T0328107

PRECISION MEDICINE

PRECISION MEDICINE

TOOLS AND QUANTITATIVE APPROACHES

Edited by

Hans-Peter Deigner
Matthias Kohl

Academic Press is an imprint of Elsevier
125 London Wall, London EC2Y 5AS, United Kingdom
525 B Street, Suite 1800, San Diego, CA 92101-4495, United States
50 Hampshire Street, 5th Floor, Cambridge, MA 02139, United States
The Boulevard, Langford Lane, Kidlington, Oxford OX5 1GB, United Kingdom

Notices
Knowledge and best practice in this field are constantly changing. As new research and experience broaden our understanding, changes in research methods, professional practices, or medical treatment may become necessary.

Practitioners and researchers must always rely on their own experience and knowledge in evaluating and using any information, methods, compounds, or experiments described herein. In using such information or methods they should be mindful of their own safety and the safety of others, including parties for whom they have a professional responsibility.

To the fullest extent of the law, neither the Publisher nor the authors, contributors, or editors, assume any liability for any injury and/or damage to persons or property as a matter of products liability, negligence or otherwise, or from any use or operation of any methods, products, instructions, or ideas contained in the material herein.

Library of Congress Cataloging-in-Publication Data
A catalog record for this book is available from the Library of Congress

British Library Cataloguing-in-Publication Data
A catalogue record for this book is available from the British Library

ISBN: 978-0-12-805364-5

Publisher: John Fedor
Acquisition Editor: Rafael Teixeira
Editorial Project Manager: Ana Claudia A. Garcia
Production Project Manager: Punithavathy Govindaradjane
Cover Designer: Victoria Pearson

Typeset by SPi Global, India

Contents

6. Workflow for Circulating miRNA Identification and Development in Cancer Research: Methodological Considerations

CHIARA M. CINISELLI, MARA LECCHI, MANUELA GARIBOLDI, PAOLO VERDERIO, MARIA G. DAIDONE

7. Analyzing the Effects of Genetic Variation in Noncoding Genomic Regions

YASMINA A. MANSUR, ELENA ROJANO, JUAN A.G. RANEA, JAMES R. PERKINS

8. Synthesis of Magnetic Iron Oxide Nanoparticles

MARCEL WEGMANN, MELANIE SCHARR

9. Magnetic Particle Imaging

ANNA BAKENECKER, MANDY AHLBORG, CHRISTINA DEBBELER, CHRISTIAN KAETHNER, KERSTIN LÜDTKE-BUZUG

10. Nanoparticles and Nanosized Structures in Diagnostics and Therapy

LISA J. JACOB, HANS-PETER DEIGNER

11. Interaction of Nanoparticles With Biomolecules, Protein, Enzymes, and Its Applications

NAVNEET PHOGAT, MATTHIAS KOHL, IMRAN UDDIN, AFROZ JAHAN

12. Modeling and Simulating Carcinogenesis

JENNY GROTEN, ANUSHA VENKATRAMAN, ROLAND MERTELSMANN

13. Sepsis: A Challenging Disease With New Promises for Personalized Medicine and Biomarker Discovery

DIDEM DAYANGAC-ERDEN, MINE DURUSU-TANRIOVER

14. Asphyxia Diagnosis: An Example of Translational Precision Medicine

JAN LÜDDECKE

15. Biomarker for Alzheimer's Disease

DOROTHEE HAAS

Contributors

Mandy Ahlborg University of Luebeck, Luebeck, Germany

Anna Bakenecker University of Luebeck, Luebeck, Germany

Chiara M. Ciniselli Fondazione IRCCS Istituto Nazionale dei Tumori; University of Milan, Milan, Italy

Maria G. Daidone Fondazione IRCCS Istituto Nazionale dei Tumori, Milan, Italy

Didem Dayangac-Erden Hacettepe University, Ankara, Turkey

Christina Debbeler University of Luebeck, Luebeck, Germany

Hans-Peter Deigner Institute of Precision Medicine, Furtwangen University, Villingen-Schwenningen; Fraunhofer EXIM/IZI, Rostock-Leipzig, Germany

Mine Durusu-Tanriover Hacettepe University, Ankara, Turkey

Manuela Gariboldi Fondazione IRCCS Istituto Nazionale dei Tumori; FIRC Institute of Molecular Oncology Foundation, Milan, Italy

Jenny Groten Albert-Ludwigs-University Freiburg, Freiburg im Breisgau, Germany

Dorothee Haas Furtwangen University, Furtwangen, Germany

Felicitas S. Holzer University of Buenos Aires, Buenos Aires, Argentina; University Paris-Sorbonne, Paris, France; CONICET (National Research Council) Argentina; FLACSO Argentina, Buenos Aires, Argentina

Lisa J. Jacob Institute of Precision Medicine, Furtwangen University, Villingen-Schwenningen, Germany

Afroz Jahan Faculty of Pharmacy, Integral University, Lucknow, India

Christian Kaethner University of Luebeck, Luebeck, Germany

Burcu Kesikli Health Institutes of Turkey-Aziz Sancar Research Center, Ankara, Turkey

Agnes Kisser Ludwig Boltzmann Institute for Health Technology Assessment, Vienna, Austria

Matthias Kohl Institute of Precision Medicine, Furtwangen University, Villingen-Schwenningen, Germany

Mara Lecchi Fondazione IRCCS Istituto Nazionale dei Tumori, Milan, Italy

Jan Lüddecke InfanDx AG, Cologne, Germany

Kerstin Lüdtke-Buzug University of Luebeck, Luebeck, Germany

Yasmina A. Mansur Department of Molecular Biology and Biochemistry, University of Malaga, Malaga, Spain

Ignacio D. Mastroleo University of Buenos Aires; CONICET (National Research Council) Argentina; FLACSO Argentina, Buenos Aires, Argentina

Roland Mertelsmann Albert-Ludwigs-University Freiburg, Freiburg im Breisgau, Germany

Meral Ozguç Center for Genomics and Rare Diseases & Biobank for Rare Diseases; Department of Medical Biology, Hacettepe University, Ankara, Turkey

James R. Perkins Research Laboratory, IBIMA-Regional University Hospital of Malaga-UMA, Malaga, Spain

Navneet Phogat Institute of Precision Medicine, Furtwangen University, Villingen-Schwenningen; University of Tübingen, Tübingen, Germany

Juan A.G. Ranea Department of Molecular Biology and Biochemistry, University of Malaga, Malaga; CIBER de Enfermedades Raras, ISCIII, Madrid, Spain

Elena Rojano Department of Molecular Biology and Biochemistry, University of Malaga, Malaga, Spain

Melanie Scharr Furtwangen University, Villingen-Schwenningen, Germany

Sowmya Srinivasan Perumbakkam Institute of Precision Medicine, Furtwangen University, Villingen-Schwenningen, Germany

Imran Uddin Aligarh Muslim University, Aligarh, UP, India

Anusha Venkatraman Albert-Ludwigs-University Freiburg, Freiburg im Breisgau, Germany

Paolo Verderio Fondazione IRCCS Istituto Nazionale dei Tumori, Milan, Italy

Stephan Vilgis Institute of Precision Medicine, Furtwangen University, Villingen-Schwenningen, Germany

Marcel Wegmann Furtwangen University, Villingen-Schwenningen, Germany

Ayse Yuzbasioglu Center for Genomics and Rare Diseases & Biobank for Rare Diseases; Department of Medical Biology, Hacettepe University, Ankara, Turkey

Preface

We are very pleased to introduce our book "Precision Medicine: Tools and Quantitative Approaches" focusing on technical prerequisites of precision medicine. With this book, we are aiming at a broad readership with background in physical or biomedical sciences interested in topics that contribute to current developments in precision medicine.

In fact, it is undeniable that medical progress is largely driven by progress in technology, for example, by DNA/RNA sequencing. In the United States, the former President Obama announced the Precision Medicine Initiative (PMI) in his State of the Union address 2015, stating that "through advances in research, technology, and policies that empower patients, the PMI will enable a new era of medicine in which researchers, providers, and patients work together to develop individualized care" (Collins and Varmus, 2015). According to the NIH, precision medicine is defined as "an emerging approach for disease treatment and prevention that takes into account individual variability in environment, lifestyle, and genes for each person" (National Research Council, 2011). In this context, precision medicine denotes an analysis and treatment approach considering individual genetic/genomic features retrievable via various "omics" techniques and environmental influences. Tools for more precise diagnosis, therapy selection, and treatment control are now increasingly becoming available; with established, continuously improved methods along with novel technological developments, we currently witness the beginning era of PM applying an integrative approach and bearing an immense potential for the future. While still in its infancy, PM is rapidly gaining momentum.

Technologies comprise current and emerging sequencing methods, metabolomics, and other high-throughput technologies as well as recent developments in imaging and optical analysis, to mention a few. The common feature of these developments or "tools," however, is that they help to deliver more conclusive data to assess, prognose, and improve the health state of an individual more precisely to the benefit of the patient.

As precise diagnosis and prognosis is based on multiple individual genomic and environmental parameters, a large body of data has to be analyzed and correlated with clinical data. Accordingly, sophisticated computational and statistical procedures representing an integral component of the PM approach were implemented into this book besides a selection of technical features affording the data required. In fact, the success of precision medicine depends on the development of accurate and reliable statistical and machine learning tools for estimating an "optimal" treatment regime given data collected from randomized experiments or observational studies.

We are delighted that we have been able to gain a number of highly renowned scientists internationally, together with ambitious and very promising junior scientists authoring our book chapters. They all did an excellent job and are the basis for the realization and success of this edition. While providing sound theoretical background on technological aspects, authors took care to demonstrate also translation into practice including recent examples.

The selection of topics addressed in this book inevitably is incomplete and biased by the view of the editors, certainly missing out other important themes. We attempted to cover topics indicating relevant progress in high-throughput technologies and also prerequisites for quality control and data analysis including topics such as biospecimen preservation and biobanking; furthermore, interesting examples related to special tools are outlined; selected ethical aspects are analyzed as well.

We thus hope that we have been successful in composing a relevant and interesting overview on areas in progress contributing to applications in current and future precision medicine.

References

Collins, F.S., Varmus, H., 2015. A new initiative on precision medicine. N. Engl. J. Med. 372 (9), 793–795.

National Research Council, 2011. Toward Precision Medicine: Building a Knowledge Network for Biomedical Research and a New Taxonomy of Disease. The National Academies Press, Washington, DC.

Hans-Peter Deigner

Institute of Precision Medicine, Furtwangen University, Villingen-Schwenningen, Germany

Fraunhofer EXIM/IZI, Rostock-Leipzig, Germany

Matthias Kohl

Institute of Precision Medicine, Furtwangen University, Villingen-Schwenningen, Germany

CHAPTER

1

Ethical Aspects of Precision Medicine: An Introduction to the Ethics and Concept of Clinical Innovation

Felicitas S. Holzer[*,†,‡,§], *Ignacio D. Mastroleo*[*,‡,§]

[*]University of Buenos Aires, Buenos Aires, Argentina [†]University Paris-Sorbonne, Paris, France [‡]CONICET (National Research Council) Argentina, Buenos Aires, Argentina [§]FLACSO Argentina, Buenos Aires, Argentina

1 INTRODUCTION

This chapter introduces the concept of clinical innovation in the context of the increasing importance of individualized therapeutic and diagnostic approaches. To start with, we want to outline three exemplary cases of what we consider clinical innovation in order to give the reader an intuition of the term "clinical innovation" that is commonly used in the current literature of research ethics and the life sciences. These exemplary cases will cover therapeutic and diagnostic interventions and a combination of both therapeutic and diagnostic procedures.

Most notably, Sugarman (2012) introduces the concept of what he calls "innovation pathway" based on the current International Society for Stem Cell Research (ISSCR) guidelines for the clinical transition of cell-based interventions. He grounds his conceptual approach to clinical innovation on the first umbilical cord blood transplantation in humans, which we identify as an exemplary case for a successful clinical innovation of a new and untested therapeutic procedure. In 1988, Mathew Farrow, a 5-year-old patient with Fanconi's anemia who had no reasonable medical alternatives for treatment, received the first successful experimental umbilical cord blood transplantation from his baby sister Alison Farrow (BBC, 2001). Since this first successful experimental transplantation of umbilical cord blood in Mathew Farrow, cord blood is now widely used as a treatment with hematopoietic stem cells for a range of different malignant and nonmalignant conditions (Gluckman et al., 1997; Taylor, 2010; Sugarman, 2012).

Precision Medicine
https://doi.org/10.1016/B978-0-12-805364-5.00001-9

However, the use of clinical innovation does not always attain the desirable results for patients deprived of validated therapeutic options. For instance, the case of Jim Gass (Kolata, 2016) provoked an outcry in international media that illustrated a growing concern about the number of stem-cell tourists worldwide. It turned out that he had several stem-cell therapies at clinics in Mexico, China, and Argentina, paying tens of thousands of dollars each time for injections in an attempt to recover from a stroke. The total cost including travel expenses reached US$300,000. Eventually, Jim Gass developed a tumor in his lower spinal column. The following tests showed that the tumor mass was made up of abnormal, primitive cells that were growing aggressively. The New York Times article (Kolata, 2016) reports that there is an increasing trend of clinics in Russia, China, Europe, and elsewhere to offer stem-cell treatments on websites with the promise to treat and cure diseases, such as muscular dystrophy, Alzheimer's disease, Parkinson's disease, spinal cord injuries, and strokes, by injecting patients with stem cells.

Successful cases such as Mathew Farrow intuitively illustrate the potential benefits of clinical innovation and are at the basis of defending this pathway for patients without a validated medical alternative. However, unfortunate cases, such as Jim Gass, also show that allowing for the unrestricted use of new and insufficiently proved interventions may have untoward consequences for both individuals and society. Consequently, clinical innovation needs to be treated with caution because patients that undergo clinical innovation, such as Mathew Farrow or Jim Gass, are neither restricted nor protected by clinical research regulations.

Examples of using new and unproved interventions outside formal clinical research are not limited to therapeutic procedures, such as umbilical cord blood transplantation or unproved stem-cell therapies, but have also been applied to diagnostic interventions. Genome-sequencing technologies, including whole-genome and whole-exome sequencing, for patients with rare diseases are recent exemplary case of a new insufficiently validated diagnostic intervention used outside formal clinical research. Patients who suffer from rare diseases often face long and burdensome diagnostic procedures over several decades. Decoding the genetic causes of disease is crucial in order to target potential therapies for rare diseases. For instance, two siblings in the United Kingdom with an unusual muscle wasting disease had to wait for 20 years until they were diagnosed at a cost of more than £14,000. Whole-exome sequencing costing approximately £1000 at this time revealed that a heterozygous mutation was likely to be disease-causing (Perdeaux, 2013; Rehm et al., 2015).

Finally, clinical innovation can be the product of combining new diagnostic and therapeutic interventions. Molly Nash was born with type-C Fanconi's anemia, a more aggressive type than the one that affected Matthew Farrow. Lacking a suitable match for a bone-marrow transplant, the Nashes conceived a baby called Adam to find a suitable donor that possibly matched with Molly. However, due to their low probability of having a baby without Fanconi's anemia, the parents had to use three different kinds of intervention, namely, in vitro fertilization, preimplantation genetic diagnosis (PGD), and umbilical cord blood transplantation, in order to attain an acceptable donor (Faison, 2005). The PGD was used to select an embryo without Fanconi's anemia and subsequently find a match for Molly. As Kahn and Mastroianni (2004) note, the chosen procedures in the Nash case had been already sufficiently validated for their standard indications in 2000 and were thus considered standard medical care. However, the combination of these interventions used to attain a promising treatment for Molly had still been considered as an "experimental" procedure at that time, which implied that the procedure was not covered by insurers (Kahn and Mastroianni, 2004, p. 92).

These exemplary cases of therapeutic and diagnostic interventions offer an overview of exemplary cases discussed in the literature on the standard concept of clinical innovation. They show that clinical innovation is a—despite its inherent risks—valuable medical activity that is different from both standard validated practice and clinical research. As Taylor notices, examples of innovations like umbilical cord blood transplantation show that clinical innovation does "[...] not follow the linear model of basic research, to translation, to clinical research, to application." Instead, clinical innovation "[...] come[s] from thinking backwards from the patient's circumstances, and forward from deep knowledge of how the body functions, to challenge the limits of current mechanisms for [...]" therapeutic, preventive, and diagnostic interventions (Taylor, 2010, p. 286, edited).

Having hinted on an intuitive definition of clinical innovation in the outline of the exemplary cases, we now want to answer the following fundamental questions:

(1) What is clinical innovation? (definition of clinical innovation)
Our aim will be to look for a sound concept and a new taxonomy of the term clinical innovation that better suits the application of new, untested interventions in individualized and precision medicine. We then want to address further questions to view the concept of clinical innovation in light of current ethics regulations for new treatments and diagnostics with a special focus on individualized therapeutic and diagnostic approaches in medicine.
(2) What is the ethical justification of clinical innovation? (justification of clinical innovation)
(3) How should we regulate clinical innovation? (ethical regulation of clinical innovation)
(4) What is the importance of clinical innovation in precision medicine?

The chapter will follow these questions as a baseline for the analysis and development of the concept and ethics of clinical innovation in the context of the emerging field of individualized approaches in medicine. First, we will clarify the definition of clinical innovation in the clinical setting. We will defend that the concept of clinical innovation is better understood by the concept of new nonvalidated practice. To show this, we will refer to major guideline papers for the conduct of human health research and contemporary concepts by Achim Rosemann, Jeremy Sugarman, Nancy King, and Alex J. London. However, most notably, we will follow Levine's (1979, 2008) conceptual interpretation of the Belmont Report (National Commission, 1979). At the heart of this first section, we thus intend to establish a changed taxonomy of the term "clinical innovation" within given standard approaches to medical practice, innovation, and clinical research. Second, we will answer the question of whether clinical innovation is ethically justifiable. Here, we will address the regulatory burdens of clinical research that currently apply to innovative interventions. We will argue that the lack of reasonable medical alternatives and the substantial health need of individuals justify establishing clinical innovation with ethically permissible concept alongside clinical research and medical practice. Finally, we will briefly present the main challenges of ethics regulations to clinical innovation and shed light on the importance of clinical innovation within the emerging field of individualized therapeutic and diagnostic approaches in medicine.

2 WHAT IS CLINICAL INNOVATION?

In the following, we will define clinical innovation as *new and insufficiently validated interventions used for the benefit of individual patients* or, in short, *new nonvalidated* practice. We will thus argue that clinical innovation is characterized by its novelty, nonvalidation, and benefit to individual patients. Our concept of clinical innovation is thought to capture the core meaning of innovation referred to in the literature and in common regulatory guidelines, such as the ISSCR guidelines.

2.1 Clinical Innovation Is Neither Clinical Research nor Validated Medical Practice

According to Beauchamp and Saghai's (2012) interpretation of the Belmont Report, the National Commission for the Protection of Human Subjects of Biomedical and Behavioral Research (hereinafter National Commission, 1979) has established two classes of activities labeled "research" and "practice" as categories for medical interventions that are logically distinguishable from each other (although they may coexist).[1] On the one hand, the commission states that

> [F]or the most part, the term "practice" refers to interventions where: (P1) the purpose of an intervention is "to provide diagnosis, preventive treatment, or therapy"; (P2) the intervention is "designed solely to enhance the well-being of an individual patient or client" (though benefit to other persons is sometimes the goal); (P3) the intervention has "a reasonable expectation of success." *(National Commission, 1979, as quoted in Beauchamp and Saghai, 2012, p. 52)*

On the other hand, the National Commission (1979) defines research as follows:

> To qualify as research two conditions are central. The first is not a necessary condition for all forms of research, but the second is a necessary condition: (R1) there is (in pertinent research methods) a formal protocol- controlled design to test a hypothesis; (R2) there is an organized design "to develop or contribute to generalizable [scientific] knowledge." *(National Commission, 1979, as quoted in Beauchamp and Saghai, 2012, p. 52)*

The National Commission (1979) recommends that if an activity has an organized design "to develop or contribute to generalizable (scientific) knowledge," it should undergo a research review to protect human subjects, irrespective of the fact that the intervention or procedure is intended to provide direct health benefit for the individual patient-subject (Levine, 2008, p. 217). This is a precautionary measure to prevent researchers from taking advantage of a loophole in the oversight system by applying investigational interventions under the label of "practice" to avoid the review process (Beauchamp and Saghai, 2012, p. 43). In order to be considered practice, the activity should be designed *solely* to enhance the well-being of an individual patient or client.

However, despite the distinction between clinical research and medical practice through the establishment of two logical, although nonmutually exclusive, domains, the National Commission already gives credit to interventions labeled as "innovation":

[1] In order to understand how research and practice may coexist, see the concept of a component analysis developed by Weijer (2000) and Weijler and Miller (2004).

> When a clinician departs in a significant way from standard or accepted practice, the innovation does not, in and of itself, constitute research. The fact that a procedure is "experimental" in the sense of new, untested or different does not automatically place it in the category of research.[2] *(National Commission, 1979, edited)*

As stated in the Belmont Report, innovation is understood as departing significantly from standard accepted practice. However, innovation does not, in itself, constitute research. Hence, the Belmont Report does not introduce a third category for innovation, but considers it as some sort of medical practice because it is designed solely to benefit the individual patient, as stated above in (P2). This has important practical consequences because considering an innovative intervention as a clinical practice implies that the intervention underlies the general regulations of medical practice and not those of clinical research (Mastroianni, 2006). Nevertheless, since the innovative intervention does not show sufficient evidence, it cannot be considered validated medical practice and hence does not fulfill the standard interpretation of condition (P3).

2.2 Clinical Innovation Is New Nonvalidated Practice

We will now specify our concept of clinical innovation as *new and insufficiently validated interventions used for the benefit of individual patients*, which we will use synonymously with *new nonvalidated practice*.

As we show in the previous section, the Belmont Report states that "practice" refers to diagnostic, preventive, or therapeutic interventions designed solely to enhance the well-being of an individual with reasonable expectations of success. In contrast, "research" is designed to test hypotheses and permit conclusions that generate generalizable knowledge, irrespective of the use of interventions or procedures intended to provide direct health benefit for the individual patient-subject.[3] However, Levine points out that there is one particular class of "medical practice" that he describes as "nonvalidated practices." Here, he notices that the Belmont Report (National Commission, 1979) refers to "nonvalidated practice" in various ways by using terms such as "innovative therapies" or "experimental design" and defines it as follows:

> *Nonvalidated practices.* A class of procedures performed by physicians conforms to the definition of "practice" to the extent that these procedures are "designed solely to enhance the well-being of an individual patient or client." However, they may not have been tested sufficiently often or sufficiently well to meet the standard of having "a reasonable expectation of success." *(Levine, 1979, p. 22)*

[2] The unedited complete quotation of the paragraph on innovation in Belmont Report is the following: "When a clinician departs in a significant way from standard or accepted practice, the innovation does not, in and of itself, constitute research. The fact that a procedure is 'experimental,' in the sense of new, untested or different, does not automatically place it in the category of research. Radically new procedures of this description should, however, be made the object of formal research at an early stage in order to determine whether they are safe and effective. Thus, it is the responsibility of medical practice committees, for example, to insist that a major innovation be incorporated into a formal research project" (National Commission, 1979).

[3] However, Levine already argues that there are plenty of medical practices including public health practices or practices for the benefits of others that overlap with the aims of research. In turn, research also takes into consideration the well-being of a research subject to some extent (Levine, 2008).

Comparing this definition of "nonvalidated practice" with the Belmont Report's definition of medical practice presented above, we note that "nonvalidated practice" meets the first two conditions of validated medical practice, namely, the purpose and scope of interventions (P1) and the design solely to benefit an individual (P2). However, the definition does not comply with the third condition of "a reasonable expectation of success" (not P3) due to a lack of sufficient evidence of the intervention. Hence, medical practice can be classified as validated or nonvalidated practice. This is in line with the reality of medical practice where not all interventions used to enhance the well-being of patients are sufficiently validated. Consequently, the term "practice" in the Belmont Report is primarily characterized by the aim to enhance an individual patient's well-being and not by its evidence.

In the previous paragraphs, we have introduced and clarified the standard definitions of research, practice, and innovation in the Belmont Report. It will be important to highlight that Levine's definition of innovation that is grounded on the Belmont Report is nonvalidated practice, while our definition is *new* nonvalidated practice. As we will argue, our definition is narrower than the one stated by Levine, but we believe it better captures the activity of introducing individualized approaches in medicine. However, in order to justify our reformulated definition, we need to analyze the three core elements of our definition of "clinical innovation": novelty, validation, and benefits for individual patients.

2.2.1 Novelty

As outlined above, Levine states that the most crucial characteristic of nonvalidated medical practice "[...] is the lack of suitable validation of the safety or efficacy of the practice" (Levine, 1979, p. 22).[4] However, according to our definition, clinical innovation does not refer to all kinds of nonvalidated interventions but only to the subset of *new* nonvalidated or insufficiently validated interventions. Hence, we want to point out that novelty and validation are two different characteristics of an intervention. On the one hand, we will stipulate that novelty refers to the recent or the first use of an intervention in the context of medical practice. In that sense, novelty is logically distinct from the long-standing use of an untested or insufficiently tested intervention. On the other hand, validation refers to the level of evidence of an intervention. An intervention can be either sufficiently or insufficiently validated (here used as a synonym of "nonvalidated") within a given practical context that Taylor (2010, p. 286) refers to as "the right circumstances" for the application of an intervention. Therefore, by introducing these two characteristics, we obtain four different categories under which an intervention can fall (see Table 1). Here, we define the term "clinical innovation" as one subset of nonvalidated practices, namely, as nonvalidated practices that are new.

We believe that our classification model substantially adds to Levine's definition of innovation as "nonvalidated practices" because it points to the distinction between new and long-standing nonvalidated medical practice, which has several practical implications (outlined in the following sections).

[4] See also "This class of activities is most commonly called innovative therapy; proposed that it should be called non-validated practice because the defining attribute was not novelty; it was lack of validation (demonstration of safety and efficacy) and the Commission's reasoning about how to deal with such practices applies to diagnostic and preventive measures, not only therapies" (Levine, 2008, p. 218).

TABLE 1 Medical Practice Characterized by Validation and Novelty

Validation	Novelty	
	New Practice	**Long-Standing Practice**
Nonvalidated practice	*New insufficiently validated interventions (i.e., clinical innovation).* Interventions recently used in medical practice without sufficient evidence *Example*: first umbilical cord blood transplant for Fanconi's anemia and other stem-cell therapies (Ballen et al., 2013)	*Long-standing insufficiently validated interventions.* Interventions used for a long time in medical practice without sufficient evidence *Example*: routine episiotomy for vaginal birth (Carroli and Mignini, 2009)
Validated practice	*New and validated interventions.* Interventions recently used in medical practice with sufficient evidence *Example*: imatinib for chronic myeloid leukemia after approval (Druker, 2009)	*Long-standing and validated interventions.* Interventions used for a long time in medical practice with sufficient evidence *Example*: amoxicillin for bacterial infections (Sutherland et al., 1972)

The current literature on clinical innovation often conflates novelty and nonvalidation (King, 2011). We however think that this would be a hasty generalization. If we associate novelty of an intervention with insufficient evidence, we may incorrectly believe that all new interventions used in practice are nonvalidated. This is apparently not the case. For instance, imatinib for chronic myeloid leukemia is an exemplary case of an intervention introduced in the context of medical practice in the early 2000s as new validated intervention after rational drug design and validation from formal clinical research (Druker, 2009). Unlike the exemplary cases of clinical innovation, imatinib followed "the linear model of basic research, to translation, to clinical research, and eventually to application" (Taylor, 2010, p. 286). Hence, it is thus distinct from the first umbilical cord blood transplantation that has been introduced in the late 1980s as a "last chance" alternative for Mathew Farrow, a patient who had no reasonable validated medical alternative at that time. Long-standing insufficiently validated interventions used in medical practice, such as routine episiotomy for vaginal birth (Carroli and Mignini, 2009), fall under Levine's definition of innovation. However, the practical problems entailed by long-standing interventions are different from the ones present in our exemplary cases of new interventions (i.e., Jim Gass, Mathew Farrow, and Molly Nash).

2.2.2 Validation

At this point, due to reasons of stipulating with the scope of this chapter, we cannot dwell on what Beauchamp and Saghai call the "epistemic problem" in the clinical context, which refers to the conditions under which medical interventions are sufficiently validated. However, following Beauchamp and Saghai (2012, pp. 49 and 50), we can note the following three points. First, a significant number of nonvalidated or insufficiently validated interventions in medical practice fall short of the validation standards set by randomized clinical trials (RCTs). Second, not all members of the scientific community accept the view that the validation of an intervention is obtained solely by RCTs. Thus, there is a strong disagreement among researchers regarding the different methods of validation at stake. In fact, "RCTs have never monopolized medical knowledge production" (Bothwell et al., 2016). Moreover, there seem to be good arguments not to consider RCTs as a universal gold standard, but rather to adopt a case-by-case approach applying different research methods and methodologies to untested and new interventions when necessary (Cartwright, 2007). Third, the fact that an intervention is accepted in practice does not provide grounds to believe that hypotheses about the safety and efficacy of that intervention are supported by sufficient evidence (e.g., routine episiotomy for vaginal birth).

Thus, instead of offering an epistemically unwavering account for the evidence of an unproved intervention, we interpret the Belmont Report's criterion (P3) of "reasonable expectations of success" as an epistemic standard relative to a specified body of information and a specified range of reasons. As we will argue, a certain level of evidence may not be sufficient to call a medical activity validated practice. However, the level of evidence may be enough to consider the activity as "clinical innovation."

In the case of innovative interventions, the limited clinical and/or preclinical evidence of effectiveness and safety may not amount to the full validation of an intervention relative to the appropriate epistemic standard. In these circumstances, physicians have good reasons to restrict the use of a new nonvalidated intervention (e.g., cord blood transplant) to individual patients. This means the intervention should not immediately be applied at a population level (i.e., all children with Fanconi's anemia) due to lacking "reasonable expectations of success." However, even if the probabilities of success are low or uncertain regarding the use of an innovation for the broader population, a new and insufficiently validated intervention may still show a minimum level of evidence to be a reasonable option in a limited number of cases in which individual patients do not have medical alternatives, as shown by the cases of Mathew Farrow or Molly Nash (below, we will refer to this as the argument of "lacking reasonable alternatives"). That is the case because even if an intervention has not been tested sufficiently often or sufficiently well, this does not imply that the intervention has no evidence at all (Levine, 1979, p. 22). Thus, it seems reasonable to distinguish between (i) a minimum level of evidence necessary to be a reasonable option for individual patients with no other medical alternatives and demonstrated by preclinical tests and (ii) the criterion of "reasonable expectations of success," which is associated with validated and established medical practices relative to an appropriate epistemic standard, as stated in (P3) of the Belmont Report.

Nevertheless, even if an intervention lacks (ii) sufficient evidence to reach the appropriate standard of validated medical practice, it has to reach (i) a minimum level of evidence to be a reasonable option for individuals who lack other reasonable alternatives (clinical innovation). Furthermore, without a minimum level of evidence in support of the efficacy and safety of the intervention, it would be irrational for physicians to prescribe and for individuals to undergo the intervention as a last chance. This explains why the ISSCR guidelines for stem-cell research and clinical translation demand for clinical innovation ("innovative care" in their terms), a sound scientific rationale and justification given the lack of alternatives for the patient. The assessment of the expected success of a clinical intervention includes any preclinical evidence for safety and efficacy (ISSCR, 2016, recommendation 3.4).[5]

[5] As a side note, we want to draw attention to the wording "reasonable chances of success" within the ISSCR framework (2016). The ISSCR guidelines state that clinical innovative interventions have to show "reasonable chance of success" to be acceptable. However, the Belmont Report (1979, condition (P3)) uses the same phrase to refer to validated and approved medical practice. Here, we want to follow this notion of reasonable chances of success. What we demand in the case of clinical innovation is a reasonable minimum of evidence to be legitimately applicable.

2.2.3 *Benefit for the Individual*

Finally, one important component of our definition of clinical innovation is that new and nonvalidated interventions are used as medical practice, such as umbilical cord blood transplants or next-generation sequencing for rare diseases. This means that the primary aim of clinical innovation is (P2) the "well-being of the patient" and not (R2) the generation of or contribution to "generalizable (scientific) knowledge" as in clinical research (National Commission, 1979; Levine, 1979).

There are two important practical implications of defining clinical innovation as a type of medical practice related to the risk-benefit evaluation and responsibility. Unlike other definitions of "clinical innovation," our definition does not state that innovative interventions fall into a "gray zone" between research and practice. Following the traditional definition of Levine and the Belmont Report, we define clinical innovation as medical practice and not as research. Thus, clinical innovation refers to the use of an intervention that falls under medical practice, understanding "practice" as a practical context where physicians apply therapies and diagnostic means with the primary aim of benefiting the individual patient. Here, we want to stress that our definition of "innovation" ought to be understood as a concept that revolves around the attendance to individual patients. We therefore call our approach "clinical" innovation, distinct from other sorts of innovation, such as technological innovation following a linear model of technological development, or public health innovation where the implementation of new practices is realized, for example, through cluster randomized trials.

At this point, we also want to clarify that our definition of clinical innovation, even though emerging from empirical observations within the context of clinical routine, is a normative category. This means that we set the category of clinical innovation within clinical practice and as decisively distinct from clinical research due to epistemic and ethical reasons and not because this approach may be more convenient in the everyday routine of medical practitioners and researchers. Thus, whether an intervention used in a specific practical situation should be labeled as a standard practice, clinical innovation or clinical research is a question beyond our conceptual approach to develop an epistemic and ethical taxonomy for clinical innovation. Nonetheless, determining the appropriate category for an intervention in a specified context has practical implications regarding the ethical permissibility of the intervention.

One important aspect of the evaluation of innovative interventions is an adequate risk-benefit analysis. Interventions considered as clinical innovation should be subject to a risk-benefit profile evaluation according to the standards of medical practice, not research (Levine, 1979, p. 22). According to Weijler and Miller (2004) and as stated by King and Churchill (2008), the risk-benefit analysis in clinical research is considerably distinct from the evaluation in medical practice. While therapeutic and nontherapeutic risks (that are not directly associated with therapeutic procedures) occur in the context of clinical research, all risks directly relate to the individual in the case of medical practice. In the case of research, nontherapeutic risks that occur in the context of nontherapeutic procedures and that are unrelated to the intended improvement of a patients' well-being procedures are balanced with potential therapeutic benefits to the research subjects *and* ought to be in reasonable relation to the knowledge gain and social benefit to society (Weijler and Miller, 2004; Emanuel et al., 2000).[6] Hence, benefits

[6] As a side note, it remains questionable whether the benefits to society should override the well-being of an individual under any circumstances (Martin et al., 1995).

to society are an additional factor in the assessment of research and the threshold for potential benefits to individuals might be considerably lower than in the case of medical practice (King and Churchill, 2008).

When we apply the aforementioned considerations on risk-benefit analyses to the concept of clinical innovation as new nonvalidated practice, we argue that the ethical evaluation of a risk-benefit profile under clinical innovation should be in accordance with the ethical evaluation of medical practice. If the nonvalidated intervention is the only reasonable intervention for diseases with potentially life-threatening or strongly life-impairing characteristics primarily designed to enhance the well-being of patients, high risks can be reasonably accepted.

Nevertheless, the risks that occur in the setting of clinical innovation ought to be in accordance with general guiding frameworks about risks. For instance, King and Churchill (2008) explain that there are some risks that can never be justified, and all significant risks of serious harms must have *unassailable justification*. The risks of harm should always be minimized.

3 WHAT IS THE ETHICAL JUSTIFICATION OF CLINICAL INNOVATION?

In the previous section, we defined clinical innovation as a new intervention that is insufficiently validated and used in the context of medical practice for the benefit of individual participants who have no other reasonable alternatives. In this section, our aim is to answer the second fundamental question we posed at the beginning of this chapter, namely, "what is the ethical justification of clinical innovation?" This justification is intended to attenuate the debunkers of clinical innovation who sustain that insufficiently validated interventions should only be accessed through the research pathway, a standpoint sustained by many authors. For instance, Emanuel (2013) holds this position by proposing a revised principle for the use of nonvalidated interventions in the Declaration of Helsinki:

> Research of Unproven, "Last Ditch" Treatments: In the treatment of a patient, where proven interventions do not exist or have been ineffective, the physician, after seeking expert advice, with informed consent from the patient or a legally authorized representative, may use an unproven intervention to promote the patient's health or well-being, but only if it is undertaken as a research study designed to evaluate its safety and efficacy. Repeated uses of an unproven intervention can only be justified as part of a research study that fulfils all the protections in this Declaration *(Emanuel, 2013, supplementary appendix)*

Emanuel's (2013) revision of the principle for the use of unproved "last-ditch" interventions has never been incorporated into ethics guidelines, such as the Declaration of Helsinki. Nevertheless, to reply to possible objections to our definition, we will present an explicit ethical justification for the use of clinical innovation as a medical practice based on an accepted formulation of this principle in the Declaration of Helsinki.

The Declaration of Helsinki in 2000 (WMA, 2000, paragraph 30) states that physicians with informed consent from the patient must be free in choice to apply unproved or new preventive, diagnostic, and therapeutic measures if they are considered potentially life-saving, reestablish health, or alleviate suffering for patients without reasonable alternatives. This principle has already been acknowledged in the first version of the declaration (WMA, 1964) under the section "clinical research combined with professional care." Here, the Declaration (1964) states in paragraph 1 (section II) that "[I] in the treatment of the sick person the doctor must be free to use a new therapeutic measure if in his judgment it offers hope of saving life, re-establishing health, or alleviating suffering." This applies to all cases in which proved diagnostic and therapeutic methods do not exist and are rare or not acceptable. It also includes conditions for the compassionate use of drugs. For instance, a physician may decide that it is reasonable to offer an individual patient an unproved intervention outside clinical trials when she is not eligible for a research study. Likewise, physicians may conclude that an individual patient can profit from innovative interventions, such as those ones that occur in the context of individualized diagnostics and therapies and that cannot be tested under a rigid and methodologically sound clinical trial. Importantly, next to assigning responsibilities to physicians, the Declaration of Helsinki gives credit to individual patients' health conditions that go along with substantial suffering. Thus, the application of potentially risky innovative interventions is not justified if the patient has no substantial health need to undergo the new and unproved intervention.

The ethical principle in the Declaration of Helsinki (WMA, 2000, paragraph 30) for the use of "last-ditch" unproved interventions, applied after a careful assessment by physicians seeking expert advice, has remained with minor changes in the revised versions of the Declaration in 2008 (WMA, 2008, paragraph 35) and in 2013 (WMA, 2013, paragraph 37). Following the latest version, the principle has been labeled as the principle for the appropriate use of "unproved interventions in clinical practice." However, clinical innovation must respect certain conditions to be considered as appropriate activity:

> In the treatment of [1] an individual patient, where proven interventions do not exist or other known interventions have been ineffective, the physician, [2, 3] after seeking expert advice, [4] with informed consent from the patient or a legally authorised representative, may use an unproven intervention if [5] in the physician's judgment it offers hope of saving life, re-establishing health or alleviating suffering. [6] This intervention should subsequently be made the object of research, designed to evaluate its safety and efficacy. [7] In all cases, new information must be recorded and, where appropriate, made publicly available. *(WMA, 2013, paragraph 37, edited)*

The Declaration of Helsinki reveals certain core principles that are also recognized by the ISSCR (2016, recommendation 3.4.1). The principle of "provision of innovative care" guides physicians when applying clinical innovation to patients, which ought to comply with the following ethical requirements: (1) exhausting circumstances and limited number of patients, (2) scientific validity, (3) independent review, (4) informed consent, (5) priority of patient well-being, (6) contribution to generalizable knowledge through clinical research, and (7) publication of results.[7] This list of ethical requirements provides an initial ethical framework to balance the aim to attain generalizable knowledge through research with the aim to attend the needs of patients with no reasonable alternatives.

Moreover, recognizing appropriate limits to clinical innovation strengthens our argument that new and insufficiently validated interventions should not always be subject to research, which goes hand in hand with restrictive regulations, precisely in all cases in which patients lack reasonable medical alternatives. We call this argument the "argument of lacking reasonable alternatives." This argument sustains that physicians can and maybe sometimes should use unproved interventions in exceptional situations within the context of medical practice.

[7] Some of the principles show similarities to the ones commonly referred to in research ethics. However, their content can differ substantially, as pointed out by Taylor (2010).

As stated by Sugarman (2012), the lack of reasonable medical alternatives for an individual goes along with a changed evaluation of the risk-benefit profile of an unproved intervention. The mere fact that individuals cannot access any alternative treatment makes even a low expectation of success of the unproved therapy ex ante a rational, reasonable, and acceptable choice, given that the intervention reaches a minimum threshold of sound evidence. Surely, we should point to the responsibilities physician have in this context. The evaluation of the reasonability and acceptability of an unproved intervention is part of a cautious balancing and judging process by the physician. It would be ethically wrong for a physician to make a patient undergo a nonvalidated intervention outside the context of clinical research in the case of available clinical trials or validated medical alternatives.

Coming back to the examples we introduced at the beginning of our chapter—except the case of Jim Gass, which we will discuss in the next section—we clearly notice that the concept of clinical innovation as new and insufficiently validated practice is necessary when physicians do not have a reasonable medical alternative at hand. In the case of the first successful transplantation of umbilical cord blood in a patient with Fanconi's anemia, the rationale had been to use a reasonable experimental alternative treatment to transplantations with hematopoietic stem cells from donors who were not available for this patient (Ballen et al., 2013). This is a paradigmatic case for the argument of lacking medical alternatives in cases of patients with life-threatening and urgent conditions.

In the diagnostic case example of genomic sequencing in patients with rare diseases, physicians use genomic-sequencing methods to search for new and not yet established genetic variants to better understand the underlying genetic disease patterns and to potentially target treatment options. Patients who often have to undergo long-lasting and burdensome diagnostic procedures at very high expenses infrequently enroll in formal research projects, such as the "Rare Diseases Genomes Project" in the United Kingdom, because there is not always a research project at hand. Thus, physicians act on rational grounds when they perform genomic sequencing in order to better understand the genetic causes of disease, which potentially leads to a better health outcome through targeted therapies (Bradley, 2013).

Thus, in view of the argument of lacking medical alternatives, clinical innovation is ethically permissible as a new reasonable approach under the rationale of medical practice that is distinct from clinical research. Clinical innovation is a justified and necessary alternative for the controlled application of new health interventions, including diagnostic and therapeutic procedures that cannot be fully developed within clinical trials.

4 HOW SHOULD WE REGULATE CLINICAL INNOVATION?

Regarding the third question we raised at the beginning of this chapter, namely, how to regulate clinical innovation, we want to shortly touch upon the debate on possible regulatory proposals. The question about regulations is of major importance because if we accept clinical innovation as a necessary and ethically justified subcategory of clinical practice, we will need to address two further questions.

First, we need to ask how to guarantee the safe access to innovative interventions for people with no reasonable medical alternatives. According to broader theories of justice and health, such as Daniels' (2008, p. 145) approach to just health care, we have to design our institutions in a way that they promote, maintain, or restore peoples' health. We argue that this would include the access to interventions under clinical innovation.

Second, we need to ask how to design regulations for clinical innovation in order to avoid harm to others or, in this case, to foster and to maintain the "public good of research." Even though the ethical justification of clinical innovation states that that it is rational for physicians and their individual patients to take risks in order to increase their expected health outcome ex ante, it is also rational for all to accept and set limits to clinical innovation, precisely when we foresee that the uncoordinated and unlimited aggregation of individual behavior would harm others. Rawls refers to individual liberties, such as the interest to restore, maintain, and promote one's own health, by placing the individual demand within a cooperative scheme: "Each person is to have an equal right to the most extensive system of equal basic liberty compatible with a similar system for all" (Rawls, 1999, p. 220). This means in terms of the application of clinical innovation that regulatory frameworks need to address the problem that innovation potentially undermines the generation of the "public good of research" created through sound research trials if unproved interventions are provided outside clinical trials. If clinical innovations were accessible to everyone without restrictions, some worry that people would fail to register for trials if they have the option to receive an unproved intervention in 100% of the cases (rather than having only a 50% chance of receiving it in a randomized control trial) (Daniels and Sabin, 1997). The ISSCR guidelines (2016, p. 3.4) answer this question by stating that clinical innovation must remain the exceptional case. In order avoid undermining public trust and delaying formal clinical trials, interventions under clinical innovation should only be applied to at most a very small number of patients outside clinical trials (ISSCR, 2016, p. 3.4). As soon as a new and insufficiently proved intervention is promising enough to be provided to a larger number of patients, clinical innovation ought to be replaced by a methodologically sound clinical trial. As seen above, the Declaration of Helsinki (WMA, 2013) acknowledges this rationale in the last section of paragraph 37 by stating that "This (unproven) intervention should subsequently be made the object of research, designed to evaluate its safety and efficacy. In all cases, new information must be recorded and, where appropriate, made publicly available."

Furthermore, as shown by various examples, such as some types of clinical stem-cell research in China (Rosemann, 2013), and the case study of Jim Gass introduced at the beginning of this chapter, clinical innovation as insufficiently validated medical practice can be conducted under the false pretenses of beneficial purposes even though the intervention may lack transparency or a minimum level of sound evidence and scientific validity. This sheds new light on the permissibility of innovative interventions and the question about its appropriate ethical regulations (WMA, 2013; Sugarman, 2012). In view of Rosemann's (2013) examples of miscarried innovative medical interventions, we sustain that clinical innovation needs to underlie regulations to offer to individuals a safe access to new and insufficiently validated interventions. We tend to support the viewpoint that regulations, similar to the Declaration of Helsinki (2013, paragraph 37) or the more elaborate ISSCR (2016) guidelines that address the provision of innovative care with cell-based interventions, should be specifically formulated for individualized therapeutic and diagnostic interventions under clinical innovation and considered independent from clinical research and validated medical practice (see Taylor, 2010 and Sugarman, 2012). Yet, there are only few regulatory guidelines of this type. Thus, there is a need to develop guidelines tailored to what Sugarman (2012) calls "responsible innovation."

With the development of independent guidelines for clinical innovation, innovative interventions would not need to follow any longer the restrictive review of clinical research. In contrast to the standard clinical research that often requires extraordinary funding, strong methodological design, and time, the review of clinical innovation could be conceptualized less burdensome than the current research regulations (Sugarman, 2012; ISSCR, 2016). In the case of independent guidelines for clinical innovation, it would be necessary to specify minimal regulatory frameworks outside the established research ethics guidelines.

However, we do not intend to dwell on the discussion about the appropriate regulation of clinical innovation, which should be a subject to a separate discussion. This discussion would follow the establishment of the category of clinical innovation interpreted as new nonvalidated practice, a subcategory of interventions that fall under clinical practice and as distinct from formal clinical research. Thus, if our conceptual work turns out to be sound, it can build the basis for a further assessment of the regulatory framework for innovative interventions.

5 WHAT IS THE IMPORTANCE OF CLINICAL INNOVATION IN PRECISION MEDICINE?

So far, this chapter presented a conceptual framework to establish "clinical innovation" within the larger category of medical practice. We intended to support our approach by arguing that this classification is necessary from an epistemic and ethical stance. In particular, it is ethically necessary and reasonable to consider clinical innovation, even if insufficiently validated, as medical practice and not as research in order to align with the urgent health needs of patients with no reasonable medical alternatives. If our previous arguments are sound, clinical innovation in the appropriate circumstances has the potential to benefit individual patients and society. Hence, we should be cautious about treating clinical innovation as an inferior kind of activity in comparison with research. As Taylor reminds us,

> That innovative therapy can be, and continues to be, so positively transformative in the right circumstances, ought to make us cautious, I think, about treating it as a presumptively flawed and inferior activity that requires the corrective guidance of the research paradigm. Each is legitimate in a certain sphere; each has different goals; and, as I shall argue, each has distinct oversight needs. *(Taylor, 2010, p. 286)*

This position would most probably have a substantive impact on the norms and practices of regulating innovative interventions in the clinical context.

In this last section, we want to put emphasis on the importance to develop a valid concept of clinical innovation in the light of the emerging field of individualized therapy and diagnostics. To name one important example for an emerging field in personalized medicine, "translational cancer research" or "precision oncology" is currently a pioneering field in oncology, combining individualized molecular diagnostics with individualized therapies, such as immunotherapy, specified radiation oncology, or cancer-prevention programs (National Center for Tumor Diseases Heidelberg (NCT), 2017). Also, research programs for whole-genome and whole-exome analyses, such as the Human Genome Project (Duke Medicine, 2011), provide another recent and pressing example for the personalized medicine initiative. One of the hopes of the Human Genome Project has been to pinpoint specific genes that cause common diseases. Even though scientists have increasingly noticed that the phenotype-genotype relations are very complex, as usually caused by multiple factors, the information gathered from the genome project has had the potential to transform health care. "Many believe that genome-based medicine [...] is the future of healthcare—the next logical step in a world in which more is known about human genetics, disease, and wellness than ever before" (Duke Medicine, 2011).

Of all the scientific and social promises that stem from advances in our understanding of the human genome, the prospect of examining a person's entire genome in order to make individualized risk predictions and treatment decisions nourishes hope within research. This discussion goes along with the hope to find the genomic basis of personalized medicine and explore its potential for good and its possible risks. As already shown by our introductory case, whole-genome and whole-exome sequencing is already applied outside large epidemiological studies that are conceptualized at a population level and focus on diagnostics of individual patients in the clinical context.

In our view, both examples, precision oncology and genome sequencing for the diagnostics or phenotype-genotype correlation in individual patients, clearly fall under the category of "clinical innovation" due to their novelty and insufficient validation and because of their primary aim to foster an individual patient's well-being. Nonetheless, precision medicine initiatives[8] are most commonly labeled as "research" projects. In view of our conclusions drawn in this chapter, we may be tempted to argue that we should not hastily put new and yet insufficiently proved interventions on the same level with research enterprises that are thought to attain generalizable knowledge. Nevertheless, precision medicine project could clearly benefit from distinguishing between different practical contexts of standard practice, clinical innovation, and sound clinical research. These different practices occurring in large projects, such as the "Precision Oncology Project" or the "Human Genome Project," are frequently interrelated and not adequately distinguished from each other. As we have argued, the disentanglement and categorization of these different practices however potentially impact on their ethics and regulation. Our taxonomy indeed intends to give individual physicians and medical teams a clearer grasp of their responsibilities toward their patients or research subjects in different circumstances and within different roles as providers of standard (practice) or innovative care (clinical innovation) and as researchers (research). We believe that precision medicine itself is context-dependent although the boundaries between the different practical contexts may sometimes be blurred. Deliberating about how to categorize and link different conducts with different medical contexts (standard validated practice, clinical innovation and research) and making this deliberation publicly available show respect for patients and foster public trust.

These considerations go along with Taylor's (2010) notion of the different practical contexts of standard care, clinical innovation, and research. According to him, these contexts are not opposed, but interact in different fields. In particular, he compares how clinical innovation and research interact in the two exemplary fields of oncology and surgery:

> Innovative therapy develops in diverse ways. Some fields, like oncology, have progressed through a close alliance between innovative therapy and research. With this model, innovative therapies are suggested by

[8] We will follow the NIH (US National Institutes of Health 2017) definition on precision medicine. According to this definition, the precision medicine initiative aims to gain better insights into the biological, environmental, and behavioral factors that drive diseases. Precision medicine is an emerging approach for disease treatment and prevention that takes into account individual variability in environment, lifestyle, and genes for each person.

> understanding where current therapies fail, and combining that knowledge with hypotheses about how changes in somewhat-known compounds might affect their action on other diseases. In clinical trials, toxic compounds of uncertain risks are offered to large cohorts, and the results, if positive, are incorporated in medical practice. Other fields, like surgery, take a very different tack: innovative therapy to address an unexpected anomaly intraoperatively can be suddenly required, and predetermined adherence to a protocol might be a fatal rigidity. Only later, after multiple experiences—and perhaps never—may a surgeon test the novel procedure through clinical research, and on many patients. Even then it may be difficult to "reduce" a patient to a randomized research subject if the surgeon is convinced that the novel procedure is better, or that choosing the best care option requires patient-specific judgment. In between oncology and surgery, of course, are many other forms of innovative therapy, including new models of health care delivery that improve care outcomes, or patient-oriented, nursing-led changes in the environment of care to promote quality and sensitivity from a patient's perspective. *(Taylor, 2010, p. 287)*

As Taylor shows, clinical research and innovative therapy are always somewhat interrelated practices that occur in different medical fields. Most notably, the fields of precision medicine and/or personalized medicine make the category of clinical innovation even more salient. As we have previously shown, undergoing new and insufficiently tested interventions outside clinical trials can be ethically justified in all cases in which individuals lack reasonable medical alternatives or when appropriate trials cannot be made available. In the case of personalized therapies, we enter the category of clinical innovation, precisely because individualized therapies and diagnostic methods cannot be offered outside "$N=1$ trials" (Sedgwick, 2014), which means that clinical studies cannot be designed including a sufficiently large number of research subjects to attain sound statistical results. Also, individuals undergo individualized therapies and/or diagnostic measures because they lack standardized or validated interventions that would otherwise be provided in the context of clinical practice.

6 CONCLUSION

The aim of this chapter has been to propose a new taxonomy and a revised definition of clinical innovation for new and yet unproved interventions in medical practice, including diagnostic and therapeutic health interventions and technologies. We have argued that clinical innovation understood as a subcategory of clinical practice is useful and even ethically required for the use and application of new and yet insufficiently validated health interventions.

When elaborating on the concept of clinical innovation, we construed the term innovation in view of Sugarman's (2012) and Taylor's (2010) approaches to translate new interventions into medical practice. We then argued that clinical innovation should be characterized as "new and insufficiently validated practice" that can be considered a special type of medical practice due to the characteristics of its novelty, its lack of sufficient validation, and the aim to directly benefit the patient. We have argued that new and insufficiently validated interventions, provided by physicians or medical teams, can be considered acceptable medical options when no reasonable alternatives are available ("argument of lacking reasonable alternatives") and when the interventions show a minimum level of evidence. However, new and not yet sufficiently validated practice departs in a significant way from standard medical practice because innovative diagnostic or therapeutic procedures are yet unproved in terms of safety and efficacy according to an appropriate epistemic standard. Thus, we have shown that clinical innovation understood as a new and insufficiently validated practice is a useful category itself in terms of ethics oversight and regulation and should also be distinguished from formal clinical research.

This claim finally finds the most promising application in the emerging field of precision medicine and individualized therapy and diagnostics, most notably in oncology and genetics. Together with the conclusions drawn from the previous paragraphs, the concept of clinical innovation and the demand for a change in regulations impact on the emerging field of personalized medicine. We suggested that precision medicine projects would gain from a more accurate and precise taxonomy that are distinguished between different practical contexts of standard practice, clinical innovation, and clinical research. This classification potentially offers to individual physicians or medical teams a clearer grasp of what their responsibilities toward patients in different circumstances are. Although there is no clear division between different medical conducts and practices in precision medicine projects where components of validated practice, clinical innovation, and clinical research interact, we still ought to rethink and deliberate on valuable taxonomies for these medical conducts due to the ethical and regulatory impact a changed classification may have. An appropriate regulation of clinical innovation has not been discussed in this chapter, but is urgently needed and should address the decision-making process, the way and need for publication of results, and the questions of payment by health insurance carriers or other stakeholders.

Acknowledgments

We are extremely grateful to Ariella Binik, Nina Hallowell, Angeliki Kerasidou, Gulamabbas Lakha, Roland Mertelsmann, Florencia Luna, Michael Morrison, Achim Rosemann, Georg Starke, Tobias Schönwitz, and Federico Vasen for their insightful comments or discussions on some of the ideas presented in this article. We also want to express our gratitude to the participants of the Oxford Bioethics Conference (November 2015), the participants of the MPhil seminar in philosophy of science at the University of Cambridge (April 2016), the participants of the FLACSO Fogarty seminar in Buenos Aires (April 2016), the participants of IMBS Symposium "Science, Ethics and Arts" at the University of Freiburg (October 2016), and the participants of the conference organized by the Ethox Center at the University of Oxford (October 2016) for their helpful comments and suggestions for this chapter.

This contribution was also made possible by financial support from CONICET (Argentina), DAAD (Germany), and the Manuel Velasco Suárez Award for Excellence in Bioethics (2014–15) awarded by the Pan American Health Organization (PAHO) Foundation and sponsored by the Ministry of Health of Mexico and UBACyT 20020150100193BA (principle of autonomy, popular sovereignty, and theory of democracy) from the University of Buenos Aires. The views expressed in this paper are personal and do not necessarily reflect the policies of the institutions mentioned above.

References

Ballen, K.K., Gluckman, E., Broxmeyer, H.E., 2013. Umbilical cord blood transplantation: the first 25 years and beyond. Blood 122 (4), 491–498.

BBC, 2001. Life Blood Podcast. Available from, http://www.bbc.co.uk/science/horizon/2001/lifebloodtrans.shtml. Accessed April 17, 2017.

Beauchamp, T.L., Saghai, Y., 2012. The historical foundations of the research-practice distinction in bioethics. Theor. Med. Bioeth. 33 (1), 45–56.

Bothwell, L.E., Greene, J.A., Podolsky, S.H., Jones, D.S., 2016. Assessing the gold standard—lessons from the history of RCTs. N. Engl. J. Med. 374 (22), 2175–2181.

Bradley, J., 2013. In: Ströck, M. (Ed.), New Initiative Will Sequence 10,000 Whole Genomes of People With Rare Genetic Diseases. Available from, http://www.cam.ac.uk/research/news/new-initiative-will-sequence-10000-whole-genomes-of-people-with-rare-genetic-diseases. Accessed April 17, 2017.

Carroli, G., Mignini, L., 2009. Episiotomy for vaginal birth. In: Cochrane Database of Systematic Reviews. John Wiley and Sons. Available from, http://onlinelibrary.wiley.com/doi/10.1002/14651858.CD000081.pub2/abstract. Accessed April 17, 2017.

Cartwright, N., 2007. Are RCTs the gold standard? BioSocieties 2 (1), 11–20.

Daniels, N., 2008. Just Health: Meeting Health Needs Fairly. Cambridge University Press, Cambridge.

Daniels, N., Sabin, J., 1997. Limits to health care: fair procedures, democratic deliberation, and the legitimacy problem for insurers. Philos. Public Aff. 26 (4), 303–350.

Druker, B.J., 2009. Perspectives on the development of imatinib and the future of cancer research. Nat. Med. 15 (10), 1149–1152. Available from, https://doi.org/10.1038/nm1009-1149. Accessed April 17, 2017.

Duke Medicine, 2011. Personalized Medicine. Available from, http://health.usnews.com/health-conditions/cancer/personalized-medicine/overview. Accessed April 17, 2017.

Emanuel, E.J., 2013. Reconsidering the Declaration of Helsinki. Supplementary appendix. Lancet 381, 1532–1533.

Emanuel, E.J., Wendler, D., Grady, C., 2000. What makes clinical research ethical? JAMA 283 (20), 2701–2711.

Faison, A., 2005. The miracle of Molly. 5280 Mag. Available from, http://www.5280.com/magazine/2005/08/miracle-molly. Accessed April 17, 2017.

Gluckman, E., Rocha, V., Boyer-Chammard, A., Locatelli, F., Arcese, W., Pasquini, R., Fernandez, M., 1997. Outcome of cord-blood transplantation from related and unrelated donors. N. Engl. J. Med. 337 (6), 373–381.

International Society for Stem Cell Research (ISSCR), 2016. Guidelines for Stem Cell Research and Clinical Translation. Available from, http://www.isscr.org/docs/default-source/guidelines/isscr-guidelines-for-stem-cell-research-and-clinical-translation.pdf?sfvrsn=2. Accessed April 17, 2017.

Kahn, J.P., Mastroianni, A.C., 2004. Creating a stem cell donor: a case study in reproductive genetics. Kennedy Inst. Ethics J. 14 (1), 81–96.

King, N.M.P., 2011. The line between clinical innovation and human experimentation. Seton Hall Law Rev. 33 (2), 473–582.

King, N.M.P., Churchill, L.R., 2008. Assessing and comparing potential benefits and risks of harm. In: Emanuel, E.J., Grady, C.C., Crouch, R.A., Lie, R.K., Miller, F.G., Wendler, D.D. (Eds.), The Oxford Textbook of Clinical Research Ethics. Oxford University Press, Oxford, pp. 514–526.

Kolata, G., 2016. A cautionary tale for 'stem cell tourism'. NY Times. Article published June 22, 2016. Available from, https://www.nytimes.com/2016/06/23/health/a-cautionary-tale-of-stem-cell-tourism.html?_r=1. Accessed April 17, 2017.

Levine, R.J., 1979. Clarifying the concepts of research ethics. Hast. Cent. Rep. 9 (3), 21–26.

Levine, R.J., 2008. The nature, scope, and justification of clinical research. In: Emanuel, E.J., Grady, C.C., Crouch, R.A., Lie, R.K., Miller, F.G., Wendler, D.D. (Eds.), The Oxford Textbook of Clinical Research Ethics. Oxford University Press, Oxford, pp. 211–221.

Martin, D.K., Meslin, E.M., Kohut, N., Singer, P.A., 1995. The incommensurability of research risks and benefits: practical help for research ethics committees. IRB: Ethics Human Res. 17 (2), 8–10.

Mastroianni, A.C., 2006. Liability, regulation and policy in surgical innovation: the cutting edge of research and therapy. J. Law Med. 16 (2), 351–440.

National Centre for Tumor Diseases (NCT) Heidelberg, 2017. Clinical and Translational Reserach Groups. Available from, https://www.nct-heidelberg.de/forschung/nct-clinical-and-translational-research-programs-groups/molecular-diagnostics.html. Accessed April 17, 2017.

National Commission, 1979. The Belmont Report: Ethical Principles and Guidelines for the Protection of Human Subjects of Research. US Government Printing Office, Washington, DC.

NIH (US National Institutes of Health), 2017. About the 'All of Us' Research Programs. Available from, https://www.nih.gov/research-training/allofus-research-program. Accessed April 17, 2017.

Perdeaux, L., 2013. The Rare Diseases Genomes Project and Genomics England: by the NHS, for the NHS. Available from, http://www.bhdsyndrome.org/forum/bhd-research-blog/the-rare-diseases-genomes-project-and-genomics-england-by-the-nhs-for-the-nhs/. Accessed April 17, 2017.

Rawls, J., 1999. A Theory of Justice. Harvard University Press, Cambridge, MA.

Rehm, H.L., Berg, J.S., Brooks, L.D., Bustamante, C.D., Evans, J.P., Landrum, M.J., Plon, S.E., 2015. ClinGen—the clinical genome resource. N. Engl. J. Med. 372 (23), 2235–2242.

Rosemann, A., 2013. Medical innovation and national experimental pluralism: insights from clinical stem cell research and applications in China. BioSocieties 8 (1), 58–74.

Sedgwick, P., 2014. What is an 'n-of-1' trial? Br. Med. J. 348. Available from, http://www.bmj.com/content/348/bmj.g2674. Accessed April 17, 2017.

Sugarman, J., 2012. Questions concerning the clinical translation of cell-based interventions under an innovation pathway. J. Law Med. Ethics 40 (4), 945–950.
Sutherland, R., Croydon, E.A.P., Rolinson, G.N., 1972. Amoxycillin: a new semi-synthetic penicillin. Br. Med. J. 3, 13–16.
Taylor, P.L., 2010. Overseeing innovative therapy without mistaking it for research: a function-based model based on old truths, new capacities, and lessons from stem cells. J. Law Med. Ethics 38 (2), 286–302.
Weijer, C., 2000. The ethical analysis of risk. J. Law Med. Ethics 28 (4), 344–361.
Weijler, C., Miller, P., 2004. When are research risks reasonable in relation to anticipated benefits. Nat. Med. 10 (6), 570–573.
Word Medical Association, 2008. Declaration of Helsinki: ethical principles for medical research involving human subjects. Available from, http://www.wma.net/en/30publications/10policies/b3/17c.pdf. Accessed April 17, 2017.
World Medical Association, 1964. Declaration of Helsinki: ethical principles for medical research involving human subjects. Available from, https://www.google.de/url?sa=t&rct=j&q=&esrc=s&source=web&cd=4&cad=rja&uact=8&ved=0ahUKEwj7vpWgl6rTAhWBxpAKHafgCSUQFgg6MAM&url=http%3A%2F%2Fwww.upenn.edu%2Fregulatoryaffairs%2FDocuments%2FDeclaration_Helsinki.doc&usg=AFQjCNGnUd3ulQnn3wl-J46kX9YUfsQH6MA. Accessed April 17, 2017.
World Medical Association, 2000. Declaration of Helsinki: ethical principles for medical research involving human subjects. J. Am. Med. Assoc. 284, 3043–3045.
World Medical Association, 2013. Declaration of Helsinki: ethical principles for medical research involving human subjects. Available from, http://www.wma.net/en/30publications/10policies/b3/index.html. Accessed April 17, 2017.

Further Reading

London, A.J., 2006. Cutting surgical practice at the joints: individualizing and assessing surgical procedures. Reltsma and Moreno, supra note 19, 19–52.
Scanlon, T., 2000. What We Owe to Each Other. Harvard University Press, Cambridge, MA.

CHAPTER

2

Issues and Challenges in the Systematic Evaluation of Biomarker Tests

Agnes Kisser

Ludwig Boltzmann Institute for Health Technology Assessment, Vienna, Austria

1 INTRODUCTION

With the sequencing of the human genome in 2001 and the fast development of the omics technologies, expectations are high, that biological processes including disease progression and treatment reactions can be measured accurately and even predicted on a molecular level (Schleidgen et al., 2013). Association studies yield vast numbers of potential biomarker candidates, but so far, there is no definitive consensus on the evidentiary requirements for the evaluation of biomarker tests for clinical routine. Diagnostic accuracy represents an important but not by itself sufficient characteristic of biomarker tests. Common challenges in the evaluation of biomarkers and medical tests are the assessment of multiple steps in a clinical path (test and treatment), the methodological challenges in systematic reviews of diagnostic accuracy and prognostic tests, the lack of direct evidence from randomized clinical trials, and the complex assessment of the applicability of the test and treatment strategy.

In this chapter, we analyze the approaches proposed in methodological guidelines by leading HTA institutes (Gartlehner, 2009; Chang and AHRQ, 2012; Deeks et al., 2010; Derksen and CVZ, 2011; Brozek et al., 2009; IQWIG, 2013; Medical Services Advisory Committee, 2005; Merlin et al., 2013; National Institute of Health and Clinical Excellence, 2011; Institute of Medicine (IOM), 2010; Nachtnebel, 2010; Centre for Reviews and Dissemination, 2009) and provide a procedural guidance for the evaluation of research questions involving biomarkers and medical tests in clinical routine. The structure of the chapter corresponds to the sequence of steps involved in the assessment of a medical test from formulating the research question to the synthesis of the evidence. In the first section, we provide definitions and clarifications of the often varying terminology used in the field. This is followed by guidance for the development of the analytic framework and the formulation of the PICO question. New study designs

Precision Medicine
https://doi.org/10.1016/B978-0-12-805364-5.00002-0

likely to be encountered in reviews of biomarker tests are presented together with alternative evidence hierarchies. Finally, guidance is provided on how to assess bias and applicability in alternative study designs and how to grade the available evidence.

2 TERMINOLOGY AND CLASSIFICATIONS

2.1 Definition: Biomarker

According to the definition of the National Institute for Health (NIH), a biomarker is "a characteristic that is objectively measured and evaluated as an indicator of normal biological processes, pathogenic processes, or pharmacological responses to a therapeutic intervention" (Biomarkers Definitions Working Group, 2001). This broad definition allows for a large variety of biomarkers, from the analysis of small molecules up to the examination of physical parameters such as blood pressure. Table 1 presents an overview of the analytic technologies used in biomarker research and the biological processes, entities, and examples of applications associated with each technology.

These technological categories however do align neither with functional differentiation of biomarkers nor with specific diseases (despite certain associations such as cancer (genomics) or diabetes (metabolomics)). The applications widely vary with regard to their stage of development: while a number of genomic, imaging, and whole body biomarkers are already in use in clinical practice, many other areas are still in the stage of early basic research and hypothesis generation in association studies.

2.2 Differentiation of Functions of Biomarkers

The methodology of HTA evaluation of a biomarker test is independent of the technological category of the biomarker with one exception: genetic/genomic biomarkers require a particular consideration as health and non-health-related (e.g., ethical, social, and legal) effects may affect more than one generation. To identify the outcomes relevant for the evaluation in a specific context of use, it is useful to differentiate biomarkers by the type of information they provide and the context of their use in clinical practice, detailed below (Biomarkers Definitions Working Group, 2001; Febbo et al., 2011; Teutsch et al., 2009). As a principle, any biomarker is only useful if the test results are associated with appropriate differential treatment options.

2.2.1 Differentiation by Type of Information

Diagnostic biomarkers are used to identify patients with a particular health condition and to differentiate it from other conditions with similar symptoms, requiring differential treatment. Tests for diagnostic biomarkers may be used as replacement for time-consuming, expensive, or invasive diagnostic procedures (e.g., replacement of echocardiogram by testing levels of brain natriuretic peptides to rule out heart disease) or as an add-on to existing methods to further refine the diagnosis. Diagnostic biomarkers can be used as screening markers to identify persons with an underlying disease in a screening population. (e.g., elevated blood glucose concentration for the diagnosis of diabetes mellitus).

TABLE 1 Analytic Technologies in Biomarker Research—Overview of Research Areas, Biological Entities, and Processes Studied—Present and Future Applications

Research Area	Biological Entity	Biological Process	Examples of Applications
Genomics	Gene characteristics (variations in sequence, copy number, epigenetic modification)	Gene expression Gene function Gene regulation	Pharmacogenomics/ pharmacogenetics, response to medication Nutrigenomics, effects of food and food constituents Epigenomics, effects of epigenetic modifications Metagenomics, study of communities of microbial organism (e.g., gut flora) Immunomics, study of immune system regulation and responses to pathogens
Transcriptomics	mRNA characteristics (variations in sequence, expression levels, processing, splicing, editing)	Gene expression	Metatranscriptomics, study of communities of microbial organism (e.g., gut flora)
Proteomics	Proteins (structure, posttranslational modifications)	Protein expression Protein-protein interactions Protein function and activity Protein secretion	Immunoproteomics, study of proteomes in immune response Secretomics, analysis of the secreted proteins in a cell
Metabolomics	Metabolites (small molecules) Metabolic profile	Biochemical pathways	Toxicology, metabolic profiling of the response to toxic insult of a chemical or drug Nutrigenomics (see above) Lipidomics, analysis of lipid species within a cell or tissue
Imaging	Cell Tissue Organ	Growth rate, metabolic rate, plaque formation, inflammation	Pharmacokinetics, analysis of time course of drug absorption, distribution, metabolism, and excretion Pharmacodynamics, biochemical and physical effects of a drug on the body
Physical or physiological measurements	Whole body	Various	Biomarkers of aging (e.g., muscle mass and muscle strength) Multiple sclerosis biomarkers (e.g., walking capacity)

Prognostic biomarkers predict the likely course of a disease in patients regardless of the treatment given. They can be used for risk stratification (triage) of patients based on their risk of disease progression to avoid expensive or invasive treatments and to ensure an optimal distribution of resources. Prognostic biomarkers may also be used in a screening population to identify people at risk to develop a specific health condition; these so-called risk biomarkers are indicative of a changed physiological state that is associated with a risk of disease. Several prognostic markers can be combined in a prediction model (e.g., fibrinogen for prognosis of primary stroke (van Holten et al., 2013)).

Predictive biomarkers (theranostic biomarker) predict the response of the patients to a specific treatment in comparison with the standard treatment, placebo, or observation only (the biomarker (information) vs treatment (effect) interaction). They are used for treatment stratification and guide the choice of treatment (Sargent et al., 2005). In vitro companion diagnostics are defined as "in vitro diagnostics that provide information that is essential for the safe and effective use of a corresponding therapeutic" (FDA Definition, Food and Drug Administration, 2014): here, the use of a specific treatment is obligatorily preceded by a test for this predictive biomarker (e.g., trastuzumab/HER2 testing for breast cancer) (Table 2).

2.2.2 Differentiation by the Context of Use

Screening markers are used to detect disease in asymptomatic or presymptomatic persons and are used for prevention. They may provide diagnostic or prognostic information.

Etiologic markers are similar to prognostic markers in that both are risk factors for a specific outcome but should be differentiated by the population they relate to: In prognosis, all the population has the same disease/condition; in etiology, the marker serves to differentiate between individuals with/without the condition and to identify causal risk factors for a specific condition.

Monitoring biomarkers are used for surveillance of the response to treatment (e.g., CEA and PSA for monitoring of tumor status during therapy and between image evaluations (Beachy and Repasky, 2008)).

Surrogate end points are laboratory measurements or physical signs used in therapeutic trials as a substitute for a clinically meaningful end point. "A surrogate end point is expected to predict clinical benefit (or harm or the lack of benefit or harm) based on epidemiological,

TABLE 2 Differentiation of Biomarker Types by Impact on Health Outcomes

		Biomarker +		Biomarker −	
	Outcome	**Drug A**	**Drug B**	**Drug A**	**Drug B**
Predictive	Overall survival	60%	30%	30%	30%
	RR	2.0		1.0	
Prognostic	Overall survival	70%	60%	35%	30%
	RR	1.2		1.2	
Both	Overall survival	60%	40%	20%	20%
	RR	1.5		1.0	

therapeutic, pathophysiological, or other scientific evidence" (Biomarkers Definitions Working Group, 2001).

ⓘ For a more in-depth overview and discussion of definitions, refer to Surrogate biomarkers (Institute of Medicine (IOM), 2010, Chapter 1, p. 17ff)

Diagnostic/Prognostic/Predictive biomarkers: Biomarkers Definitions Working Group (2001)

2.3 Medical Tests

By definition, not only a biomarker must represent a specific biological process, but also it must be objectively measurable. Thus, biomarkers are always associated with a corresponding medical test. The test must be applicable in clinical routine and reproducible and accurately translate the biomarker into a measurement parameter. Often, several tests are available to measure the same biomarker—in this case, not only evaluation takes into account the specific test performance characteristics, but also costs and handling of the test might become decisive factors.

Medical tests can be differentiated by test methodology:

- Imaging

  This category comprises radiology (classic x-Ray), sonography (ultrasound), x-ray, computed tomography, magnetic resonance imaging, angiography, positron emission tomography, and other methods of nuclear medicine.
- Analysis of body fluids or smears

  Most frequently analyzed body fluids are blood, urine, and cerebrospinal fluid; less common are synovial fluid, sweat, saliva, and gastric juices. The analysis involves chemical or molecular biologic assays and cytological examinations of cell smears or suspensions.
- Endoscopy

  This consists in the investigation of interiors of organs or body cavities with an endoscope (viewing tube) that is introduced through a small incision or an existing body orifice (the nose, mouth, anus, urethra, and vagina).
- Measurement of body functions

  Examples for this category are the measurement of blood pressure and the measurement of the electric activity of the heart (electrocardiography, ECG) or of the brain (electroencephalography, EEG).
- Examination of biopsies

  This consists in the removal of a tissue sample followed by the histological examination of the tissue. The analysis of the biopsy may also involve chemical or molecular biologic tests.

2.4 Evaluation Framework

The purpose of a systematic review of any medical test is to identify and present evidence of its clinical utility: the health outcomes associated with its use (Chang and AHRQ, 2012; Derksen and CVZ, 2011; National Institute of Health and Clinical Excellence, 2011; Febbo et al., 2011; Morrison et al., 2012). Unlike the outcomes of therapeutic interventions, the clinical outcomes of medical tests are only in part directly induced by the test procedure but most of them indirectly by patient management decisions and treatments initiated according to the test results. The majority of the studies evaluating medical tests cover only segments of this path.

A number of frameworks have already been proposed for the evaluation of medical tests (Morrison et al., 2012; Lijmer et al., 2009). Organizing frameworks can and should be "used to categorize key questions and suggest which types of studies would be most useful for the review" (Chang and AHRQ, 2012, pp. 2–6). These parameters can be mapped to a three-step evaluation process (Institute of Medicine (IOM), 2010). The three tiers are interrelated: they can be assessed individually but need to be correlated to come to a final evaluation. A change in technical parameters will require a reevaluation of patient-relevant outcomes, and it specifically includes the contextual analysis and the assessment with regard to the specific context of use. Especially for diagnostics, clinical practice of testing algorithms may strongly vary between countries.

The first step, analytic validation, describes the ability of a test to reliably and accurately measure a biomarker of interest, including limits of detection, limits of quantitation, reference value cutoff concentration, reliability, and reproducibility.

The qualification step comprises the actual evidentiary assessment of the association between biomarker and disease states. In the case of surrogate biomarkers, the qualification step includes in addition assessment of evidence that interventions targeting the biomarker have an impact on health outcomes. Evidence on the impact on health outcomes can be provided by direct or by indirect evidence.

The third step of biomarker evaluation—utilization—consists in the analysis of the evidence "with regard to the proposed use of the biomarker." In this step, evaluators should take into consideration the specific context of use of the biomarker with regard to target population, setting, and purpose of the biomarker.

Table 3 gives an overview of the characteristics evaluated in each step of the biomarker evaluation, with example parameters. Where varying terms were used in the literature, they are listed in brackets below each of the evaluation steps.

TABLE 3 Steps in the Evaluation of Biomarkers

Steps in Biomarker Evaluation	Biomarker Characteristics Evaluated	Parameters (Examples)
Analytic validation (technical efficacy, analytic validity)	Ability of a test/chemical assay to quantitate a biomarker of interest	Technical quality of a radiological image, reproducibility, repeatability
Qualification 1 (test accuracy—diagnostic/prognostic accuracy, clinical validity)	Ability of a test to classify a patient into a disease, phenotype, or prognosis category	Sensitivity, specificity, SROC curve
Qualification 2 (clinical utility, therapeutic efficacy, patient outcome efficacy)	Ability of a test to improve patient outcomes	Changes in patient management, mortality, morbidity
Utilization (societal aspects, economic aspects)	Contextual analysis and risk-benefit assessment with regard to the proposed use	Opportunity costs

ⓘ For a more in-depth overview, refer to three steps in the biomarker evaluation process (Institute of Medicine (IOM), 2010, p. 5ff)

Review of different types of analytic frameworks: Lijmer et al. (2009)

3 CONTEXT FOR ANALYSIS

3.1 Developing an Analytical Framework

A systematic review is the method of choice for the evaluation of medical tests. The same quality criteria as for systematic reviews of therapeutic interventions apply (Gartlehner, 2009; Chang and AHRQ, 2012; Nachtnebel, 2010). A specific challenge in the evaluation of medical tests is that more time needs to be dedicated to a careful definition of the review question to avoid ambiguity during literature search and literature selection and the evaluation of the selected studies.

Due to the indirect influence on patient outcomes, a systematic review of a medical test starts with clarifying the embedment of the test in clinical routine and the implications resulting thereof. A useful tool is to create an analytic framework, including decision-making, further tests and treatments, patient outcomes, and their surrogates (Harris et al., 2001) (Fig. 1). This might include liaising with relevant experts and also by taking into account the indications of the manufacturer. Often, there are several scenarios possible with varying positions of the test in the diagnostic chain, and in the first step, the scenario(s) relevant to the commissioner of the study need(s) to be clarified.

3.2 Formulation of Review Question With PICO

3.2.1 Population

The review should clarify the target population of the test, i.e., to which patients the test is planned to be applied to. This includes information on the following:

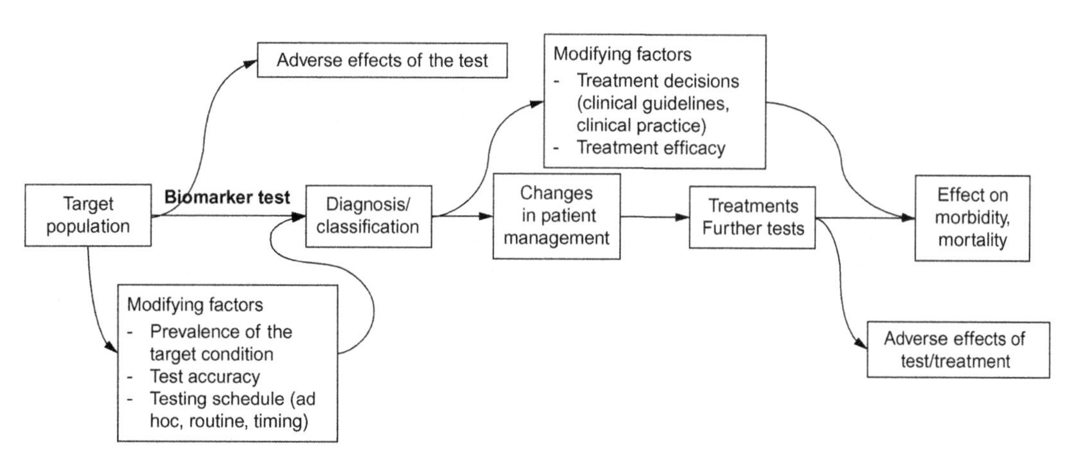

FIG. 1 Schematic representation of an analytic framework for the evaluation of a biomarker test.

- Demographic characteristics (age and sex)
- Medical history (prior diseases and treatments and comorbidity)
- The clinical setting in which the test will be used (inpatient, ambulant/outpatient, doctor's office, self-administration, etc.)
- The prevalence of the target disease in the target population
- For prognostic markers—the observed probability (i.e., the observed proportion of an event in a given time period (Tripepi et al., 2010a,b)) of the outcome being predicted

From the prevalence of the disease, the reviewer may deduce the pretest probability of the target health condition in the population. In combination with the diagnostic accuracy parameters, the pretest probability allows to calculate the posttest probabilities.

The detailed description of the target population is further required for the assessment of the available studies with regard to the applicability of their results to the review question.

3.2.2 *Intervention*

An essential characteristic of medical tests is that they are usually not stand-alone interventions, but used within in a testing and treatment strategy, often comprising several sequential or parallel diagnostic tests and various treatment options according to specific combinations of test results. In the review question, therefore, the intervention under review should be defined as the entire sequence of tests and treatments in which the test under review is included.

Bossuyt et al. identified three options on how to integrate a medical test in an existing testing and treatment strategy depicted in Fig. 2—replacement, triage, or add-on (Bossuyt et al., 2006). The purpose of a replacement test is usually to maintain the same test performance (sensitivity and specificity) as the comparator while increasing cost-effectiveness or reducing adverse events due to invasive procedures. Triage tests serve to avoid invasive or expensive procedures by decreasing unnecessary referrals. They should maintain the same sensitivity as the comparator, since test negatives of the new test will not be tested by the existing test

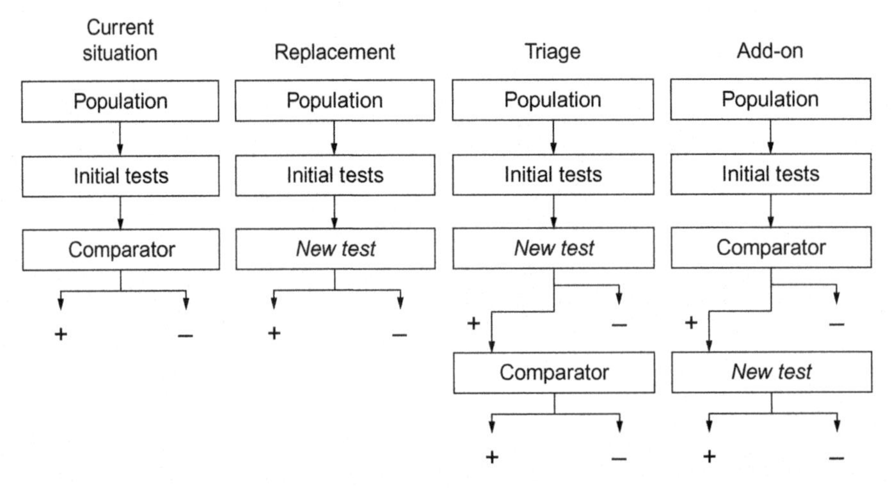

FIG. 2 Different options to integrate a medical test in an existing test and treatment strategy comprising initial tests and a comparator test against which a new test is evaluated. *Bossuyt, P.M., Irwig, L., Craig, J., Glasziou, P., 2006. Comparative accuracy: assessing new tests against existing diagnostic pathways. BMJ 332 (7549), 1089–1092.*

but may have lower specificity. Add-on tests finally serve to refine a diagnosis with the goal to improve treatment decisions and, thus, outcomes (Lord et al., 2009).

Initial tests and patient management and treatment decisions need to be clarified with relevant experts and guidelines to adjust to the national context.

Description of the intervention should further include variants of the test, a definition of the cutoff point(s), and the timing of the application of the test (follow-up). Moreover, the description of the test should include an assessment of infrastructure and processes necessary for implementation of the test in practice.

In diagnostic accuracy studies, the test whose performance is evaluated is called "index test."

3.2.3 *Comparator*

Considering the current test and treatment strategy, i.e., the sequence of tests, patient management decisions, and treatments, the comparator is the test currently used to provide a classification of the disease for diagnosis, to identify clinically relevant subgroups and to make a management decision. This may vary in dependence of the intended placement of the new test in the current management strategy. The choice of the most appropriate comparator for the review question should therefore be clarified by involving clinical guidelines and possibly experts. (Lord et al., 2011).

DIFFERENTIATION BETWEEN REFERENCE STANDARD AND COMPARATOR

In diagnostic accuracy studies, a reference standard is needed to define the target condition (Lord et al., 2011). Naive estimates of diagnostic accuracy are calculated based on the assumption that the reference standard is "perfect," i.e., capable of identifying "true" cases of disease and nondisease with 100% accuracy. This is problematic in circumstances where, in fact, the reference standard is poor and in cases where no reference standard is available. In these cases, estimates of diagnostic accuracy are likely to be biased and are unsuitable to substitute valid measurements of clinical outcomes (Trikalinos and Balion, 2012). Also, it is not possible to use a diagnostic accuracy study if the new test under evaluation is expected to have superior diagnostic accuracy to the current reference standard. In these cases, the best way to determine diagnostic effectiveness would be a trial.

Diagnostic accuracy studies allow a direct comparison of two tests against a same reference standard in fully paired direct comparisons, where all samples are tested with the index test, one or more comparator tests, and the reference standard. As an alternative, participants may be randomly allocated to receive the index or the comparator test (randomized direct comparison). If the estimates of diagnostic accuracy from different study populations are compared, the comparison is indirect (Takwoingi et al., 2013).

3.2.4 *Outcomes*

Accurate diagnosis is a prerequisite for a successful therapy, but it should not be seen in isolation. Instead, the benefit to patients resulting from diagnosis should be measured in patient-relevant outcomes (Hsu et al., 2011), such as survival (mortality), clinical events, adverse events, patient-reported outcomes (health-related quality of life), activity, and function.

Based on the analytic framework developed, the reviewer should first explore all outcomes resulting from embedment of the test in the testing and treatment strategy in comparison with clinical practice without the test (as described in "intervention" and "comparator"). Reviewers should then make a careful selection of the relevant outcomes both to the process of testing and to the results of the test (Segal, 2012) by mapping them according to the following categories (Segal, 2012; Bossuyt and McCaffery, 2009):

- Clinical management effects due to testing
- Direct health effects of testing
- Emotional, social, cognitive, behavioral responses to testing
- Legal and ethical effects of testing

A decision that outcomes are relevant for a review depends on the type of test under review and on the needs of the stakeholders of the study (Segal, 2012). A review assessing the inclusion of a test in the benefit catalog of health insurances or hospital interventions might be more restricted on outcomes directly affecting the patient, while a review serving medical guideline development or the choice of a screening algorithm needs to take into account the outcomes on a societal level. As a consequence, the prioritization of the relevant outcomes should involve the commissioner of the study. The outcomes are decisive only if they differ between current and new testing and treatment strategy.

The outcomes should explicitly be rated by importance a priori (Hsu et al., 2011):

- Clinical management effects due to testing

  This category describes the clinical consequences (in patient-relevant outcomes—mortality, morbidity, and quality of life) that the use of a test will induce. This includes expected consequences based on a negative (true negative, TN, and false negative, FN) or positive (true positive, TP, and false positive, FP) test result, as supported by primary literature on therapy decisions and outcomes in the particular setting under review. Further consequences might be induced by unexpected findings: they are particularly prominent in imaging methods.

  Clinical management effects might be of particular importance in the evaluation of diagnostic and prognostic tests (Segal, 2012).

  The desired management effects define which test performance parameters are of greatest importance: A test used to rule out the presence of a disease or a high risk for a disease should have high sensitivity and negative predictive value, NPV; a test used as "add-on" to refine diagnosis should have high specificity or positive predictive value, PPV.
- Direct health effects of testing

  This category relates to the health effects that are directly induced by the test procedure and might be particularly relevant for invasive procedures or procedures that involve radiation, while other forms of testing (e.g., a vaginal swab) most likely will not have any health consequences.
- Emotional, social, cognitive, behavioral responses to testing

  Emotional responses might include relief or anxiety as a consequence of a test result—outcomes of this type might be particularly relevant in screening or prognostic tests. Test results might induce a change in behavior in the testees—for example, opting for a healthier lifestyle to compensate a high-risk prognosis or sustaining an unhealthy lifestyle as a consequence of a low-risk prognosis. Emotional responses might also be related to the test procedure itself—for example, psychological symptoms following colonoscopy (Berzin et al., 2010). The effects might also extend to family members. Social issues such as stigmatization, discrimination, and privacy/confidentiality should also be considered. Genetic tests might have complex impact on behavior, e.g., regarding family planning.

- Legal and ethical effects

  Legal consequences might arise in the case of reportable diseases (to be reported by the health-care provider) or diseases representing a safety threat in certain professions (pilots, surgeons, and employees in gastronomy), which warrant disclosure to the employer. Besides disclosure or reporting requirements, legal issues might involve consent and ownership of data and/or samples, patents, licensing, and proprietary testing.

  Genetic tests have a special status with regard to the ethical and legal effects because of the possible impact on family members. This depends on whether the testing is for inherited or acquired genetic mutations and the inheritance pattern of the trait.

  Based on the inventory of outcomes, the reviewer will need to choose the relevant outcomes for the review, depending on time and resources available and the intended purpose of the review.

ⓘ A checklist for assessing the context of submission and the proposed impact of a biomarker test/technology on current clinical practice is proposed in Merlin et al. (2013).

The ACCE model proposes a checklist of targeted questions specifically for genetic testing (Haddow and Palomaki, 2003).

---

4 STUDY DESIGNS AND STUDY SELECTION

To identify the evidence base relevant to a particular research question, a systematic literature search is conducted to identify all studies relevant to the research question, and the evidence base is established based on the quantity and quality of the studies included (Hillier et al., 2011).

4.1 Literature Search

Several common challenges are associated with literature search for medical tests with the following key points (Relevo, 2012):

Due to still underdeveloped indexing and reporting of studies of diagnostic tests, literature search should not rely (exclusively) on diagnostic search filters; in particular, these filters are inappropriate for systematic review of clinical effectiveness.

If the name of the diagnostic test(s) relevant for the research question is not known, search strategies should capture the "concept of diagnostic tests," e.g., diagnosis OR diagnose OR diagnostic OR di(sh) OR "gold standard" OR "ROC" OR "receiver operating characteristic" OR sensitivity and specificity(mh) OR likelihood OR "false positive" OR "false negative" OR "true positive" OR "true negative" OR "predictive value" OR accuracy OR precision.

To identify all studies for a systematic review, searches should include text words (not subject headings alone) and be combined with hand search including additional sources of information: specialized databases, citation tracking, and regulatory documents (Relevo, 2012).

ⓘ A thorough description of "effective search strategies for systematic reviews of medical tests" is given by Relevo et al. in the AHRQ Methods Guide for Medical Test Reviews (Relevo, 2012, p. 4–1ff).

4.2 Hierarchies of Evidence

For assessments of therapeutic effectiveness, a hierarchy of evidence has been established and is widely accepted based on the degree of bias associated with observational and nonrandomized studies in comparison with randomized controlled trials (Samson and Schoelles, 2012; Benson and Hartz, 2000; Kunz and Oxman, 1998; Concato et al., 2000): this hierarchy attributes the highest level of evidence (level I) to systematic reviews and meta-analyses of RCT and level II to evidence obtained from at least one (properly designed) RCT. Levels III and IV subsequently refer to nonrandomized comparative studies and case series, respectively (Table 4). Similarly, according to Grading of Recommendations Assessment, Development and Evaluation (GRADE), only RCTs are a priori considered to provide high-quality evidence about treatment effects (Guyatt et al., 2011a).

Study designs of biomarker studies may vary from classical intervention studies. This section is therefore meant to help reviewers to categorize the studies identified during literature search.

TABLE 4 Evidence Hierarchies by Research Questions

Evidence Level		Intervention	Diagnostic Accuracy	Prognosis	Screening	Etiology
Highest	I	Systematic review of level II studies	Systematic review of level II studies	Systematic review of level II studies	Systematic review of level II studies	Systematic review of level II studies
High	II	RCT	Diagnostic accuracy study Independent blinded comparison Valid reference standard Consecutive patient sample Defined clinical presentation	Prospective (inception) cohort studies (phase 2 or phase 3 explanatory studies)	RCT	Prospective cohort study
Moderate	III	Nonrandomized controlled trial/ cohort/follow-up study	Diagnostic accuracy study not meeting the criteria for level II, diagnostic case-control study	Cohort study (phase 1 explanatory study) or control arm of randomized trial	Non randomized controlled trial/cohort/ follow-up study	Retrospective cohort study or case-control study
Low	IV	Case series, case-control, or historically controlled studies	Diagnostic accuracy study with poor reference standard, study of diagnostic yield	Case series or case-control studies	Case series, case-control, or historically controlled studies	Cross-sectional study or case series

Following the evaluation framework, in order to fulfill the qualification step, evaluators perform first an evidentiary assessment on the causal relationship between biomarker and disease pathogenesis and second an assessment of the evidence that interventions targeting the (surrogate) biomarker impacting the health outcome of interest (Institute of Medicine (IOM), 2010).

In principle, the most appropriate study design to evaluate the impact of a biomarker test on clinical management effects and health outcomes is an RCT (Merlin et al., 2009; OCEBM Levels of Evidence Working Group, n.d.). However, the assessment of biomarkers does not necessarily include a classical intervention research question (therapeutic effectiveness), but instead may include questions on prognosis, etiology, or diagnostic accuracy, depending on the context of use. For some of these questions, the only evidence feasible and/or ethical will be from observational studies, and different evidence hierarchies may apply (Institute of Medicine (IOM), 2010; Merlin et al., 2009; OCEBM Levels of Evidence Working Group, n.d.). To reflect this, revised evidence hierarchies have been elaborated (Merlin et al., 2009; OCEBM Levels of Evidence Working Group, n.d.), which should be considered when prioritizing available evidence (Table 3).

4.3 Diagnostic Accuracy Studies

Diagnostic accuracy studies are cross-sectional by nature. Study designs are differentiated in "single-gate" (diagnostic cohort study) and "two-gate" (diagnostic case-control studies) studies.

In "single-gate" studies, all study participants are first tested with an index test and then with the reference standard. Provided all participants undergo both tests, the sequence of testing may be reversed (reversed-flow design), without influencing estimates of diagnostic accuracy (Rutjes et al., 2005). "Two-gate" studies make comparisons between participants with confirmed disease/condition and healthy participants. The "two-gate" design is intrinsically prone to spectrum bias, potentially leading to inflated estimates of the diagnostic accuracy (Rutjes et al., 2005, 2006). Quality assessment should identify if a "two-gate" study represents only a limited spectrum of disease and nondisease and if so omit the study from the meta-analysis (if done) (Bossuyt and Leeflang, 2008).

The identification of sources of bias in diagnostic accuracy studies requires a detailed assessment of the study design, including criteria on reference standard, study population, and blinding.

ⓘ When can diagnostic accuracy be used as surrogate for health outcomes? (Samson and Schoelles, 2012; Lord et al., 2006)

- The index test has similar sensitivity as the comparator test but has other positive attributes such as higher specificity, lower costs, fewer adverse events, or being less invasive; the value of the test corresponds to the benefits of avoiding adverse events or costs associated with the comparator test.
- The index test has higher sensitivity and similar specificity than the comparator test, and the extra cases detected by the new, more sensitive index test represent the same spectrum of disease (size, grade, and severity) or the same definition of disease for which treatment response is known. This condition may, e.g., be fulfilled if in clinical diagnostic routine, test cases are subsequently confirmed by the reference standard (linked-evidence approach).
- The index test has higher sensitivity and similar specificity than the comparator test, and treatment response of the extra cases detected by the index test has been shown in trials (linked-evidence approach).

If the index test is less sensitive or less specific than the comparator test but has other positive attributes, assessing the trade-off of benefits and harms of using the index test will require direct evidence from an RCT (Lord et al., 2006).

RCTs are also needed if no or only a poor reference standard is available to determine diagnostic accuracy or if the index test is expected to perform better than the current reference standard, which by definition cannot be demonstrated in diagnostic accuracy studies.

4.4 Prognostic Studies

A prognostic test is used to predict a patient's likelihood to experience a medical event (disease development or progression), within a defined time interval and using the observed proportion of the population experiencing this event as reference.

Biomarkers are indicative of a physiological state and therefore not necessarily causal. Cross-sectional studies do not allow for causal inferences to be made since in these studies, biomarker-disease measurements occur simultaneously. In order to show causality and, hence, prognostic value, prospective cohort studies are required that follow health outcomes over time in a population characterized by the levels of the biomarker (Institute of Medicine (IOM), 2010).

Huguet et al. recently proposed an adaptation of the GRADE framework to research on prognostic factors in which they suggest to consider the phase of investigation in the ranking of evidence: a high level of evidence for prognosis would be provided by prospective or retrospective cohort studies that test a fully developed hypothesis and conceptual framework on the underlying processes for the prognosis of a health condition (Lord et al., 2006). Studies in an early stage of investigation, to generate hypotheses, would be attributed as a moderate level of evidence.

4.5 Randomized Clinical Trial Designs

The evaluation of biomarkers used for treatment selection necessarily requires a randomized design to isolate the effect of the marker on therapeutic efficacy from all the other factors influencing a treatment choice (Mandrekar and Sargent, 2009). To this end, new trial designs have been proposed, with substantial variability in the labeling of the trial designs (Tajik et al., 2013). Trial designs can be classified according to the patient flow in the studies; each category allows assessing different effects (Table 5) (Freidlin et al., 2010).

- Targeted or enrichment designs (Tajik et al., 2013; Freidlin and Korn, 2014)

  Patients are screened for the presence or absence of the marker, and only the subgroup of patients defined by a specific marker status is studied (e.g., HER2/trastuzumab trial (Andre et al., 2014)). The study is powered to detect a clinically meaningful effect in the marker-positive subgroup, but does not provide information on the treatment benefit in the marker-negative group. It therefore does not allow to answer the question, whether costs and inconvenience associated with biomarker-based treatment allocation are worthwhile (Freidlin and Korn, 2014). Furthermore, causality of any association of treatment effect with biomarker status may not be established, as there is no comparison with a biomarker-negative group.

TABLE 5 List of Effects That Can Be Assessed and Questions That Can Be Answered by the Trials of Each Design Category

		Marker by Treatment	**Biomarker Strategy**		
Questions trial can answer	Enrichment	With randomization stratified by biomarker status	With treatment randomization in the control arm	RCT of discordant results	Double RCT
Treatment effects					
Q1. How does the experimental treatment compare with the control treatment in biomarker positives?	✓	✓	✓	–	✓
Q2. How does the experimental treatment compare with the control treatment in biomarker negatives?	–	✓	✓	–	✓
Q3. How does the experimental treatment compare with the control treatment in overall study population?	–	✓	✓	–	✓
Biomarker effects					
Q4. Is the biomarker status associated with the outcome in the standard of care group? (is the biomarker prognostic?)	–	✓	✓	–	✓
Q5. Is the biomarker status associated with the outcome in the experimental treatment group?	–	✓	✓	–	✓
Biomarker by treatment effect					
Q6. Is the biomarker status associated with a benefit of experimental treatment? (is the biomarker predictive?)	–	✓	✓	–	✓

TABLE 5 List of Effects That Can Be Assessed and Questions That Can Be Answered by the Trials of Each Design Category—cont'd

		Marker by Treatment	Biomarker Strategy		
Strategy effects			✓		✓
Q7. How does the biomarker-based treatment strategy compare with the control treatment in the overall study population?	–	✓ Indirect	✓	✓	✓
Q8. How does the biomarker-based treatment strategy compare with the experimental treatment in the overall study population?	–	✓ Indirect	✓	✓	✓

Alternatively, marker-negative patients may be assigned a control treatment—this form of hybrid design was used for example in the TAILORx trial for the evaluation of Oncotype DX (Zujewski and Kamin, 2008).

- Marker by treatment interaction design or tests at baseline in RCT (Sargent et al., 2005; Lijmer and Bossuyt, 2009; Janatzek, 2011)

All patients are randomly assigned to treatments based on a randomization that may or may not be stratified based on biomarker status. This is the most efficient trial design in situations in which there are two or more existing treatment options with no definitive evidence for one being preferred in a given population (Freidlin et al., 2010).

This design allows the embedment of the evaluation of various diagnostic strategies in classical intervention studies. To this end, either before or after randomization, all study participants are tested with one or more diagnostic test strategies (e.g., biomarker-based and non-marker-based/standard). Then, all participants are randomized to either of two treatments: this randomization should ideally be blinded and independent of test results. In principle, this study design may also be performed retrospectively in archived samples of a classical intervention study of the treatment under consideration, provided that the standard diagnostic treatment decision can be determined retrospectively (Janatzek, 2011). Retrospective stratification, however, involves a higher risk of bias by confounding. A stratified randomization by biomarker will ensure appropriate power of the study and minimize selection bias. Because all patients are randomized to both treatment and control, this study design is also designated "randomize all."

With randomization stratified according to biomarker status, this study design allows to assess the relationship between the test (biomarker) and the treatment, i.e., whether the biomarker is predictive (modifier of treatment effect) or prognostic (favorable outcome in marker-positive patients regardless of treatment).

- Biomarker strategy designs (Sargent et al., 2005; Mandrekar et al., 2005)
  In this design, marker status is first determined in all patients, and then, patients are randomized to either a biomarker-based or a biomarker-independent strategy for allocation of treatment. The biomarker-independent strategy may be a standard diagnostic pathway or a randomized treatment allocation. This type of trial design allows to directly compare the outcome of all patients in the marker-based arm with the outcome of the patients in the non-marker-based arm. The efficiency of this trial design is limited by the fact that in many cases, treatment choices by either diagnostic strategy would be the same for a large number of the patients, reducing the potential observable differences between the groups and thus increasing the required participant number (Mandrekar and Sargent, 2009). This study design allows to directly compare the potential of two diagnostic strategies to differentiate between patients likely to profit and patients unlikely to profit of a specific treatment. This study design does not allow to assess the relationship between the test (biomarker) and the treatment, i.e., whether the biomarker is predictive (modifier of the treatment effect) or prognostic (favorable outcome in marker-positive patients regardless of treatment) (Freidlin and Korn, 2010).
- RCT of discordant test results (Lijmer and Bossuyt, 2009; Janatzek, 2011)
  This is an adaptation of the marker-based strategy design, enriching on the fraction of participants that receive discordant results using the two diagnostic strategies and sparing a treatment randomization to those in which both diagnostic strategies come to the same treatment decision.
  All participants are tested with both strategies; only those with discordant test results are randomized to the treatment options under consideration. In both groups (new +/ standard −) and (new −/standard +), superiority (or noninferiority) of the treatment chosen by the biomarker-based strategy over the treatment option chosen by standard diagnosis needs then to be demonstrated (Janatzek, 2011).
- Double-randomized controlled trial
  This design combines a randomization to the testing strategy, similar to the biomarker-strategy design, with a randomization to the treatment in both biomarker-positive and biomarker-negative groups to allow to explain the biomarker-drug relationship (i.e., predictive vs prognostic factor) (Merlin et al., 2013). This design poses practical limitations, especially if the biomarker is uncommon. Ethical challenges might arise if a new treatment is tested as a replacement for an old therapy and an effect is only plausible in biomarker-positive patients: in this case, it might be considered unethical to randomize biomarker-negative patients to a treatment, where no effect is to be expected while forgoing an effective treatment (Merlin et al., 2013), which is also of concern for the marker by treatment interaction design.

5 CRITICAL APPRAISAL-SPECIFIC ISSUES IN BIOMARKER STUDIES

5.1 Diagnostic Accuracy Studies

Diagnostic accuracy relates to the strength of association between the results of the index test and a reference standard (or "gold standard," meaning the best available diagnostic

TABLE 6 Tabular Presentation of the Changes in Classification Induced by a New Test Compared With Standard Strategy—For Example, p16/Ki-67 Triage Compared With Direct Referral to Colposcopy

Putative Benefit of New Test	Sensitivity	Specificity	TP	TN	FP	FN	Inconclusive Results
Simpler, less time	Lower	Higher	↘	↗	↘↘	↗	Unknown
Benefits and harms from changes in classification			Benefit from treatment	Benefit from avoiding unnecessary tests	Anxiety and morbidity from unnecessary additional testing and treatment	Possible detriment from delayed diagnosis	Benefit from treatment, anxiety and morbidity from unnecessary testing and treatment
Certainty/uncertainty of the benefits/harms			No uncertainty	Major uncertainty (not clear if clinicians would trust test results)	No uncertainty	Major uncertainty (impact of negative result on screening attendance unclear)	No uncertainty (all cases will be confirmed by colposcopy +/− biopsy)

approach). The presumed consequences of classification in each of the categories (FP, TP, FN, and TN) need to be defined (see example in Table 6) and the directness of the link assessed. Benefits of correct classification should outweigh the harms of misclassification. This assessment should also include consequences of inconclusive results (Hsu et al., 2011).

All sources of bias particularly common to diagnostic accuracy studies have been comprehensively reviewed by Whiting et al. (2004), and based on these results, the QUADAS-2 checklist has been developed to assess the main empirically validated sources of bias in diagnostic accuracy studies (Whiting et al., 2011).

Studies with a "two-gate" or "diagnostic case-control" design are prone to spectrum bias, because they possibly omit those cases that are difficult to diagnose and thereby would overestimate the diagnostic accuracy. To assess this potential bias, reviewers should assess the patient selection in these studies and estimate if despite the case-control design the full spectrum of disease and nondisease is represented.

The appropriate interval between index test(s) and reference standard needs to be defined in context with the review question. There is risk of bias if

- interventions have been initiated between index test and reference standard and it cannot be excluded that they have (already) influenced the condition of the participant at the time of conduct of the reference standard,
- it cannot be excluded that the condition of the participant has significantly improved (e.g., clearance of an infection) or worsened (e.g., tumor progression) in the time interval chosen,

- it cannot be ascertained that the reference standard can detect the condition after the time interval chosen (e.g., motor symptoms of multiple sclerosis).

Per definition diagnostic accuracy is inferred from the comparison of the index test with a reference test, which for this purpose is considered "ideal," i.e., able to detect the "true" presence or absence of disease. As a consequence of this assumption, estimates of diagnostic accuracy are not possible or potentially biased if

- no reference standard is available,
- available reference standards are known to be imperfect,
- not all participants receive the same reference standard (differential verification bias),
- not all participants receive a reference standard (partial verification bias),
- interpretation of the results of index test or reference test is not blinded to the results of the other test, respectively (review bias).

There is no consensus on how to process indeterminate results. Often, these results are either excluded or classified as positive—both procedures may lead to overestimation of test performance. It is recommended to classify indeterminate results by "intention to diagnose,"i.e., false negative if the reference standard result is positive and false positive if the reference standard result is negative (Schuetz et al., 2012; Shinkins et al., 2013).

Following the quality assessment of the single studies, the same rules as for grading of interventions apply to the overall grading of study limitations in a body of evidence using the GRADE framework (Guyatt et al., 2011b).

Directness. An important challenge is to decide on whether diagnostic accuracy alone, a linked-evidence approach, or direct evidence from RCT is needed to answer the specific review question (Singh et al., 2012). It is tempting to use diagnostic accuracy as a surrogate outcome for patient outcomes; the linkage between diagnostic accuracy and clinical outcomes, however, is in many cases indirect and challenging to establish.

If diagnostic accuracy is used as intermediate outcome in a "linked-evidence" approach, separate grading of the body of evidence for each link in the chain (based on the analytic framework) is required.

If a poor reference standard, the expected performance of a new test or other (e.g., clinical and organizational) factors lowers the inference of clinical outcomes from diagnostic accuracy, evidence from studies on diagnostic accuracy would need to be downgraded for indirectness. According to GRADE, indirectness includes an assessment of applicability (of the results). For diagnostic accuracy studies, this is evaluated using the QUADAS-2 tool (Whiting et al., 2011). Investigators need information on inclusion and exclusion criteria, settings and locations of data collections, methods of patient recruitment, and sampling to decide whether evidence about the test is valid, clinically relevant, and applicable to specific patient groups or individuals (Rutjes et al., 2005).

Diagnostic and prognostic biomarkers are not codependent with one specific treatment. Nevertheless, they cannot provide clinical benefit without inducing patient management changes and treatment decisions compared with the established test and treatment strategy. Specific challenges here are to identify (pragmatic) RCT reflecting the population and clinical practice in the country/health system in which the biomarker is intended to be introduced (Merlin et al., 2013).

The evaluation of inconsistency according to GRADE applies to binary/dichotomous outcomes and relative, not absolute measures of effect (Guyatt et al., 2011c). Instead of forest plots, the most common representation format of diagnostic test performance to allow detection of inconsistency is a summary receiver operating characteristic (ROC) curve (Eng, n.d.), displaying sensitivity and specificity results from various studies. This can be complemented by a bubble plot of true-positive versus false-positive rates spread in ROC space. (Singh et al., 2012). Reviewers should seek to resolve inconsistency through a critical consideration of possible explanations (differences in populations, interventions, or outcomes) for any detected inconsistency.

Imprecision and publication bias are defined and assessed as for intervention studies (Guyatt et al., 2011d,e). The impact of imprecision on clinical outcomes may be determined by calculating posttest probabilities. In contrast to clinical trials, there is no register for diagnostic accuracy studies. In adaptation of an approach for prognostic studies, one possibility is to generally assume that publication bias is present, except if a diagnostic test has been studied in a large number of diagnostic accuracy studies (Huguet et al., 2013).

A dose-response association might be observed for tests with continuous outcomes and multiple cutoffs. Inconsistency therefore may arise from test thresholds/cutoffs for positive/negative categorization varying across test accuracy studies.

Plausible unmeasured confounders. The strength of evidence is increased if, despite plausible confounders that would decrease the diagnostic accuracy, the diagnostic accuracy measured is high.

In intervention studies, strength of association may lead to upgrading of the strength of evidence, if an observed association is large enough that it cannot have occurred solely as a result of bias from confounding factors (Singh et al., 2012). For diagnostic accuracy studies, this domain can be applied for upgrading of the evidence when diagnostic accuracy of an index test is measured with an imperfect reference standard and hence may be underestimated (Singh et al., 2012).

5.2 Prognosis Studies

An adaptation of the GRADE framework for prognostic studies has been proposed by Huguet et al. (2013).

In a systematic review, Hayden et al. (2006) have identified quality items used in assessing six sources of bias in prognosis studies: study participation, study attrition, measurement of prognostic factors, measurement of and controlling for confounding variables, measurement of outcomes, and analysis approaches. Based on this work, a team of epidemiologists, statisticians, and clinicians developed the quality in prognosis studies (QUIPS) tool for assessing bias in studies of prognostic factors that demonstrated acceptable reliability (Hayden et al., 2006), and so far, it is the only tool available for this purpose.

Reviewers should downgrade for inconsistency if estimates of the prognostic factor association with the outcome vary in direction and there is no or minimal overlap of the confidence intervals. If a meta-analysis is conducted, before downgrading for inconsistency, a subgroup analysis in a priori defined subgroups (e.g., differences in population, duration of follow-up, and study methods) should be performed (Huguet et al., 2013).

Indirectness may be present if the study population does not represent the population of the review question (e.g., patients of a headache clinic vs general population) (Huguet et al.,

2013). Similarly, downgrading for indirectness would be justified if the prognostic factor or the outcomes assessed would not represent the full bandwidth of the review question.

Of particular importance in the review of prognostic studies is publication bias. In contrast to clinical trials, there is no register for prognostic research studies. One possibility therefore is to generally assume that publication bias is present, except if a prognostic factor has been studied in a large number of cohort studies (Huguet et al., 2013).

Other domains are assessed as for intervention studies, with the exception of plausible confounders: in contrast to intervention studies, the effects of inadequate control of confounding on the study effects are unclear and as such cannot be taken into account to estimate the accuracy of the effect estimate. Risk of bias by confounding, however, is covered by the quality appraisal of the single studies (Huguet et al., 2013).

5.3 Linked-Evidence Evaluation of Co-dependent Technologies

Decisions on reimbursement and/or implementation in clinical practice of new technologies require evidence on the clinical benefit in terms of patient health outcomes.

Biomarker tests are not stand-alone technologies but need to be assessed with regard to patient management changes and treatment incited by the test results. This poses several evidentiary challenges, notably a lack of (direct) evidence from randomized controlled trials as discussed, for example, in Khoury et al. (2010). To reduce decision-making uncertainty in the absence of direct evidence, several institutions propose linked-evidence approaches for the assessment of medical tests (Medical Services Advisory Committee, 2005; Chang and AHRQ, 2012; Derksen and CVZ, 2011; Nachtnebel, 2010; Freidlin et al., 2010), while other institutions do not include it as an option (IQWIG, 2013). Linked evidence describes a chain of arguments linking different types of evidence, provided that (a) data are transferrable across different parts of the linkage and (b) for each element, evidence is gathered systematically and transparently and is considered internally valid.

A framework for evaluating evidence on the clinical benefit of codependent technologies for reimbursement decisions has recently been developed and is summarized in the following (Merlin et al., 2013). Predictive biomarkers guide treatment choices: A drug A is expected to perform better than a drug B in biomarker-positive patients, while no differences of treatment effect are expected in biomarker-negative patients.

In the first step, the biological plausibility of the relationship between the drug and the biomarker is evaluated: evidence must be presented that the biomarker is predictive (treatment effect modifier) or prognostic (indicative of disease progression independent of treatment) or both. Only double-randomized trials or trials with a marker by treatment interaction design allow an answer to this question; the relationship cannot be clarified by a marker-strategy design. If the biomarker is a prognostic factor, other treatments will also be likely to have a favorable outcome in the marker-positive subgroup and should be included in the comparison.

Highest level of evidence on the clinical benefit of a test and treatment strategy is provided by a double-randomized controlled trial, with one randomization to the testing strategy and a second randomization to treatment and control. Direct evidence, albeit of a lower level, can also be provided by trials with a marker-strategy design or a biomarker by treatment design (Merlin et al., 2013). If direct evidence is not available, there are various options to provide linked evidence:

- A marker by treatment design is linked to a randomized controlled trial of the same treatment/control in an untested population, thereby allowing comparing the relative effectiveness of treatment versus control in a biomarker-stratified with a biomarker-unstratified population.
- Only patients with discordant results in a standard testing strategy versus a biomarker-based strategy are randomized to treatment/control and compared with the RCT in an untested population.
- Patients in the standard and the treatment arm of an RCT are retrospectively analyzed for marker status, or tests are performed in archived samples. Relative effectiveness of treatment versus control is then compared with a biomarker-positive population in an enrichment design. Test accuracy serves as an additional link in this option.

References

Andre, F., O'Regan, R., Ozguroglu, M., Toi, M., Xu, B., Jerusalem, G., et al., 2014. Everolimus for women with trastuzumab-resistant, HER2-positive, advanced breast cancer (BOLERO-3): a randomised, double-blind, placebo-controlled phase 3 trial. Lancet Oncol. 15 (6), 580–591. Epub 2014/04/20. Available from: http://www.ncbi.nlm.nih.gov/pubmed/24742739.

Beachy, S.H., Repasky, E.A., 2008. Using extracellular biomarkers for monitoring efficacy of therapeutics in cancer patients: an update. Cancer Immunol. Immunother. 57 (6), 759–775. Epub 2008/01/12. Available from: http://www.ncbi.nlm.nih.gov/pubmed/18188561.

Benson, K., Hartz, A.J., 2000. A comparison of observational studies and randomized, controlled trials. Am J. Ophthalmol. 130 (5), 688. Epub 2000/11/18. Available from: http://www.ncbi.nlm.nih.gov/pubmed/11078861.

Berzin, T.M., Blanco, P.G., Lamont, J.T., Sawhney, M.S., 2010. Persistent psychological or physical symptoms following endoscopic procedures: an unrecognized post-endoscopy adverse event. Dig. Dis. Sci. 55 (10), 2869–2873. Epub 2010/04/16. Available from: http://www.ncbi.nlm.nih.gov/pubmed/20393877.

Biomarkers Definitions Working Group, 2001. Biomarkers and surrogate endpoints: preferred definitions and conceptual framework. Clin. Pharmacol. Ther. 69 (3), 89–95. Epub 2001/03/10. Available from: http://www.ncbi.nlm.nih.gov/pubmed/11240971.

Bossuyt, P., Leeflang, M.M., 2008. Developing criteria for including studies. In: Cochrane Handbook for Systematic Reviews of Diagnostic Test Accuracy. The Cochrane Collaboration. Version 0.4 [updated Sept 2008]. Available from: http://srdta.cochrane.org/. (Chapter 6).

Bossuyt, P.M., McCaffery, K., 2009. Additional patient outcomes and pathways in evaluations of testing. Med. Decis. Mak. 29 (5), E30–8. Epub 2009/09/04. Available from: http://www.ncbi.nlm.nih.gov/pubmed/19726782.

Bossuyt, P.M., Irwig, L., Craig, J., Glasziou, P., 2006. Comparative accuracy: assessing new tests against existing diagnostic pathways. BMJ 332 (7549), 1089–1092. Epub 2006/05/06. Available from: http://www.ncbi.nlm.nih.gov/pubmed/16675820.

Brozek, J.L., Akl, E.A., Jaeschke, R., Lang, D.M., Bossuyt, P., Glasziou, P., et al., 2009. Grading quality of evidence and strength of recommendations in clinical practice guidelines: part 2 of 3. The GRADE approach to grading quality of evidence about diagnostic tests and strategies. Allergy 64 (8), 1109–1116. Epub 2009/06/06. Available from: http://www.ncbi.nlm.nih.gov/pubmed/19489757.

Centre for Reviews and Dissemination, 2009. Systematic Reviews. Chapter 2: Clinical Tests. Available from: http://www.york.ac.uk/inst/crd/pdf/Systematic_Reviews.pdf.

Chang, S., AHRQ, 2012. Methods Guide for Medical Test Reviews. AHRQ Publication N012-EHC017. Available from: http://www.effectivehealthcare.ahrq.gov/reports/final.cfm.

Concato, J., Shah, N., Horwitz, R.I., 2000. Randomized, controlled trials, observational studies, and the hierarchy of research designs. N. Engl. J. Med. 342 (25), 1887–1892. Epub 2000/06/22. Available from: http://www.ncbi.nlm.nih.gov/pubmed/10861325.

Deeks, J.J., Bossuyt, P.M., Gatsonis, C. (Eds.), 2010. Cochrane Handbook for Systematic Reviews of Diagnostic Test Accuracy Version 0.9.0. The Cochrane Collaboration. Available from: http://srdta.cochrane.org/.

Derksen, J., CVZ, 2011. Medical tests (assessment of established medical science and medical practice). CVZ Report 293. Available from: http://www.zorginstituutnederland.nl/publicaties.

Eng, J. (n.d.). ROC analysis: web-based calculator for ROC curves. Retrieved 18.10.2013, from http://www.jrocfit.org.

Febbo, P.G., Ladanyi, M., Aldape, K.D., De Marzo, A.M., Hammond, M.E., Hayes, D.F., et al., 2011. NCCN task force report: Evaluating the clinical utility of tumor markers in oncology. J. Natl. Compr. Cancer Netw. 9 (Suppl 5), S1–32. quiz S3. Epub 2011/12/22. Available from: http://www.ncbi.nlm.nih.gov/pubmed/22138009.

Food and Drug Administration. [4 November 2014]; Available from: http://www.fda.gov/MedicalDevices/ProductsandMedicalProcedures/InVitroDiagnostics/ucm301431.htm.

Freidlin, B., Korn, E.L., 2010. Biomarker-adaptive clinical trial designs. Pharmacogenomics 11 (12), 1679–1682. Epub 2010/12/15. Available from: http://www.ncbi.nlm.nih.gov/pubmed/21142910.

Freidlin, B., Korn, E.L., 2014. Biomarker enrichment strategies: matching trial design to biomarker credentials. Nat. Rev. Clin. Oncol. 11 (2), 81–90. Epub 2013/11/28. Available from: http://www.ncbi.nlm.nih.gov/pubmed/24281059.

Freidlin, B., McShane, L.M., Korn, E.L., 2010. Randomized clinical trials with biomarkers: design issues. J. Natl. Cancer Inst. 102 (3), 152–160. Epub 2010/01/16. Available from: http://www.ncbi.nlm.nih.gov/pubmed/20075367.

Gartlehner, G., 2009. Internes Manual. Abläufe und Methoden. Teil 2, 2. Aufl. Ludwig Boltzmann Institut für Health Technology Assessment, Wien. Available from: http://eprints.hta.lbg.ac.at/713/.

Guyatt, G., Oxman, A.D., Akl, E.A., Kunz, R., Vist, G., Brozek, J., et al., 2011a. GRADE guidelines: 1. Introduction-GRADE evidence profiles and summary of findings tables. J. Clin. Epidemiol. 64 (4), 383–394. Epub 2011/01/05. Available from: http://www.ncbi.nlm.nih.gov/pubmed/21195583.

Guyatt, G.H., Oxman, A.D., Vist, G., Kunz, R., Brozek, J., Alonso-Coello, P., et al., 2011b. GRADE guidelines: 4. Rating the quality of evidence—study limitations (risk of bias). J. Clin. Epidemiol. 64 (4), 407–415. Epub 2011/01/21. Available from: http://www.ncbi.nlm.nih.gov/pubmed/21247734.

Guyatt, G.H., Oxman, A.D., Kunz, R., Woodcock, J., Brozek, J., Helfand, M., et al., 2011c. GRADE guidelines: 7. Rating the quality of evidence—inconsistency. J. Clin. Epidemiol. 64 (12), 1294–1302. Epub 2011/08/02. Available from: http://www.ncbi.nlm.nih.gov/pubmed/21803546.

Guyatt, G.H., Oxman, A.D., Kunz, R., Brozek, J., Alonso-Coello, P., Rind, D., et al., 2011d. GRADE guidelines 6. Rating the quality of evidence—imprecision. J. Clin. Epidemiol. 64 (12), 1283–1293. Epub 2011/08/16. Available from: http://www.ncbi.nlm.nih.gov/pubmed/21839614.

Guyatt, G.H., Oxman, A.D., Montori, V., Vist, G., Kunz, R., Brozek, J., et al., 2011e. GRADE guidelines: 5. Rating the quality of evidence—publication bias. J. Clin. Epidemiol. 64 (12), 1277–1282. Epub 2011/08/02. Available from: http://www.ncbi.nlm.nih.gov/pubmed/21802904.

Haddow, J.E., Palomaki, G.E., 2003. ACCE: a model process for evaluating data on emerging genetic tests. In: Khoury, M., Little, J., Burke, W. (Eds.), Human Genome Epidemiology: A Scientific Foundation for Using Genetic Information to Improve Health and Prevent Disease. Oxford University Press, New York, pp. 217–233.

Harris, R.P., Helfand, M., Woolf, S.H., Lohr, K.N., Mulrow, C.D., Teutsch, S.M., et al., 2001. Current methods of the US Preventive Services Task Force: a review of the process. Am. J. Prev. Med. 20 (3 Suppl), 21–35. Epub 2001/04/18. Available from: http://www.ncbi.nlm.nih.gov/pubmed/11306229.

Hayden, J.A., Cote, P., Bombardier, C., 2006. Evaluation of the quality of prognosis studies in systematic reviews. Ann. Intern. Med. 144 (6), 427–437. Epub 2006/03/22. Available from: http://www.ncbi.nlm.nih.gov/pubmed/16549855.

Hillier, S., Grimmer-Somers, K., Merlin, T., Middleton, P., Salisbury, J., Tooher, R., et al., 2011. FORM: an Australian method for formulating and grading recommendations in evidence-based clinical guidelines. BMC Med. Res. Methodol. 11, 23. Epub 2011/03/02. Available from: http://www.ncbi.nlm.nih.gov/pubmed/21356039.

van Holten, T.C., Waanders, L.F., de Groot, P.G., Vissers, J., Hoefer, I.E., Pasterkamp, G., et al., 2013. Circulating biomarkers for predicting cardiovascular disease risk; a systematic review and comprehensive overview of meta-analyses. PLoS ONE 8 (4), e62080. Epub 2013/05/01. Available from: http://www.ncbi.nlm.nih.gov/pubmed/23630624.

Hsu, J., Brozek, J.L., Terracciano, L., Kreis, J., Compalati, E., Stein, A.T., et al., 2011. Application of GRADE: making evidence-based recommendations about diagnostic tests in clinical practice guidelines. Implement. Sci. 6, 62. Epub 2011/06/15. Available from: http://www.ncbi.nlm.nih.gov/pubmed/21663655.

Huguet, A., Hayden, J.A., Stinson, J., McGrath, P.J., Chambers, C.T., Tougas, M.E., et al., 2013. Judging the quality of evidence in reviews of prognostic factor research: adapting the GRADE framework. Syst. Rev. 2 (1), 71. Epub 2013/09/07. Available from: http://www.ncbi.nlm.nih.gov/pubmed/24007720.

Institut für Qualität und Wirtschaftlichkeit im Gesundheitswesen—IQWIG, 2013. Allgemeine Methoden Version 4.1. Available from: https://www.iqwig.de/download/IQWiG_Methoden_Version_4-1.pdf.

Institute of Medicine (IOM), 2010. Evaluation of Biomarkers and Surrogate Endpoints in Chronic Disease. The National Academies Press, Washington, DC. Available from: http://www.iom.edu/Reports/2010/Evaluation-of-Biomarkers-and-Surrogate-Endpoints-in-Chronic-Disease.aspx.

Janatzek, S., 2011. The benefit of diagnostic tests—from surrogate endpoints to patient-relevant endpoints. Z. Evid. Fortbild. Qual. Gesundhwes. 105 (7), 504–509. Epub 2011/10/01. Nutzen diagnostischer Tests - vom Surrogat zur Patientenrelevanz. Available from: http://www.ncbi.nlm.nih.gov/pubmed/21958609.

Khoury, M.J., Coates, R.J., Evans, J.P., 2010. Evidence-based classification of recommendations on use of genomic tests in clinical practice: dealing with insufficient evidence. Genet. Med. 12 (11), 680–683. Epub 2010/10/27. Available from: http://www.ncbi.nlm.nih.gov/pubmed/20975567.

Kunz, R., Oxman, A.D., 1998. The unpredictability paradox: review of empirical comparisons of randomised and non-randomised clinical trials. BMJ 317 (7167), 1185–1190. Epub 1998/10/31. Available from: http://www.ncbi.nlm.nih.gov/pubmed/9794851.

Lijmer, J.G., Bossuyt, P.M., 2009. Various randomized designs can be used to evaluate medical tests. J. Clin. Epidemiol. 62 (4), 364–373. Epub 2008/10/24. Available from: http://www.ncbi.nlm.nih.gov/pubmed/18945590.

Lijmer, J.G., Leeflang, M., PMM, B., 2009. Proposals for a Phased Evaluation of Medical Tests. Med. Tests-White Paper Ser.. Available from: http://www.ncbi.nlm.nih.gov/pubmed/21290784.

Lord, S.J., Irwig, L., Simes, R.J., 2006. When is measuring sensitivity and specificity sufficient to evaluate a diagnostic test, and when do we need randomized trials? Ann. Intern. Med. 144 (11), 850–855. Epub 2006/06/07. Available from: http://www.ncbi.nlm.nih.gov/pubmed/16754927.

Lord, S.J., Irwig, L., Bossuyt, P.M., 2009. Using the principles of randomized controlled trial design to guide test evaluation. Med. Decis. Mak. 29 (5), E1–E12. Epub 2009/09/24. Available from: http://www.ncbi.nlm.nih.gov/pubmed/19773580.

Lord, S.J., Staub, L.P., Bossuyt, P.M., Irwig, L.M., 2011. Target practice: choosing target conditions for test accuracy studies that are relevant to clinical practice. BMJ 343, d4684. Epub 2011/09/10. Available from: http://www.ncbi.nlm.nih.gov/pubmed/21903693.

Mandrekar, S.J., Sargent, D.J., 2009. Clinical trial designs for predictive biomarker validation: theoretical considerations and practical challenges. J. Clin. Oncol. 27 (24), 4027–4034. Epub 2009/07/15. Available from: http://www.ncbi.nlm.nih.gov/pubmed/19597023.

Mandrekar, S.J., Grothey, A., Goetz, M.P., Sargent, D.J., 2005. Clinical trial designs for prospective validation of biomarkers. Am. J. Pharmacogenomics 5 (5), 317–325. Epub 2005/10/04. Available from: http://www.ncbi.nlm.nih.gov/pubmed/16196501.

Medical Services Advisory Committee, 2005. Guidelines for the Assessment of Diagnostic Technologies. Commonwealth of Australia, Canberra. Available from: http://www.msac.gov.au/.

Merlin, T., Weston, A., Tooher, R., 2009. Extending an evidence hierarchy to include topics other than treatment: revising the Australian 'levels of evidence'. BMC Med. Res. Methodol. 9, 34. Epub 2009/06/13. Available from: http://www.ncbi.nlm.nih.gov/pubmed/19519887.

Merlin, T., Farah, C., Schubert, C., Mitchell, A., Hiller, J.E., Ryan, P., 2013. Assessing personalized medicines in Australia: a national framework for reviewing codependent technologies. Med. Decis. Mak. 33 (3), 333–342. Epub 2012/08/17. Available from: http://www.ncbi.nlm.nih.gov/pubmed/22895559.

Morrison, A., Boudreau, R., CADTH, 2012. Evaluation Frameworks for Genetic Tests [Environmental Scan issue 36]. Canadian Agency for Drugs and Technologies in Health, Ottawa.

Nachtnebel, A., 2010. Evaluation von Diagnostika—Hintergrund, Probleme, Methoden. HTA Projektbericht Nr 36. Available from: http://eprints.hta.lbg.ac.at/898/.

National Institute of Health and Clinical Excellence, 2011. Diagnostics Assessment Programme Manual. Available from: http://www.nice.org.uk/media/8A3/34/DAP_programme_manual_final_for_upload_22_Dec_11.pdf.

OCEBM Levels of Evidence Working Group. "The Oxford 2011 Levels of Evidence". Oxford Centre for Evidence-Based Medicine.; Available from: http://www.cebm.net/index.aspx?o=5653.

Relevo, R., 2012. Effective Search Strategies for Systematic Reviews of Medical Tests. Methods Guid. Med. Test Rev. Available from: http://www.ncbi.nlm.nih.gov/pubmed/22834020.

Rutjes, A.W., Reitsma, J.B., Vandenbroucke, J.P., Glas, A.S., Bossuyt, P.M., 2005. Case-control and two-gate designs in diagnostic accuracy studies. Clin. Chem. 51 (8), 1335–1341. Epub 2005/06/18. Available from: http://www.ncbi.nlm.nih.gov/pubmed/15961549.

Rutjes, A.W., Reitsma, J.B., Di Nisio, M., Smidt, N., van Rijn, J.C., Bossuyt, P.M., 2006. Evidence of bias and variation in diagnostic accuracy studies. CMAJ 174 (4), 469–476. Epub 2006/02/16. Available from: http://www.ncbi.nlm.nih.gov/pubmed/16477057.

Samson, D., Schoelles, K.M., 2012. Developing the topic and structuring systematic reviews of medical tests: utility of PICOTS, analytic frameworks, decision trees, and other frameworks. Methods Guid. Med. Test Rev. Available from: http://www.ncbi.nlm.nih.gov/pubmed/22834028.

Sargent, D.J., Conley, B.A., Allegra, C., Collette, L., 2005. Clinical trial designs for predictive marker validation in cancer treatment trials. J. Clin. Oncol. Off. J. Am. Soc. Clin. Oncol. 23 (9), 2020–2027. Epub 2005/03/19. Available from: http://www.ncbi.nlm.nih.gov/pubmed/15774793.

Schleidgen, S., Klingler, C., Bertram, T., Rogowski, W.H., Marckmann, G., 2013. What is personalized medicine: sharpening a vague term based on a systematic literature review. BMC Med. Ethics 14, 55. Epub 2013/12/24. Available from: http://www.ncbi.nlm.nih.gov/pubmed/24359531.

Schuetz, G.M., Schlattmann, P., Dewey, M., 2012. Use of 3x2 tables with an intention to diagnose approach to assess clinical performance of diagnostic tests: meta-analytical evaluation of coronary CT angiography studies. BMJ 345, e6717. Epub 2012/10/26. Available from: http://www.ncbi.nlm.nih.gov/pubmed/23097549.

Segal, J.B., 2012. Choosing the important outcomes for a systematic review of a medical test. Methods Guid. Med. Test Rev. Available from: http://www.ncbi.nlm.nih.gov/pubmed/22834029.

Shinkins, B., Thompson, M., Mallett, S., Perera, R., 2013. Diagnostic accuracy studies: how to report and analyse inconclusive test results. BMJ 346, f2778. Epub 2013/05/18. Available from: http://www.ncbi.nlm.nih.gov/pubmed/23682043.

Singh, S., Chang, S.M., Matchar, D.B., Bass, E.B., 2012. Grading a body of evidence on diagnostic tests. In: Chang, S.M., Matchar, D.B., Smetana, G.W., Umscheid, C.A. (Eds.), Methods Guide for Medical Test Reviews. Agency for Healthcare Research and Quality, Rockville, MD.

Tajik, P., Zwinderman, A.H., Mol, B.W., Bossuyt, P.M., 2013. Trial designs for personalizing cancer care: a systematic review and classification. Clin. Cancer Res. 19 (17), 4578–4588. Epub 2013/06/22. Available from: http://www.ncbi.nlm.nih.gov/pubmed/23788580.

Takwoingi, Y., Leeflang, M.M., Deeks, J.J., 2013. Empirical evidence of the importance of comparative studies of diagnostic test accuracy. Ann. Intern. Med. 158 (7), 544–554. Epub 2013/04/03. Available from: http://www.ncbi.nlm.nih.gov/pubmed/23546566.

Teutsch, S.M., Bradley, L.A., Palomaki, G.E., Haddow, J.E., Piper, M., Calonge, N., et al., 2009. The Evaluation of Genomic Applications in Practice and Prevention (EGAPP) Initiative: methods of the EGAPP Working Group. Genet Med 11 (1), 3–14. Epub 2008/09/25. Available from: http://www.ncbi.nlm.nih.gov/pubmed/18813139.

Trikalinos, T.A., Balion, C.M., 2012. Chapter 9: options for summarizing medical test performance in the absence of a "gold standard". J. Gen. Intern. Med. 27 (Suppl 1), S67–75. Epub 2012/06/08. Available from: http://www.ncbi.nlm.nih.gov/pubmed/22648677.

Tripepi, G., Jager, K.J., Dekker, F.W., Zoccali, C., 2010a. Statistical methods for the assessment of prognostic biomarkers (part II): calibration and re-classification. Nephrol. Dial. Transplant. 25 (5), 1402–1405. Epub 2010/02/20. Available from: http://www.ncbi.nlm.nih.gov/pubmed/20167948.

Tripepi, G., Jager, K.J., Dekker, F.W., Zoccali, C., 2010b. Statistical methods for the assessment of prognostic biomarkers (Part I): discrimination. Nephrol. Dial. Transplant. 25 (5), 1399–1401. Epub 2010/02/09. Available from: http://www.ncbi.nlm.nih.gov/pubmed/20139066.

Whiting, P., Rutjes, A.W., Reitsma, J.B., Glas, A.S., Bossuyt, P.M., Kleijnen, J., 2004. Sources of variation and bias in studies of diagnostic accuracy: a systematic review. Ann. Intern. Med. 140 (3), 189–202. Epub 2004/02/06. Available from: http://www.ncbi.nlm.nih.gov/pubmed/14757617.

Whiting, P.F., Rutjes, A.W., Westwood, M.E., Mallett, S., Deeks, J.J., Reitsma, J.B., et al., 2011. QUADAS-2: a revised tool for the quality assessment of diagnostic accuracy studies. Ann. Intern. Med. 155 (8), 529–536. Epub 2011/10/19. Available from: http://www.ncbi.nlm.nih.gov/pubmed/22007046.

Zujewski, J.A., Kamin, L., 2008. Trial assessing individualized options for treatment for breast cancer: the TAILORx trial. Future Oncol. 4 (5), 603–610. Epub 2008/10/17. Available from: http://www.ncbi.nlm.nih.gov/pubmed/18922117.

Further Reading

Balshem, H., Helfand, M., Schunemann, H.J., Oxman, A.D., Kunz, R., Brozek, J., et al., 2011. GRADE guidelines: 3. Rating the quality of evidence. J. Clin. Epidemiol. 64 (4), 401–406. Epub 2011/01/07. Available from: http://www.ncbi.nlm.nih.gov/pubmed/21208779.

Hayden, J.A., van der Windt, D.A., Cartwright, J.L., Cote, P., Bombardier, C., 2013. Assessing bias in studies of prognostic factors. Ann. Intern. Med. 158 (4), 280–286. Epub 2013/02/20. Available from: http://www.ncbi.nlm.nih.gov/pubmed/23420236.

Trikalinos, T.A., Balion, C.M., 2012b. Options for summarizing medical test performance in the absence of a "Gold Standard". Methods Guid. Med. Test Rev. Available from: http://www.ncbi.nlm.nih.gov/pubmed/22834025.

US Food and Drug Administration, Devices @FDA Catalog. Available from: http://www.accessdata.fda.gov/scripts/cdrh/devicesatfda/.

CHAPTER

3

Statistical Learning in Precision Medicine

Sowmya Srinivasan Perumbakkam, Matthias Kohl

Institute of Precision Medicine, Furtwangen University, Villingen-Schwenningen, Germany

1 INTRODUCTION

Statistical learning is the combination of machine learning with statistics. Similar to machine learning, it studies computer algorithms for learning with the goal to make predictions, where the same algorithms as in machine learning are applied, but in addition, it involves statistical models and the assessment of their uncertainty. The learning always requires some sort of data such as direct experience or instruction and is performed automatically, i.e., without human intervention or assistance. This perfectly fits to precision medicine, which aims to use all information available from a patient to guide decisions made with regard to the prevention, diagnosis, and treatment of diseases to achieve an optimal therapeutic effect. Hence, we could say that the success of precision medicine at least in parts will depend on the development and application of accurate and reliable statistical (machine) learning models for estimating the optimal treatment regime from all the data that are available. In particular, data from omics experiments are expected to lead to significant advances in precision medicine, i.e., patients can be treated according to their own molecular characteristics (Tian, 2015; Chen and Snyder, 2013).

More and more research confirms that statistical (machine) learning is a key technique for precision medicine. Menden et al. (2013) for instance have developed machine learning models to predict the response of cancer cell lines to drug treatment, based on both the genomic features of the cell lines and the chemical properties of the considered drugs. Yu et al. (2016) have found that a machine learning approach for identifying critical disease-related features was able to accurately differentiate between two types of lung cancers—adenocarcinoma and squamous cell carcinoma—and also predict patient survival times better than pathologists, who classify tumors by grade and stage. The authors believe that applying machine learning techniques to other types of cancer could also prove effective (Yu et al., 2016).

In the sequel, we describe the most important components and steps involved in statistical learning for patient data.

Precision Medicine
https://doi.org/10.1016/B978-0-12-805364-5.00003-2

2 DATA

Everything starts with the data. In statistics and machine learning, data are usually categorized as either numerical or categorical. *Numerical data* are any information that is measurable in numbers such as height, weight, blood pressure, or gene expression of a particular person. *Categorical data* represent nominal data such as a person's gender, marital status, or any sort of data in the dichotomous yes/no format and ordinal data such as grades, medical scores, or semiquantitative measurements (e.g., low-medium-high concentration). The data can further be divided into (independent) *exploratory variables/features* that predict or affect the (dependent) *response variables/features*; see, for instance, Section 1.3 in Whitlock and Schluter (2015).

As the quality of the data ultimately determines the quality and reproducibility of the results, we strongly recommend a thorough and detailed quality control of the data before statistical procedures are applied. The quality of the data can be assessed by summary statistics such as mean, median, standard deviation, and interquartile range. And visualizations such as scatter plots, bar charts, box-and-whisker plots, and histograms. The importance of data quality is well known in predictive analytics and should be regarded of similar high importance in precision medicine (Jugulum, 2016). In addition, the quality control contributes to the "understanding" of the data at hand and often provides clues on the best possible analysis of the data.

3 TYPES OF STATISTICAL LEARNING ALGORITHMS

The algorithms in statistical (machine) learning can be categorized based on the way the data are used to learn. In *supervised learning*, the procedures generate a function that maps the input data (exploratory features) to desired outputs (response features); special cases are classification and regression. In classification, the response variable is categorical, whereas in regression, it is usually numerical. In *unsupervised learning*, there is no response but only exploratory variables, and the aim is to understand the relationships between the features or between the observations; see, for example, Section 2.1 in James et al. (2013).

Semisupervised learning combines both "labeled" (response known) and "unlabeled" (response unknown) observations and tries to generate an appropriate function or classifier. *Transduction* is related to the field of semisupervised learning, but does not explicitly construct a function. Instead, it tries to predict new outputs based on given inputs and outputs and on new inputs. In *reinforcement learning*, the algorithms learn a policy of how to act in a certain environment by maximization of a reward (resp. minimization of a punishment); that is, every action has some impact on the environment, and the environment provides feedback that guides the learning algorithm. Finally, *learning to learn*—also called meta learning—consists of self-improving machine learning algorithms (Taiwo Oladipupo, 2010).

In the sequel, we will focus on supervised learning, more precisely, on the typical steps involved in the development of predictive models for precision medicine.

4 STEPS IN SUPERVISED LEARNING

A supervised learning algorithm analyzes the available (so-called training) data and produces a predictive model, which is also called a classifier (categorical response) or a regression model (numerical response). Ideally, the predictive model gives the correct output value (response) for any valid input object. In this section, we give an overview of the most important steps involved in the development and identification of a best possible predictive model.

4.1 Preparation of Data

A major part of the preparation of the data is called preprocessing, which refers to processes performed on the raw data to make them better suitable for further analysis steps. If the exploratory features have very different units and ranges, the computations may become numerically unstable, and the interpretation of the resulting models may be difficult. In such a setting, we can often improve the situation by centering and scaling/standardizing such features. The *centering* of a feature is performed by subtracting some measure of location such as mean or median from the data. A feature is *scaled/standardized* by dividing its (centered or uncentered) values by a measure of scale such as standard deviation or interquartile range. When mean and standard deviation are applied, one also speaks of *z scores* in allusion to the standard normal distribution (sometimes also called z distribution) having a mean of zero and a standard deviation of one; see, for instance, pages 143–146 in (Dancey et al., 2012).

In addition to this simple centering and scaling, one also applies other variance stabilizing and normalizing transformations. This especially holds for all kind of omics data, where such transformations form an integral part of the preprocessing. From a statistical point, variance stabilizing and normalizing transformations harmonize the data and thus increase the reliability and reproducibility of the results (Quackenbush, 2002; De Bruyne et al., 2007; Kohl, 2010; Griffiths et al., 2010).

Other ways of transforming the exploratory features are multivariate methods such as independent component analysis (ICA), principal component analysis (PCA), or the "spatial sign" transformation. The ICA finds new features as linear combinations of the original features such that the components are independent (Hyvärinen and Oja, 2016). PCA can be applied to transform the data to a smaller subspace where the new features are uncorrelated; see, for example, Section 10.2 in James et al. (2013). The "spatial sign" transformation projects the data to the unit circle in p dimensions, where p is the number of features (Serneels et al., 2006). As such transformations may also reduce the dimension of the data, they can also be seen as feature selection procedures; for more details, see Section 4.4.

A key for preparing the data is also the appropriate integration of data from different sources and domains with the goal to provide a unified representation of the data usable for precision medicine (Barash et al., 2015).

4.2 Handling Imbalanced Datasets

In classification, it often happens that one response category (majority class) in the dataset has a (clearly) higher number of instances than others. In such a situation, which is called imbalanced, unbalanced, or skewed, most of the classification algorithms give results that are biased toward the majority class, i.e., they tend to ignore the underrepresented classes (Longadge et al., 2013). The problem is especially common with clinical databases, where disease prevalence is typically fairly low among screening populations (García-Pedrajas et al., 2012; Li et al., 2010; Mazurowski et al., 2008). Imbalance is also of particular relevance in precision medicine, since the final aim is to make reliable predictions for small groups or even single patients. It is therefore of primary importance in precision medicine to recognize and to properly address the imbalance issue. The problem of imbalance learning is relatively new and has attracted increasing research interest over the last 15 years (Chawla et al., 2004; He and Garcia, 2009; López et al., 2013).

There are three major techniques to address the issue: algorithmic cost-sensitive approaches, data preprocessing approaches, and feature selection approaches. Various empirical studies have shown that in some application domains, including certain specific imbalanced learning domains, cost-sensitive learning is superior to data preprocessing methods (Liu and Zhou, 2006; McCarthy et al., 2005; Zhou and Liu, 2006). These cost-sensitive algorithms however take for granted the availability of a cost matrix and its associated cost items, whereas in many situations, an explicit description of the misclassification costs is difficult or even impossible (Galar et al., 2012). If misclassification costs are unknown, which is typically the case in medical practice, we recommend informed resampling, an up-to-date preprocessing approach, addressing most of the shortcomings of simple resampling while keeping the computational costs within reasonable limits. The oldest example of informed undersampling is one-sided sampling (OSS) (Kubat and Matwin, 1997), a method to identify and delete the borderline and noisy observations from the majority class. Neighborhood cleaning rule (NCR) (Laurikkala, 2001) is another approach that applies the edited nearest neighbor rule (ENNR) (Wilson and Martinez, 2000) to wisely choose the samples to remove from the majority class. More recently, the easy ensemble and balance cascade algorithms (He and Garcia, 2009; Liu et al., 2009) have shown good results by employing ensemble methods to clever undersample the data. As NCR, the synthetic minority oversampling technique (SMOTE) algorithm also exploits the k-nearest neighbor technique to selectively oversample the minority class and randomly introduce new synthetic samples between the closest neighbors from this class (Chawla et al., 2002). SMOTE has gained popularity (He and Ma, 2013; Batista et al., 2004; Napierała et al., 2010), and further extensions well reviewed by Maciejewski and Stefanowski (2011) have been proposed to release questionable assumptions behind the method (Maciejewski and Stefanowski, 2011).

4.3 Imputation of Missing Data

The presence of missing values in a dataset not only reduces the statistical power of the study but also impedes the implementation of mining methods that require a complete dataset such as many machine learning methods (Sterne et al., 2009). Although regulatory guidance on the design and conduct of clinical trials intend to reduce the rate of missing values in the generated data, the amount of missing cases always is pretty high in medical datasets (Little et al., 2012).

Rubin (1976) distinguishes between three types of missingness (Rubin, 1976). In case of *missing completely at random* (MCAR), cases with missing values can be thought of as a random sample of all the cases. This however occurs rarely in practice. *Missing at random* (MAR) means that conditioned on all the data we have, any remaining missingness is completely random; that is, it especially does not depend on some unrecorded variables. Hence, missingness can be modeled using the observed data. Then, we can use specialized missing data analysis methods on the available data to correct for the effects of missingness. Finally, *missing not at random* (MNAR) is neither MCAR nor MAR. It is difficult to handle, since it requires knowledge or strong assumptions about the patterns of missingness.

Nowadays, *multiple imputation* methods such as multivariate imputation via chained equations (MICE) (Azur et al., 2011) or Amelia (Zhang, 2016) are known to work well under the MAR assumption and are regarded as state of the art. These algorithms generate multiple values for each missing value leading to multiple datasets. Each of the datasets is analyzed separately, and finally, the results are pooled.

4.4 Feature Selection

Often, datasets especially high-dimensional datasets (e.g., omics datasets) include redundant or irrelevant features (Liu et al., 2014). Such features reduce the efficiency of the learning and the comprehensibility of the learned results. They also favor the generation of predictive models that are overfitting the training data; that is, the models fit well to the training data but perform clearly worse on new unseen data (Blum and Langley, 1997; Guyon and Elisseeff, 2003). The process of identifying and selecting the features that are required to build an accurate model should therefore be considered unavoidable in the learning process. The three main approaches of feature selection are filter, wrapper, and embedded methods. *Filters* select subsets of variables in a preprocessing step, independently of the chosen machine learning algorithm. They assign scores to each feature based on some statistical measure such as p values, information gain, or fold changes and select a user-specified number of features with the best scores that will then be used for building the predictive model (Guyon and Elisseeff, 2003). The basic idea behind *wrappers* is to repeatedly fit the model with different sets of exploratory features and compute the performance of the resulting predictive model. The feature set and the respective model giving the best performance are selected (Guyon and Elisseeff, 2003). In case of some learning algorithms, the search for the best set of exploratory features is embedded in the learning procedure. Such *embedded methods* report on feature importance and are highly efficient (Guyon and Elisseeff, 2003).

4.5 Learning Algorithms

A large number of algorithms have been proposed for classification and regression, ranging from classical approaches such as linear discriminant analysis or linear regression to modern approaches such as support vector machines or deep neural networks (Harrell, 2015; Hastie et al., 2009; Goodfellow et al., 2016). All algorithms have advantages and disadvantages in terms of computational complexity, performance, interpretability, robustness, etc. As Dudoit et al. (2002) show, simple methods may work equally well as complex methods. This is also confirmed by the MAQC-II study, which found that the performance of the developed predictive models largely depended on the end point and that different approaches led to models of similar performance (MAQC Consortium, 2010).

4.6 Evaluating the Performance

For a reliable evaluation of the performance of a predictive model on new data, a test dataset is required, which includes data that have not been used in the training of the model. This is usually done by randomly splitting the given data into a training and a test dataset. By applying resampling methods such as cross validation or bootstrap, a realistic estimate of the performance of the predictive model on new data can be obtained. In case of *k-fold cross validation* ($k \geq 2$), the dataset is randomly split into *k*-subsets (drawing without replacement). Each subset is held out, while the model is trained on all other subsets. This process is completed when performance is determined for each observation in the dataset and an overall performance estimate is provided. It is recommended to repeat *k*-fold cross validation several times to get a final performance estimate (Kohavi, 1994). *Bootstrap* was invented by Efron (1979, 1982) and is in detail described in Efron and Tibshirani (1993). Nowadays, there are many variants of bootstrap (Chernick and LaBudde, 2011). The standard bootstrap approach works by drawing with replacement a random sample out of the original dataset and uses this random sample for training the predictive model. The performance of the model is then evaluated on the remaining data. This procedure is usually repeated many times (e.g., 1000 times) (Kohavi, 1994).

Of course, the evaluation of the performance of a predictive model requires also the selection of a performance measure. Many measures have been proposed in literature including standard measures such as R^2 and AIC for regression or sensitivity and specificity in binary classification (Harrell, 2015; Kuhn and Johnson, 2013; Steyerberg et al., 2010; Parker, 2011; Kohl, 2012). The selection of the most appropriate performance measure for the situation at hand is of crucial importance as different measures usually lead to different optimal models.

4.7 Improving Model Performance

Many learning algorithms include parameters that have to be specified by the user, where the user does not know, which set of parameters will give the best result. Finding this set of parameters is called *model tuning*. Typically, the tuning of a model is performed by an inner cross validation or bootstrap loop; that is, each training dataset by resampling is further split into training and test sets, and the set of parameters giving the best performance in this inner loop are selected (Kuhn and Johnson, 2013).

Another very popular approach that aims at improving model performance is *model averaging* or *ensemble learning* (Yang et al., 2010). Instead of just using the result of one learning algorithm, the results of multiple learning algorithms are combined to a so-called ensemble model. Ensemble classifiers were shown to perform better than any of the single ones (Freund and Schapire, 1995). This is of particular interest when working with medical datasets that often have high dimensionality but may only have a few strongly predictive features of the target response and many weak ones. In this case, the effects of weak features may have to be combined to achieve good predictions. Averaging the predictions of a number of models that contain different sets of features aggregates the predictive power of their possibly weak features in the final prediction (Schapire, 1990).

5 SKETCH OF THE WORKFLOW

We start with the raw data, which means that all available data have already been imported to some computer program and have been integrated and combined into one dataset, where the observations and features are represented by rows and columns, respectively. In very high-dimensional settings, it is often just the other way round, i.e., the features are included in the rows. After the quality control of the raw data, the data will be preprocessed if necessary. However, in some cases, for instance, in case of microarray data, the preprocessing procedure uses information from all observations. In this particular case, the observations get connected, and it would no longer be possible to separate an independent test set; that is, the preprocessing of the data must be postponed and should be done after training and test set have been separated. Next, it often makes sense to perform a quality control of the preprocessed data, as the preprocessing may also lead to unexpected side effects, since a selected preprocessing method might not work for all kind of data, e.g., applying the logarithm to a feature that contains nonpositive values. If the given dataset is large enough, one sometimes generates a holdout sample; that is, a part of the data is randomly selected and excluded from all further analysis. This dataset may act as independent test set, where the performance of the final predictive model can be evaluated and the results can be compared with the results obtained by resampling. Now, the outer resampling loop starts; that is, one either starts the first run of k-fold cross validation or generates the first bootstrap sample leading to the first pair of training and test sets. In case of class imbalance, the next step consists of balancing the training set. In case of missing values, the (balanced) training set will undergo a multiple imputation leading to multiple imputed training sets, where the subsequent steps have to be performed for each of these imputed datasets. If a filter method is used for selecting the best features, the (balanced and imputed) training set may now be filtered, i.e., uninformative features (columns) are removed from the dataset. If a learning algorithm is used that needs to be tuned, another loop will start now. In this loop, the (balanced and imputed) training set will be split into training and test sets by some resampling procedure, and for each parameter set, the respective model will be trained and its performance evaluated. Now, there may be even one more inner loop, if variable selection is performed by a wrapper. That is, for each parameter set, the model has to be trained for various sets of features, whose performance could again be compared by using resampling. This will finally lead to the optimal set of features for a given parameter set, which then will give an optimal predictive model for a given parameter set. By comparing the resampled results of all parameter sets, the best parameter set and the corresponding best predictive model can be identified. Now, either the next imputed dataset is used or the next iteration of the outer resampling loop starts. By aggregating the results of the outer resampling loop, the expected performance of the predictive model that is generated by applying the selected algorithms to the whole dataset is obtained. In each of the steps, there is no single best method, but there are many algorithms and combinations of algorithms that are available, even the order of the steps may be interchanged, and one should at least try a few of them. That is, the above procedure should be repeated for different preprocessing methods, different balancing methods, different multiple imputation methods, different feature selection methods, and different learning algorithms; see also Supplementary Document 5 of (MAQC Consortium, 2010), which includes the standard operating procedures (SOPs) and methods and analysis for MAQC-II (balancing is missing!).

6 CONCLUSION

As there is no single best procedure in statistical learning, the development of optimal predictive models in precision medicine is a difficult task involving a huge number of computations and should be done by a researcher experienced in statistical learning. Moreover, the complexity of the data in precision medicine requires that there is also a close collaboration between the data analyst and researchers from various other fields such as medicine, biology, and biochemistry.

By the growing amount of all kinds of patient data, e.g., from patient data management systems and omics databases, and the growing computational power, we can expect that in the near future, more and more reliable and robust predictive models for various diseases will be developed.

Acknowledgment

The authors acknowledge the financial support by the Federal Ministry of Education and Research of Germany in the projects PATIENTS and MultiFlow (project numbers 03FH019PX3 and 03FH046PX4).

References

Azur, M.J., Stuart, E.A., Frangakis, C., Leaf, P.J., 2011. Multiple imputation by chained equations: what is it and how does it work? Int. J. Methods Psychiatr. Res. 20 (1), 40–49.

Barash, C.I., Elliston, K.O., Faucett, W.A., Hirsch, J., Naik, G., Rathjen, A., Wood, G., 2015. Harnessing big data for precision medicine: a panel of experts elucidates the data challenges and proposes key strategic decisions points. Appl. Transl. Genom. 4, 10–13.

Batista, G.E.A.P.A., Prati, R.C., Monard, M.C., 2004. A study of the behavior of several methods for balancing machine learning training data. SIGKDD Explor. Newsl. 6 (1), 20–29.

Blum, A.L., Langley, P., 1997. Selection of relevant features and examples in machine learning. Artif. Intell. 97 (1), 245–271.

Chawla, N.V., Bowyer, K.W., Hall, L.O., Kegelmeyer, W.P., 2002. Smote: synthetic minority over-sampling technique. J. Artif. Intell. Res. 16, 321–357.

Chawla, N.V., Japkowicz, N., Kotcz, A., 2004. Editorial: special issue on learning from imbalanced data sets. ACM SIGKDD Explor. Newsl. 6 (1), 1–6.

Chen, R., Snyder, M., 2013. Promise of personalized omics to precision medicine. Wiley Interdiscip. Rev. Syst. Biol. Med. 5 (1), 73–82.

Chernick, M.R., LaBudde, R.A., 2011. An Introduction to Bootstrap Methods with Applications to R. John Wiley & Sons.

Dancey, C.P., Reidy, J.G., Rowe, R., 2012. Statistics for the Health Sciences. Sage.

De Bruyne, V., Al-Mulla, F., Pot, B., 2007. Methods for microarray data analysis. Methods Mol. Biol. 382, 373–391.

Dudoit, S., Fridlyand, J., Speed, T.P., 2002. Comparison of discrimination methods for the classification of tumors using gene expression data. J. Am. Stat. Assoc. 97 (457), 77–87.

Efron, B., 1979. Bootstrap methods: another look at the jackknife. Ann. Stat. 7, 1–26.

Efron, B., 1982. In: The jackknife, the bootstrap, and other resampling plans. Society of Industrial and Applied Mathematics CBMS-NSF Monographs. vol. 38.

Efron, B., Tibshirani, R., 1993. An Introduction to the Bootstrap. Chapman & Hall.

Freund, Y., Schapire, R.E., 1995. A decision-theoretic generalization of on-line learning and an application to boosting. In: Computational Learning Theory. Springer, pp. 23–37.

Galar, M., Fernandez, A., Barrenechea, E., Bustince, H., Herrera, F., 2012. A review on ensembles for the class imbalance problem: bagging-, boosting-, and hybrid-based approaches. IEEE Trans. Syst. Man Cybern. C: Appl. Rev. 42 (4), 463–484.

García-Pedrajas, N., Pérez-Rodríguez, J., García-Pedrajas, M., Ortiz-Boyer, D., Fyfe, C., 2012. Class imbalance methods for translation initiation site recognition in DNA sequences. Knowl.-Based Syst. 25 (1), 22–34.

Goodfellow, I., Bengio, Y., Courville, A., 2016. Deep Learning. MIT Press.

Griffiths, W.J., Koal, T., Wang, Y., Kohl, M., Enot, D.P., Deigner, H.P., 2010. Targeted metabolomics for biomarker discovery. Angew. Chem. Int. Ed. 49 (32), 5426–5445.

Guyon, I., Elisseeff, A., 2003. An introduction to variable and feature selection. J. Mach. Learn. Res. 3, 1157–1182.

Harrell, F.E., 2015. Regression Modeling Strategies, second ed. Springer.
Hastie, T., Tibshirani, R., Friedman, J., 2009. The Elements of Statistical Learning, second ed. Springer.
He, H., Garcia, E.A., 2009. Learning from imbalanced data. IEEE Trans. Knowl. Data Eng. 21 (9), 1263–1284.
He, H., Ma, Y., 2013. Imbalanced Learning: Foundations, Algorithms, and Applications. John Wiley & Sons.
Hyvärinen, A., Oja, E., 2016. Independent component analysis: algorithms and applications. Neural Netw. 13 (4–5), 411–430.
James, G., Witten, D., Hastie, T., Tibshirani, R., 2013. An Introduction to Statistical Learning. Springer.
Jugulum, R., 2016. Importance of data quality for analytics. In: Sampaio, P., Saraiva, P. (Eds.), Quality in the 21st Century. Springer.
Kohavi, R., 1994. In: A study of cross-validation and bootstrap for accuracy estimation and model selection. Proceedings of the 14th International Joint Conference on Artificial Intelligence—Vol. 2 (IJCAI'95). Morgan Kaufmann Publishers Inc., San Francisco, CA, pp. 1137–1143.
Kohl, M., 2010. Development and validation of predictive molecular signatures. Curr. Mol. Med. 10 (2), 173–178.
Kohl, M., 2012. Performance measures in binary classification. Int. J. Stat. Med. Res. 1 (1), 79–81.
Kubat, M., Matwin, S., 1997. In: Addressing the curse of imbalanced training sets: one-sided selection. Proceedings of the Fourteenth International Conference on Machine Learning, pp. 179–186.
Kuhn, M., Johnson, K., 2013. Applied Predictive Modeling. Springer.
Laurikkala, J., 2001. Improving Identification of Difficult Small Classes by Balancing Class Distribution. Springer.
Li, D.C., Chiao-Wen Liu, C.W., Hu, S.C., 2010. A learning method for the class imbalance problem with medical data sets. Comput. Biol. Med. 40 (5), 509–518.
Little, R.J., D'Agostino, R., Cohen, M.L., Dickersin, K., Emerson, S.S., Farrar, J.T., Frangakis, C., Hogan, J.W., Molenberghs, G., Murphy, S.A., Neaton, J.D., Rotnitzky, A., Scharfstein, D., Shih, W.J., Siegel, J.P., Stern, H., 2012. The prevention and treatment of missing data in clinical trials. N. Engl. J. Med. 367 (14), 1355–1360.
Liu, X.Y., Zhou, Z.H., 2006. In: The influence of class imbalance on cost-sensitive learning: an empirical study. Sixth International Conference on Data Mining (ICDM'06), Hong Kong. pp. 970–974.
Liu, X.Y., Wu, J., Zhou, Z.H., 2009. Exploratory undersampling for class-imbalance learning. IEEE Trans. Syst. Man Cybern. B: Cybern. 39 (2), 539–550.
Liu, Q., Ribeiro, B., Sung, A.H., Suryakumar, D., 2014. In: Mining the big data: the critical feature dimension problem. 2014 IIAI 3rd International Conference on Advanced Applied Informatics (IIAIAAI). IEEE, pp. 499–504.
Longadge, R., Dongre, S., Malik, L., 2013. Class imbalance problem in data mining; review. Int. J. Comput. Sci. Netw. 2 (1), 83–87.
López, V., Fernández, A., García, S., Palade, V., Herrera, F., 2013. An insight into classification with imbalanced data: empirical results and current trends on using data intrinsic characteristics. Inf. Sci. 250, 113–141.
Maciejewski, T., Stefanowski, J., 2011. In: Local neighbourhood extension of SMOTE for mining imbalanced data. IEEE Symposium on Computational Intelligence and Data Mining (CIDM), Paris. 2011, pp. 104–111.
MAQC Consortium, 2010. The MicroArray quality control (MAQC)-II study of common practices for the development and validation of microarray-based predictive models. Nat. Biotechnol. 28 (8), 827–838.
Mazurowski, M.A., Habas, P.A., Zurada, J.A., Lo, J.Y., Baker, J.A., Tourassi, G.D., 2008. Training neural network classifiers for medical decision making: the effects of imbalanced datasets on classification performance. Neural Netw. 21 (2–3), 427–436.
McCarthy, K., Zabar, B., Weiss, G., 2005. In: Does cost-sensitive learning beat sampling for classifying rare classes? Proceedings of the 1st International Workshop on Utility-Based Data Mining (UBDM '05). ACM, New York, pp. 69–77.
Menden, M.P., Iorio, F., Garnett, M., McDermott, U., Benes, C.H., Ballester, P.J., et al., 2013. Machine learning prediction of cancer cell sensitivity to drugs based on genomic and chemical properties. PLoS ONE 8 (4), e61318.
Napierała, K., Stefanowski, J., Wilk, S., 2010. In: Szczuka, M., Kryszkiewicz, M., Ramanna, S., Jensen, R., Hu, Q. (Eds.), Learning from imbalanced data in presence of noisy and borderline examples. Proceedings of the 7th International Conference on Rough Sets and Current Trends in Computing (RSCTC'10). Springer-Verlag, Berlin, Heidelberg, pp. 158–167.
Parker, C., 2011. In: An analysis of performance measures for binary classifiers. 2011 IEEE 11th International Conference on Data Mining, Vancouver, BC, Canada, pp. 517–526.
Quackenbush, J., 2002. Microarray data normalization transformation. Nat. Genet. 32 (Suppl), 496–501.
Rubin, D.B., 1976. Inference and missing data. Biometrika 63 (3), 581–592.
Schapire, R.E., 1990. The strength of weak learnability. Mach. Learn. 5 (2), 197–227.

Serneels, S., De Nolf, E., Van Espen, P.J., 2006. Spatial sign preprocessing: a simple way to impart moderate robustness to multivariate estimators. J. Chem. Inf. Model. 46 (3), 1402–1409.

Sterne, J.A., White, I.R., Carlin, J.B., Spratt, M., Royston, P., Kenward, M.G., Wood, A.M., Carpenter, J.R., 2009. Multiple imputation for missing data in epidemiological and clinical research: potential and pitfalls. BMJ 338, b2393.

Steyerberg, E.W., Vickers, A.J., Cook, N.R., Gerds, T., Gonen, M., Obuchowski, N., Pencina, M.J., Kattan, M.W., 2010. Assessing the performance of prediction models: a framework for some traditional and novel measures. Epidemiology (Cambridge, Mass) 21 (1), 128–138.

Taiwo Oladipupo, A., 2010. Types of machine learning algorithms. In: Zhang, Y. (Ed.), New Advances in Machine Learning. InTech.

Tian, G., 2015. Precision medicine: how the genomics can change medical mode. Life World 41, 42–45.

Whitlock, M.C., Schluter, D., 2015. The Analysis of Biological Data, 2nd ed. Roberts and Company Publishers.

Wilson, D.R., Martinez, T.R., 2000. Reduction techniques for instance based learning algorithms. Mach. Learn. 38 (3), 257–286.

Yang, P., Yang, Y.H., Zhou, B.B., Zomaya, A.Y., 2010. Curr. Bioinforma. 5 (4), 296–308.

Yu, K.H., Zhang, C., Berry, G.J., Altman, R.B., Ré, C., Rubin, D.L., Snyder, M., 2016. Predicting non-small cell lung cancer prognosis by fully automated microscopic pathology image features. Nat. Commun. 7, 12474.

Zhang, Z., 2016. Multiple imputation for time series data with Amelia package. Ann. Transl. Med. 4 (3), 56.

Zhou, Z.H., Liu, X.Y., 2006. Training cost-sensitive neural networks with methods addressing the class imbalance problem. IEEE Trans. Knowl. Data Eng. 18 (1), 63–77.

CHAPTER

4

Biobanks as Basis of Individualized Medicine: Challenges Toward Harmonization

Ayse Yuzbasioglu,†, Burcu Kesikli‡, Meral Ozguç*,†*

*Center for Genomics and Rare Diseases & Biobank for Rare Diseases, Hacettepe University, Ankara, Turkey †Department of Medical Biology, Hacettepe University, Ankara, Turkey ‡Health Institutes of Turkey-Aziz Sancar Research Center, Ankara, Turkey

1 INTRODUCTION

Biobanks are defined as infrastructures for collections of human biological samples and associated data such as medical, familial, social, and genetic (Public Population Project in Genomics (P3G) (http://www.p3g.org/)) (http://www.coe.int/t/dg3/healthbioethic/Activities/10_Biobanks/biobanks_for_Europe.pdf). As a crucial cornerstone of research and development in life sciences, future diagnostics, and therapy, they have a major role in providing samples used for biotechnological applications such as drug design and genetic test development. Availability of samples to allow for analysis of diseases at the molecular level plays a role in the development of new therapies.

Biobanks can also be built to contain nonhuman material from environment animals, plants, and microorganisms and thus foster the protection of biodiversity and the environment (http://www.oecd.org/sti/biotech/oecdbestpracticeguidelinesforbiologicalresourcecentres.htm) (https://esbb.org/).

Life sciences research always depended on biological samples usually acquired for a specific project and kept in the possession of the researcher initiating the study as "collections." However, today, biobanks are facilitators of biomedical research by providing samples to a wide community of researchers.

Spanish regulation about biobanks (Royal Decree 1716/2011) makes a distinction between collections and biobanks. Collections can only be used by an investigator in a particular research project as specified in the original consent, whereas biobanks are described as platforms that provide high-quality samples for scientific community in a broad sense (Arias-Diaz et al., 2013).

Precision Medicine
https://doi.org/10.1016/B978-0-12-805364-5.00004-4

There are two categories of human sample biobanks, either for research or for nonresearch purposes (http://www.oecd.org/sti/biotech/guidelinesforhumanbiobanksandgeneticresearch-databaseshbgrds.htm).

Storing tissues and organs for therapeutic purposes especially for transplantation is a well-known type of nonresearch biobank.

As an example to biobanks for therapeutic purposes, cord blood banks are well-known. They can be affiliated with an academic establishment such as Cambridge Blood and Stem Cell Biobank, or they can be private and/or commercial structures (http://www.haem.cam.ac.uk/cambridge-blood-and-stem-cell-biobank/) (Petrini, 2010).

In the United States, Coriell Institute for Medical Research (www.coriell.org), a private nonprofit organization, is an example of a provider of a large sample collection. By sustaining and providing this wide range of samples worldwide to researchers, it contributes to medical research and has started to play an active role in personalized medicine (Coriell Personalized Medicine Collaborative (CPMC)).

Many countries have national legislation for obtaining and processing human tissue. For example, in the United Kingdom, obtaining and storing human tissue are governed by a specific legislation as Human Tissue Act 2004 (Weale and Lear, 2007).

In the United States, US Food and Drug Administration (FDA) (www.fda.gov) is the organization that is responsible for governing the use of tissues for transplantation since 1993. FDA furthered its regulatory basis by forming Human Tissue Task Force (HTTF), that is, a network of different offices: the Center for Biologics Evaluation and Research (CBER), the Office of Regulatory Affairs (ORA), and the Office of the Commissioner (OC). The regulatory action is toward assessing risks in human cells, tissues, and cellular- and tissue-based products (HCT/Ps) (https://www.fda.gov/downloads/BiologicsBloodVaccines/SafetyAvailability/TissueSafety/UCM114829.pdf).

For research biobanks, there are yet no overarching global regulatory approaches. In some countries, there are specific laws or regulations, but mostly, biobanking research is governed by guidelines issued by scientific and international organizations.

When there are no specific guidelines for biobanking, countries can refer to *Convention for the Protection of Human Rights and Dignity of the Human Being with Regard to the Application of Biology and Medicine (1997)* by Council of Europe. This law has been ratified by many non-European countries as well.

Tables 1–3 contain some examples of governance models for biobank research:

(Calzolari et al., 2013; Scott et al., 2012; Salvaterra et al., 2008; Macilotti et al., 2008).

Biobanks have different purposes for establishment and have different structures.

Biobanks differ in parameters, such as ownership of the collections. A survey revealed that in Europe, 39% of biobanks are affiliated with universities, 39% with national agencies, 19% with nonprofit organizations, and only 3% are privately owned (http://ipts.jrc.ec.europa.eu/publications/).

Each biobank with a defined purpose can hold a variety of different biological samples. Cells can be obtained from tumors, biopsy samples, and tissues. Some biobanks house only DNA samples or DNA and whole blood and body fluids such as plasma, serum, and urine (Peakman and Elliott, 2010).

According to a survey study, among 56 European biobanks, 38 of them stored DNA, 37 of them stored tissue (66%), 25 of them stored cell lines (45%), and 20 of them stored serum (36%) samples (Arampatzis et al., 2016).

Research biobanks can be organizations with a specific emphasis on a particular disease or a specific population (Asslaber and Zatloukal, 2007).

TABLE 1 Countries With a Specific Biobank Legislation

Country	Law, Regulation
Estonia	Human Genes Research Act (2001)
The United Kingdom	Human Tissue Act (2004)
The Netherlands	Civil code, Article 467 (1994)
Iceland	Act on Biobanks No. 110 (2000)
Denmark	Law on Biobanks No. 312 (2003)
Sweden	Law No. 297 (2005)
Norway	Act on Biobanks (2003)
Finland	The Finnish Parliament Biobank Law (688/2012)
Spain	Royal Decree (411/1996) (R)
The United States	Human Tissue Task Force (2007)

TABLE 2 Guidelines by International Organizations

Organization	Guideline
World Health Organization (WHO)	Guideline for Obtaining Informed Consent for the Procurement and Use of Human Tissues, Cells, and Fluids in Research (2003)
United Nations Educational, Scientific, and Cultural Organization (UNESCO)	Universal Declaration on Bioethics and Human Rights (2005)
	International Declaration on Human Genetic Data (2003)
	Universal Declaration on the Human Genome and Human Rights (1997)
Human Genome Organization	Statement on DNA Sampling: Access and Control (1998)
Organization for Economic Cooperation and Development (OECD)	OECD Guidelines on the Protection of Privacy and Transborder Flows of Personal Data (2013)
	OECD Guidelines on Human Biobanks and Genetic Research Databases (2009)
	OECD Best Practice Guidelines for Biological Resource Centers (2007)
The Council for International Organizations of Medical Sciences (CIOMS)	The International Ethical Guidelines for Epidemiological Studies (2009)
The Council for International Organizations of Medical Sciences CIOMS/WHO	International Ethical Guidelines for Biomedical Research Involving Human Subjects (2016)
World Medical Association (WMA)	Helsinki Declaration on Medical Research Involving Human Subjects (2013)
	WMA Declaration of Taipei on Ethical Considerations Regarding Health Databases and Biobanks (2016)
Nuffield Council on Bioethics	The collection, linking and use of data in biomedical research and health care: ethical issues (2015)

TABLE 3 Countries With Guidelines for Biobanks

Country	Guideline
The Netherlands	Code for Proper Secondary Use of Human Tissue in The Netherlands (2002)
Australia	Guidelines for Human Biobanks, Genetic Research Databases and Associated Data—Government of West Australia Department of Health (2010)
	National Statement on Ethical Conduct in Human Research (2007)
France	Ethical Issues Raised by Collections of Biological Materials and Associated Data: "Biobanks," "Biolibraries"—National Consultative Bioethics Committee for Health and Life Sciences (2003)
Germany	Biobanks for Research—National Ethics Council Opinion (2004)
Italy	Biobanks and Research on Human Biological Material—National Bioethics Committee Opinion (2006)
	Guideline for Clinical Protocols of Genetic Research—Italian Society of Human Genetics (2006)
	Guideline for Genetic Biobanks—Telethon (2003)
	Guideline for the Establishment and Accreditation of Biobanks (2006)
Japan	Ethical Guidelines for Analytical Research on the Human Genome/Genes (2001)
Switzerland	Biobanks: Obtainment, Preservation and Utilization of Human Biological Material (2006)
The United Kingdom	Human Tissue and Biological Samples for Use in Research—Medical Research Council (2001)

Disease-specific biobanks are usually situated in hospital clinics with access to patients and their clinical data.

In the case of disease-specific biobanks, rare diseases (80% are of genetic origin) that affect a small population size benefit largely from sample collections at biobanks. Biobanks for rare diseases are crucial to attain the right size of sample size for clinical trials and also for the development of new tests and therapies. There are well-known biobanks for rare diseases.

The Italian Telethon Network of Genetic Biobanks (http://www.telethon.it/en) was established in 2008. A successful collaboration with patient organizations enabled a large number of sample collection from almost 900 rare genetic diseases (Baldo et al., 2016).

DNA and Cell Bank of Genethon (http://www.genethon.fr/en/rd-2/dna-and-cell-bank/) in France is another example of a genetic biobank. It was established in 1990 and is one of the largest rare disease biobanks in Europe holding more than 250,000 biological samples mostly from neuromuscular diseases.

An important group of disease-specific biobanks are tumor/cancer repositories. A well-known example is the Canadian Tumor Repository Network (CTRNet) (www.ctrnet.ca) established and funded by the Canadian Institute of Health Sciences (CIHR).

There is a long list of medical centers worldwide that house many different disease-specific collections, one example being the Biobank of Nephrological Diseases at University Medical Centers (UMCs) supported by the Dutch government and by National

Parelsnoer-String of Pearls Initiative (PSI) (www.parelsnoer.org and www.string-of-pearls.org) (Navis et al., 2014).

Population biobanks are collections of samples from donors to conduct prospective research on mostly complex diseases. These are longitudinal studies with a large number of samples to attain statistical validity. Long-term periodic medical control of participants and follow-up on lifestyle parameters can lead to assessment of environmental input for susceptibility to specific diseases in the population. Such biobank studies help to organize health-care policy and in the near future will be instrumental in precision medicine initiatives.

UK Biobank (www.ukbiobank.ac.uk) is an example of population biobank that was initiated as a prospective national study to investigate disease incidence from a large cohort (about 500,000 participants/ages 40–49). The sample acquisition is done through the health-care system where medical- and health-related data are linked to the samples that are then processed in a central facility (Elliott et al., 2008).

In a centralized system, sample transportation to the main laboratory is a parameter that has to be well controlled.

In other cases, samples may be acquired and processed in a local system such as American Cancer Society Cancer Prevention Study (http://www.cancer.org/Research). Then, the bottleneck is quality management and harmonization of sample collection and processing at each local laboratory.

Estonia started its National Genome Project (EGP) based on a public/private partnership model in 2001. EGP Foundation was supported by public funding, and EGeen International Corporation (EGI) as a private entity gave initial support at the establishment phase. However, the project eventually was supported solely by the state. The biobank samples were collected through the health-care system by integration of general practitioners (Leitsalu et al., 2015a,b).

Iceland is an early example that established population biobank by funding through deCODE Genetics, a private company (https://www.decode.com). The main mission was to discover genetic variations causing susceptibility to diseases in the Icelandic population (160,000 volunteers).

A well-known and established type of biobanks for especially cancer research is tissue banks. They also differ in size and composition and have different goals. Tissues of high quality are highly instrumental for biomarker discovery, pathway analysis, and drug design studies in the field of oncology (Zatloukal and Hainaut, 2010).

Samples in tissue biobanks are either disease or organ specific or routinely collected through pathology departments (Riegman et al., 2007).

A unique example for tissue specific biobanks are brain biobanks for neuroscience research. Some examples are BrainNet Europe, UK Brain Banks Network, Australian Brain Bank Network, and, more recently, the NIH NeuroBioBank. Samples are either provided from autopsy centers or through preregistration donor programs of some biobanks (Palmer-Aronsten et al., 2016).

Samples accrued for diagnostic purposes in clinical pathology and genetic testing laboratories are also very valuable samples that can be turned into research material if proper ethical procedures can be met.

Residual tissues describe archived pathology specimens obtained for diagnostic purposes and there is an obligation to keep them in storage for a defined period of time. This is usually around ten years not to risk reduction in sample size for possible need that may occur for re-diagnostic purposes.

Leftover tissues are materials that are described as medical wastes that remain after a diagnostic or a surgical procedure.

Secondary use of residual material for future research has its own ethical challenges, the main issue being a model for a proper consent type (Tasse, 2016).

In some jurisdictions, consent may be waived if the identity of the residual tissue is not known to the researcher (Guidance on Research Involving Coded Private Information or Biological Specimens (2008) (Office for Human Research Protections)) (https://www.hhs.gov/ohrp/regulations-and-policy/guidance/research-involving-coded-private-information/index.html; https://www.hta.gov.uk/policies/human-tissue-act-2004).

For leftover material, a type of consent proposed is precautionary consent described as *consent sought for research use of residual materials where the specifics of the future research projects are unknown* (Gefenas et al., 2012).

Another type of consent for leftover material is "presumed consent." It is a way of assuming that patients allow for their samples to be used for research after the procedures for their diagnostic workup is completed. Unless they opt out, the leftover material even if it is not anonymized can be regarded as research material.

For example, in Belgium and in the Netherlands, patients are informed with various means about the possible use of their leftover biomaterial for research upon their initial visit to the health-care center. Then, the informed patients have an option to agree or disagree to the use of their sample for future research (Riegman et al., 2008; Riegman and Van Veen, 2011).

2 GOVERNANCE

The activities of a biobank involve a close net of relations between various stakeholders: operators of the biobank, donors of samples/participants, funding agencies, research ethics committees (RECs), and researchers (Master et al., 2015).

Governance is described not only as a regulatory activity but also as *a strategy for patterning a network of interaction* (Gottweis and Lauss, 2012).

WMA states the following principles for governance of databases and biobanks:

Protection of individuals: Governance should be designed, so the rights of individuals prevail over the interests of other stakeholders and science.

Transparency: Any relevant information on health databases and biobanks must be made available to the public.

Participation and inclusion: Custodians of health databases and biobanks must consult and engage with individuals and their communities.

Accountability: Custodians of health databases and biobanks must be accessible and responsive to all stakeholders (https://www.wma.net/policies-post/wma-declaration-of-taipei-on-ethical-considerations-regarding-health-databases-and-biobanks/).

At a biobank, the network is running at various defined stages of the operation: acquisition of samples, processing of samples, maintenance of samples (archiving)/management of sample related data and IT infrastructure, and access to the samples by researchers (Yuzbasioglu and Ozguc, 2013).

At all stages, there are two major domains of concern, ethical and scientific/technical.

Ethical and legal guidelines set the principles for individuals' rights and benefits thus applies to the activities of the biobank related to donors' rights such as autonomy, privacy, benefit, and harm.

3 CONSENT

Informed consent is an ethical principle that became a part of biomedical research process after the Nuremberg trials (the Nuremberg Code (1947)) (http://unesdoc.unesco.org/images/ 0017/001781/178124e.pdf).

WHO describes informed consent as "a decision to participate in research, taken by a competent individual who has received the necessary information; who has adequately understood the information; and who, after considering the information, has arrived at a decision without having been subjected to coercion, undue influence or inducement, or intimidation" (2011).

In a study to assess consent documents used by German biobanks (30 registered biobanks) (http://.www.biobanken.de), it was concluded that further harmonization is needed to improve networking among biobanks to the benefit of donors. As shown by the study, even within one single country, there are a large variety of unharmonized issues in the informed consent documents; it is obvious that it will be even more difficult to have a global template for biobank networking on the international scale (Hirschberg et al., 2013).

In research ethics, informed consent is a binding document governing basic principles of human dignity, human rights, individuals' right to privacy, and above all the principle of autonomy. Traditionally, it is obtained for a specific project for which the participant gives consent (specific consent) to take part. In biobanking research, the main mission is to facilitate biomedical research by future projects that are not specified at the time of sample collection. In this case, the donor of the biological sample cannot be truly "informed" about basic issues such as the content, identity of the principle investigator, and duration of the project that are only foreseen in the future.

Further issues related to biobanking consent are the indefinite duration of storage of biological samples, access to medical records, recontract of participants, data sharing, commercial developments, access to research results, and protection of privacy and confidentiality (Beskow et al., 2010).

There are some examples to basic definitions of consent forms in Table 4 (Master et al., 2015; Steinsbekk et al., 2013).

To facilitate biobanking, research and sharing of sample's blank or broad consent are the most practical form. In this way, participants need not be contacted in case of each new research project involving their samples. Blank consent puts no limitations on the scope of research where biologicals samples will be used in the future (Lunshof et al., 2008).

Whereas in broad consent, for all future projects, ethical review needs to be obtained, and a very important point is that the participant can use his/her right to withdraw from the project (Hansson et al., 2006).

Thus, broad consent is not a blanket consent where samples may be used for nonbiological-medical purposes such as parenthood testing or forensic use (Hansson, 2009; Strech et al., 2016).

TABLE 4 Types of Consent Used for Biobanking

Consent	Definition
Specific	Permission for a given defined study only
Broad	Permission given for future research unforeseen at the time of donation but related to original purpose of sample collection
Tiered	Permission given for a specific use and reconsent is required for further studies
Blank/general	Permission given to any type of future research
Dynamic	Interactive internet-based consent that keeps donor informed about each new project
Presumed	Unless donors opt out, it is presumed that their samples can be used for research

A consent form for donating samples to a biobank should have the following minimum set of information:

- Purpose and scope of the biobank
- Types of material that are stored
- Funding
- Duration for storage
- Type of data collected (personal, medical/health, genetic, imaging, etc.)
- Provisions for protection of confidentiality and security of the samples
- Risk and benefit to the participant
- Possibility of recontract and/or resampling
- Ownership of the samples
- Right to withdraw and possible consequences
- Possible options for commercialization
- Access policy to samples and data

(http://www.bbmri-eric.eu), (http://www.isber.org), (http://www.p3g.org), and (http://genomicsandhealth.org).

It should be kept in mind that respect for individual autonomy embedded in the principle of consent is a culture-dependent parameter (Turner, 2005; Grady, 2015).

People in various communities depend on their families and/or communities for decisions involving taking part in research or receiving medical treatment. In some cases, agreement from legal representative of the community can be obtained; however, this is not a substitution for individual's informed consent (http://portal.unesco.org/en/ev.php-URL_ID=31058&URL_DO=DO_TOPIC&URL_SECTION=201.html).

In our own experience in Rare Diseases Biobank at Hacettepe University (http://www.hubigem.hacettepe.edu.tr/english/), we also encounter cultural aspects that need to be taken into consideration when obtaining consent. The society has a long tradition of high reverence for medical doctors, and we experience that parents seek the advice/decision of the referring medical doctor for consenting to donate samples from their children. Most families cannot differentiate between giving samples for future research that requires long-term storage before any result can be reached and donating for diagnostic purposes. They are anxious to reach a result in a short time toward the diagnosis of their child. Thus, for example, most of the parents' choice is positive for the return of results. Families with a rare disease usually want the information to be shared within the family, and the concern for privacy is replaced by concern for well-being of other family members who may need genetic counseling. Obtaining consent from illiterate parents is another issue where written consent forms need to be supplemented by oral information, and this information needs to be simple, accurate, and comprehensible. So in every society, obtaining a consent needs to pay attention to the needs of the donors/families, and safeguarding them should be kept ahead of legal protection of the biobank management.

4 VULNERABLE GROUPS

4.1 Minors

Since consent is related to the autonomy of the individual to make responsible decisions, a special attention must be given to the group of persons who are limited in their capacity to consent. Minors, mentally disabled individuals, and people with learning difficulties are vulnerable groups who will not be able to evaluate the circumstances for being involved in a research project.

Consent from minors are obtained from parents or legal guardians provided that there is no harm but direct benefit, and if no direct benefit is obtained, then there is minimal risk involved (Borry et al., 2008).

Regarding risk to the participant, Ethical considerations for clinical trials on medicinal products conducted with the pediatric population (http://ftp.cordis.europa.eu/pub/fp7/docs/ethical-considerations-paediatrics_en.pdf) states that arterial puncture, peripheral venous lines, and skin punch biopsies involve more than minimal risk and are not permissible if there is no direct benefit.

With regard to parental consent, reconsenting the child who reaches an age of competence is an issue. There is an evolving concept that when minors reach mental maturity, they should be asked to assent or dissent for research (Hens et al., 2009). Then, there is discussion about determining a fix biological age to consult the child, yet this is a parameter that needs to be evaluated on an individual basis parameter (Ashcroft et al., 2003; Wendler and Shah, 2003; Giesbertz et al., 2016).

4.2 Deceased Persons

The principles of autonomy and protection of privacy formulated in consent become irrelevant in case of death of a research participant. Especially in large population biobanks with long-term storage for longitudinal studies, data already acquired and continuing storage of the samples raise issues that are not commonly addressed in research consent forms. For continuing use of samples, can family members give consent? Another question is the return of results from already conducted studies to the family members. Again, the principles of autonomy and privacy come into question. Especially in genetic studies in returning results to the family members, harm and benefit should be weighed, and discrimination for the family members should be considered (Tasse, 2011).

The issue is also discussed in some international documents (https://www.wma.net/ http://en.unesco.org/ http://www.who.int/en/).

In the United States, US Department of Health and Human Services policy (45 CFR 46 19,916) describes "human subjects" as living persons and thus does not require that consent be taken for use of samples from deceased persons. However, if consent was taken for research participation when the person was alive, then the scope of the consent needs to be honored after death.

In Canada, ethical review is required for research with samples from a deceased person, and consent given prior to death is honored. In case there is no consent, then a third party is allowed to give consent for the use of samples.

Two biobanks have specific policy for deceased persons' samples and data:

UK Biobank has a policy that states that only a research participant can withdraw consent, and family members cannot withdraw consent after death of a research participant (http://www.ukbiobanka.uk/faq).

CARTaGENE project of Canada states that unless specified by the living will of a research participant, upon death neither samples nor data will be withdrawn from the study (http://www.cartagene.qc.ca/en/home)

5 FROM INDIVIDUAL RIGHTS TO SOLIDARITY

In biobanking research, the question of proper consent types from broad to restricted runs parallel to comparing the weight of individual's rights and privacy with the good and benefit of society (Forsberg et al., 2011).

There are recent trends balancing individualistic approaches in consent versus a new type of right called "solidarity rights."

Rights in an individualist context are mainly for civil and political rights (International Covenant on Civil and Political Rights (1946)).

International Covenant on Economic and Social and Cultural Rights (1996) emphasizes the right to work, health, housing, and education and as such needs governmental policies for implementation.

Solidarity rights furthermore need a global approach since they include the right to peace, healthy environment, development, communication, and ownership of common heritage of humanity (Wellman, 2000; Meslin and Garba, 2011; Prainsack and Buyx, 2013; Chadwick and Berg, 2001).

According to Williams (2005), the recent development of large-scale population biobanks needs to address the collectivist nature of the process rather than the individual's rights protected by the issue of consent. Knoppers defines a new niche for especially longitudinal biobanking projects as public health projects that concern the whole public and thus should be governed by solidarity rights (Knoppers, 2005b).

Solidarity rights are linked to the definition of the human genome as a heritage of humanity (http://portal.unesco.org/en/ev.php-URL_ID=13177&URL_DO=DO_TOPIC&URL_SECTION=201.html) and linked to this definition; it is suggested that genetic databases are a domain that should be regarded as public good (Knoppers and Fecteau, 2003).

6 PRIVACY

Protection of the privacy of an individual is a very basic ethical principle in biomedical research. International organizations issuing declarations and guidelines emphasize the primacy of privacy in biomedical research.

UNESCO Universal Declaration on Bioethics and Human Rights in Article 9 states the following:

> The privacy of the persons concerned and the confidentiality of the personal information should be respected. To the greatest extent possible such information should not be used or disclosed for purposes other than those for which it was collected or consented to, consistent with international law, in particular international human rights law.

The article emphasizes the role of consent to protect privacy and indicates that protection of individual privacy is an issue embedded in human rights law.

However, today in biobanking research, we are experiencing that to reach adequate sample sizes for research, it is becoming a common practice to join forces at national or international level by sharing samples and data.

While protecting the privacy of an individual during research, UNESCO Universal Declaration on Bioethics and Human Rights in Articles 13, 15, and 24 states "solidarity," "sharing of benefits," and "international cooperation" as agencies toward the benefit of the whole human kind.

To protect privacy yet be able to share samples, *anonymization* is a method that can be used to unlink personal identifiers from the research data (Wallace, 2016).

Total anonymization of data makes it irretrievable; thus, it has a disadvantage for the donor if consent wants to be withdrawn or the donor needs to be recontracted. Also, any research result cannot be returned to the donor (https://www.wma.net/policies-post/wma-declaration-of-taipei-on-ethical-considerations-regarding-health-databases-and-biobanks/).

Pseudoanonymization allows the sample to be coded, yet the personal data are not irretrievably unlinked. This type of anonymization is a preferred method in biobank research.

There is still need for harmonized definition in relation to anonymization and reidentification of personal data (Knoppers, 2005a).

Especially with the new EU legislation (General Data Protection Regulation (EU 2016/679)), it will be beneficial to be able to use harmonized terminologies and guidelines. This will enable to be able to share samples yet to safeguard privacy (Sariyar and Schlunder, 2016).

With genomics research and use of WGS techniques with biobank samples, the question of re identification comes up even if the original sample has been anonymized (Fullerton et al., 2010; Jacobs et al., 2009).

7 ACCESS TO SAMPLES AND DATA

To be able to share samples and data while protecting the privacy of donors and right of researchers, "access" policy of biobanks creates transparency in the operation of the biobank and protects the rights of all partners. Some biobanks may have designated committees to review applications for access to the samples; in others, the director/operator of the biobank assumes this responsibility.

In designing an access policy, some core elements summarized below may be found in "P3G Sample and Data Access: Core Elements" (www.p3gobservatory.org/download/P3G/sample):

- Allow access to samples and/or data.
- Indicate the time covered by the granted access.
- An ethics committee review document of the project should accompany the application.
- No access by third parties and no reidentification.
- Use of samples only for the stated purpose.
- Remaining samples may be returned to the biobank or destroyed after the completion of the project.
- The question of ownership of the samples is critical, but it is the general consensus that the samples belong to original source.

Material and Data Access Agreements can be used to share samples after an access policy has been put in place by biobanks. These agreements are signed by the principle investigator and an authorized representative of the receiving institute. Also the type of material that is being transferred has to be well defined. If only data are being transferred, it is a license agreement to the use of the data. In agreement forms, liability clauses should be added so that all responsibility of handling the samples should lie with the PI receiving them. It is the usual practice that biobanks not guarantee the quality of the samples being transferred in such agreements. Intellectual property rights and breach of contract clauses are also points that need to be well defined within the access policy and need to be reflected in the agreement forms accordingly (Material and Data Access Agreements: Core Elements. Saminda Pathmasiri and Bartha Maria Knoppers—P3G Ethics and Policymaking Core (www.p3gobservatory.org)).

7.1 Public Trust

Besides the stakeholders in the biobanking activities, another very important parameter is the public and the perception of the public about biobanks. Trust of the public that includes the donors themselves in biobanking increases the utility of biobanks and their input to scientific discoveries (Gaskell et al., 2013).

A successful example of public engagement is displayed by UK Biobank.

Interim Advisory Group (IAG) was formed in 2003 to provide ethical input into governance of the project. It served as an advisory group to the funders, and the group worked to create awareness about biobanks and tried to create public trust and confidence in the project. Societal acceptance of biobanks is a very crucial input for sustainability of the projects not only socially but also financially (Tutton et al., 2004).

8 SUSTAINABILITY

Both disease-specific biobanks and population-based biobanks for longitudinal studies house large number of samples for long periods of time and share them with researchers, and the scope is moving from being national/regional to global. To maintain such resources with high running costs, one big challenge is the question of economic sustainability.

Many public biobanks lack the managerial background in economics to make sound analysis of running costs. It is advised that biobanks, to ensure economic maintenance, need to develop a business plan (Baiden et al., 2005) with well-described mission and organizational scheme, market analysis, and financial planning with provisions for running costs (Simeon-Dubach and Henderson, 2014; Henderson et al., 2017).

Funding for biobanks can come from different sources: private funds, support from the hosting institute, public funding, research grants, and donations. When considering costs for running a biobank, a study on a survey of biobanks from literature and results of questionnaires from individual biobanks in France and the Netherlands indicates that major costs in running a research biobank is of human resources, sample handling and processing, and equipment costs for laboratories and IT infrastructure (Yuille et al., 2017).

Biobanks that are supported by public funds most of the time do not need to generate revenues, but for long-term sustainability, a cost recovery model needs to be put in place, and a plan to offer services such as consultation can also be included in the business plan (Clement et al., 2014; Gonzalez-Sanchez et al., 2013; Albert et al., 2014).

However, revenue generated by cost recovery is usually a very small contribution to the finances of biobanks (Vaught et al., 2011).

Biobanks not only are *containers* of biospecimens but also must be utilized with maximum capacity for research or health. Biobanks that make transparent access policies and that ensure high utility of the samples with established guidelines for safety and quality of the samples and ethical guidelines for protection of donor privacy transparent have better public trust that helps toward increased utility, and this in turn catalyzes sustainability (Daniel et al., 2015).

Besides sustainability, high quality of specimens is a very important contributing factor for networking of biobanks, and harmonization of quality assurance programs is vital to increase global collaborations and *interoperability* (Kiehntopf and Krawczak, 2011).

9 QUALITY ASSURANCE MANAGEMENT (QAM)

QA management is mandatory to meet the standards for physical capacity and personnel capability required for maintenance of high-quality samples for biomedical research. At biobanks, there has to be a division of labor and responsibilities.

Director or biobank manager is the person who administrates all processes at the biobank. He/she is responsible for hiring of staff, training and education activities, development of access policy, and other policy issues.

All biobanks have separate managerial positions for specific domains such as IT and quality control (QC).

Designated senior technical personnel run the everyday procedures such as sample acquisition and storage, writing of SOPs upgrading, and running the QC systems.

QAM applied for each step of biobanking process starts at the preanalytic stage to ensure the stability and integrity of samples and continues at each step of sample handling up to the distribution stage. Type of transport vials, time, and temperature of transportation are some preanalytic parameters contributing to sample quality. They need to be properly recorded and accompany the sample itself to the biobank and evaluated before the storage stage. Transport documents need to be signed by the depositor and the responsible person in biobank to confirm that the state of sample is suitable for acceptance and processing (Betsou et al., 2010).

10 SECURITY

Security issue is a main component of the quality assurance management. Security guidelines have to be established for physical infrastructures and also for samples and data. The following are a set of minimum requirements:

Physical infrastructure

- Restricted entry only by dedicated personnel
- Use of generators and constant power supplies
- Alarm systems in case of power failure
- Constant liquid nitrogen supply systems

Samples

- Assurance of traceability
- Backup storage systems optimally in a separate location
- Processing by dedicated personnel

Data

- Maintenance of IT system in a separate location
- Restricted access by dedicated personnel
- Hard copy backup for the data and software systems
- Use of appropriate coding/anonymization systems

Quality control systems QCS

- Periodic maintenance and calibration of the instruments
- Membership to proficiency testing programs
- Development of in-house quality controls

Standard operating procedures (SOPs) are written documentation of all the methods at each step of sample handling from acquisition to transfer to external facilities. This needs to be written in a clear and concise way, and all changes and upgrading need to be supervised and approved by the biobank manager. Changes need to be circulated among the lab personnel, and necessary training must be organized (Grizzle et al., 2015).

Best practice guidelines can be found from international organizations such as OECD (Best Practice Guidelines for Biological Resource Centers) (www.oecd.org/dataoecd/7/13/38777417.pdf) and societies such as International Society for Biological and Environmental Repositories (ISBER) (www.isber.org/Pubs/BestPractices2008.pdf).

As part of a quality management system, certification through International Organization for Standardization (ISO) is available, and ISO 9001 is the most commonly used standard for biobanks (Davis et al., 2012).

In France, there is a specific accreditation system for biobanks, French Standard NF S 96-900 (Di Donato, 2015).

11 BIOBANKS FOR PRECISION MEDICINE

We are experiencing a paradigm shift in medicine today. The concept of "right medicine at right dose for the right person" describes the concept of personalized/precision medicine. Such an approach emerging in health care means that instead of treating an advanced disease, prevention can take place at an early stage before onset and populations can be screened to identify individuals with disease susceptibility provided there is an available and reliable biomarker and treatments are tailored to the individual molecular profiles. Pharmacogenetic testing can give information about an individual's response or adverse reaction to a drug, thus leading to better individualized drug prescription. Patient rights in health care for therapeutic decisions are taken into consideration rather than a paternalistic approach, and the new trend is symbolized in predictive, preventive, personalized, participatory medicine (PPPPM) (Khoury et al., 2012).

Sensitive molecular profiling and stratification of individuals are becoming possible following the development and emergence of new "omics" technologies. Genomics technologies such as SNP profiling in genome-wide association studies (GWAS) and whole-exome (WES) or whole-genome sequencing (WGS) are being applied in biomedical research into molecular mechanisms in common and rare monogenic diseases (Ginsburg, 2013).

Today, genetic profiling is considered as a main component leading to personalized health care. There is hope that genomics will soon be translated into clinical applications for better and individually tailored diagnostics and prognostics in cancer and other common diseases. Personal genome profiling has the potential to reveal if the individual falls within a risk group for a specific disease susceptibility. However, since the result is based on prediction of a risk and the contribution of each genome variation to disease is rather small, genomic profiling may not be considered yet as diagnostic in common diseases where environmental factors also play a major role (Burke and Psaty, 2007).

In rare monogenic diseases, identification of causative genome variants has the potential for the development of genetic tests and also for annotation of pathways as drug targets. All "omics" platforms used in a holistic approach for analysis of transcriptome, proteome, and metabolome have the high potential to develop biomarkers and to develop targeted therapies. Rare diseases are a very good example for applications of precision medicine with rare diseases biobanks playing an important role as essential stakeholders (Gulbakan et al., 2016).

Especially for rare diseases where number of patients at one center may be a limiting factor, collaborative biobanking on an international level may be foreseen for studies such as gene identification, investigation for new biomarkers, and clinical trials (Gainotti et al., 2016).

Biobanks, as essential infrastructure for personalized medicine, have been rightfully included in the platform chains that lead from basic to clinical studies linking early epidemiological data, disease modeling, biomarker discovery and validation, and development of diagnostic tools and drugs (Hewitt, 2011; Yan, 2008; Golubnitschaja et al., 2014).

11.1 Return of results

As genomics becomes a tool for precision medicine, a large number of samples in biobanks are being utilized in whole-genome sequencing (WGS) projects that yield a very high content of information (big data). These data sets potentially contain genome variance information with yet unknown clinical validity and utility, and these are defined as incidental or secondary findings (Wolf, 2013; Wolf et al., 2012).

Even if the debate is still unsolved about return of incidental findings, considering WGS data's high potential for individual's health outcome, there is a growing tendency that these findings should be returned. Yet, there are a variety of handicaps indicated for not being able to return results to the participants of research:

- Legal, societal or organizational (the lack of genetic counselors, guidelines, and awareness on the part of research participants),
- financial (the lack of specific funding for feedback process) need for harmonized guidelines,
- the lack of general public awareness,
- REC input to modulate return of results, and
- education of the health-care personnel (Budin-Ljosne et al., 2016).

11.2 Networking activities

11.2.1 *BBMRI-ERIC*

Science and technological tools for biobanking are becoming more sophisticated, and collaborative research is becoming more common to overcome fragmentation and sharing of expertise for more sustainable projects. European Commission has developed the program called European Strategy Forum on Research Infrastructures (EFRSI) to define research infrastructures that will have strategic impact for Europe. One of the infrastructure projects that was selected is Biobanking and BioMolecular Resources Research Infrastructure (BBMRI) (Reichel et al., 2014).

Furthermore, in 2009, a commission regulation (EC-723/2009) was developed to allow the development of European Research Infrastructure Consortium (ERIC).

Today, BBMRI-ERIC has 16 states and one international organization (IARC/WHO) as members (Austria, Belgium, Czech Republic, Estonia, Finland, France, Germany, Greece, Italy, Latvia, Malta, the Netherlands, Norway, Poland, Sweden, and the United Kingdom) and three member states with observer status (Cyprus, Switzerland, and Turkey) (http://www.bbmri-eric.eu/national-nodes/).

BBMRI also has a directory that holds data from 515 European biobanks. This directory is a useful tool to identify partners or access to samples, and thus, it fosters networking activity (Holub et al., 2016).

11.2.2 *EuroBioBank*

Eurobiobank network for rare diseases (Fullerton et al.) was established in 2001 through funding from the EC fifth framework program. The major aim of the network is to provide access to samples (DNA, cells, and tissues) and data to facilitate research efforts toward identification of new genes and finding treatment for rare diseases where only a very small number of diseases have a treatment.

Eurobiobank is the only biobanking network in Europe for rare diseases and has 25 members from 11 countries. The number of biological samples in the catalog is about 140,000 as of 2016 (http://www.eurobiobank.org/).

Eurobiobank is a collaborator of BBMRI and a partner of RD-Connect (http://rd-connect.eu), an FP7 program aiming at coordinating the activities of RD patient registries and biobanking activities.

IRDiRC is an international consortium founded by NIH and EC in 2009 (http://www.irdirc.org) with a wide mission of policy development in rare diseases research with a global outlook joining both European and American rare disease communities. Both Eurobiobank and RD-Connect contribute to the aims and working programs of IRDiRC, and with such a wide network, it is hoped that set goals for RD research will be reached in a short period of time.

12 FUTURE

To gain maximum benefit from precision medicine approaches in health care in the future, international cooperation of biobanks will be an enabling factor. Harmonization of biobanking activities around technical and ethical guidelines will facilitate interoperability on an international level. Using of shared SOPs in sample handling and using of matching data processing and exchange formats will help the sharing of samples and data (Harris et al., 2012).

A framework has been organized to guide harmonization of sharing genomic and clinical data by Global Alliance for Genomic and Health (GA4GH) (Knoppers, 2014).

The framework sets the goals for biobanks:

- Transparency
- Accountability
- Engagement
- Data quality and security
- Privacy, data protection, and confidentiality
- Risk-benefit analysis
- Recognition and attribution
- Sustainability
- Education and training
- Accessibility and dissemination

Biobanking has been an activity that is mostly associated with health research in the United States and Europe and other developed countries. However, in recent years, biobanks are beginning to be established in developing countries as well (Chen and Pang, 2015). The following are some examples of these biobanks:

H3Africa—to initiate genomic studies in the African population (http://h3africa.org) (Davies et al., 2017).

China Kadoorie Biobank—to study genetics of common diseases (http://www.ckbiobank.org/site/).

Gambian National DNA Bank—for genetic analysis of infectious diseases.

13 CONCLUSION

In conclusion, biobanking is becoming an essential partner for precision medicine initiatives, and with fast acceleration in genomics technology and decrease in the cost of analysis such as WES and WGS, establishment of biobanks will spread globally as genomics for public health will improve health care. Both for clinical and research use of biobank samples and to foster international collaboration, public awareness of biobanks is crucial. Furthermore, biobanking activities that are transparent gain the trust of the public that can lead to sustainable funding. Biobanking needs to be a multidisciplinary activity; pathology, biology, computer sciences, bioinformatics, ethics, and law are only some of the disciplines involved.

To usher countries in the era for precision medicine, more education is needed at various levels. Literacy in genomics needs to increase among health-care personnel, and medical school curricula and postgraduate studies need to be adjusted to this end (Ozguc, 2014).

Biobanking as a specific area in biolaw would contribute to harmonization on the international level as biobanking is gaining momentum on the global scale.

Not only sharing of samples and data but also sharing of technology and science of biobanking more globally especially with less developed countries would help improve global health and solidarity between countries.

For harmonization at the global level, solidarity and working toward a common good—better health for all—should be the driving force for sharing of the benefits of biobanks for biomedical research. The following steps can foster global biobanking:

Sharing of technology and know-how.

Training of personnel.

Diffusion of awareness raising activities.

Capacity building for RECs—for international projects involving biobanks, equal reviewing in each participating countries should be fostered.

Harmonization of terminology leading to a common glossary.

Harmonized IT and laboratory information management systems (LIMS).

Use of common disease ontologies.

Harmonized quality assurance and accreditation systems.

Use of common models for consent and MTA forms.

References

Albert, M., Bartlett, J., Johnston, R.N., Schacter, B., Watson, P., 2014. Biobank bootstrapping: is biobank sustainability possible through cost recovery? Biopreserv. Biobank. 12, 374–380.

Arampatzis, A., Papagiouvanni, I., Anestakis, D., Tsolaki, M., 2016. A classification and comparative study of European biobanks: an analysis of biobanking activity and its contribution to scientific progress. Arch. Med. 8 (3), 6.

Arias-Diaz, J., Martin-Arribas, M.C., Garcia del Pozo, J., Alonso, C., 2013. Spanish regulatory approach for biobanking. Eur. J. Hum. Genet. 21, 708–712.

Ashcroft, R., Goodenough, T., Williamson, E., Kent, J., 2003. Children's consent to research participation: social context and personal experience invalidate fixed cutoff rules. Am. J. Bioeth. 3, 16–18.

Asslaber, M., Zatloukal, K., 2007. Biobanks: transnational, European and global networks. Brief Funct. Genom. Proteom. 6, 193–201.

Baiden, F., Baiden, R., Williams, J., Akweongo, P., Clerk, C., Debpuur, C., Philips, J., Hodgson, A., 2005. Review of antenatal-linked voluntary counseling and HIV testing in sub-Saharan Africa: Lessons and options for Ghana. Ghana Med. J. 39, 8–13.

Baldo, C., Casareto, L., Renieri, A., Merla, G., Garavaglia, B., Goldwurm, S., Pegoraro, E., Moggio, M., Mora, M., Politano, L., Sangiorgi, L., Mazzotti, R., Viotti, V., Meloni, I., Pellico, M.T., Barzaghi, C., Wang, C.M., Monaco, L., Filocamo, M., 2016. The alliance between genetic biobanks and patient organisations: the experience of the telethon network of genetic biobanks. Orphanet. J. Rare Dis. 11, 142.

Beskow, L.M., Friedman, J.Y., Hardy, N.C., Lin, L., Weinfurt, K.P., 2010. Developing a simplified consent form for biobanking. PLoS ONE 5, e13302.

Betsou, F., Lehmann, S., Ashton, G., Barnes, M., Benson, E.E., Coppola, D., DeSouza, Y., Eliason, J., Glazer, B., Guadagni, F., Harding, K., Horsfall, D.J., Kleeberger, C., Nanni, U., Prasad, A., Shea, K., Skubitz, A., Somiari, S., Gunter, E., 2010. International Society for Biological and Environmental Repositories (ISBER) Working Group on Biospecimen Science, Standard preanalytical coding for biospecimens: defining the sample PREanalytical code. Cancer Epidemiol. Biomark. Prev. 19, 1004–1011.

Borry, P., Stultiens, L., Goffin, T., Nys, H., Dierickx, K., 2008. Minors and informed consent in carrier testing: a survey of European clinical geneticists. J. Med. Ethics 34, 370–374.

Budin-Ljosne, I., Mascalzoni, D., Soini, S., Machado, H., Kaye, J., Bentzen, H.B., Rial-Sebbag, E., D'abramo, F., Witt, M., Schamps, G., Katic, V., Krajnovic, D., Harris, J.R., 2016. Feedback of individual genetic results to research participants: is it feasible in Europe? Biopreserv. Biobank. 14, 241–248.

Burke, W., Psaty, B.M., 2007. Personalized medicine in the era of genomics. JAMA 298, 1682–1684.

Calzolari, A., Napolitano, M., Bravo, E., 2013. Review of the Italian current legislation on research biobanking activities on the eve of the participation of national biobanks' network in the legal consortium BBMRI-ERIC. Biopreserv. Biobank. 11, 124–128.

Chadwick, R., Berg, K., 2001. Solidarity and equity: new ethical frameworks for genetic databases. Nat. Rev. Genet. 2, 318–321.

Chen, H., Pang, T., 2015. A call for global governance of biobanks. Bull. World Health Organ. 93, 113–117.

Clement, B., Yuille, M., Zaltoukal, K., Wichmann, H.E., Anton, G., Parodi, B., Kozera, L., Brechot, C., Hofman, P., Dagher, G., Biobanks, E.-U.E.G.O.C.R.I., 2014. Public biobanks: calculation and recovery of costs. Sci. Transl. Med. 6, 261fs45.

Daniel, N., Fraticelli, F., Franceschilli, S., Zinno, F., 2015. More than just a "container": centrality and versatility of biobanks in the era of scientific challenges. Int. J. Lab. Med. Res. 1, 106.

Davies, J., Abimiku, A., Alobo, M., Mullan, Z., Nugent, R., Schneidman, M., Sikhondze, W., Onyebujoh, P., 2017. Sustainable clinical laboratory capacity for health in Africa. Lancet Glob. Health 5, e248–e249.

Davis, E., Hampson, K., Bray, C., Dixon, K., Ollier, W., Yuille, M., 2012. Selection and implementation of the ISO9001 standard to support biobanking research infrastructure development. Biopreserv. Biobank. 10, 162–167.

Di Donato, J.H., 2015. Biobanking for rare diseases-impact on personalised medicine. In: Ozguc, M. (Ed.), Rare Diseases Integrative PPPM Approach as the Medicine of the Future. In: Golubnitschaja, O. (Ed.), Advances in Predictive, Preventive and Personalized Medicine, Vol. 6. Springer, Dordrecht, pp. 23–31.

Elliott, P., Peakman, T.C., Biobank, U.K., 2008. The UK biobank sample handling and storage protocol for the collection, processing and archiving of human blood and urine. Int. J. Epidemiol. 37, 234–244.

Forsberg, J.S., Hansson, M.G., Eriksson, S., 2011. Biobank research: who benefits from individual consent? Br. Med. J. 343, .

Fullerton, S.M., Anderson, N.R., Guzauskas, G., Freeman, D., Fryer-Edwards, K., 2010. Meeting the governance challenges of next-generation biorepository research. Sci. Transl. Med. 2, 15cm3.

Gainotti, S., Turner, C., Woods, S., Kole, A., McCormack, P., Lochmuller, H., Riess, O., Straub, V., Posada, M., Taruscio, D., Mascalzoni, D., 2016. Improving the informed consent process in international collaborative rare disease research: effective consent for effective research. Eur. J. Hum. Genet. 24, 1248–1254.

Gaskell, G., Gottweis, H., Starkbaum, J., Gerber, M.M., Broerse, J., Gottweis, U., Hobbs, A., Helen, I., Paschou, M., Snell, K., Soulier, A., 2013. Publics and biobanks: Pan-European diversity and the challenge of responsible innovation. Eur. J. Hum. Genet. 21, 14–20.

Gefenas, E., Dranseika, V., Serepkaite, J., Cekanauskaite, A., Caenazzo, L., Gordijn, B., Pegoraro, R., Yuko, E., 2012. Turning residual human biological materials into research collections: playing with consent. J. Med. Ethics 38, 351–355.

Giesbertz, N.A., Bredenoord, A.L., Van Delden, J.J., 2016. When children become adults: should biobanks re-contact? PLoS Med. 13, e1001959.

Ginsburg, G.S., 2013. Realizing the opportunities of genomics in health care. JAMA 309, 1463–1464.

Golubnitschaja, O., Kinkorova, J., Costigliola, V., 2014. Predictive, preventive and personalised medicine as the hardcore of "Horizon 2020": Epma position paper. EPMA J. 5, 6.

Gonzalez-Sanchez, M.B., Lopez-Valeiras, E., Morente, M.M., Fernandez Lago, O., 2013. Cost model for biobanks. Biopreserv. Biobank. 11, 272–277.

Gottweis, H., Lauss, G., 2012. Biobank governance: heterogeneous modes of ordering and democratization. J. Commun. Genet. 3, 61–72.

Grady, C., 2015. Enduring and emerging challenges of informed consent. N. Engl. J. Med. 372, 855–862.

Grizzle, W.E., Gunter, E.W., Sexton, K.C., Bell, W.C., 2015. Quality management of biorepositories. Biopreserv. Biobank. 13, 183–194.

Gulbakan, B., Ozgul, R.K., Yuzbasioglu, A., Kohl, M., Deigner, H.P., Ozguc, M., 2016. Discovery of biomarkers in rare diseases: Innovative approaches by predictive and personalized medicine. EPMA J. 7, 24.

Hansson, M.G., 2009. Ethics and biobanks. Br. J. Cancer 100, 8–12.

Hansson, M.G., Dillner, J., Bartram, C.R., Carlson, J.A., Helgesson, G., 2006. Should donors be allowed to give broad consent to future biobank research? Lancet Oncol. 7, 266–269.

Harris, J.R., Burton, P., Knoppers, B.M., Lindpaintner, K., Bledsoe, M., Brookes, A.J., Budin-Ljosne, I., Chisholm, R., Cox, D., Deschenes, M., Fortier, I., Hainaut, P., Hewitt, R., Kaye, J., Litton, J.E., Metspalu, A., Ollier, B., Palmer, L.J., Palotie, A., Pasterk, M., Perola, M., Riegman, P.H., Van Ommen, G.J., Yuille, M., Zatloukal, K., 2012. Toward a roadmap in global biobanking for health. Eur. J. Hum. Genet. 20 (11), 1105.

Henderson, M.K., Goldring, K., Simeon-Dubach, D., 2017. Achieving and maintaining sustainability in biobanking through business planning, marketing, and access. Biopreserv. Biobank. 15, 1–2.

Hens, K., Nys, H., Cassiman, J.J., Dierickx, K., 2009. Biological sample collections from minors for genetic research: a systematic review of guidelines and position papers. Eur. J. Hum. Genet. 17, 979–990.

Hewitt, R.E., 2011. Biobanking: the foundation of personalized medicine. Curr. Opin. Oncol. 23, 112–119.

Hirschberg, I., Knuppel, H., Strech, D., 2013. Practice variation across consent templates for biobank research. A survey of German biobanks. Front. Genet. 4, 240.

Holub, P., Swertz, M., Reihs, R., Van Enckevort, D., Muller, H., Litton, J.E., 2016. Bbmri-Eric directory: 515 biobanks with over 60 million biological samples. Biopreserv. Biobank. 14, 559–562.

Jacobs, K.B., Yeager, M., Wacholder, S., Craig, D., Kraft, P., Hunter, D.J., Paschal, J., Manolio, T.A., Tucker, M., Hoover, R.N., Thomas, G.D., Chanock, S.J., Chatterjee, N., 2009. A new statistic and its power to infer membership in a genome-wide association study using genotype frequencies. Nat. Genet. 41, 1253–1257.

Khoury, M.J., Gwinn, M.L., Glasgow, R.E., Kramer, B.S., 2012. A population approach to precision medicine. Am. J. Prev. Med. 42, 639–645.

Kiehntopf, M., Krawczak, M., 2011. Biobanking and international interoperability: Samples. Hum. Genet. 130, 369–376.

Knoppers, B.M., 2005a. Biobanking: International norms. J. Law Med. Ethics 33, 7–14.

Knoppers, B.M., 2005b. Of genomics and public health: Building public "goods"? CMAJ 173, 1185–1186.

Knoppers, B.M., 2014. Framework for responsible sharing of genomic and health-related data. HUGO J. 8, 3.

Knoppers, B.M., Fecteau, C., 2003. Human genomic databases: A global public good? Eur. J. Health Law 10, 27–41.

Leitsalu, L., Alavere, H., Tammesoo, M.L., Leego, E., Metspalu, A., 2015a. Linking a population biobank with national health registries-the estonian experience. J. Pers. Med. 5, 96–106.

Leitsalu, L., Haller, T., Esko, T., Tammesoo, M.L., Alavere, H., Snieder, H., Perola, M., Ng, P.C., Magi, R., Milani, L., Fischer, K., Metspalu, A., 2015b. Cohort profile: Estonian biobank of the Estonian genome center, University of Tartu. Int. J. Epidemiol. 44, 1137–1147.

Lunshof, J.E., Chadwick, R., Vorhaus, D.B., Church, G.M., 2008. From genetic privacy to open consent. Nat. Rev. Genet. 9, 406–411.

Macilotti, M., Izzo, U., Pascuzzi, G., Barbareschi, M., 2008. Legal aspects of biobanks. Pathologica 100, 86–115.

Master, Z., Campo-Engelstein, L., Caulfield, T., 2015. Scientists' perspectives on consent in the context of biobanking research. Eur. J. Hum. Genet. 23, 569–574.

Meslin, E.M., Garba, I., 2011. Biobanking and public health: is a human rights approach the tie that binds? Hum. Genet. 130, 451–463.

Navis, G.J., Blankestijn, P.J., Deegens, J., De Fijter, J.W., Homan Van Der Heide, J.J., Rabelink, T., Krediet, R.T., Kwakernaak, A.J., Laverman, G.D., Leunissen, K.M., Van Paassen, P., Vervloet, M.G., Wee, P.M., Wetzels, J.F., Zietse, R., Van Ittersum, F.J., Investigators, B.-N., 2014. The biobank of Nephrological diseases in the Netherlands cohort: the string of pearls initiative collaboration on chronic kidney disease in the university medical centers in the Netherlands. Nephrol. Dial. Transplant. 29, 1145–1150.

Ozguc, M., 2014. Education for all in the era of personalized medicine. EPMA J. 5 (Supp 1), A139.

Palmer-Aronsten, B., Sheedy, D., Mccrossin, T., Kril, J., 2016. An international survey of brain banking operation and characterization practices. Biopreserv. Biobank. 14, 464–469.

Peakman, T., Elliott, P., 2010. Current standards for the storage of human samples in biobanks. Genome Med. 2, 72.

Petrini, C., 2010. Umbilical cord blood collection, storage and use: ethical issues. Blood Transfus. 8, 139–148.

Prainsack, B., Buyx, A., 2013. A solidarity-based approach to the governance of research biobanks. Med. Law Rev. 21, 71–91.

Reichel, J., Lind, A.S., Hansson, M.G., Litton, J.E., 2014. Eric: a new governance tool for biobanking. Eur. J. Hum. Genet. 22, 1055–1057.

Riegman, P.H., Van Veen, E.B., 2011. Biobanking residual tissues. Hum. Genet. 130, 357–368.

Riegman, P.H., Dinjens, W.N., Oosterhuis, J.W., 2007. Biobanking for interdisciplinary clinical research. Pathobiology 74, 239–244.

Riegman, P.H., Morente, M.M., Betsou, F., De Blasio, P., Geary, P., Marble Arch International Working Group on Biobanking for Biomedical Research, 2008. Biobanking for better healthcare. Mol. Oncol. 2, 213–222.

Salvaterra, E., Lecchi, L., Giovanelli, S., Butti, B., Bardella, M.T., Bertazzi, P.A., Bosari, S., Coggi, G., Coviello, D.A., Lalatta, F., Moggio, M., Nosotti, M., Zanella, A., Rebulla, P., 2008. Banking together. A unified model of informed consent for biobanking. EMBO Rep. 9, 307–313.

Sariyar, M., Schlunder, I., 2016. Reconsidering anonymization-related concepts and the term "identification" against the backdrop of the European legal framework. Biopreserv. Biobank. 14, 367–374.

Scott, C.T., Caulfield, T., Borgelt, E., Illes, J., 2012. Personal medicine—the new banking crisis. Nat. Biotechnol. 30, 141–147.

Simeon-Dubach, D., Henderson, M.K., 2014. Sustainability in biobanking. Biopreserv. Biobank. 12, 287–291.

Steinsbekk, K.S., Kare Myskja, B., Solberg, B., 2013. Broad consent versus dynamic consent in biobank research: is passive participation an ethical problem? Eur. J. Hum. Genet. 21, 897–902.

Strech, D., Bein, S., Brumhard, M., Eisenmenger, W., Glinicke, C., Herbst, T., Jahns, R., Von Kielmansegg, S., Schmidt, G., Taupitz, J., Troger, H.D., 2016. A template for broad consent in biobank research. Results and explanation of an evidence and consensus-based development process. Eur. J. Med. Genet. 59, 295–309.

Tasse, A.M., 2011. Biobanking and deceased persons. Hum. Genet. 130, 415–423.

Tasse, A.M., 2016. A comparative analysis of the legal and bioethical frameworks governing the secondary use of data for research purposes. Biopreserv. Biobank. 14, 207–216.

Turner, L., 2005. From the local to the global: bioethics and the concept of culture. J. Med. Philos. 30, 305–320.

Tutton, R., Kaye, J., Hoeyer, K., 2004. Governing Uk biobank: the importance of ensuring public trust. Trends Biotechnol. 22, 284–285.

Vaught, J., Rogers, J., Myers, K., Lim, M.D., Lockhart, N., Moore, H., Sawyer, S., Furman, J.L., Compton, C., 2011. An NCI perspective on creating sustainable biospecimen resources. J. Natl. Cancer Inst. Monogr. 2011, 1–7.

Wallace, S.E., 2016. What does anonymization mean? Datashield and the need for consensus on anonymization terminology. Biopreserv. Biobank. 14, 224–230.

Weale, A.R., Lear, P.A., 2007. Organ transplantation and the human tissue act. Postgrad. Med. J. 83, 141–142.

Wellman, C., 2000. Solidarity, the individual and human rights. Hum. Rights Q. 22, 639–657.

Wendler, D., Shah, S., 2003. Should children decide whether they are enrolled in nonbeneficial research? Am. J. Bioeth. 3, 1–7.

Williams, J.R., 2005. UNESCO's proposed declaration on bioethics and human rights—a bland compromise. Dev. World Bioeth. 5, 210–215.

Wolf, S.M., 2013. Return of individual research results and incidental findings: facing the challenges of translational science. Annu. Rev. Genomics Hum. Genet. 14, 557–577.

Wolf, S.M., Crock, B.N., Van Ness, B., Lawrenz, F., Kahn, J.P., Beskow, L.M., Cho, M.K., Christman, M.F., Green, R.C., Hall, R., Illes, J., Keane, M., Knoppers, B.M., Koenig, B.A., Kohane, I.S., Leroy, B., Maschke, K.J., Mcgeveran, W., Ossorio, P., Parker, L.S., Petersen, G.M., Richardson, H.S., Scott, J.A., Terry, S.F., Wilfond, B.S., Wolf, W.A., 2012. Managing incidental findings and research results in genomic research involving biobanks and archived data sets. Genet. Med. 14, 361–384.

Yan, Q., 2008. The integration of personalized and systems medicine: bioinformatics support for pharmacogenomics and drug discovery. Methods Mol. Biol. 448, 1–19.

Yuille, M.M., Feller, P.I., Georghiou, L., Laredo, P., Welch, E.W., 2017. Financial sustainability of biobanks: from theory to practice. Biopreserv. Biobank. 15, 85–92.

Yuzbasioglu, A., Ozguc, M., 2013. Biobanking: sample acquisition and quality assurance for "omics" research. New Biotechnol. 30, 339–342.

Zatloukal, K., Hainaut, P., 2010. Human tissue biobanks as instruments for drug discovery and development: impact on personalized medicine. Biomark. Med 4, 895–903.

CHAPTER

5

Sequencing in Precision Medicine

Stephan Vilgis[*], *Hans-Peter Deigner*[*,†]
[*]Institute of Precision Medicine, Furtwangen University, Villingen-Schwenningen, Germany
[†]Fraunhofer EXIM/IZI, Rostock-Leipzig, Germany

1 SEQUENCING TECHNOLOGIES

Thinking about DNA, most people know that the DNA is the carrier of the genetic information and the fundamental of nearly all living organisms. This information, however, is by no means sufficient for modern science to comprehend the complex contexts in a living being. The discovery of DNA dates back to the year 1869, where Friedrich Miescher succeeded in extracting a substance from the cell nuclei of leukocytes (Dahm, 2005), which he called nuclein. Twenty years later, in 1889, Richard Altmann had found a possibility of isolating proteins and nucleic acid from this nuclein (Altmann, 1889) before Albrecht Kossel discovered the four bases adenine, cytosine, thymine, and guanine in 1896 (Kossel, 1896). Finally, in 1953, the scientists Crick and Watson discovered the DNA double helix (Watson and Crick, 1953) and were awarded the 1962 Nobel Prize in Medicine "for their discoveries concerning the molecular structure of nucleic acids and its significance for information transfer in living material."

The research of Crick and Watson marked a milestone in the history of science and gave rise to modern molecular biology, starting the era of genomics concerned with questions how genes control biochemical processes within cells.

Particularly, in the field of genomics, sequencing of nucleic acids plays one of the main roles because of the key importance of DNA. Thus, knowledge of DNA sequences is mandatory or at least useful in almost any area of biological research.

2 MEANING OF SEQUENCING

DNA sequencing is the process of determining the nucleotide order of a given DNA polynucleotide chain containing the information for the hereditary and biochemical properties of terrestrial life. Put simply, sequencing means "reading the DNA"; therefore, the ability to determine the length and the composition of such sequences is imperative to biological research.

Precision Medicine
https://doi.org/10.1016/B978-0-12-805364-5.00005-6

Since the introduction of the first-generation sequencing technologies (FGSTs), the technological demands have been growing, especially in high-throughput and massively parallel sequencing possibilities. These requirements were initially covered by the development of next-generation sequencing technologies (NGSTs). These technologies also come to their limits as soon as it is about sequencing in real time or one tries to prepare the sample by as little manipulation as possible to reduce biases.

Therefore, further developments such as the third-generation sequencing technologies (TGSTs) seem to be ideal candidates covering at least partially all the desired demands.

3 FIRST-GENERATION SEQUENCING

The first two methods for the sequencing of DNA were developed between 1975 and 1977, and these are, namely, the method of Maxam-Gilbert and the method of Sanger (Maxam and Gilbert, 1977; Sanger and Coulson, 1975).

In general, first-generation sequencing consists of three basic phases by means of sample preparation, physical/chemical sequencing, and reassembly. The first step involves cutting off the nucleotide chains into several small fragments and amplification of the fragments if the amount of sample is too low. The second step, the sequencing process itself, uses physical or chemical detection to determine the base call and the last step to reassemble the fragments to obtain the contiguous sequences.

3.1 Maxam-Gilbert "Base-Specific Chemical Cleavage" Sequencing

The Maxam-Gilbert method was developed in 1977 by Allan Maxam and Walter Gilbert.

Their technology is based on the base-specific chemical cleavage of the DNA by suitable reagents and subsequent separation of the fragments by SDS-PAGE.

For this, the double-stranded DNA (dsDNA) is fragmented, denatured into small fragments of single-stranded DNA (ssDNA), and labeled with ^{32}P at the 5′ end of the DNA fragment by a kinase.

The labeled ssDNA fragments are further used for four different reactions each made by different cleavage reagents, where within each reaction the base-specific cleavage takes place (Maxam and Gilbert, 1977).

3.1.1 *Chemical Cleavage of Specific Bases*

Purines and pyrimidines make up the two groups of nucleotide bases (cytosine and thymine out of pyrimidines and adenine and guanine out of purines).

The combination of various chemicals will lead to a breakage of the glycoside bond between the ribose and the base of the nucleotides resulting in a base-specific cleavage as shown below.

Dimethylsulfate + piperidine:	→ cleavage of guanines
Dimethylsulfate + piperidine in folic acid	→ cleavage of guanines and adenines
Hydrazine + piperidine:	→ cleavage of thymines
Hydrazine + piperidine in sodium chloride	→ cleavage of cytosines

By this, the DNA fragments are cut into different sizes dependent on the sequence. An example is shown in Fig. 1.

The labeled fragments of each reaction are subsequently separated by their size using, e.g., a high-resolution sodium dodecyl sulfate polyacrylamide gel electrophoresis (SDS-PAGE) and subsequently made visible using radioautography where the smallest fragments occur at the bottom of the gel.

By comparison of the bands from the purine reactions (A+G with G) and by comparison of the bands from the pyrimidine reactions (C+T with C), the bases can be identified.

For example, if the identical band is visible at A+G and G, the base must be G, whereas if the band is only visible at A+G, the base must be A, same with the pyrimidines.

Finally, the nucleotide sequence can be read from the bottom to the top of the gel.

Due to the hazardous chemicals needed for Maxam-Gilbert sequencing and the complexity for automatization, this technology was readily replaced by the breakthrough of the Sanger sequencing technology.

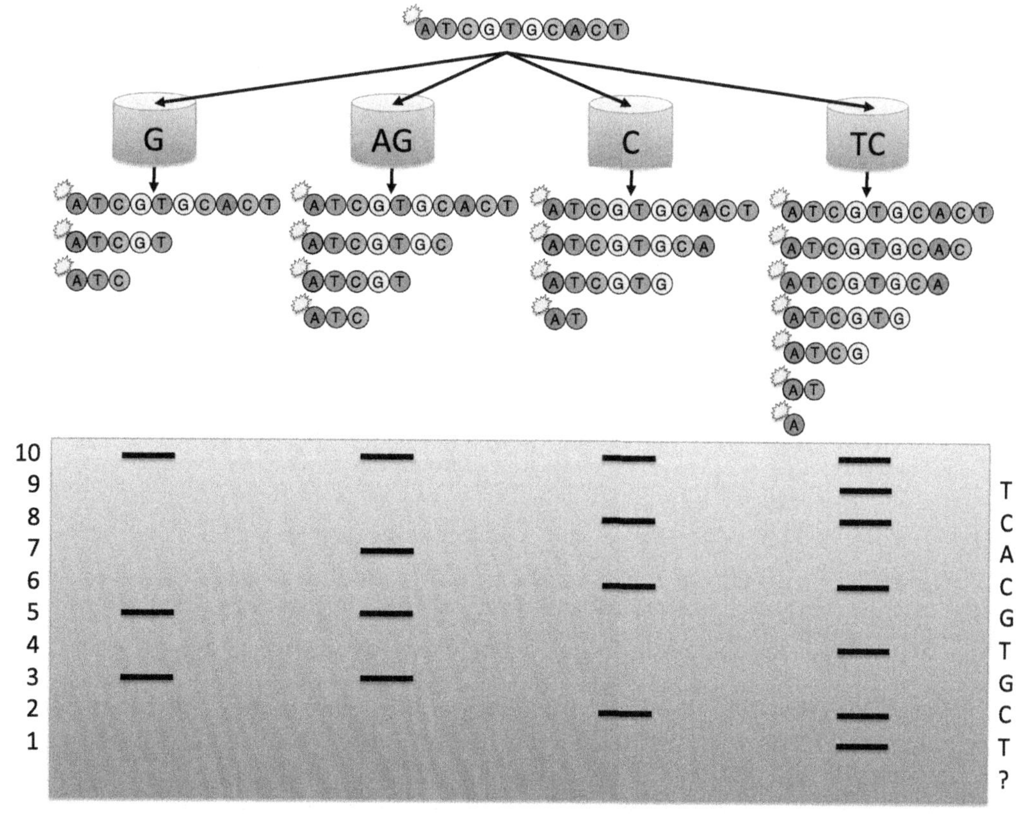

FIG. 1 Maxam-Gilbert sequencing principle. The labeled DNA fragments undergo four different chemical reactions and within the specific bases are cleaved resulting in fragments with different sizes. The fragments are separated on a gel and visualized with autoradiography.

3.2 Sanger Chain-Termination Sequencing

In 1977, Fred Sanger and colleagues developed a related technique based on the detection of radiolabeled partial digestion of DNA fragments (Sanger and Coulson, 1975; Sanger et al., 1977).

In this process, the double-stranded DNA (dsDNA) gets digested by restriction enzymes into short fragments, and these subsequently are denatured into single-stranded DNA (ssDNA).

The ssDNA fragments are further processed within four different reactions containing a radio- or fluorescence-labeled DNA primer, a DNA polymerase, the dNTPs, and a small amount of one of the four ddNTPs.

3.2.1 *dNTPS and ddNTPS*

Dideoxyribonucleoside triphosphates (ddNTP) are artificial DNA nucleotides constructed like the deoxyribonucleoside triphosphates (dNTPs). However, the ribose (sugar) is deoxidized at position 2′ and 3′. By this modification, it is achieved that no further base can be incorporated during strand synthesis, since the missing hydroxyl group is required for the phosphodiester bond in the DNA backbone (Fig. 2).

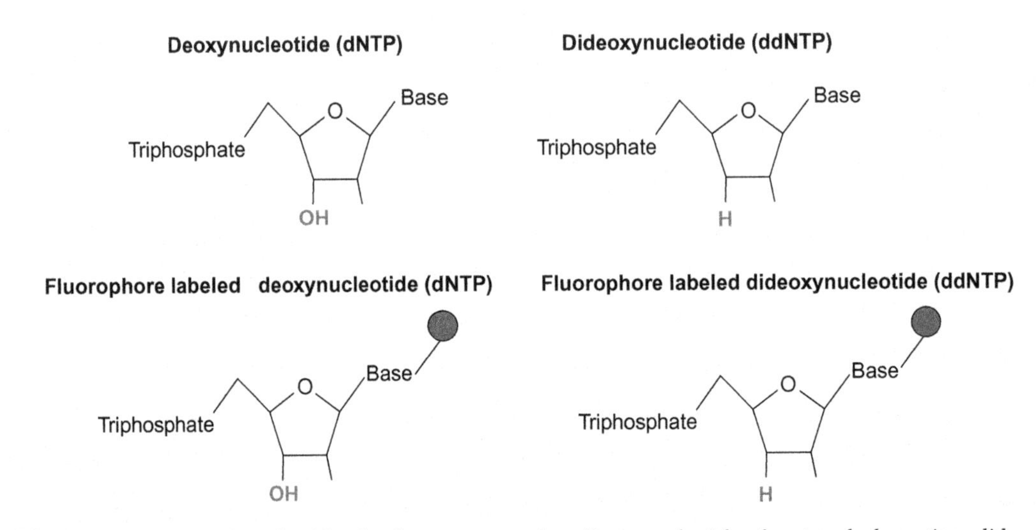

FIG. 2 Typically used nucleotides for Sanger sequencing. Deoxynucleotides for strand elongation, dideoxynucleotides for chain termination, fluorophore-labeled dNTP enabling detection during strand elongation, and fluorophore-labeled ddNTP for (reversible) chain termination.

The labeled DNA primer binds to the complementary sequence of the 3′ end of the ssDNA, and the polymerase starts strand synthesis. This reaction takes place in each of the four vials, each containing one different ddNTP (ddATP, ddTTP, ddGTP, or ddCTP).

If the polymerase incorporates one ddNTP by chance, neither further dNTP nor a ddNTP can be incorporated in the next step leading to chain termination. Thus, several labeled fragments of different sizes are created within each vial. The labeled fragments of each vial are subsequently separated by their size using, e.g., a high-resolution sodium dodecyl sulfate polyacrylamide gel electrophoresis (SDS-PAGE), where the smallest fragments occur at the bottom of the gel. The sequence then is determined by "reading" the gel from the bottom to the top (Fig. 3).

In a more recent adaptation of this principle, each ddNTP (A, C, T, or G) nowadays carries a unique, fluorescent molecule; by this approach, the strand synthesis is terminated and at the same time labeled at the terminal end of the fragment.

The sequence thus can be determined using capillary electrophoresis where each fluorophore at the terminal end of the fragment emits a specific wavelength.

The ABI Prism 310 genetic analyzer introduced in 1995 by Applied Biosystems is available today as version ABI Prism 3730xl still uses this technology and still applied in reliable genetic diagnostics since the accuracy is about 99.999% (Liu et al., 2012).

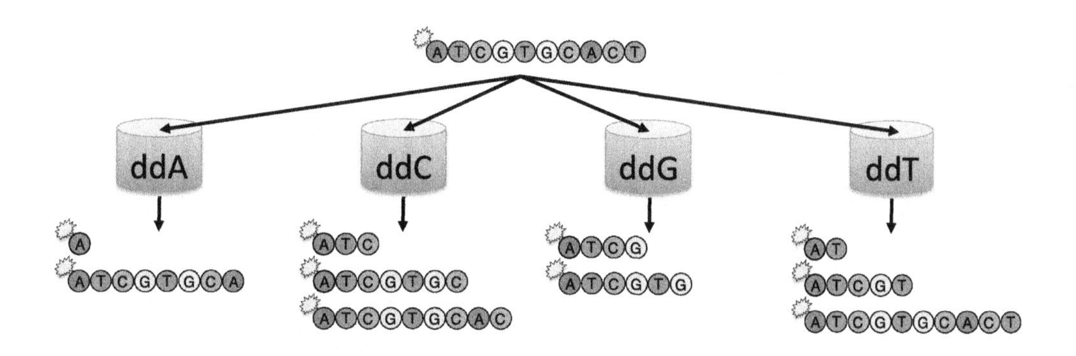

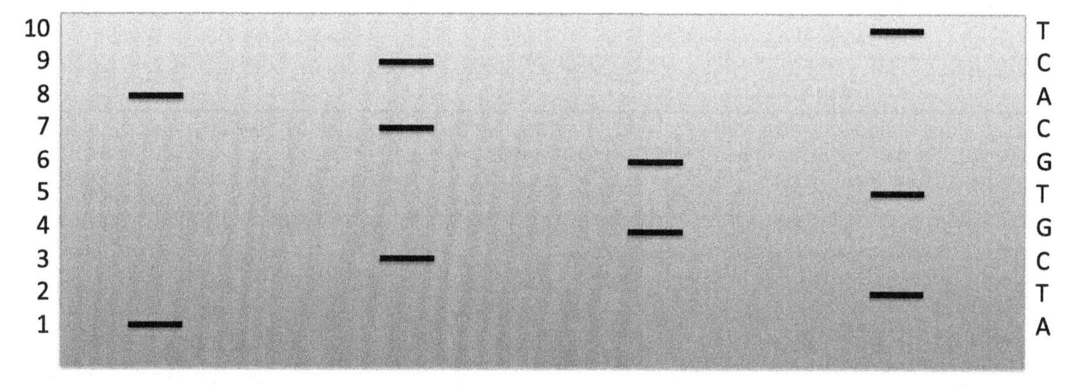

FIG. 3 Sanger sequencing principle. The labeled DNA fragments undergo four different chemical reactions and within the specific bases are blocked resulting in chain termination and fragments with different sizes. The fragments are separated on a gel and visualized with radiography.

4 NEXT-GENERATION SEQUENCING

The request for high-throughput technologies led to the introduction of the second-generation sequencing technologies since the first-generation sequencing technologies are not capable of performing high throughput although having a lower error rate. The Roche/454 platform (Margulies et al., 2005) was the first available next-generation sequencing technology (NGST) introduced in 2005 employing pyrosequencing, followed by the Illumina/Solexa system (Bentley et al., 2008) in 2006 using also sequencing by synthesis, Applied Biosystems SOLiD system (Valouev et al., 2008) in 2007 using sequencing by ligation, and Ion Torrent system (Rothberg et al., 2011) in 2010, also based on sequencing by synthesis. A comparison of commonly used NGSTs is shown in Table 1.

TABLE 1 Comparison of Next-Generation Sequencing Technologies

Device	Roche 454 GS FLX	Illumina HiSeq 2000	Applied Biosystems SOLiDv4	Applied Biosystems Sanger 3730xl	Life Technologies Ion Torrent PGM 318
Sequencing mechanism	Pyrosequencing	Sequencing by synthesis	Ligation and two-base coding	Dideoxy chain termination	Ion semiconductor sequencing
Read length	700 bp	50 bp single end, 50 bp paired end	50 + 35 bp or 50 + 50 bp	400–900 bp	400 bp
Accuracy	99.9%	98%	99.94% (raw data)	99.999%	99.4%
Reads	1 M	3 G	1200–1400 M	–	5.5 M
Output data/run	0.7 Gb	600 Gb	120 Gb	1.9–84 kb	4 Gb
Time/run	24 h	3–10 days	7 days for single end 14 days for paired end	20 min–3 h	2–4 h
Advantage	Read length, fast	High throughput	Accuracy	High quality, long read length	Large output
Disadvantage	Error rate with polybase >6, high cost, low throughput	Short-read assembly	Short-read assembly	High cost, low throughput	Low throughput

Data from Liu, L., Li, Y., Li, S., Hu, N., He, Y., Pong, R., Lin, D., Lu, L., Law, M., 2012. Comparison of next-generation sequencing systems. J. Biomed. Biotechnol. 2012, 251364, https://doi.org/10.1155/2012/251364. and Life Technologies.

These commercially available high-throughput sequencing platforms share three crucial steps to create a so-called sequencing library:

(1) Fragmentation
(2) Adapter ligation
(3) Amplification

These three steps are the fundamentals enabling massively parallel, high-throughput sequencing by reducing the sequencing time since the fragments can be analyzed in parallel and by cluster generation immobilizing the fragments on a surface using adapters followed by amplification to improve the signal detection since each cluster consists of the same sequences. Depending on the platform, each NGST uses a variety of fluidic and optic technologies to perform and record the molecular sequencing reactions.

Compared with Sanger sequencing, these NGSTs are cheaper and less time-consuming since most NGSTs are capable of performing multiplex sequencing in a high-throughput manner, generating up to billions of reads. Nevertheless, all NGSTs are restricted in their read length and their low throughput capacity or poor read-assembly capacity (Liu et al., 2012).

After their launch, these technologies have been continuously refined and improved, and many novel capabilities have been achieved within a relatively short period of time (Sanger and Coulson, 1975; Margulies et al., 2005; Bentley et al., 2008; Valouev et al., 2008; Rothberg et al., 2011).

NGSTs use various chemistries and detection mechanisms (such as luminescence, fluorescence, or hydrogen detection). The common principle of all NGSTs is their need for clonally amplified molecules to increase the signal during the sequencing process since they are not able to detect single molecules (Liu et al., 2012).

Clonal amplification and reverse transcription seem to be adequate methods to solve this problem; however, it affects sample properties, for example, by changing the sample length, by means of the necessary adapter ligation. Furthermore, the amplification using PCR introduces errors and increases the bias of sequencing results (Linsen et al., 2009).

As sequencing cycles progress, the clonal strands get out of phase due to chemistry inefficiencies. The result is a limited read length, higher error rates (compared with Sanger sequencing), and suboptimal sequencing efficiency (Linsen et al., 2009).

4.1 Roche 454 Pyrosequencing

Pyrosequencing is based on sequencing by synthesis, where the pyrophosphate production during DNA polymerization is detected by light. The principle was first demonstrated in 1987 by Nyren (2007).

Pyrosequencing starts with the amplification of the DNA strand using PCR, where the PCR primer contains a biotin at the 5′ end, leading to a biotinylated amplified product that binds to magnetic beads coated with streptavidin. The DNA becomes denatured leaving only the template strand on the bead. By applying a magnetic field, the beads, containing the strand to be sequenced are kept in a well having the same size as the bead itself, whereas the unwanted by-products are washed away, before a primer and a polymerase are bound to the template strand.

The sequencing process itself takes place in an emulsion consisting of the beads and further enzymes such as sulfurylase, luciferin, and apyrase.

During sequencing, the dNTPS are sequentially dispensed into the reaction well. If the complement dNTP gets incorporated, an inorganic pyrophosphate (PP_i) and a hydrogen (H^+) are released.

The PP_i reacts with adenosine 5′ phosphosulfate (APS) catalyzed by ATP sulfurylase to adenosine triphosphate (ATP) and anionic sulfate (SO_4^{2-})

$$PP_i + APS \rightarrow ATP + SO_4^{2-} \rightarrow \text{catalyzed by sulfurylase}$$

The obtained ATP drives the luciferase reaction requiring O_2, oxidizing luciferin into oxyluciferin, AMP, PP_i, CO_2, and energy (E_{Photon}) in form of a detectable light signal (hf):

$$ATP + O_2 + \text{luciferin} \rightarrow AMP + PP_i + \text{oxyluciferin} + CO_2 + hf$$
$$(h = \text{Planck's constant};\ f = \text{frequency})$$

The emission of the light signal during the addition of dNTPs is proportional to the amount of PPi and therefore proportional to the amount of the incorporated dNTP. Since the four different dNTPs are sequentially dispensed, each incorporation can be checked one after another by having a signal or not. These signals can be recorded in a so-called program, where the height of one peak is proportional to the measured energy. This means the peak increases within a series of identical bases (homopolymer) since all dNTPs of a homopolymer sequence are incorporated almost simultaneously. If no reaction takes place, the dNTPS are degraded by the apyrase enzyme (Harrington et al., 2013). An overview of the Roche 454 Pyrosequencing principle is shown in Fig. 4.

4.2 Illumina/Solexa Sequencing

The Illumina sequencing principle, also known as Solexa, shows similarity to the Sanger sequencing approach since both methods include chain termination and both technologies use DNA primer, DNA polymerase, and deoxynucleoside triphosphates (dNTPs) but differ in chain termination reaction (Sanger and Coulson, 1975; Bentley et al., 2008).

4.2.1 Basic Sequencing Principle

The identification of the nucleotides relies on the attachment of randomly fragmented and amplified ssDNA to a planar, optically transparent surface (flow cell). The bound fragments are subsequently clonally amplified by bridge amplification, building so-called clusters. The detection of the nucleotides is then achieved by a process comprising sequential washing and scanning operations using modified fluorescence label dNTP, which is also known as reversible terminators. Thus, one base was incorporated at the time and excited by a laser, leading to a specific signal to determine the base call (Magi et al., 2016). This so-called sequencing by synthesis method gives the scientists the power to sequence several gigabytes of data in a single run, and this has made Illumina the leading company for NGST applications (Magi et al., 2016).

4.2.2 Reversible Terminators

The classical chain termination method (Sanger) requires a single-stranded DNA template, a DNA primer, a DNA polymerase, normal deoxynucleoside triphosphates (dNTPs), and modified radioactively or fluorescently labeled dideoxynucleotide triphosphates (ddNTPs).

The ddNTPs irreversibly terminates the DNA strand elongation during synthesis, due to a missing 3′-OH group, required for the formation of a phosphodiester bond between two nucleotides as shown in Fig. 5 (Sanger and Coulson, 1975; Bentley et al., 2008).

The Illumina technology uses a similar chemistry consisting of a DNA primer, a DNA polymerase, normal dNTPs, and modified dNTPS instead of ddNTPS.

These modified dNTPs containing so-called reversible terminators differ from naturally occurring ones in having an initially blocked 3′-OH group and a cleavable fluorophore at the base. Compared with using dNTPS instead of ddNTPS, the synthesis is briefly interrupted (not terminated) and continued after laser excitation (Bentley et al., 2008). Thus, Illumina sequencing technology is capable of generating data in a high-throughput manner, based on the two main parts after the sample preparation: the "cluster generation" and the "sequencing-by-synthesis technology."

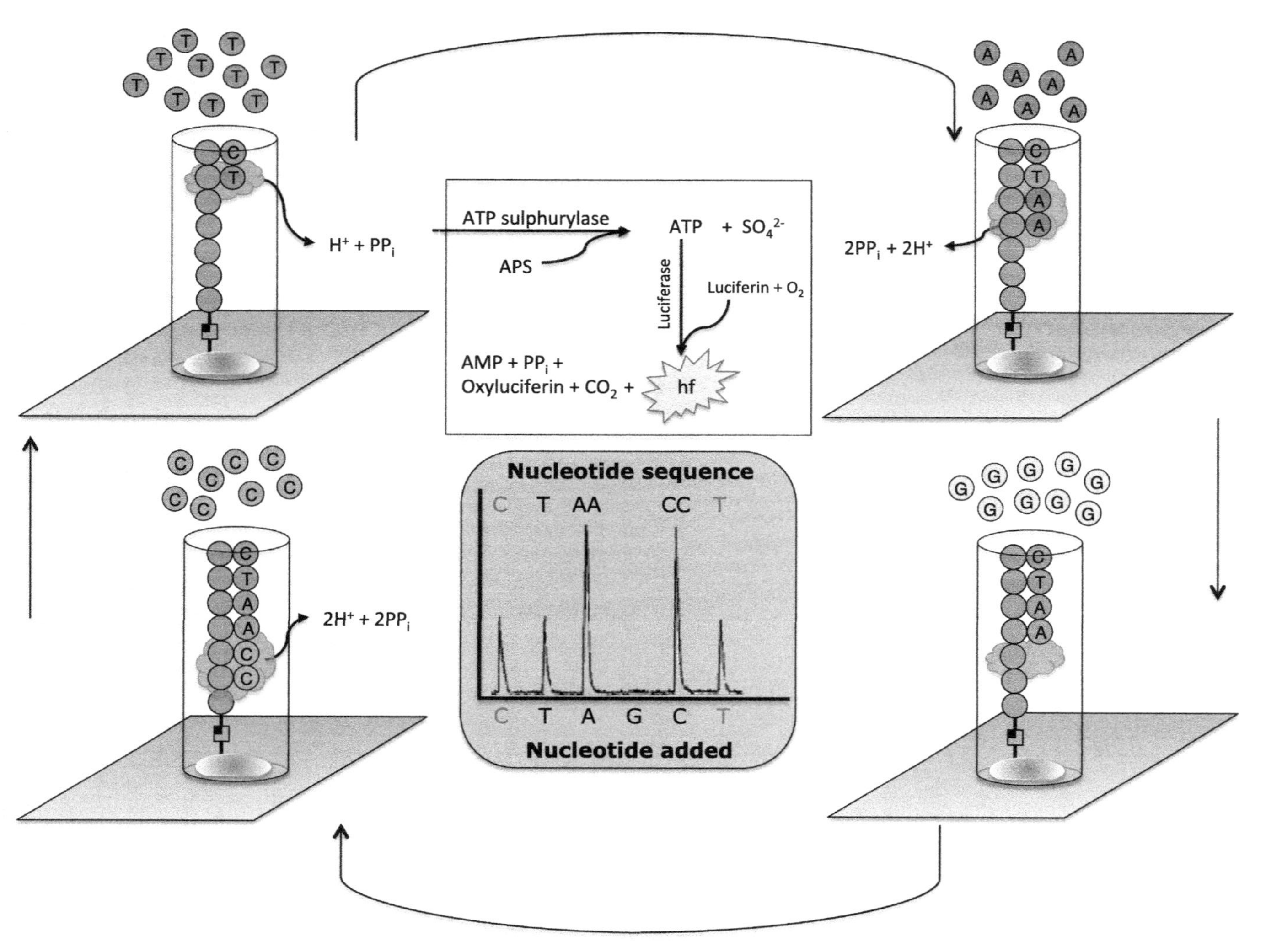

FIG. 4 Principle of pyrosequencing. The nucleotides are sequentially dispensed into the reaction; incorporation of the nucleotide inside the emulsion (not shown) leads to the release of PP_i and finally to a detectable light signal (hf). The intensity is proportional to the number of incorporated nucleotides.

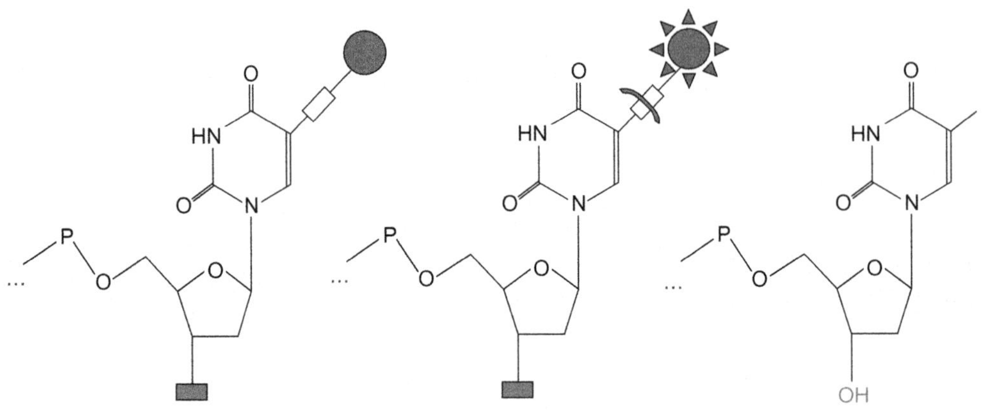

FIG. 5 Reversible terminators with blocked 3′ ribose groups (left and center, bottom) and unblocked 3′ group of the ribose (right, bottom) allowing the next base incorporation. The fluorophore (left, top) becomes cleaved after laser excitation (center, top).

4.2.3 *Cluster Generation*

Cluster generation is a process required for amplification of the strands of interest since the sensitivity of single nucleotide detection is too low. The prepared cDNA strands, therefore, containing the adapters for binding to the flow cell, become denatured prior to loading on a flow cell.

For cluster generation, these ssDNA fragments have to be immobilized on the flow cell. The fragments are randomly bound, with their 3′ end, in lanes on the flow cell, on which each lane represents a channel, containing two different types of oligonucleotides. On the first type of oligonucleotide, the adapter region of the ssDNA samples may hybridize with the complementary oligonucleotide on the surface, before a polymerase creates the complement of the hybridized strands. After polymerization, the double-stranded molecules are denatured, and the original strands are washed away.

The newly synthesized strand may now hybridize with its free end of the complementary second type of immobilized oligonucleotides in the flow cell, building a single-stranded bridge.

A polymerase creates the complementary strand of the folded molecule, building a double-stranded bridge. The double-stranded bridge is subsequently denatured resulting in a forward and a reverse single-stranded copy of the molecule bound at the flow cell oligonucleotides.

These steps are repeated several times occurring simultaneously for each fragment on the flow cell, resulting in hundreds of identical copies of the strands, forming a cluster. Finally, the reverse strands were cleaved and washed away, and the free oligonucleotides on the flow cell were blocked to prevent unwanted priming, resulting in clusters that only contain the forward single-stranded molecules. The process of cluster generation is shown in Fig. 6 (Illumina, 2010).

4.2.4 *Sequencing by Synthesis*

Sequencing begins by adding a primer on the primer sequence near the open end of the forward strand, prior to the addition of modified dNTPs.

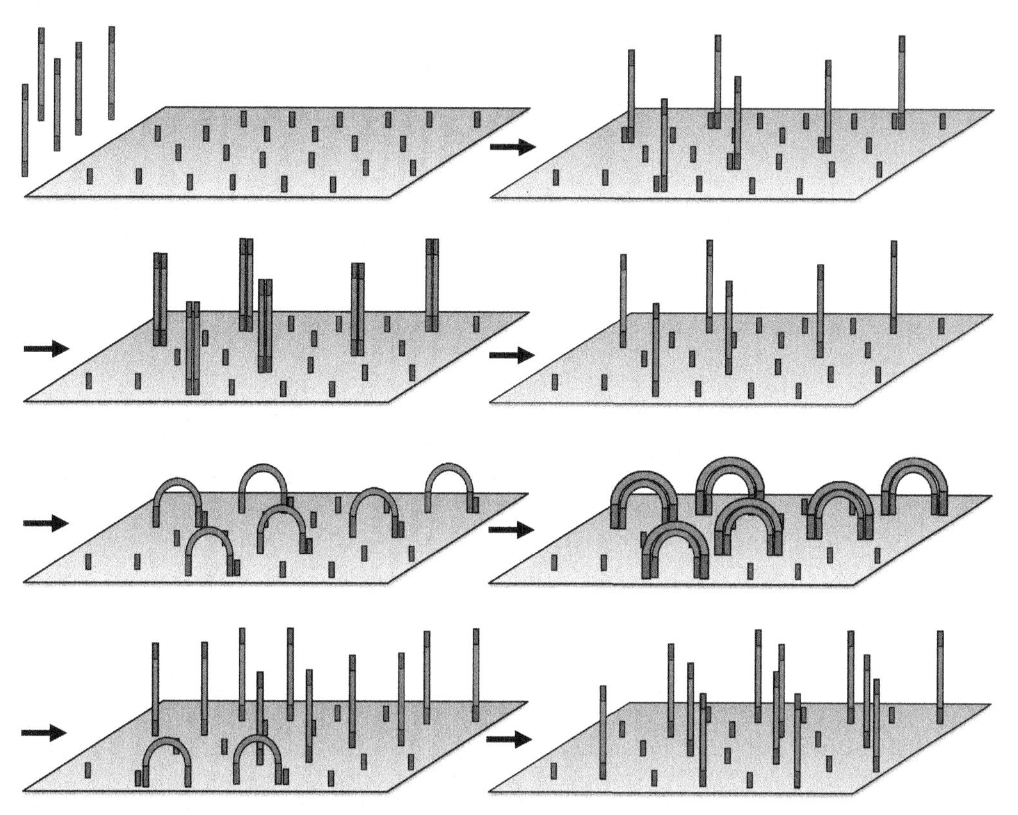

FIG. 6 Bridge amplification, key feature used by Illumina (Solexa) sequencing: The single stranded DNA fragments are randomly bound to the flow cell surface prior a polymerase creates the complementary strand and the original strand is cleaved. The newly synthesized strands fold to the complementary nucleotide on the surface building a bridge. A polymerase synthesizes a double-stranded bridge before the strands are denatured resulting in two single-stranded fragments on the flow cell. This clonal amplification occurs simultaneously on every fragment building so-called clusters.

In every cycle, the four different modified dNTPs compete to bind on the forward strand, where only the complement base can bind. After each binding of the nucleotide, the nucleotide is excited by a light source (laser), which results in a characteristic emitted signal. This laser excitation occurs simultaneously in the complete cluster, leading to a measurable fluorescence signal generated by hundreds of fluorophores. Furthermore, the laser excitation leads to an unblocking of the initially blocked 3′-OH group. The unblocking of the 3′-OH group enables the incorporation of a complementary dNTP during the second cycle.

This process is repeated several times and is called "sequencing by synthesis" (Fig. 7). In fact, most sequencing methods utilize this principle. The read length thereby is determined by the number of cycles, while the wavelength and the intensity of the emitted signal determine the base call. All identical strands in one cluster are read simultaneously affording millions of parallel reads.

Finally, to obtain the complete sequence of the initially fragmented DNA, the short-read products are assembled and aligned against a reference sequence. These data can further be evaluated by using an appropriate bioinformatics routine (Illumina, 2010).

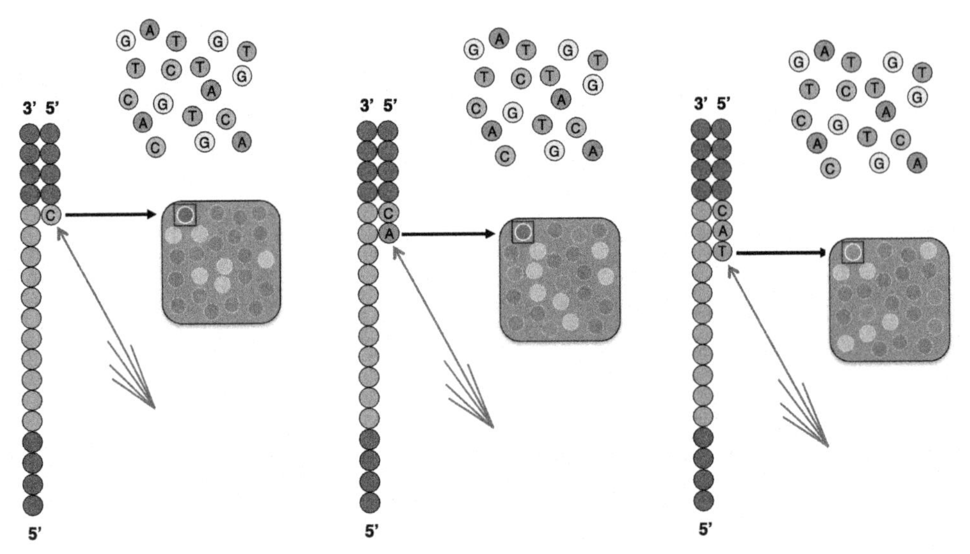

FIG. 7 Sequencing by synthesis: During sequencing, the complement reversible terminator becomes incorporated and excited by a laser resulting in a base-specific signal that can be interpreted after imaging.

4.3 SOliD (Sequencing by Oligonucleotide Ligation and Detection)

The SoliD sequencing was invented by George Church and developed by Applied Biosystems in 2008 (Nair, 2012). The process involves sample preparation, PCR, ligation reactions, and data analysis. During sample preparation, the sample becomes fragmented using nebulization or sonication or digestion. Nebulization is carried out using compressed nitrogen to force the DNA through a small hole leading to shear forces that break the DNA into small fragments. Sonication uses ultrasonic waves that create gas bubbles leading to resonance vibration at the inside of the DNA shearing the sample into small fragments. For digestion of the DNA, restriction enzymes are used to cut the DNA into small fragments.

After the fragmentation, two different adapters are ligated to the 5′ (P1 adapter) and the 3′ end (P2 adapter) of each fragment followed by PCR to amplify the DNA fragments. The PCR is done using PCR beads coated with complementary strands to the P1 adapter so that the P1 adapter may hybridize to the bead. Each fragment binds to its own bead, since the number of beads is higher than the number of fragments. After performing PCR, each bead consists of millions of copies of the same fragment, so-called polonies. To separate the polonies from the empty beads, additional polystyrene beads containing a sequence complementary to the P2 adapter may now bind the polonies. Due to a low density of the bead-to-bead structure, the empty beads can be centrifuged out, leaving only the polonies in the supernatant, which can be immobilized on a glass surface.

Sequencing begins with the addition of primers, fluorescently labeled probes, and ligase to append the probes to the primers. The primers bind to the template, followed by the hybridization and ligation of the probe. Fluorescence is measured prior to the cleavage of the fluorophore; steps then are repeated. In detail, the probes consist of eight nucleotides, where the first two nucleotides correspond to the fluorophore attached to the 3 terminal nucleotides of the 5′ end. After the primer has annealed to the complementary adapter, a probe may anneal to the template strand and becomes ligated to the nascent strand. By laser excitation, the fluorophore signal can be detected. The 3 terminal dye nucleotides are then cleaved, enabling the ligation of the next probe during the next step. Since the probe consists of five nucleotides after cleavage, only every fifth color can be detected. Accordingly, this step is repeated four times with an offset of one base. To determine the sequence, all five offset sequences are overlaid one over another, resulting in a color schema for every base. To decode the entire sequence, a so-called two-base color encoding is used. Here, each nucleotide of the template strand can be determined by comparison of the recorded color with a color chart. Starting with the first known base (e.g., thymine) as part of the P1 adapter, this base will line up with the first base (e.g., adenine) of the complementary probe with a given color (e.g., yellow). The next (second) base can be determined using a color chart by comparing the known (first) base with the resulting color (Fig. 8). In short, the known base and the resulting color determine the next base in the sequence. This technology has advantages in the sequencing of single nucleotide polymorphisms, where palindromic sequences cannot be determined correctly (Applied Biosystems, 2008).

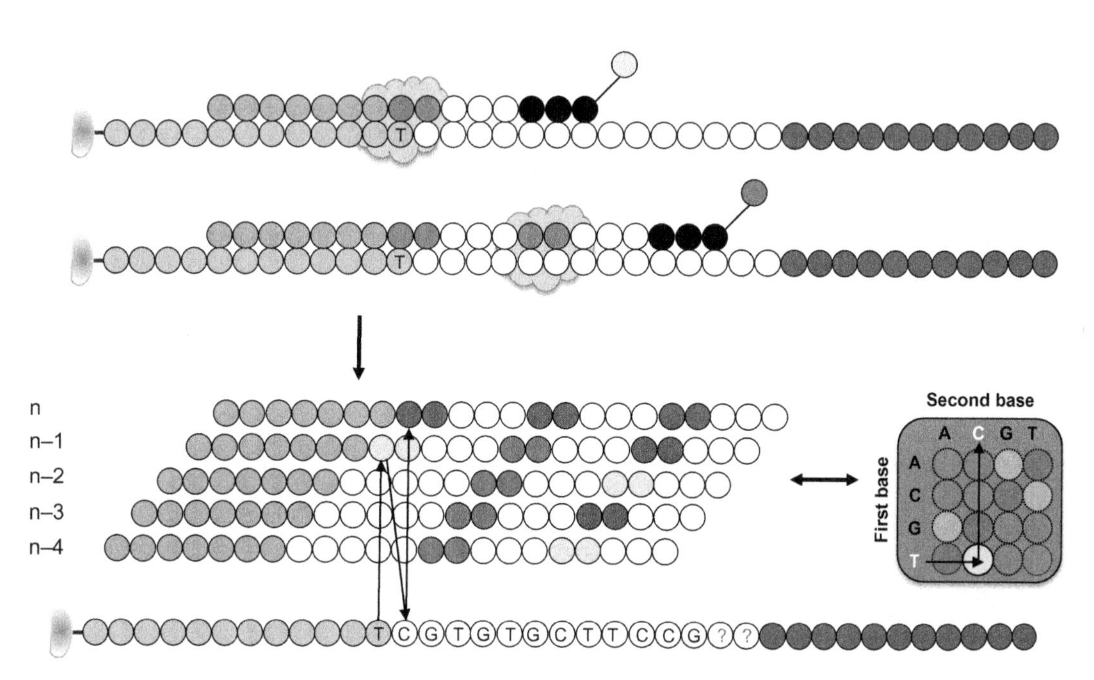

FIG. 8 Basic principle of SOLiD sequencing: DNA to be sequenced is attached to a bead via an adapter containing a thymine at the terminal end, where the sample is ligated. A primer binds to the adapter, where a ligase attaches different fluorophore labeled probes to the primer. The first two nucleotides of the probe determine the fluorophore. After binding, the three terminal nucleotides containing the fluorophore are cleaved. By this approach, each fifth base can be marked. To determine the base-call the fragments are overlaid with an offset of 1, by this the base-call can be decoded using the two-base-color-encoding scheme.

4.4 Ion Torrent (Ion Semiconductor) Sequencing

Ion Torrent uses a semiconductor chip covered with millions of small wells capturing chemical information (proton release) during the sequencing process. The chemical information can be translated into digital information and the base call determined.

Like in all NGSTs, the DNA first must be fragmented and ligated with adapters for the subsequent bead binding. The adapters allow binding of each fragment to its own bead where all fragments are amplified using emulsion PCR; each bead is covered with millions of identical single-stranded fragments.

The beads now flow across a semiconductor chip, where each single bead is deposited into a single well.

The wells are sequentially flooded with nucleotides. Whenever a nucleotide gets incorporated by a polymerase, a hydrogen ion is released, resulting in a pH change. This change can be transferred into a digital signal by which the base call can be determined. Similar to the pyrosequence technology, this signal increases if several identical nucleotides are incorporated simultaneously in homopolymers (Fig. 9) (Quail et al., 2012).

5 THIRD-GENERATION SEQUENCING

Despite the advances in NGSTs, the goal of the rapid, comprehensive, and unbiased sequencing of RNA has not been achieved yet. The ability to quickly and accurately quantify and decode the sequence of all cellular RNA molecules directly without reverse transcription or other manipulations has been desired for long. The initial attempts on determining RNA sequence in the 1970s by Maxam and Gilbert did not involve intermediate complementary DNA (cDNA) synthesis or other steps and relied on the tendency of various RNAses to cut RNA molecules at certain nucleotides (Donis-Keller et al., 1977). Subsequent technological evolutions such as Sanger or NGSTs required several molecules for signal amplification due to the lack of single-molecule sensitivity. To fill this gap, all available technologies to sequence RNA molecules rely on multiple manipulations converting natural RNA into cDNA detectable using electric or chemical signals and various sensing mechanisms. It is, however, clear by now that each manipulation introduced to nucleotide strands causes artifacts and inaccuracies in measurements (Linsen et al., 2009).

Novel technologies that are under development today, such as single-molecule real-time (SMRT) systems like the PacBio RS II from Pacific Biosciences (2012) or as nanopore-based systems like the Oxford Nanopore Technologies (ONT) MinION (Mikheyev and Tin, 2014), are now designed in ways requiring only minimal or no manipulation of natural nucleotides. A comparison of these two TGSTs is shown in Table 2.

5.1 Pacific Biosciences "Single-Molecule Real-Time" Sequencing

Pacific Biosciences has developed the single-molecule real-time (SMRT) sequencing, a method that enables direct observation of a single molecule in real time. This progress was achieved by using a nanophotonic visualization chamber called zero-mode waveguide (ZMW). Several thousand ZMWs make up the so-called SMRT cell. Each ZMW is a small

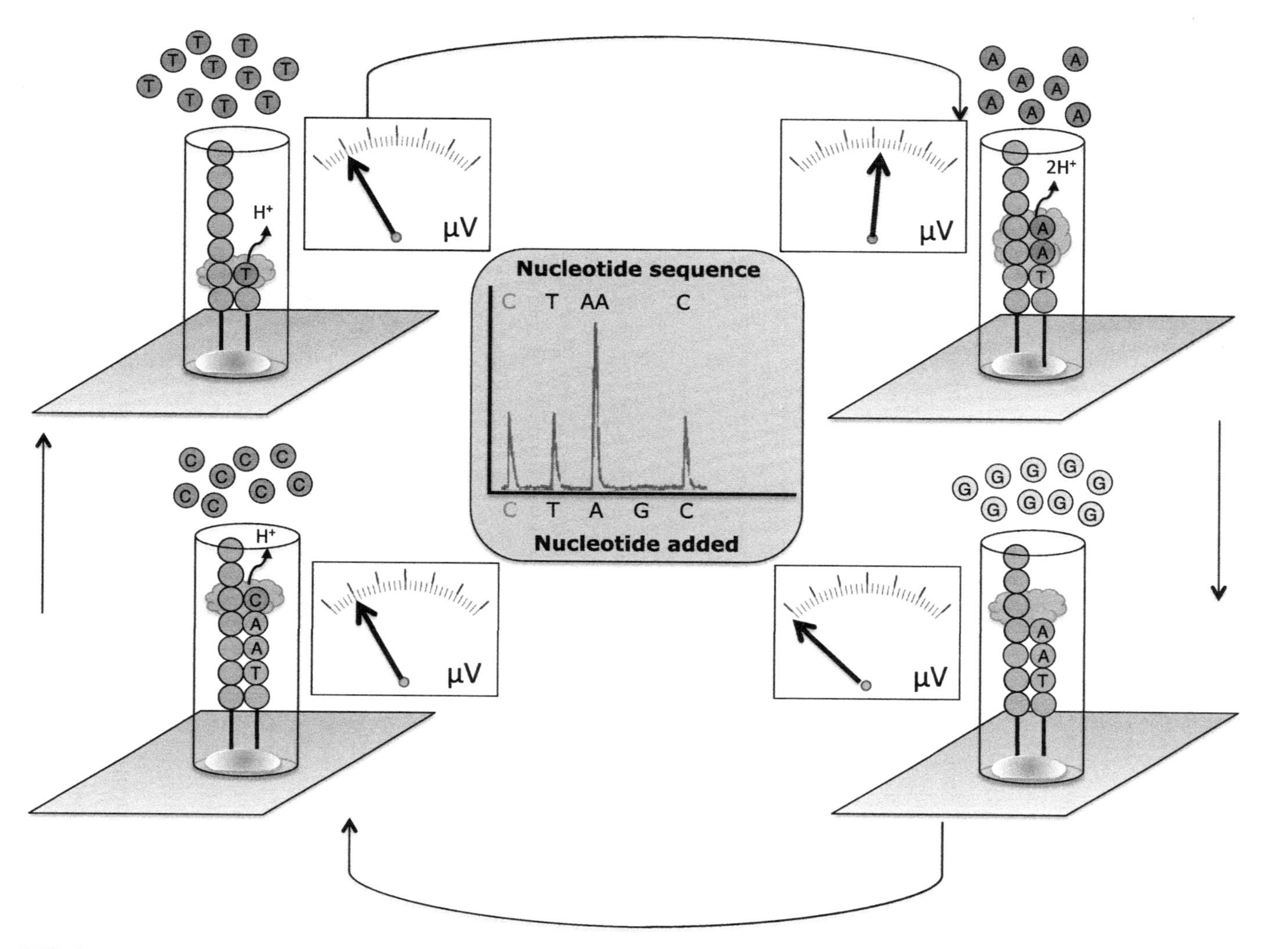

FIG. 9 Principle of Ion Torrent sequencing. The nucleotides are sequentially dispensed into the reaction. The incorporation of the nucleotide inside the emulsion (not shown) leads to a release of H^+ and thereby to an increased pH. The signal is translated into electric signals and recorded as voltage changes where the peak height is proportional to the number of incorporated nucleotides.

TABLE 2 Comparison of Third-Generation Sequencing Technologies

Sequencer	Pacific Biosciences PacBio RS II	Oxford Nanopore Technologies MinION Mk1B
Sequencing mechanism	Single-molecule, real-time sequencing	Nanopore sequencing
Read length	Average > 10,000 bp, some reads >60,000 bp (according to manufacturer)	Average > 5000 bp > 10,000 bp, some reads (according to manufacturer)
Accuracy	85%–89%	65%–88%
Reads	365.000 (according to manufacturer)	–
Output data/run	7.6 Gb (according to manufacturer)	> 20 Gb (according to manufacturer)
Time/run	From 30 min to 6 h per SMRT cell	Controlled by user (up to 48 h)
Advantage	No amplification, read length, high throughput, palindromes, homopolymers, low diversity	Amplification not mandatory, read length, palindromes, low diversity, homopolymers, high throughput, rapid sample preparation, prize
Disadvantage	Still in development, high error rate	Still in development, high error rate

Data from Pacific Biosciences/Oxford Nanopore Technologies and Quail, M.A., Smith, M., Coupland, P., Otto, T.D., Harris, S.R., Connor, T.R., Bertoni, A., Swerdlow, H.P., Gu, Y., 2012. A tale of three next generation sequencing platforms: comparison of Ion Torrent, Pacific Biosciences and Illumina MiSeq sequencers. BMC Genomics, 13, 341.

chamber, about 70 nm in diameter, fabricated in a thin metal film about 100 nm deposited on a glass substrate, where a polymerase is affixed at the bottom. The extremely small size of the ZMW prevents visible laser light, from passing through the ZMW, meaning that the light decays after 20–30 nm as it enters the ZMW resulting in a detection volume of about $20 * 10^{-21}$ L. This setting is required to illuminate solely the bottom of the ZMV when a light source is applied to it and to obtain fluorescence emission of the very nucleotide just been attached to the growing strand.

Pacific Biosciences uses four fluorescently labeled nucleotides, which generate distinct emission spectrums. A distinctive feature of the used dNTPS is that distinct (four different) fluorophores are linked to the phosphate group of the very base unlike most other sequencing technologies utilizing fluorophores.

The phospholinked dNTPs may diffuse in and out of the ZMW. When the affixed polymerase encounters the correct dNTP, the dNTP becomes incorporated, and the fluorophore excited by a laser emits a detectable light signal prior to the cleavage of the phosphodiester bond along with the release of the fluorescent tag as shown in Fig. 10 (Quail et al., 2012).

5.2 Oxford Nanopore Technologies Nanopore Sequencing

Nanopore sequencing is based on tiny holes with a size of about 1.4 nm diameter (Liu et al., 2012), inserted into a membrane with high electric resistance. The membrane is immersed in an electrolyte solution, and a potential difference is applied across the membrane, generating a current through the nanopore. If an analyte interacts with the

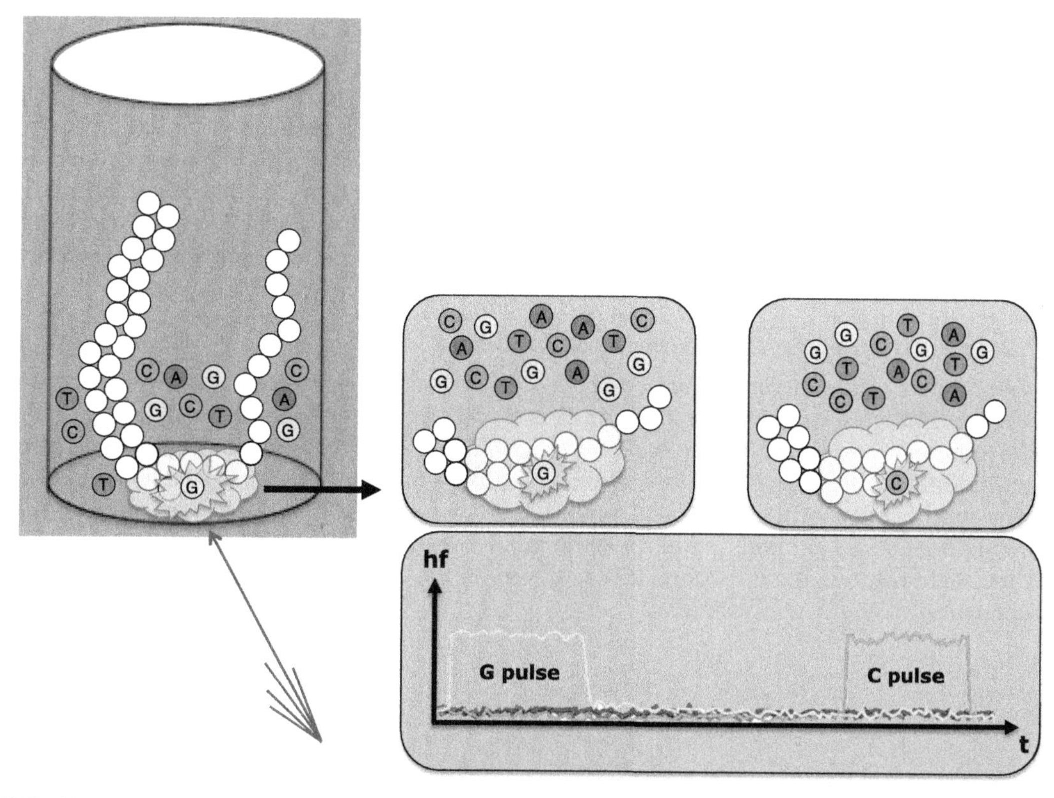

FIG. 10 Principle of Pacific Biosciences sequencing: A DNS strand diffuses into a ZMW, and the adaptor binds to a polymerase immobilized at the bottom. Each of the four nucleotides (G, C, T, and A) is labeled with a different fluorescent dye (indicated also in red (A), yellow (G), green (T), and blue (C)), so that they have distinct emission spectrums. As a nucleotide is held in the detection volume by the polymerase, a light pulse is produced that identifies the base. The fluorescence output of the color corresponding to the incorporated base (here for base G) is elevated. The dye-linker-pyrophosphate product is cleaved from the nucleotide and diffuses out of the ZMW, ending the fluorescence pulse. The polymerase translocates to the next position. The next nucleotide associates with the template in the active site of the polymerase, initiating the next fluorescence pulse, which corresponds to base C here.

pore by passing through it, a small current disruption can be detected. This gives the possibility to produce a massively parallel continuous read length of over >100 kb, which, theoretically, is not limited (Mikheyev and Tin, 2014; Jain et al., 2016; Johnson et al., 2017; Smith et al., 2017). Another advantage of this technology is to circumvent the use of polymerase chain reaction (PCR), thereby opening the possibility of direct RNA/DNA sequencing without (in the case of RNA) requiring preceding reverse transcription (Smith et al., 2017).

In 2014, the MinION, the first sequencer using nanopore technology, was released by Oxford Nanopore Technologies (ONT) and successfully employed by several independent genomic laboratories (Jain et al., 2016). The device was made commercially available in 2015. The MinION identifies nucleotide bases by measuring current disruptions generated by nucleotide strands passing through a biological nanopore.

5.2.1 Basic Sequencing Principle

ONT nanopore sequencing is based on biological nanopores inserted into a synthetic polymer membrane. The membrane, immersed in an electrolyte solution, exhibits a very high electric resistance, allowing ions only passing the nanopore if a suitable potential is applied across the membrane; the potential generates an ionic current through the nanopore. Single molecules entering the nanopore cause characteristic disruptions in the current (Fig. 11). Measuring this disruption, DNA or RNA molecules can be characterized (Kasianowicz et al., 1996), not relying on a measurement of base incorporations as most employed by current major technologies (Jain et al., 2016; Johnson et al., 2017; Smith et al., 2017).

5.2.2 Interaction of the Pore With the DNA Strand

Although in vitro experiments already have shown that single-stranded DNA fragments can be driven through a nanopore by an appropriate potential applied (Kasianowicz et al., 1996), DNA fragments and the nanopore had to be slightly modified to achieve a controlled passage of the biopolymer through the nanopore for generating a measurable signal (Astier et al., 2006).

The modification of the DNA is a prerequisite to fulfilling two crucial functions while bringing the DNA fragments in reach to the nanopore and controlling the passage speed of the fragment through the nanopore; it is achieved by the ligation of the so-called sequencing adapters (also known as Y-adapters).

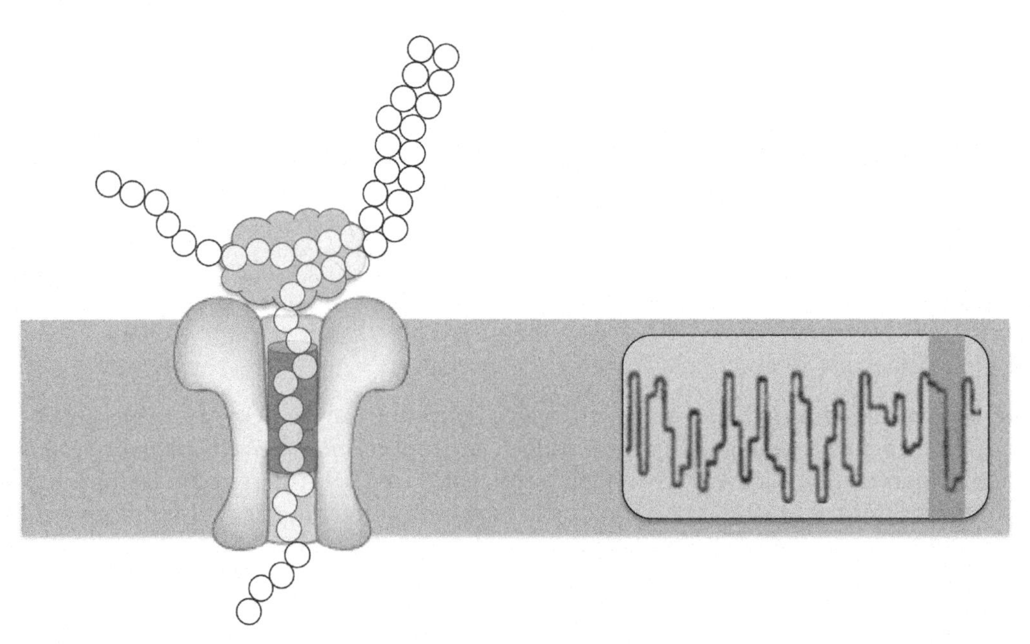

FIG. 11 Principle of nanopore sequencing. A biological nanopore is embedded in a membrane with high electric resistance immersed in an electrolyte solution. Ions may only pass through the nanopore if a suitable potential is applied across the membrane generating an ionic current. If an analyte enters the pore and passes the "sensing zone," a disruption of the current can be detected.

5.2.3 *Sequencing Adapters*

The adapters are ligated to prior end-repaired and dA-tailed DNA fragments during library preparation. The end-repair step is needed to convert fragmented DNA to repaired DNA with 5′ phosphorylated and 3′ dA-tailed blunt ends, allowing the ligation of the adapter. The ligation is assisted by hybridization of the A and T overhangs of the DNA fragments and adapters, respectively, resulting in fragments adapted at both ends.

The sequencing adapter contains a loaded enzyme (also known as motor protein), controlling the passage of the DNA through the nanopore and a specific nucleotide sequence for subsequent tether binding, required to attach the DNA fragment to the membrane.

The tethers have a dual function of tethering the DNA strand to the membrane and reducing the diffusion of the DNA strands from three to two dimensions. Furthermore, the DNA remains mobile in the bilayer. Together, these functions improve DNA capturing by approximately three orders of magnitude (20.000× higher sensitivity) compared with capture without the tethers (Brown, 2016).

5.2.4 *Biological Nanopores for Sequencing*

Today, the technique of utilizing biological nanopores for DNA sequencing is based on the studies of Kasianowicz and colleagues showing the possibility to translocate polynucleotides through a nanopore and to characterize the length distribution of individual polynucleotides using current disruptions in biological nanopores. It has been shown that each nucleotide passing through the pore causes a short-term disruption in the ion-flow proportional to the size and chemical property of the nucleotide base (Kasianowicz et al., 1996).

In 1996, Kasianowicz and colleagues used the naturally occurring Hla toxin from *Staphylococcus aureus* resulting in a self-assembled, heptameric nanopore (leading to the denomination R7) with an inner diameter of about 2 nm embedded a planar lipid bilayer membrane (Kasianowicz et al., 1996).

A problem, however, which had to be solved, was the sensing resolution, detecting each single nucleotide in a DNA strand, since the sensing zone on the inside of natural occurring α-hemolysin nanopores is about 5 nm in length, much longer than the base-to-base distance (340 pm). Dealing with this problem, the idea was taken up by Bayley and colleagues, and this led to "the breakthrough that one free nucleotide gives a distinguishable signal" (Tim Harris, from the applied physics and instrumentation group at the Howard Hughes Medical Institute's Janelia Farm Research Campus in Ashburn, Virginia, 2008) achieved by two alterations concerning the nanopore in 2006.

Bayley and colleagues used an engineered α-hemolysin nanopore with an inserted cyclodextrin plug in the pore channel leading to a prolonged passage velocity (Astier et al., 2006).

Cyclodextrin is a ring-shaped molecule that narrows the neck of the nanopore, acting as a molecular recognition agent (Gyurcsanyi, 2008). The nucleotides, passing through the pore, must be squeezed through this plug, where a phosphate group on the nucleotide briefly binds to the cyclodextrin, thus blocking the pore. The duration a nucleotide binds to the cyclodextrin is dependent on its size, giving characteristic readouts for each nucleotide (Astier et al., 2006; Gyurcsanyi, 2008).

Furthermore, to direct the ssDNA through the nanopore, an exonuclease enzyme had been attached to the entry of the nanopore, serving each base at a time while directing the strand into the hole.

By this, it was achieved that the dsDNA becomes single stranded by the exonuclease and is directed through the nanopore one base at a time.

The upside, using α-hemolysin nanopores for long, single-stranded DNA (ssDNA) sequencing, is, besides a linear translocation velocity, a homogeneous and self-assembled structure of the nanopore with an inside diameter of about 2nm. Nevertheless, it had also been revealed by Stoddart et al. in 2009 that the hitherto used pores have three regions within the pore, where a nucleotide generates a greater influence in the ionic flow than at other positions, leading to higher complexity (Stoddart et al., 2009). Up to the year 2012, over 160 enzymes coming into consideration for use as motor protein >1000 pores and their mutants have been tested (Brown, 2016).

Although it seems feasible that each nucleotide is base-called at a time, the nucleotides are read in short "words," in the so-called sensing zone, depending on a distinct k-mer length. This k-mer length (currently 5-mer (Turner, 2016)) is mainly determined by the pore length and by the size of the cyclodextrin plug (Japrung et al., 2010), resulting in $4^5 = 1024$ possible states/signals (Turner, 2016).

As the ssDNA strand moves through the pore, the combination of nucleotides in the strand being processed in the sensing region creates a characteristic disruption in the electric current (Olasagasti et al., 2013) that enables real-time sequencing of the ssDNA strand (theoretically) regardless of the length of the strand.

In general, biological nanopores can be produced out of different proteins secreted by pathogenic bacteria, such as the α-hemolysin (Hla) toxin of *S. aureus*, the porin MspA of *Mycobacterium smegmatis*, or the α-hemolysin (HlyA) toxin of uropathogenic *Escherichia coli* (UPEC), which, however, all bear different pore properties (Japrung et al., 2010).

At present, ONT uses a nonameric lipoprotein (denomination of the R9.x flow cell) from an engineered Curlin sigma s-dependent growth Gene (CsgG) from *E. coli* (*E. coli*) for pore formation, where it acts as part of a protein secretion machinery carrying peptides through the membrane (Brown, 2016).

5.2.5 *One Directional Read, Two Directional Read and One Directional Squared Read*

At the moment, Oxford Nanopore Technologies offers three different read options for strand sequencing, with each method, however, influencing the sequencing confidence.

The simplest method is the one-directional (1-D) read, where only one strand (template strand) is read whether the sample is single stranded or double stranded.

In the case of a two-directional (2-D) read, the sample must be double stranded or become double stranded during the library preparation. In comparison with a 1-D protocol where only the template strand of the sample is base-called, in a 2-D protocol, the complementary strand is also base-called since a hairpin adapter is attached to one end of the double-stranded DNA. This method enables sequencing of both strands during one read due to the hairpin adapter. By generating a characteristic current disruption, caused by the hairpin nucleotide composition, the read can bioinformatically divide in a forward and a reverse read using appropriate bioinformatics and data processing.

A combination of 1-D and 2-D read represents the newly available one-directional squared (1-D^2) read; as for the 2-D read, the sample must be or become double stranded. In this case, however, no hairpin adapter is needed; the so-called 1-D^2 adapters within the sample and the sequencing adapter, however, are still required. Here, the template strand is read followed by the complementary strand since the motor protein is now able to attach to opposite adapter on the complementary strand resulting directly in a forward and a reverse read. Compared with a 1-D read, the 2-D read and the 1-D^2 read increase the sequencing confidence, since both strands are base-called and bioinformatically compared (Goodwin et al., 2015).

6 ISSUES TO BE IMPROVED FOR OBTAINING AN OPTIMAL SEQUENCING TECHNOLOGY

With the introduction and the continuous improvements of the NGSTs, science became closer, but it is still far away from the ultimate goal of the rapid, comprehensive, and unbiased sequencing of nucleotides; this is due to the need of sample amplification and especially due to limited read length (Liu et al., 2012; Linsen et al., 2009).

While de novo genome assemblies can be produced from NGS data, assembly continuity is often relatively poor due to the limited suitability of short reads for handling long repeats. Assembly quality, however, can be greatly improved by using TGS long reads since repetitive regions can be easily expanded into using longer sequencing lengths, despite bringing higher error rates at the base level (Linsen et al., 2009).

One of the main problems in improving error rates is the lack of high-quality and full-length reference sequences deposited in reference databases; this is associated with the currently prevailing high-throughput method of Illumina (Schloss et al., 2016) (Solexa) sequencing using shorter fragments.

An optimal sequencing technology, therefore, should

i. fully sequence natural DNA/RNA fragments of various sizes (from few bases up to several hundred megabases) with no/minimal errors,
ii. require minimal sample manipulation,
iii. operate in real time,
iv. minimize the hands-on time and the time for sequencing and data analysis,
v. generate no sequence-dependent or other biases for optimal quantification and genotyping performance,
vi. require a minimum amount of input material,
vii. have high efficiency in all sequencing steps,
viii. directly detect modified nucleotides (e.g., methylated cytosines),
ix. operate without any special external environment requirements (e.g., temperature and humidity),
x. be sufficiently affordable and economically viable to allow worldwide adoption and operation in research and clinical settings.

Nanopore technology holds the promise to meet these requirements and to overcome current shortcomings; it, however, still requires further improvement to reach the ultimate desired sequencing goals. It can be expected, however, that we will soon experience the great power of game-changing nanopore technology in sequencing as a key tool in precision medicine and exhibiting tremendous impact on diagnosis, prognosis, therapy selection, and control. Roche has already announced a new nanopore-based technology for the near future, apparently applying a similar technology with biological nanopores like ONT, however, using phosphate-tagged nucleotides enabling single-molecule DNA sequencing.

References

Altmann, R., 1889. Ueber Nucleinsäuren. Archiv für Anatomie und Physiologie. Physiologische Abteilung. Leipzig, S. 524–536.

Applied Biosystems, 2008. Principles of Di-Base Sequencing and the Advantages of Color Space Analysis in the SOLiD System. Application Note 04/2008 139AP10-01.

Astier, Y., Braha, O., Bayley, H., 2006. Toward single molecule DNA sequencing: direct identification of ribonucleoside and deoxyribonucleoside 5′-monophosphates by using an engineered protein nanopore equipped with a molecular adapter. J. Am. Chem. Soc. 128 (5), 1705–1710.

Bentley, D.R., Balasubramanian, S., Swerdlow, H.P., et al., 2008. Accurate whole human genome sequencing using reversible terminator chemistry. Nature 456 (7218), 53–59.

Brown, C., 2016. No thanks, I've already got one. Oxford Nanopore Webinar 03/2016.

Dahm, R., 2005. Friedrich Miescher and the discovery of DNA. Dev Biol. 278 (2), 274–288.

Donis-Keller, H., Maxam, A.M., Gilbert, W., 1977. Mapping adenines, guanines, and pyrimidines in RNA. Nucleic Acids Res. 4 (8), 2527–2538.

Goodwin, S., Gurtowski, J., Ethe-Sayers, S., Deshpande, P., Schatz, M.C., WR, M.C., 2015. Oxford Nanopore sequencing, hybrid error correction, and de novo assembly of a eukaryotic genome. Genome Res. 25 (11), 1750–1756. https://doi.org/10.1101/gr.191395.115.

Gyurcsanyi, R.E., 2008. Chemically-modified nanopores for sensing. Trends Anal. Chem. 27.

Harrington, C.T., Lin, E.I., Olson, M.T., Eshleman, J.R., 2013. Fundamentals of pyrosequencing. Arch. Pathol. Lab. Med. 137.

Illumina, 2010. Illumina Sequencing Technology. Illumina Technology Spotlight.

Jain, M., et al., 2016. The Oxford Nanopore MinION: delivery of nanopore sequencing to the genomics community, genome biology. Genome Biol. 17 (1), 239. https://doi.org/10.1186/s13059-016-1103-0.

Japrung, D., Henricus, M., Li, Q., Maglia, G., Bayley, H., 2010. Urea facilitates the translocation of single-stranded DNA and RNA through the alpha-hemolysin nanopore. Biophys. J. 98 (9), 1856–1863.

Johnson, S.S., et al., 2017. Real-Time DNA Sequencing in the Antarctic Dry Valleys Using the Oxford Nanopore Sequencer, pp. 2–7. https://doi.org/10.7171/jbt.17-2801-009.

Kasianowicz, J.J., Brandin, E., Branton, D., Deamer, D.W., 1996. Characterization of individual polynucleotide molecules using a membrane channel. Proc. Natl. Acad. Sci. U. S. A. 93 (24), 13770–13773.

Kossel, A., 1896. Ueber die basischen Stoffe des Zellkerns. Zschr. physiol. Chem. 22, 176.

Linsen, S.E., de Wit, E., Janssens, G., et al., 2009. Limitations and possibilities of small RNA digital gene expression profiling. Nat. Methods 6 (7), 474–476.

Liu, L., Li, Y., Li, S., Hu, N., He, Y., Pong, R., Lin, D., Lu, L., Law, M., 2012. Comparison of next-generation sequencing systems. J Biomed Biotechnol 2012, 251364. https://doi.org/10.1155/2012/251364.

Magi, A., Giusti, B., Tattini, L., 2016. Characterization of MinION nanopore data for resequencing analyses. Brief. Bioinform. (Epub ahead of print).

Margulies, M., Egholm, M., Altman, W.E., et al., 2005. Genome sequencing in microfabricated high-density picolitre reactors. Nature 437 (7057), 376–380.

Maxam, A., Gilbert, W., 1977. A new method for sequencing DNA. Proc. Natl. Acad. Sci. U. S. A. 74 (2), 560–564.

Mikheyev, A.S., Tin, M.M.Y., 2014. A first look at the Oxford Nanopore MinION sequencer. Mol. Ecol. Resour. 14 (6), 1097–1102. https://doi.org/10.1111/1755-0998.12324.

Nair, P., 2012. Profile of George M. Church. Proc. Natl. Acad. Sci. U. S. A. 109 (30), 11893–11895. https://doi.org/10.1073/pnas.1204148109.

Nyren, P., 2007. The history of pyrosequencing. Methods Mol. Biol. 373, 1–14.

Olasagasti, F., Lieberman, K.R., Benner, S., Cherf, G.M., Dahl, J.M., Deamer, D.W., Akeson, M., 2013. Replication of individual DNA molecules under electronic control using a protein Nanopore. Nat. Nanotechnol. 5 (11), 798–806.

Quail, M.A., Smith, M., Coupland, P., Otto, T.D., Harris, S.R., Connor, T.R., Bertoni, A., Swerdlow, H.P., Gu, Y., 2012. A tale of three next generation sequencing platforms: comparison of Ion Torrent, Pacific Biosciences and Illumina MiSeq sequencers. BMC Genomics 13, 341.

Rothberg, J.M., Hinz, W., Rearick, T.M., et al., 2011. An integrated semiconductor device enabling nonoptical genome sequencing. Nature 475 (7356), 348–352.

Sanger, F., Coulson, A.R., 1975. A rapid method for determining sequences in DNA by primed synthesis with DNA polymerase. J. Mol. Biol. 94 (3), 441–448.

Sanger, F., Nicklen, S., Coulson, A.R., 1977. DNA sequencing with chain-terminating inhibitors. Proc. Natl. Acad. Sci. U. S. A. 74 (12), 5463–5467.

Schloss, P.D., Jenior, M.L., Koumpouras, C.C., Westcott, S.L., Highlander, S.K., 2016. Sequencing 16S rRNA gene fragments using the PacBio SMRT DNA sequencing system. Peer J. 4 (March), e1869.

Smith, A.M., et al., 2017. Reading Canonical and Modified Nucleotides in 16S Ribosomal RNA Using Nanopore Direct RNA Sequencing. bioRxiv. Available at http://biorxiv.org/content/early/2017/04/29/132274.abstract.

Stoddart, D., Heron, A.J., Mikhailova, E., Maglia, G., Bayley, H., 2009. Single-nucleotide discrimination in immobilized DNA oligonucleotides with a biological nanopore. Proc. Natl. Acad. Sci. U. S. A. 106 (19), 7702–7707.

Turner, D., 2016. In: MinION applications: microbial genomics and direct RNA sequencing. 2nd Annual Microbiology & Immunology 2016 Virtual Conference.

Valouev, A., Ichikawa, J., Tonthat, T., et al., 2008. A high-resolution, nucleosome position map of *C. elegans* reveals a lack of universal sequence-dictated positioning. Genome Res. 18 (7), 1051–1063.

Watson, J.D., Crick, F.H.C., 1953. A Structure for Deoxyribose Nucleic Acid. Nature 171, 737–738.

Further Reading

McCarthy, A., 2010. Third generation DNA sequencing: pacific biosciences single molecule real time technology. Chem. Biol. 17. https://doi.org/10.1016/j.chembiol.2010.07.004.

CHAPTER

6

Workflow for Circulating miRNA Identification and Development in Cancer Research: Methodological Considerations

Chiara M. Ciniselli[*,†], *Mara Lecchi*[*], *Manuela Gariboldi*[*,‡], *Paolo Verderio*[*], *Maria G. Daidone*[*]

[*]Fondazione IRCCS Istituto Nazionale dei Tumori, Milan, Italy [†]University of Milan, Milan, Italy [‡]FIRC Institute of Molecular Oncology Foundation, Milan, Italy

1 INTRODUCTION

In the recent era of personalize medicine, the identification of biomarkers measurable from body fluids represents a unique opportunity to discover and develop noninvasive biomarkers for diagnosis, prognosis, and monitoring of specific disease conditions, as well as for predicting the response to a specific therapeutic strategy. In addition, the availability of high-throughput technologies for the analysis of circulating biomarkers, including the novel next-generation sequencing one, opened new prospectives for guiding diagnostic and therapeutic strategies especially in the oncological scenario. Despite the huge number of research studies highlighting new promising biomarkers, a limited number are then translated and fully applied in the real clinical practice. As reported in the commentary by Poste (2011), <0.1% of the discovered biomarkers are then used in the routine clinical setting also due to a lack of shared and standardized conditions for specimen collection, handling, and storage. A successful implementation of biomarkers in the clinical routine practice is however a prerequisite for the progression of personalized medicine. Thus, a comprehensive evaluation of the whole process for biomarker development is necessary, starting from the early biomarker discovery phase, through the biomarker development and validation, and finally to the biomarker implementation into in vitro diagnostic tests including external quality control for their analytical validation (Verderio et al., 2010).

Precision Medicine
https://doi.org/10.1016/B978-0-12-805364-5.00006-8

In this chapter, we will focus on circulating microRNAs (miRNAs), with a particular attention on the aspect related to their identification and validation, also considering the methodological-statistical aspects of their combination in a composite score. An example, in colorectal cancer (CRC) setting, will be finally provided as propaedeutic application.

2 BIOMARKERS

2.1 Definitions and Types

According to the recently published Biomarkers, EndpointS, and other Tools (BEST) Resource (https://www.ncbi.nlm.nih.gov/books/NBK338448/), biomarker is "a defined characteristic that is measured as an indicator of normal biological processes, pathogenic processes, or responses to an exposure or intervention, including therapeutic interventions." Biomarker can have molecular, histological, radiographic, or physiological characteristics. According to their role and the BEST Resource classification, cancer biomarkers can be mainly classified in different categories: *diagnostic*, used to detect or confirm the presence of a disease or condition of interest or to identify individuals with a subtype of the disease; *predictive*, used to identify individuals who are more likely than similar individuals without the biomarker to experience a favorable or unfavorable effect from exposure to a medical product or an environmental agent; or *prognostic*, used to identify likelihood of a clinical event, disease recurrence, progression, or death in patients with the disease or the medical condition of interest. A complete description of the terms and definitions related to the biomarkers context can be found at the BEST Resource webpage (https://www.ncbi.nlm.nih.gov/books/NBK338448/).

2.2 Workflow for Biomarker Identification and Development

The development of a new cancer biomarker is a process that begins with biomarker discovery, followed by a rigorous definition and evaluation of the whole process of its determination (i.e., analytical validation), and it continues by establishing if the biomarker test reliably measures or predicts the clinical question of interest (i.e., clinical validation). The ultimate step of this process consists in evaluating the benefits and risks for patients in using the developed biomarker test (Verderio et al., 2010). Briefly, biomarker discovery begins with the identification of promising biomarkers that are usually validated in independent set(s) of subjects (Goossens et al., 2015). After their identification, candidate biomarkers should be evaluated in terms of analytical validation, by assessing the performance characteristics of the biomarker assay and its optimal analytical setting to guarantee a satisfactory level of reproducibility and accuracy, and in terms of clinical validation (i.e., evaluation whether the test results are robust and reliable with respect to the outcome of interest) (Verderio et al., 2010; Goossens et al., 2015). The ultimate goal of the analytical validation is to reduce the number of promising biomarkers that fail in the clinical setting as a result of the lack of robust analytical validations (Verderio et al., 2010). Table 1 summarizes the phases that should be followed for the analytical validation of a promising biomarker including the implementation of internal and external quality assessment scheme. This scheme tries to cover the need of

TABLE 1 Phases for the Analytical Validation of a Biomarker

Phase	Description
I—Operating procedure setting up	Definition of the operating procedures for biomarker determination
II—Operating procedure standardization	Validation of the operating procedures in terms of precision and accuracy according to the defined standards
III—Internal quality control	Evaluation of the validated standards within laboratory
IV—External quality assessment	Between-laboratory comparison and assessment of biomarker accuracy

developing, as much as possible, standardized operating procedures for the whole analytical process of biomarker measurement including their monitoring. In this context, assays for cancer biomarkers should be analytically validated by appropriately implementing external quality assurance (EQA) schemes both before their fully implementation into routine laboratory testing in order to generate clinically useful information and to monitor their accuracy during time (Verderio et al., 2010).

3 CIRCULATING miRNAs

miRNAs are short (19–25 nucleotides) noncoding RNA molecules that regulate gene expression at the posttranscriptional level (Calin and Croce, 2006). miRNAs are involved in different biological processes from cell proliferation, differentiation, metabolism, and embryogenesis to inflammation, aging, and programmed cell death (Becker and Lockwood, 2013; de Planell-Saguer and Rodicio, 2013). Accordingly, the detection of miRNAs has attracted enormous interest since their discovery, as highlighted by the huge number of scientific publication on that topic in different clinical scenarios, from the oncological one to cardiovascular disease (Becker and Lockwood, 2013; de Planell-Saguer and Rodicio, 2013). In addition, following the first investigation of Chen et al. (2008)—who highlighted the feasibility of detecting miRNAs from plasma and serum—many independent research studies have confirmed the role of circulating miRNAs as potential noninvasive biomarkers for diagnosis and monitoring of different human cancers (Cortez et al., 2011; Zanutto et al., 2014; Verma et al., 2015). However, as reported in the recent years by many authors (Jarry et al., 2014; Singh et al., 2016), the overlap between specific miRNA signatures also within the same cancer type is poor, consequently suggesting the need to define shared workflows for their evaluation (Singh et al., 2016). A possible explanation may be due to the challenge of the technical issues of their detection and evaluation, from the preanalytical and analytical workflow to the postanalytical ones including the aspect of data normalization and signature building (Jarry et al., 2014; Verderio et al., 2016).

3.1 Preanalytical Factors in miRNA Analysis

Preanalytical factors are a common source for erroneous laboratory test results (Becker and Lockwood, 2013) that account for 46%–68% of the errors observed during the total testing process (https://www.mlo-online.com/ebook/201405/resources/index.htm). Preanalytical variables are defined as factors of a test that can affect the composition of the samples of interest, including those related to the sample collection and processing, transport and storage, and nucleic acid extraction methods for molecular biology (Jarry et al., 2014).

In the field of miRNA analysis, general guidelines for specimen collection and handling have not been universally implemented, and many variables are still under investigation (Becker and Lockwood, 2013). An extensive review of most of the preanalytical factors involved in the evaluation of circulating and tissue miRNAs was reported by Becker and Lockwood (2013). Specifically, for circulating miRNAs, the type of starting material (i.e., plasma or serum), the collection methods, the sample processing (i.e., protocols for plasma/serum separation), the storage conditions, and ultimately the extraction methods (i.e., guanidine/phenol/chloroform-based protocols or columns or bead methods) are known preanalytical factors involved in the identification and quantification of circulating miRNAs (Becker and Lockwood, 2013; Butz and Patocs, 2015; Tiberio et al., 2015).

Moreover, the expression of circulating miRNAs can be influenced by another important factor, the hemolysis, which can occur during blood collection or plasma/serum processing (Tiberio et al., 2015). Hemolysis, recognizable by eye due to pink discoloration of serum or plasma, derives from the release of the red blood cell (RBC) contents into the fluids, including miRNAs (Kirschner et al., 2011, 2013; Pritchard et al., 2012; Yamada et al., 2014). Many published studies investigated the effect of hemolysis revealing its substantial impact on the expression of certain miRNAs, especially for miR-16 and miR-456 that are the most abundant miRNAs in RBCs (Kirschner et al., 2011, 2013; Pritchard et al., 2012; Yamada et al., 2014). Another study highlighted a similar condition also for miR-451 and miR-92a, a proposed colon cancer biomarker (Kirschner et al., 2011). This means that altered levels of these miRNAs may reflect blood-cell-based phenomena rather than the presence of cancer, if used for diagnostic purpose (Pritchard et al., 2012; Kirschner et al., 2013).

Of note, the expression of some miRNAs could be influenced by hemolysis at levels not assessable by eye. Accordingly, various hemolysis indexes—based on specific spectrophotometric absorbance measurements—have been suggested, such as the absorbance peaks at 414 (Kirschner et al., 2011), the hemolysis ratio (Zanutto et al., 2014; Fortunato et al., 2014), the hemolysis score (Appierto et al., 2014), and the hemoglobin concentration obtained by the Harboe method (MacLellan et al., 2014). Once identified the hemolyzed samples, irrespectively from the index method used, the strategy of removing that samples from the analysis could raise concerns especially if the hemolyzed samples come from patients who suffer from the disease of interest (Yamada et al., 2014) or are submitted to treatment approaches that—by themselves—induce hemolysis. Thus, a suitable alternative could be to evaluate the influence of hemolysis on the miRNAs of interest implementing ad hoc in vitro controlled hemolysis experiments by artificially introducing different percentages of RBCs in a hemolysis-free plasma sample. Details on this approach are reported in the paper of Pizzamiglio et al. (2017).

3.2 Analytical Factors in miRNA Analysis

The detection of miRNAs, in both tissue and body fluids, is also a challenge due to several intrinsic characteristics of these molecules, such as the small size and the high similarity within miRNA families (de Planell-Saguer and Rodicio, 2013; Tiberio et al., 2015). Several techniques are currently available for assessing miRNA levels in body fluids, such as miRNA microarrays, quantitative real-time reverse-transcription PCR (qRT-PCR), and deep sequencing. According to the number of miRNAs that could be simultaneously evaluated, these technologies could be classified into high, medium, and low throughput (Tiberio et al., 2015). An overview of the different platforms available for miRNA analysis (in both tissue and body fluids) is reported by de Planell-Saguer and Rodicio (2013).

Currently, the qRT-PCR is one of the most commonly used methods for analyzing circulating miRNAs with high sensitivity and specificity (de Planell-Saguer and Rodicio, 2013). Different platforms have been developed by different companies to allow the simultaneous evaluation of a panel of miRNAs (Tiberio et al., 2015), such as microfluidic cards and arrays (i.e., TaqMan OpenArray, TaqMan TLDA microfluidic cards, mercury LNA qPCR, and miScript). All these platforms are designed to evaluate about 700 miRNAs simultaneously with high sensitivity and specificity starting from <100ng of input RNA (Tiberio et al., 2015). The output of this expression analysis is provided in terms of cycle threshold (Ct) values and thus analyzed, with the proper adjustments, according to the already available approaches, that is, relative expression (Livak and Schmittgen, 2001). Obviously, the use of these platforms allowed the evaluation of annotated miRNAs only, whereas sequencing approaches (i.e., miRNA-Seq) can be useful tools to discover novel miRNAs or IsomiRs. On the other hand, the costs and the equipments necessary for such analysis are important aspects that should be currently considered (Tiberio et al., 2015).

3.3 Workflow for miRNA-Based Signature Development

We have recently proposed (Verderio et al., 2016) a workflow covering the major essential steps involved in the cancer biomarker-signature development, from laboratory to clinical practice in the field of miRNA-based studies performed with qRT-PCR assays. This workflow, as reported in Fig. 1, summarizes all the key phases involved in biomarker studies, from biomarker discovery to their analytical and clinical validation including issues related to the development of operative procedures for their analysis. The process, as already mentioned, should start with a discovery phase, followed by a validation step, and by the clinical application of the identified biomarker signature, in case of strong evidences. Two additional assay-oriented steps could be introduced in the workflow, before and/or after the validation phase.

During the *biomarker discovery*, the aim is to identify candidate biomarkers, from large sets of tested biomarkers, which may be associated with the disease under study. During this phase, high-throughput platforms, allowing the evaluation of hundreds or thousands of molecules, are usually employed. In this initial phase, researchers have to clearly define the target population to be evaluated (together with inclusion/exclusion criteria), the biological specimens with the corresponding preanalytical procedures, and all the conditions for experimental running (analytical procedures) and for data analysis (postanalytical procedures). *Validation* phase refers instead to the evaluation of the role of the (signature) biomarkers previously identified on independent cohort(s) of subjects. The final goal of this phase is the identification of signatures to be translated in the clinical practice for subject classification and outcome prediction. Obviously, discovery and validation phases should be implemented on distinct cohorts of subjects in order to evaluate the performance of the identified biomarkers. An *assay optimization step*, before the validation phase, could be included in this workflow if

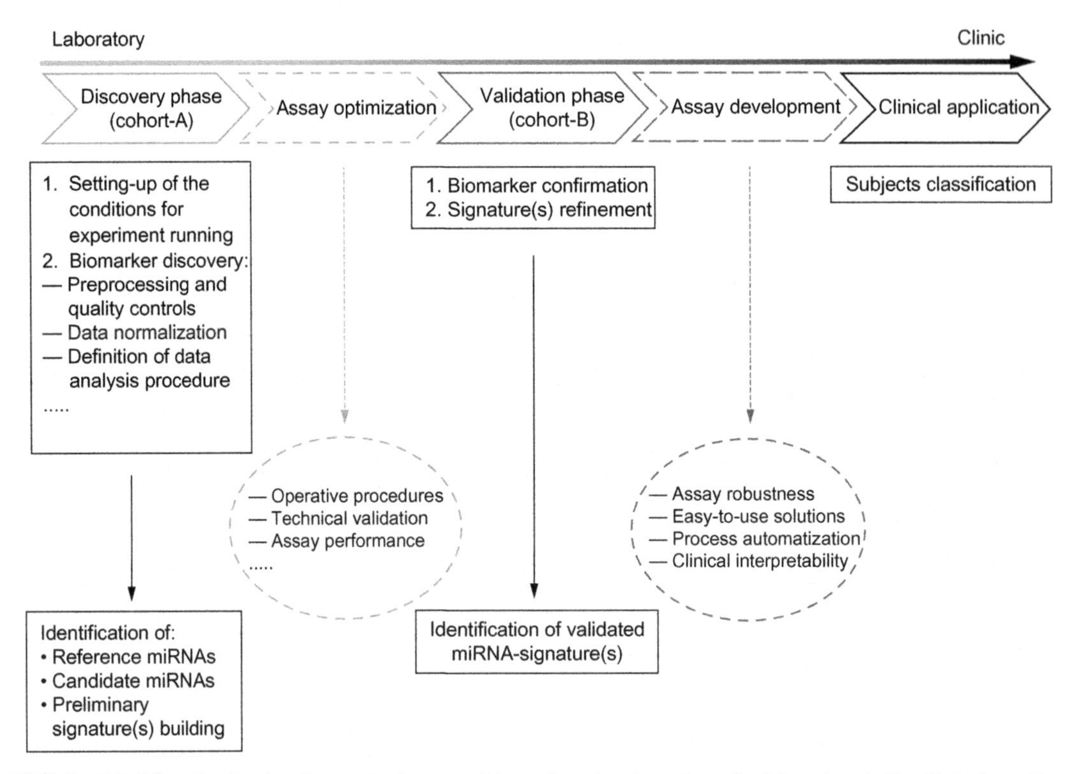

FIG. 1 Workflow for the development of a cancer biomarker signature, from the laboratory to the clinical practice.

different assays are used in the discovery and validation phases, in order to evaluate their level of reproducibility. An *assay development step* should also be added after the validation phase, to set up an easy-to-use assay to be used in the clinical setting, including the implementation of EQA schemes (Verderio et al., 2016).

From a statistical-methodological point of view, the main key issues involved in this workflow are those related to the data normalization of high-throughput qPCR data and to the building and validation of miRNA-based signatures.

4 STATISTICAL ISSUES IN miRNA-BASED SIGNATURE DEVELOPMENT

4.1 Data Normalization of High-Throughput qPCR Data

Data normalization represents a crucial preprocessing step aimed at removing experimentally induced variations and at differentiating true biological changes. Inappropriate normalization strategies can in fact induce misleading effects affecting the results of the analysis and the corresponding conclusions (Deo et al., 2011). Since in the circulating miRNA context there are not yet verified and shared *reference miRNAs* that could be used for data normalization, the preprocessing step of data normalization is really a major challenge, especially in the analysis of high-throughput qPCR data (Deo et al., 2011; Kang et al., 2012).

A common approach, especially in the last years, was to use presumed stable expressed *reference miRNAs* according to the existing data reported in literature without proper validation of their stable expressions in the specific context (Kang et al., 2012). Other authors reported the use of synthetic RNAs or miRNA molecules as spike-in controls for mRNA/miRNA expression, not only for monitoring the efficiency of RNA purification and reverse transcription but also for data normalization (Kang et al., 2012). A different strategy consists in identifying suitable endogenous controls for each study through the systematic evaluation of the expression level of a set of candidate *reference miRNAs* (Kang et al., 2012), usually identified from a pilot study with samples representative of the experimental conditions under investigation (Deo et al., 2011). A suitable (set of) endogenous control(s) should be adequately expressed in the sample specimens of interest with minimal variability in expression among samples under the investigated experimental conditions (Silver et al., 2006).

The currently most accepted and widely used method for data normalization of circulating miRNAs is the one proposed by Mestdagh et al. (2009), based on the computation of the global mean of the expressed miRNAs. Instead of using a single or a set of *reference miRNAs*, authors proposed the use of the mean expression value of all the expressed miRNAs as normalization factor. This method is obviously valid if a large number of miRNAs are profiled (i.e., discovery experiments performed in the initial phase of a study) but is almost never applicable in validation studies focused on a limited number of miRNAs. To overcome this issue, in the same paper, the authors proposed to search the set of reference miRNAs that most resemble the mean expression value of all the miRNAs and to use that set of *reference miRNAs* for data normalization in subsequent validation studies. Table 2 summarizes the principal strategies reported in literature for the identification of the best set of *reference miRNAs*.

Briefly, both the geNorm (Vandesompele et al., 2002) and NormFinder (Andersen et al., 2004) algorithms are data-driven methods aimed at identifying the best subset of *reference* miRNA starting from a set of *reference* candidates. Specifically, in geNorm approach, the pairwise variation of each candidate *reference* with all the others is computed as the standard deviation of the log-transformed ratio, and the gene-stability measure M is then calculated as the average of the pairwise variation of a particular *reference RNA* with all other *references*, and those with the lowest stability value (*M*-value) are identified as the most stable ones (Vandesompele et al., 2002). The NormFinder approach is however a model-based approach in which the inter- and intragroup variations are computed separately and then combined in a measure of stability representing the estimated systematic error: a low stability value means a low systematic error and therefore a stable expression across samples (Andersen et al., 2004). The BestKeeper software (Pfaffl et al., 2004), in addition to what implemented in geNorm, enables the identification of the *reference* and the analysis of the *target* variable.

TABLE 2 Data Normalization Strategies

Strategy Name	Stability Measure	References
geNorm	*M*-value	Vandesompele et al. (2002)
BestKeeper	BestKeeper index (BKI)	Pfaffl et al. (2004)
NormFinder	Stability value	Andersen et al. (2004)
NqA	*M*-value and stability value	Verderio et al. (2014)

In the NqA algorithm (Verderio et al., 2014), the N miRNAs expressed in all the samples are firstly used to compute the mean expression value and to identify a list of miRNAs differentially expressed between the conditions under investigation. In parallel, a subset of G candidate *reference miRNAs* is identified according to appropriate selection criteria such as variability, invariance between comparison groups (using the overall mean as normalization method), and coregulation. Subsequently, the identified G miRNAs are evaluated through both geNorm and NormFinder *R*-based function, ranked according to the stability values and forwardly combined in $G-1=S$ subsets, with at least two miRNAs, according to their stability value. Then, the distribution, within each jth set ($j=2,\ldots, S$), is compared between groups, and finally, the smallest set of reference miRNAs showing results with the highest agreement with those obtained when considering the overall mean were identified as the best subset of reference miRNAs (Verderio et al., 2014).

4.2 Composite Score Generation

Once identified the reference miRNAs, the subsequent step is represented by the combination of biomarkers in a composite score (i.e., miRNA signature). Both parametric and nonparametric combination-based methods could be used to find the *optimal* linear combination of a set of biomarkers in order to achieve greater discriminatory ability (as assessed by the area under the ROC curve) than those obtained using single biomarker alone. An overview of this method is reported in the recent paper of Yan et al. (2015). An alternative approach is the logistic regression model that allows to assess and predict the probability of the occurrence of a binary outcome (Harrell, 2001). Fig. 2 graphically summarizes the workflow we have proposed (Verderio et al., 2016) for developing miRNA-based signatures based on the logistic regression theory.

The process starts with the identification of the candidate miRNAs that should be included in the initial multivariate model. These candidates could be selected from literature, from prior evidences or be the results of previous univariate analysis within the same study. Then, during the fitting of the initial multivariate model, it is mandatory to consider the number of events per variable (EPV) to properly take into consideration the overfitting issue. This problem can in fact occur when the number of covariates is so much larger with respect to the low number of outcome events, so that the estimated model tends to capture not only the underlining process that generated the data but also the noise, thus leading to an over- or underestimation of the risk of the event in high- or low-risk patients (Pavlou et al., 2016). As a rule of thumb, it has been suggested that the model estimates are likely to be reliable when the EPV is at least 10 (Verderio et al., 2016; Pavlou et al., 2016). Accordingly, as reported in Fig. 2, when EPV is <10, the use of penalized regression strategies may represent a useful tool to handle overfitting as much as possible, whereas when EPV ≥ 10, standard regression strategies can be used (Verderio et al., 2016).

4.2.1 *Penalized Regression Models*

Penalized regression models are a useful tool to be used for limiting as much as possible the overfitting when a large number of covariates are considered in a model with respect to the number of outcome events. In the penalized regression models, the estimates of the

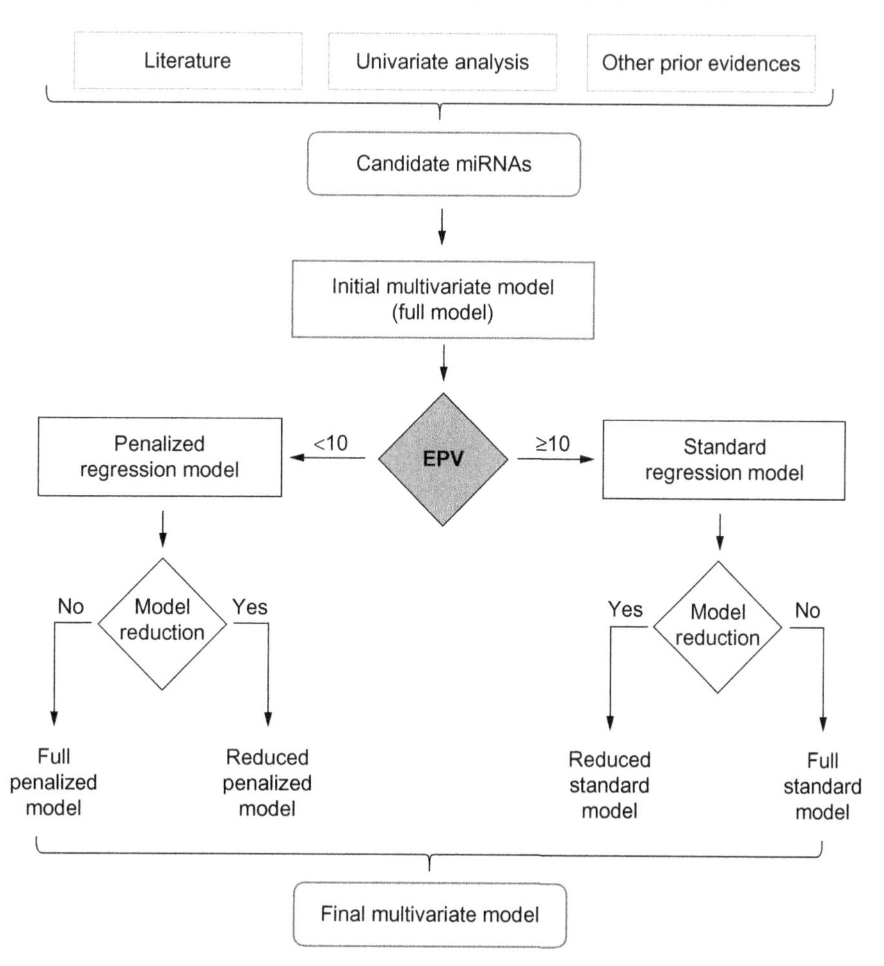

FIG. 2 Statistical analysis flowchart for the development of miRNA signatures.

beta-coefficients are obtained by maximizing the penalized log-likelihood and not the log-likelihood function, as in the standard regression models (Pavlou et al., 2016). This will allow a direct penalization of the beta estimates during model fitting. Briefly, a penalty term, corresponding to the functional form of the constraints, and a tuning parameter, corresponding to the amount of shrinkage, are applied to the beta-coefficient estimation. The functional form of the constraints means different types of penalization regression, as reported by Pavlou et al. (2016) in his recent paper. A general approach is, for example, the penalized maximum likelihood estimation (PMLE) developed for logistic regression models (Harrell, 2001; Moons et al., 2004).

4.2.2 Model Reduction

Another important theme, once identified and fitted the initial multivariate model, is the definition of the final model with the intent to obtain a more parsimonious model without a substantial loss of information; the final model could correspond to the full initial model or defined as a reduced one (see Fig. 2; Verderio et al., 2016; Moons et al., 2012a). In case of no model reduction, the final multivariate model is the initial multivariate one (full model), in which all the a priori included predictors are considered and no predictor selection is performed. In the other case, the initial multivariate model is reduced, and a final model, including fewer predictors, is obtained. Several well-established approaches for standard regression models are available, such as *backward elimination* (Moons et al., 2012a). For PMLE, a reduced model can be obtained using the R-square method (Moons et al., 2004). An alternative approach to the standard selection methods is the *all subsets regression*, which can discover from all the possible combinations of variables the ones explaining more variation in patients' outcome than those obtained by using the standard stepwise/backward algorithms (Verderio et al., 2016; Altman and Royston, 2000).

4.2.3 Model Evaluation

Once the model was developed, its performance should be assessed according to specific performance measures as reported by Steyerberg et al. (2010). One of the most common measures is discrimination that refers to the ability of the model to discriminate between individuals with the disease from those without the disease. The *c*-index or the equivalent area under the ROC curve (AUC) are the indexes used to measure it. AUC values range between 0.5 and 1.0, with values equal to 1 indicating a perfect discrimination between the two groups and values near 0.5 meaning no discriminatory ability. Another aspect to evaluate is calibration that refers to the agreement between the probability of developing (or having) the outcome of interest as estimated by the model and that observed. It is usually graphically assessed: perfect predictions should be on the 45 degrees line of the graph, and for logistic regression, smoothing techniques can be used to estimate the observed probabilities of the outcome in relation to the predicted ones (Steyerberg et al., 2010).

4.2.4 Model Validation

Validation of a model traditionally refers of assessing the performance of that model in subjects other than those used for model development. When applied to new subjects, the performance of a model is generally lower than that observed in the sample on which the model was developed. Therefore, the performance of a developed model should be accurately evaluated in new individuals before its implementation and application in clinical practice. To this purpose, two different types of model validation can be adopted: internal and external validation techniques (Moons et al., 2012a, 2012b).

Internal validation implies the properly fitting and validation of a model using only a series of samples. The simpler strategy is the splitting of the data in a training set—used for model building—and a testing set, used for model validation. Extension of the training-testing splitting that includes also resampling technique is the leave-one-out, *k*-fold, and repeated random split cross validation. An alternative is the bootstrap approach, especially when the development sample is relatively small and/or a large number of candidate predictors are studied (Moons et al., 2012a; Taylor et al., 2008). With the latter approach, all the data available are used for model development providing also information about the overfitting and optimism of the model. Even if these internal validation methods can correctly control overfitting and optimisms, they cannot substitute the external validation ones (Moons et al.,

2012b). The performance of a developed and internally validated model should in fact be tested or validated in new individuals, similar to that considered in the developed set, before its implementation and application in clinical practice. Validation on new data, such as subjects from another country (*geographic validation*) or recruited at another time (*temporal validation*), finally allows the evaluation of the generalizability of the model (Moons et al., 2012b). The aim of this phase is to apply the developed model to new individuals and to quantify the model's predictive performance. An extensive description of these approaches was reported in the recent paper by Austin et al. (2016).

4.2.5 Model Updating and Extension

A lower performance of a prediction model can be however obtained when tested on the new individuals compared with what observed in the development set. In such case, a valid alternative is to update the existing model by adjusting or recalibrating it on the new data (Moons et al., 2012b; Vergouwe et al., 2017). In this situation, the updated model combines both the information from the original model with those from the new individuals, theoretically improving the transportability to other individuals. An example is the adjustment of the baseline risk or the reestimation of all the coefficients of the covariates included in the model (i.e., *model revision*). Obviously, the performance of an updated model should be tested before applied in the routine practice.

In addition, to improve the performance of an already existing clinical model, new marker(s) can be incorporated. Different model-extension strategies can be adopted for this purpose, such as the reestimation of all the regression coefficients including the new marker (Nieboer et al., 2016). The performance of the extended model can be evaluated through calibration and discrimination or according to other developed measures that assess the added value of the new marker(s), such as the net reclassification improvement and the integrated discrimination improvement indexes (Steyerberg et al., 2010).

4.3 Assay-Oriented Steps

As reported in Fig. 2, in the proposed workflow, two assay-oriented steps could be introduced in order to set up an easier assay from the evaluation of the candidate biomarkers arising from the discovery phase. A first phase could be placed between the discovery and the validation phase, especially if different types of assays are used to profile the candidate miRNAs in the two cohorts of subjects. This phase will allow the evaluation of the level of reproducibility between the employed assays. After the validation of the biomarkers on an independent cohort of subjects and in case of implementation of the assay in the clinical routine, an assay optimization step should be considered in the workflow (Verderio et al., 2016).

4.4 Clinical Validation and Implementation

The last phase of the process is represented by the evaluation of the clinical utility of the biomarker test. This will imply the evaluation of the improvements associated with its introduction in the clinics for treatment decision-making and patient outcomes. This requires the availability of high level of evidence that corroborates the use of the marker for improving patient outcome in order to justify its incorporation into routine clinical care. An extensive discussion of this topic is reported by Febbo et al. (2011).

5 COLORECTAL CANCER

CRC is one of the major causes of cancer death in western countries (Mazeh et al., 2013). Complete surgical excision of the primary tumor is the only treatment for early CRC, and tumor stage at diagnosis remains the only independent predictor of survival (Bustin and Murphy, 2013). On the basis of the natural history of CRC progression and of the long time interval of progression from normal mucosa to invasive cancer, many efforts have been focused on the development of screening programs for CRC prevention and detection at an early stage, when cancer is most likely curable. According to these considerations and to the CRC age-associated risk, current screening guidelines recommend routine testing after the age of 50 years (Mazeh et al., 2013). Colonoscopy is the most accurate test for both detection and removal of adenomas but presents some limits due to its invasiveness and high cost (Park et al., 2010). Accordingly, strategies based on the search of human hemoglobin in stool (fecal occult blood test (FOBT) or the recent fecal immunochemical test (FIT) tests) have been proposed and implemented for large-scale population screening program (Mazeh et al., 2013; Park et al., 2010). In Italy, the majority of the CRC screening programs are based on FOBT/FIT offered every 2 years to 50- to 69-year-old people. Subjects with a negative test are invited to repeat the test after 2 years, whereas subjects with a positive result are contacted in order to perform a total colonoscopy at referral centers during dedicated sessions. According to the colonoscopy results, subjects undergo surgery or, after the neoplasm removal, are enrolled in a follow-up program. A detailed description of the CRC screening program characteristics is reported in the 11th report of the National Centre for Screening Monitoring (http://www.osservatorionazionalescreening.it).

5.1 Application

Currently, at the Fondazione IRCCS Istituto Nazionale dei Tumori in Milano, there is an ongoing study that aimed to discover circulating miRNAs able to identify the presence of CRC or adenomas in subjects that resulted positive to the FIT test (Ciniselli et al., 2016). Briefly, blood samples were collected before colonoscopy, and circulating miRNAs, extracted from plasma, were analyzed using PCR-based assays. Following the above described workflow and according to the time line of the projects, a *discovery* phase was firstly implemented with the aim to investigate the suitability of searching miRNAs in plasma from FIT+ individuals and to identify a set of *reference* and candidate miRNAs to be deeply investigated in the subsequent phases. Plasma samples of FIT+ subjects with a colonoscopy at our institute were analyzed using a high-throughput assay. Once identified the set of *reference* and candidate miRNAs through the NqA R-function, two cohorts of subjects were prospectively collected, an internal cohort (colonoscopy at our institute) and an external one (colonoscopy in one of the institutes joining the CRC screening program of the Local Health Authority of Milan).

In addition, in order to investigate if our candidate miRNAs were influenced by hemolysis, an ad hoc in vitro controlled hemolysis experiment was implemented by artificially introducing different percentages of red blood cells (%RBCs) in a hemolysis-free plasma sample. Details on the design of this experiment are reported in the paper of Pizzamiglio et al. (2017). miRNAs known as hemolysis related in literature (miR-16, miR-486, miR-451, and miR-92a) were identified as such also in our experiments, showing a significant different expression compared with the sample uncontaminated by RBCs, whereas our reference miRNAs resulted were not influenced by hemolysis. In addition, the availability for each contaminated tube of both hemolysis indexes and the known %RBC concentration allowed the building of a calibration curve useful for estimating the percentage of RBCs in new plasma samples (Pizzamiglio et al., 2017).

On the discovery cohort, by using the NqA software, we identified a subset of *reference* miRNAs for data normalization and a set of candidate miRNAs. Based on these results, a custom microfluidic card including the candidate and the reference miRNAs was designed to analyze the samples from the internal cohort. By starting from our candidate miRNAs, penalized logistic regression models were implemented using an all-subset approach. The performance of the most promising signatures on the external cohort is now under investigation, and preliminary analysis showed statistically significant results also on this cohort. The complete description and results of the entire study will be sooner released.

6 CONCLUSION

The identification of biomarkers from body fluids has attracted much interest in the research community, especially among oncologists. In such scenario, miRNAs represent a promising tool due to their involvement in primary biological processes and to their easy detectability in plasma/serum. A large number of studies have in fact been published on this topic, suggesting miRNAs as diagnostic, prognostic, or predictive biomarkers. However, some issues about the preanalytical phase are still under investigation, and fully understanding of all these aspects should be deeply addressed before its ultimate translation into the clinics (Becker and Lockwood, 2013). In addition, incomplete reporting of details about their analytical evaluation and the subsequent statistical analysis are additional aspects that could explain poor concordance among independent researches. Thus, the use of shared workflow for their evaluation and analysis is required in order to aid the application of miRNAs in the clinical patient management.

Here, we summarized the most important steps involved in the biomarker identification-validation process, by focusing both on the study design and to the statistical- methodological aspect related to their identification and combination in a composite score.

Acknowledgment

This work was supported by grant from Associazione Italiana per la Ricerca sul Cancro (AIRC, Grant 12162 to G Sozzi).

References

Altman, D.G., Royston, P., 2000. What do we mean by validating a prognostic model? Stat. Med. 19 (4), 453–473.

Andersen, C.L., Jensen, J.L., Orntoft, T.F., 2004. Normalization of real-time quantitative reverse transcription-PCR data: a model-based variance estimation approach to identify genes suited for normalization, applied to bladder and colon cancer data sets. Cancer Res. 64 (15), 5245–5250.

Appierto, V., Callari, M., Cavadini, E., Morelli, D., Daidone, M.G., Tiberio, P., 2014. A lipemia-independent NanoDrop((R))-based score to identify hemolysis in plasma and serum samples. Bioanalysis 6 (9), 1215–1226.

Austin, P.C., van Klaveren, D., Vergouwe, Y., Nieboer, D., Lee, D.S., Steyerberg, E.W., 2016. Geographic and temporal validity of prediction models: different approaches were useful to examine model performance. J. Clin. Epidemiol. 79, 76–85.

Becker, N., Lockwood, C.M., 2013. Pre-analytical variables in miRNA analysis. Clin. Biochem. 46 (10–11), 861–868.

Bustin, S.A., Murphy, J., 2013. RNA biomarkers in colorectal cancer. Methods 59 (1), 116–125.

Butz, H., Patócs, A., 2015. Technical aspects related to the analysis of circulating microRNAs. In: Igaz, P. (Ed.), Circulating microRNAs in Disease Diagnostics and their Potential Biological Relevance. Experientia Supplementum, vol. 106. Springer, Basel.

Calin, G.A., Croce, C.M., 2006. MicroRNA signatures in human cancers. Nat. Rev. Cancer 6 (11), 857–866.

Chen, X., Ba, Y., Ma, L., Cai, X., Yin, Y., Wang, K., et al., 2008. Characterization of microRNAs in serum: a novel class of biomarkers for diagnosis of cancer and other diseases. Cell Res. 18 (10), 997–1006.

Ciniselli, C.M., Verderio, P., Pizzamiglio, S., Bottelli, S., Gariboldi, M., 2016. Workflow for the identification and validation of microRNAs: the colorectal cancer experience. In: Europe Biobank Week. http://europebiobankweek.eu/wp-ontent/themes/offreWP_ebw/images/abstract_book_V5.pdf.

Cortez, M.A., Bueso-Ramos, C., Ferdin, J., Lopez-Berestein, G., Sood, A.K., Calin, G.A., 2011. MicroRNAs in body fluids—the mix of hormones and biomarkers. Nat. Rev. Clin. Oncol. 8 (8), 467–477.

de Planell-Saguer, M., Rodicio, M.C., 2013. Detection methods for microRNAs in clinic practice. Clin. Biochem. 46 (10–11), 869–878.

Deo, A., Carlsson, J., Lindlof, A., 2011. How to choose a normalization strategy for miRNA quantitative real-time (qPCR) arrays. J. Bioinforma. Comput. Biol. 9 (6), 795–812.

Febbo, P.G., Ladanyi, M., Aldape, K.D., De Marzo, A.M., Hammond, M.E., Hayes, D.F., et al., 2011. NCCN Task Force report: evaluating the clinical utility of tumor markers in oncology. J. Natl. Compr. Cancer Netw. 9 (Suppl. 5), S1–32. quiz S33.

Fortunato, O., Boeri, M., Verri, C., Conte, D., Mensah, M., Suatoni, P., et al., 2014. Assessment of circulating microRNAs in plasma of lung cancer patients. Molecules 19 (3), 3038–3054.

Goossens, N., Nakagawa, S., Sun, X., Hoshida, Y., 2015. Cancer biomarker discovery and validation. Transl. Cancer Res. 4 (3), 256–269.

Harrell Jr., F.E., 2001. Regression Modeling Strategies. Springer-Verlag, New York.

Jarry, J., Schadendorf, D., Greenwood, C., Spatz, A., van Kempen, L.C., 2014. The validity of circulating microRNAs in oncology: five years of challenges and contradictions. Mol. Oncol. 8 (4), 819–829.

Kang, K., Peng, X., Luo, J., Gou, D., 2012. Identification of circulating miRNA biomarkers based on global quantitative real-time PCR profiling. J. Anim. Sci. Biotechnol. 3 (1). 4-1891-3-4.

Kirschner, M.B., Kao, S.C., Edelman, J.J., Armstrong, N.J., Vallely, M.P., van Zandwijk, N., et al., 2011. Haemolysis during sample preparation alters microRNA content of plasma. PLoS One 6 (9), e24145.

Kirschner, M.B., Edelman, J.J., Kao, S.C., Vallely, M.P., van Zandwijk, N., Reid, G., 2013. The impact of hemolysis on cell-free microRNA biomarkers. Front. Genet. 4, 94.

Livak, K.J., Schmittgen, T.D., 2001. Analysis of relative gene expression data using real-time quantitative PCR and the 2(−Delta Delta C(T)) method. Methods 25 (4), 402–408.

MacLellan, S.A., MacAulay, C., Lam, S., Garnis, C., 2014. Pre-profiling factors influencing serum microRNA levels. BMC Clin. Pathol. 14, 27. 6890-14-27. eCollection 2014.

Mazeh, H., Mizrahi, I., Ilyayev, N., Halle, D., Brucher, B., Bilchik, A., et al., 2013. The diagnostic and prognostic role of microRNA in colorectal cancer—a comprehensive review. J. Cancer 4 (3), 281–295.

Mestdagh, P., Van Vlierberghe, P., De Weer, A., Muth, D., Westermann, F., Speleman, F., et al., 2009. A novel and universal method for microRNA RT-qPCR data normalization. Genome Biol. 10 (6). R64-2009-10-6-r64. Epub 2009 Jun 16.

Moons, K.G., Donders, A.R., Steyerberg, E.W., Harrell, F.E., 2004. Penalized maximum likelihood estimation to directly adjust diagnostic and prognostic prediction models for overoptimism: a clinical example. J. Clin. Epidemiol. 57 (12), 1262–1270.

Moons, K.G., Kengne, A.P., Woodward, M., Royston, P., Vergouwe, Y., Altman, D.G., et al., 2012a. Risk prediction models: I. Development, internal validation, and assessing the incremental value of a new (bio)marker. Heart 98 (9), 683–690.

Moons, K.G., Kengne, A.P., Grobbee, D.E., Royston, P., Vergouwe, Y., Altman, D.G., et al., 2012b. Risk prediction models: II. External validation, model updating, and impact assessment. Heart 98 (9), 691–698.

Nieboer, D., Vergouwe, Y., Ankerst, D.P., Roobol, M.J., Steyerberg, E.W., 2016. Improving prediction models with new markers: a comparison of updating strategies. BMC Med. Res. Methodol. 16 (1), 128.

Park, D.I., Ryu, S., Kim, Y.H., Lee, S.H., Lee, C.K., Eun, C.S., et al., 2010. Comparison of guaiac-based and quantitative immunochemical fecal occult blood testing in a population at average risk undergoing colorectal cancer screening. Am. J. Gastroenterol. 105 (9), 2017–2025.

Pavlou, M., Ambler, G., Seaman, S., De Iorio, M., Omar, R.Z., 2016. Review and evaluation of penalised regression methods for risk prediction in low-dimensional data with few events. Stat. Med. 35 (7), 1159–1177.

Pfaffl, M.W., Tichopad, A., Prgomet, C., Neuvians, T.P., 2004. Determination of stable housekeeping genes, differentially regulated target genes and sample integrity: BestKeeper–excel-based tool using pair-wise correlations. Biotechnol. Lett. 26 (6), 509–515.

Pizzamiglio, S., Zanutto, S., Ciniselli, C.M., Belfiore, A., Bottelli, S., Gariboldi, M., et al., 2017. A methodological procedure for evaluating the impact of hemolysis on circulating microRNAs. Oncol. Lett. 13 (1), 315–320.

Poste, G., 2011. Bring on the biomarkers. Nature 469 (7329), 156–157.

Pritchard, C.C., Kroh, E., Wood, B., Arroyo, J.D., Dougherty, K.J., Miyaji, M.M., et al., 2012. Blood cell origin of circulating microRNAs: a cautionary note for cancer biomarker studies. Cancer Prev. Res. (Phila.) 5 (3), 492–497.

Silver, N., Best, S., Jiang, J., Thein, S.L., 2006. Selection of housekeeping genes for gene expression studies in human reticulocytes using real-time PCR. BMC Mol. Biol. 7, 33.

Singh, R., Ramasubramanian, B., Kanji, S., Chakraborty, A.R., Haque, S.J., Chakravarti, A., 2016. Circulating microRNAs in cancer: hope or hype? Cancer Lett. 381 (1), 113–121.

Steyerberg, E.W., Vickers, A.J., Cook, N.R., Gerds, T., Gonen, M., Obuchowski, N., et al., 2010. Assessing the performance of prediction models: a framework for traditional and novel measures. Epidemiology 21 (1), 128–138.

Taylor, J.M., Ankerst, D.P., Andridge, R.R., 2008. Validation of biomarker-based risk prediction models. Clin. Cancer Res. 14 (19), 5977–5983.

Tiberio, P., Callari, M., Angeloni, V., Daidone, M.G., Appierto, V., 2015. Challenges in using circulating miRNAs as cancer biomarkers. Biomed. Res. Int. 2015, 731479.

Vandesompele, J., De Preter, K., Pattyn, F., Poppe, B., Van Roy, N., De Paepe, A., et al., 2002. Accurate normalization of real-time quantitative RT-PCR data by geometric averaging of multiple internal control genes. Genome Biol. 3 (7). RESEARCH0034.

Verderio, P., Mangia, A., Ciniselli, C.M., Tagliabue, P., Paradiso, A., 2010. Biomarkers for early cancer detection—methodological aspects. Breast Care (Basel) 5 (2), 62–65.

Verderio, P., Bottelli, S., Ciniselli, C.M., Pierotti, M.A., Gariboldi, M., Pizzamiglio, S., 2014. NqA: an R-based algorithm for the normalization and analysis of microRNA quantitative real-time polymerase chain reaction data. Anal. Biochem. 461, 7–9.

Verderio, P., Bottelli, S., Pizzamiglio, S., Ciniselli, C.M., 2016. Developing miRNA signatures: a multivariate prospective. Br. J. Cancer 115 (1), 1–4.

Vergouwe, Y., Nieboer, D., Oostenbrink, R., Debray, T.P., Murray, G.D., Kattan, M.W., et al., 2017. A closed testing procedure to select an appropriate method for updating prediction models. Stat. Med. 36 (28), 4529–4539. https://doi.org/10.1002/sim.7179.

Verma, A.M., Patel, M., Aslam, M.I., Jameson, J., Pringle, J.H., Wurm, P., et al., 2015. Circulating plasma microRNAs as a screening method for detection of colorectal adenomas. Lancet 385 (Suppl. 1), S100. 6736(15)60415–9.

Yamada, A., Cox, M.A., Gaffney, K.A., Moreland, A., Boland, C.R., Goel, A., 2014. Technical factors involved in the measurement of circulating microRNA biomarkers for the detection of colorectal neoplasia. PLoS One 9 (11), e112481.

Yan, L., Tian, L., Liu, S., 2015. Combining large number of weak biomarkers based on AUC. Stat. Med. 34 (29), 3811–3830.

Zanutto, S., Pizzamiglio, S., Ghilotti, M., Bertan, C., Ravagnani, F., Perrone, F., et al., 2014. Circulating miR-378 in plasma: a reliable, haemolysis-independent biomarker for colorectal cancer. Br. J. Cancer 110 (4), 1001–1007.

CHAPTER

7

Analyzing the Effects of Genetic Variation in Noncoding Genomic Regions

Yasmina A. Mansur[*,#], *Elena Rojano*[*,#], *Juan A.G. Ranea*[*,†], *James R. Perkins*[‡]

[*]Department of Molecular Biology and Biochemistry, University of Malaga, Malaga, Spain
[†]CIBER de Enfermedades Raras, ISCIII, Madrid, Spain [‡]Research Laboratory, IBIMA-Regional University Hospital of Malaga-UMA, Malaga, Spain

1 INTRODUCTION

The sequencing of the human genome gave rise to an explosion of interest in the genetic basis of diversity among individuals. Moreover, it has greatly facilitated the study of many common familial traits and complex diseases such as asthma, cancer, diabetes, and psychiatric disorders (Lowe and Reddy, 2015). Such studies have routinely found disease-associated genetic variants, such as single-nucleotide polymorphisms (SNPs), outside of protein-coding genes, that is, in the noncoding regions of the genome (Li et al., 2014).

This chapter will focus on SNPs in noncoding regions in the context of disease. After a brief definition, we will outline the different genotyping techniques available for their detection, including genome-wide association studies (GWAS), which are often used to find associations between genetic variants and specific traits. We will also explain the different biological mechanisms that can be affected by variation in noncoding regions, such as DNA methylation, histone modifications, transcription factor (TF) binding, alternative splicing, and mRNA stability. We will present examples of how such variants can lead to specific diseases such as cancer and asthma. Finally, we will provide a summary of the different tools available to annotate and classify SNPs in noncoding regions via prediction of their putative functional effects.

This chapter will be useful for researchers interested in (i) developing procedures for the identification of disease-associated SNPs in noncoding regions using GWAS, (ii) learning more

[#]Equal Contribution.

Precision Medicine
https://doi.org/10.1016/B978-0-12-805364-5.00007-X

about the potential roles of such SNPs in disease through their effects on genomic elements, and (iii) using publicly available tools and datasets to annotate and assign putative function to SNPs of interest.

2 SNPs AND DISEASE

We are all genetically different. These differences can underlie susceptibility to common diseases, human traits, and differential responses to drugs. SNPs represent an important class of genetic variant, defined as a single-base change in the DNA sequence. They occur an average of every 300 bases in our genomes (https://ghr.nlm.nih.gov/primer/genomicresearch/snp). This means that two individuals can potentially differ at 10 million different sites, although in reality most people are far more similar (Durbin et al., 2010). Although SNPs do not typically refer to large insertions or deletions within the genome, in practice, a range of variations are also referred to as SNPs, including insertions, deletions, and variations, with <1% frequency. SNPs are mainly biallelic, although tri- and tetra-allelic forms can be found (Phillips et al., 2004; Westen et al., 2009).

SNPs are not evenly distributed along the genome. In fact, fewer disease-associated SNPs have been found in protein-coding regions than in noncoding, perhaps unsurprisingly given that noncoding regions account for the majority of the genome (Zhang and Lupski, 2015). SNPs in coding regions can lead to modifications in the structure and function of encoded proteins if they are nonsynonymous (Zhao et al., 2014; i.e., they change the amino-acid sequence of the protein). Such SNPs are responsible for some of the most well-known inherited monogenic disorders and are routinely analyzed for diagnostic purposes (Syvänen, 2001). They can also contribute to common diseases. The factor V Leiden mutation and SNPs in the apolipoprotein E gene (Major et al., 2000; Schmidt et al., 2002) are examples of variants in coding regions associated with common diseases that can be used to assess risk.

SNPs in noncoding regions do not affect the structure of encoded proteins. However, they can affect gene regulation through various mechanisms (Shastry, 2009). For instance, if present in promoter or enhancer elements, TF binding can be affected. Additionally, SNPs located in introns may affect the binding of proteins that regulate alternative splicing, mRNA stability, or the splice-site recognition (Fig. 1). Interestingly, a correlation has been observed between SNPs and diseases related to histone modifications, affecting enhancer function (Ernst et al., 2011; Fareed and Afzal, 2013). Several examples of disease-associated SNPs in noncoding regions are given in Table 1. These were found using GWAS, which we will describe in greater detail below.

3 SNP GENOTYPING

The study of SNPs is crucial for medical research due to their potential involvement in disease mechanisms. Thus, many genetic studies have been performed. Such studies aim to identify SNPs that cause changes in biological processes that lead to disease states (Emahazion et al., 2001; Kwok and Gu, 1999; Schork et al., 2000; Tost and Gut, 2005). One of the most common approaches is to perform a case-control study in which large-scale SNP genotyping is performed in a patient group and healthy control population. By comparing allele frequencies between groups, we aim to find SNPs associated with the disease. This information can

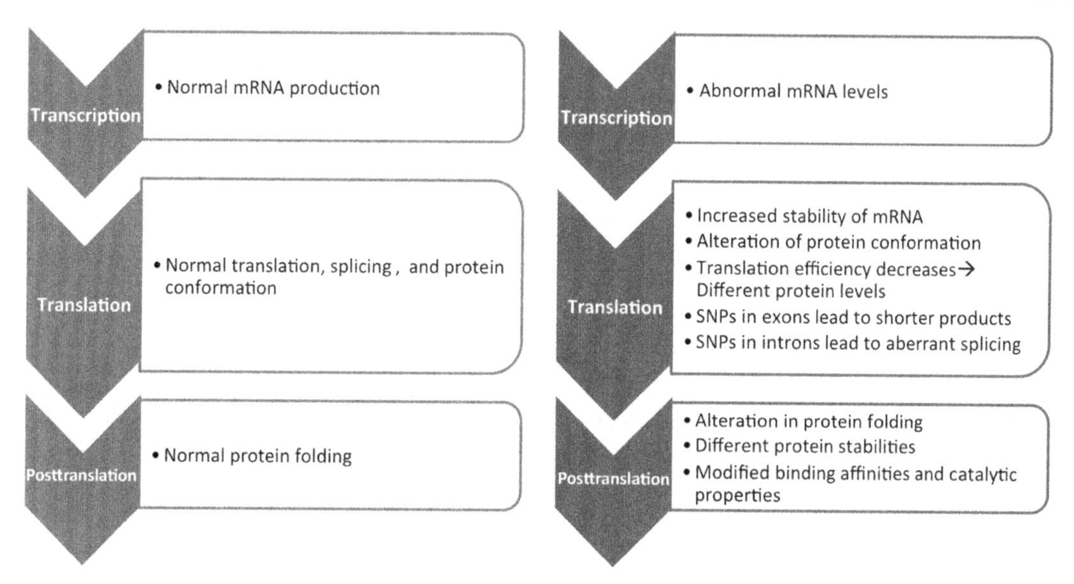

FIG. 1 The potential effects of SNPs on biological processes at the transcriptional, translational, and posttranslational levels. Neutral effects are shown on the *left*, deleterious effects on the *right*.

TABLE 1 Examples of SNPs Located in Intergenic Regions Associated With Cancer

Disease	Location	Strongest SNP-Risk Allele
Breast cancer	8q24.21	rs13281615-T
	2q35	rs13387042-A
Basal-cell carcinoma	8q24.21	rs16901979-A
		rs6983267-G
Melanoma	22q13.1	rs2284063-G
Ovarian cancer	9p22.2	rs3814113-T

Results obtained from GWAS studies.
Adapted from Fareed, M., Afzal, M., 2013. Single nucleotide polymorphism in genome-wide association of human population: a tool for broad spectrum service. Egypt. J. Hum. Genet. 14 (2) 123–134. doi:10.1016/j.ejmhg.2012.08.001.

then be used to direct further study into disease mechanisms and to find biomarkers for susceptibility, prevention and treatment, key concepts in precision medicine.

3.1 Variant Detection Techniques

The detection of a single-base change in the 3 billion base pairs present in the human genome is a complex task for which several methods have been proposed. As stated in the previous sections, SNPs can cause changes in biological processes, leading to disease. To detect such associations, a vast number of SNPs must be genotyped in large sample populations (Gunderson et al., 2006). Thus, it is vital to achieve a high level of throughput by increasing the number of SNPs that can be analyzed simultaneously.

In most cases, genotyping consists of the generation of allele-specific products for SNPs of interest followed by their detection for genotype determination. The most commonly used genotyping technologies comprise PCR-based amplification and/or hybridization.

3.1.1 *Primer Extension*

Primer extension utilizes enzyme specificity to achieve allelic discrimination via the allele-specific incorporation of nucleotides during the primer extension reaction with a DNA template. These assays can use either one single primer for detecting both alleles or specific primers for detecting each allele (Kim and Misra, 2007).

3.1.2 *Hybridization*

Hybridization-based approaches rely on differences in the thermal stability of double-stranded DNA to differ between perfectly matched and mismatched target-probe pairs. Since hybridization approaches do not require enzymatic reactions to discriminate alleles, their implementation on high-throughput platforms is possible, for example, with microarrays. Two of the most widely used technologies for this process are the GeneChip array technology from Affymetrix and the TaqMan genotyping assay from Applied Biosystems (Kim and Misra, 2007), which combines hybridization with primer extension. A related technique is dynamic allele-specific hybridization in which the generation of melting curves is used to distinguish between alleles (Russom et al., 2006).

3.1.3 *Mismatch Repair Detection*

Mismatch repair detection employs an in vivo bacterial mismatch repair system to detect polymorphisms (Faham et al., 2001). In this approach, the genomic region of interest from a reference sample is amplified by PCR, cloned in a vector containing an active *Cre* marker gene and grown in a modified *Escherichia coli* strain. The activity of the *Cre* gene is then used to determine the genotype.

3.1.4 *BeadArray and SNPlex*

Other genotyping technologies include combinations of two or more allele discrimination approaches, such as BeadArray (Illumina, California) and SNPlex (Applied Biosystems, California) assays. BeadArray consists of a high-throughput SNP genotyping platform that combines hybridization, primer extension, and ligation to generate allele-specific products, followed by PCR for amplification (Ferguson et al., 2000; Kim and Misra, 2007). SNPlex combines hybridization and ligation for generating allele-specific product and PCR for amplification (Tobler et al., 2005).

3.2 Genome-Wide Association Studies

A GWAS consists of the measurement and analysis of variation with the aim of identifying genetic risk factors for common diseases (Feero and Guttmacher, 2010). This is commonly achieved through a case-control study that compares the genotypes of patients with a given disease against healthy controls. When this is performed for variants across the entire genome, it can be considered a GWAS, that is, genome-wide. There are several important concepts to consider when designing, analyzing, and interpreting a GWAS.

3.2.1 *Minor Allele Frequency*

Within a population, two or more base-pair possibilities can exist for a given SNP, known as alleles. The frequency of an SNP is determined by the frequency of the less common allele, known as the *minor allele frequency*. The frequency of an allele within a population and the risk for complex diseases associated with that allele are key features to consider when planning a genetic study. As far as GWAS are concerned, a large panel of genetic markers and large sample sizes, consisting of unrelated individuals, are required (Bush and Moore, 2012).

3.2.2 *Linkage Disequilibrium*

The genotype data collected in the HapMap and 1000 Genomes Projects provide information related to *linkage disequilibrium* (LD), a key concept in GWAS and essential for the design of genotyping technology. LD can be considered a property of SNPs that describes the degree of correlation between them. This is related to the number of recombination events that have taken place within a population. Eventually, after a certain number of recombination events, all the alleles should in theory be completely independent to each other (*linkage equilibrium*) (Bush and Moore, 2012).

LD values depend on several factors, such as the population size and the number of generations for which the population has existed. Consequently, different human populations present different patterns of LD, and this has to be considered when designing a genetic association study (Bush and Moore, 2012).

A statistical measure of correlation, r^2, can be used to analyze LD. Other measures of LD, such as D and D', are also available. High values of r^2 between several closely occurring SNPs imply that they convey similar information. Therefore, it can be inferred that only a subset of these SNPs needs to be genotyped to capture the variation within the population (Syvänen, 2001). This has important implications for the design of arrays as it implies that not all SNPs across the genome need to be measured by the technology—instead, only a subset should be enough to capture most differences.

These SNPs, called *tag SNPs*, are specifically selected to capture variation at nearby sites in the genome that present high LD values. Tag SNPs avoid the need to genotype every SNP in the targeted chromosomal region, and thus, they simplify the identification of genetic variation and their association with phenotypes. Moreover, this reduces the costs of genotyping. However, as LD patterns are population-specific, tag SNPs for a given population may not be useful for a different one (Bush and Moore, 2012). Moreover, rarer variants will not be detected in this manner: exome and genome sequencing will be more appropriate for the detection of such variants (Lee et al., 2014).

3.2.3 *Direct vs Indirect Associations*

Genetic associations can be one of two different types: direct or indirect. The former means that the associated SNP is functional, that is, it is thought to be affecting a biological mechanism and causing the phenotype. Indirect associations can be made when the influential SNP is not directly genotyped; however, a tag SNP in high LD with the influential SNP is genotyped and statistically associated with the phenotype (Hirschhorn and Daly, 2005). Therefore, imputation and fine-mapping studies are required to home in on the functionally important SNP. A summary of the different steps in a GWAS workflow is given in Fig. 2.

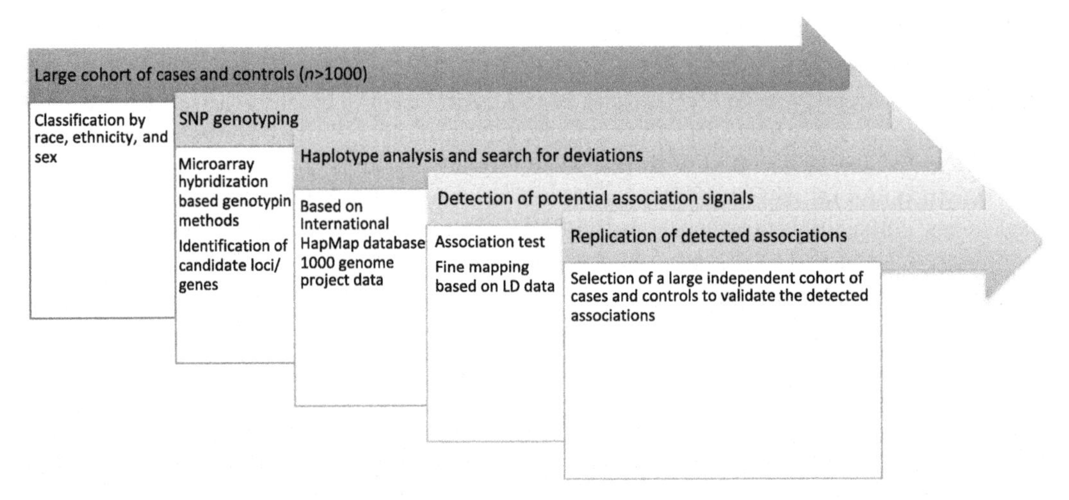

FIG. 2 Overview of a common workflow used while performing genome-wide association studies.

4 SNPs ASSOCIATED WITH GENE REGULATION

As mentioned previously, many disease-associated SNPs detected by GWAS are located outside of coding regions. These variants may be involved in a disease by affecting the regulation of gene expression. Transcription is regulated by two major mechanisms: changes in DNA sequences that are responsible for genetic regulation, typically through protein binding or affecting intron regions, and epigenetic regulation exerted by chromatin structures and DNA modifications affecting gene activity. Both mechanisms can be affected by genetic variations in different ways. Concretely, we will explain the effects of SNPs in noncoding regions on DNA methylation, histone modifications, TF binding affinity, alternative splicing, and mRNA stability, as shown in Fig. 3.

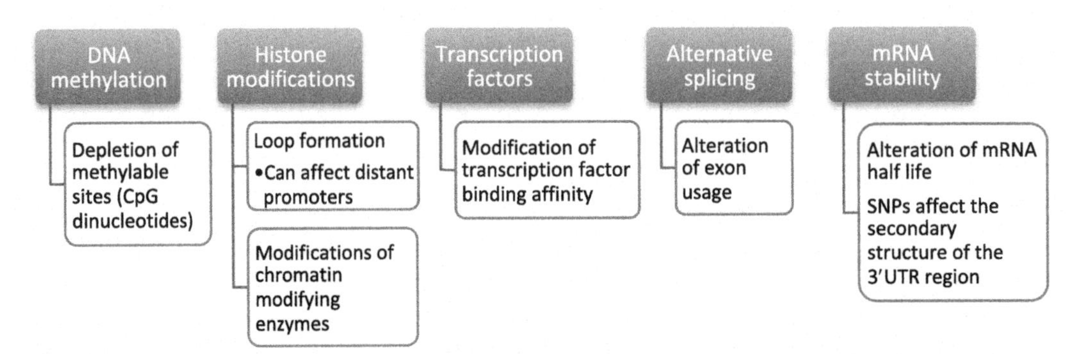

FIG. 3 Different biological mechanisms affected by the presence of noncoding SNPs. SNPs can affect DNA methylation through the depletion of CpG dinucleotides, histone modifications, transcription factor binding, alternative splicing, and mRNA stability.

4.1 DNA Methylation

DNA methylation is the best-characterized epigenetic modification associated with human disease. Such methylation consists of the addition of a methyl group to DNA by DNA methyltransferases (DNMTs) on the 5′-carbon of the pyrimidine ring in cytosine that leads to the formation of 5-methylcytosine (m^5C). DNA methylation is commonly found in regions of at least 200 bp that contain a GC content of more than 55%: CpG islands (Takai and Jones, 2002).

4.1.1 Sequence Specific Regulation of DNA Methylation

Many observations have indicated that genetic variants can exert an effect on DNA methylation, mostly by the depletion of methylable sites such as CpG dinucleotides (Zaina et al., 2010).

Examples of such events include point mutations in the human v-Ki-ras2 Kirsten rat sarcoma viral oncogene homologue (*KRAS*) gene in which G/A (C/T in the opposite strand) transitions take place (Jiang and Zhao, 2006; Mariyama et al., 1989). C/T transitions can lead to the depletion of genome CpG dinucleotides. Thus, regulatory proteins incorrectly recognize the DNA and cannot regulate transcription. Other examples include the destruction of a single CpG nucleotide by a T/C or G/A SNP and the consequent loss or gain of DNA methylation, which has been shown to greatly affect the promoter activity of the gene matrix metalloproteinase 1 (*MMP1*) and potassium-chloride cotransporter 3 (*SLC12A6*; Moser et al., 2009; Wang et al., 2008). The presence of SNPs in both genes has been associated with increased risk of preterm premature rupture of membranes and psychiatric disorders, respectively.

If a certain variant results in the depletion of a methylable site that has spreading effects on adjacent sequences, it is logical to expect the existence of an association between DNA methylation state and effects in a proximal DNA sequence. In fact, a seminal work proved that specific SNP genotypes associate with DNA methylation patterns and possibly act as modifiers of DNA methylation in *cis* (Kerkel et al., 2008). This study showed that allele-specific DNA methylation (ASM) was specifically associated with adjacent SNP genotype in several loci. In the same vein, a more recent study investigated the impact of SNPs located at CpG dinucleotides on ASM (Shoemaker et al., 2010). These results show that CpG dinucleotide-affecting SNPs can determine a major fraction of ASM and hence the disruption of CpG sites through genetic variation can be determinant of differential methylation. The next step is to determine which factors are involved in SNP-operated ASM *cis*-regulation. It has been suggested that a given CpG dinucleotide SNP can impact DNA methylation if it is included in a region normally targeted by epigenetic regulators (Shoemaker et al., 2010).

4.2 Histone Modifications

It has been observed that some of the modification patterns are associated with biological events such as transcription, histone deposition, and even mitosis/meiosis. Such modifications include acetylation, methylation, phosphorylation, B-*N*-acetylglucosamination, deimination, ADP-ribosylation, ubiquitylation, SUMOylation, proline isomerization, and histone-tail clipping. The presence of posttranslational modifications on histone tails affects their ability to bind DNA and thus modifies gene expression (Kouzarides, 2007).

Histone modifications can also be affected by genetic variation. In particular, the presence of SNPs can affect histone marks and alter the accessibility of TFs in different cellular states (McVicker et al., 2013).

4.2.1 *Loop Formation*

Gene expression can be regulated by distal regulatory elements via their interaction with promoters in the 3D structure of the genome. Chromosome conformation capture (3C) technology has helped to demonstrate that regulatory sequences are capable of controlling transcription through the formation of a loop that can bring regulatory regions of target coding genes located hundreds of kilobases away into contact (de Laat and Dekker, 2012; Hughes et al., 2014).

Such interactions have been described in a study by Wright et al., who focused on the SNP rs6983267, located intergenically and notably distal from the nearest gene, *MYC*. The results of this study revealed that rs6983267 takes part in a chromatin loop that allows long-distance interaction with the *c-MYC* promoter. Concretely, the rs6983267 G allele displays an enhancer-like histone mark that accounts for allele-specific c-*MYC* promoter activation (Wright et al., 2010). This SNP has been linked with different types of cancer due to its location (within risk loci 8q24) and its impact on c-MYC levels, such as prostate (Zheng et al., 2007), breast, and colon cancer (Ahmadiyeh et al., 2010; Grisanzio and Freedman, 2010).

4.2.2 *Noncoding SNPs Affect Chromatin-Modifying Enzymes*

Histone modifications can be affected by mutations in regulatory elements of chromatin-modifying enzymes. Thus, it has been observed that mutations play a key role in cancer and many other genetic diseases through the modulation of the chromatin microenvironment (Tsuge et al., 2005). In this study, they observed that a SNP in the regulatory region of *SMYD3* creates a third E2F binding site that results in increased *SMYD3* expression, which is related to carcinogenesis for breast, liver, and colon tissue. They found the allele leading to three binding sites to be higher in individuals suffering from breast cancer, hepatocellular carcinoma, and colorectal cancer (Tsuge et al., 2005).

4.3 Binding of TFs to Regulatory Elements

4.3.1 *TFs and Histone Modifications*

A relationship between TFs involved in altering gene expression via the recruitment of the transcriptional machinery and specific combinations of histone modifications has been observed (Ptashne, 2007, 2013). In this way, noncoding *cis*-regulatory variants that often occur at TF binding sites may affect histone marks.

Variation in noncoding regulatory sequences may affect histone modifications, and the downstream consequences can include chromatin remodeling and transcriptional effects (Kouzarides, 2007). It is still unknown whether histone modifications are a cause or a consequence of gene regulation and which DNA elements direct cell type-appropriate histone marking (Henikoff and Shilatifard, 2016; Jenuwein and Allis, 2001). To this end, studies of genetic variants disrupting TF binding sites can shed light on whether TF binding leads to histone modification or vice versa.

In a study by Rintisch et al., focused on the impact of genetic variation on histone modification patterns, they identified quantitative differences in histone trimethyllysine levels among rat recombinant inbred strains (Rintisch et al., 2014). Their results suggested that the TFs identified, MYC, MAX, and E2F1, among others, directly or indirectly recruit H3K4 methyltransferase or demethylase complexes to the promoters of the differentially modified genes in the heart and liver. However, further experiments are required to confirm the interaction between these TFs and the histone methyltransferase or demethylase complexes.

4.3.2 SNPs Affect TF Binding

As already stated, GWAS studies have found disease-associated SNPs within intergenic regions. Regulatory SNPs (rSNPs) can affect gene regulation through the modification of TF binding affinities to their target genomic sequences (Zuo et al., 2015). The identification of potential rSNPs may provide vital information regarding disease mechanisms, hence much effort has been focused on this field. For instance, Kuosmanen et al. identified the rSNP rs113067944 in the ferritin (*FTL*) promoter region, which, although rare, leads to reduced TF binding for the nuclear factor E2-related factor 2 (NFE2L2), leading to a decrease in transcription (Kuosmanen et al., 2016).

4.4 Alternative Splicing

Alternative splicing of mRNA is the process by which many gene products with distinct functions can potentially be generated from a single coding sequence (Brett et al., 2002). This process has the potential to produce protein isoforms with different biological properties, such as protein interaction partners, subcellular localization, and catalytic ability (Stamm et al., 2005). One study found mutations affecting the splicing machinery directly (Ward and Cooper, 2010); however, splicing can be affected through other mechanisms, giving rise to human diseases (Tazi et al., 2009). Initial research was focused on the impact of exonic mutations on splice-site selection (Cooper and Mattox, 1997; Mardon et al., 1987). Nonetheless, it has been observed that SNPs in both intronic and regulatory regions can alter exon usage and thus might affect alternative splicing (Lu et al., 2012).

In one study, an upstream promoter polymorphism and two intronic SNPs affecting dopamine receptor DRD2 splicing sites were identified by Zhang et al., which are related to cognitive and brain disorders, such as schizophrenia and drug addiction (Zhang et al., 2007). Similarly, Kawase et al. described an intronic SNP in the histocompatibility minor serpin-domain containing (*HMSD*) gene (Kawase et al., 2007). This SNP generates a novel allelic splice variant of *HMSD* coding for a minor histocompatibility antigen that can be targeted in immunotherapy against hematologic malignancies.

4.5 mRNA Stability

The stability of mRNA depends on *cis*-acting sequences within the mRNA (mostly in the 3′-untranslated region, UTR) and on trans-acting RNA-binding proteins (Martin and Ephrussi, 2009). UTRs consist of regulatory elements of genes that control translation and RNA decay. In this way, SNPs in the 3′-UTR may alter the binding of such trans-acting factors and thus alter the mRNA half-life. This can occur in two ways: (1) by preventing their correct function due to their location within the binding sites of these factors or (2) by altering the secondary structure of the 3′-UTR region involved in protein binding site formation.

Boffa et al. studied the effect of SNPs in the 3′-UTR regions of *CPB2* gene (coding for thrombin-activatable fibrinolysis inhibitor, *TAFI*) on mRNA stability (Boffa et al., 2008). The results of this study showed that SNPs in the 3′-UTR regions influence the abundance of mRNA by either affecting the stability of the *CPB2* transcript or altering polyadenylation site selection.

Another study by Halvorsen et al. was focused on the structural consequences that arise from disease-associated mutations in SNPs located in the 3′- and 5′-UTR of genes (Halvorsen et al., 2010). The mutations that affect RNA can have either local or global effects on structural ensembles (Waldispühl and Clote, 2007). If they belong to the latter, a relationship between the RNA structure and the molecular mechanism of the disease can be assumed. One study found that multiple SNPs that alter the mRNA structural ensemble were observed in the UTRs of genes associated with diseases such as β-thalassemia, cartilage-hair hypoplasia, retinoblastoma, chronic obstructive pulmonary disease (COPD), and hyperferritinemia-cataract syndrome. The term "riboSNitch" was proposed to define a regulatory RNA in which a determined SNP has a structural consequence that give rise to a disease phenotype (Halvorsen et al., 2010).

5 DISEASES AFFECTED BY SNPs IN NONCODING REGIONS

Hundreds of disease genetic studies have been performed to identify the genetic background behind disease susceptibility. GWAS have found many noncoding SNPs associated with common diseases such as cancer and asthma (Schaub et al., 2012). It would therefore appear that the role of SNPs in noncoding regions is potentially as important as that of coding regions. Examples of SNPs affecting gene regulation and their associated diseases are given in Table 2.

TABLE 2 Examples of SNPs Affecting Different Mechanisms Associated With Gene Regulation

Feature	Examples of SNPs	Effects	Disease	References
DNA methylation	In *MMP1* and *SLC12A6* promoters	Loss of DNA methylation, no transcriptional regulation	Premature rupture of membranes Psychiatric disorders	Moser et al. (2009) and Wang et al. (2008)
Histone modifications	rs6983267	Alters *MYC* transcription	Carcinogenesis	Wright et al. (2010)
	Regulatory regions of *SMYD3*	Increased *SMYD3* expression	Hepatocellular carcinoma Colorectal cancers	Tsuge et al. (2005)
TF affinity	rs113067944	Decreases binding to NRF2		Kuosmanen et al. (2016)
Alternative splicing	Promoter and two introns of *DRD2*		Schizophrenia and drug addiction	Zhang et al. (2007)
mRNA stability	In 3'-UTR of *CPB2*	mRNA abundance		Boffa et al. (2008)

5.1 Cancer

Cancer has become one of the leading causes of death worldwide, and much research is focused on identifying its molecular bases (Ferlay et al., 2015). GWAS have been used to search for inherited germ line and somatic variants. A summary of important noncoding SNPs related to cancer is given in Table 3.

5.1.1 Roles of Somatic Variants in Cancer

Whole-genome sequencing (WGS) of tumor genomes has provided examples in which somatic variants play a key role in tumorigenesis, such as the gain of TF binding sites. One such example is that of the telomerase reverse transcriptase (*TERT*) gene that encodes for the catalytic subunit of the telomerase enzyme, involved in avoiding apoptosis. The expression of *TERT* is repressed in normal cells; however, it is overexpressed in cancer cells. Such overexpression relies on mutations in the promoter of *TERT* that create new binding motifs for ETS family of TFs. These bind the *TERT* promoter and increase its expression. One such mutation is a disease segregating single-base change located 57bp upstream from the ATG start site, which has been associated with melanoma (Heidenreich et al., 2014).

TABLE 3 Examples of SNPs in Noncoding Regions Associated With Cancer

Type of Variation	Location	Affected Gene/ Detected SNP	Molecular Effects	Effects Upon Regulation	Type of Cancer	References
Somatic	Promoter	*TERT*	Creates new TF binding sites for ETS	Overexpression of *TERT*	Melanoma	Heidenreich et al. (2014)
Germ line	Promoter	*MDM2*	Increased binding of Sp1	Overexpression of *MDM2*: suppression of p53 pathway	Several	Bond and Levine (2007) and Bond et al. (2004)
	Enhancers	rs339331	Increase d*HOXB13* binding	Overexpression of *RFX6*	Prostate	Huang et al. (2014)
		First intron of *LMO1*	Affects binding of GATA	Modified expression of *LMO1*	Neuroblastoma	Oldridge et al. (2015)
	ncRNAs	3′-UTR of *PCM1*	May affect miRNA binding	Modified expression of *PCM1* mRNA	Ovary	Chen et al. (2015)
		miR-27a	Alters processing of pre-miR-27a Increased expression of *HOXA10*	Reduces susceptibility to cancer	Gastric	Yang et al. (2014)

5.1.2 Roles of Germline Variants in Cancer

The link between inherited germ-line variants and complex disorders such as cancer has been studied in many GWAS (Chang et al., 2014). Noncoding regions of the genome appear to be involved in cancer susceptibility, development, and growth (Chen et al., 2014; Maurano et al., 2012). Such noncoding variants have been observed in different regulatory elements, such as promoters, enhancers, and in ncRNAs.

PROMOTER MUTATIONS

The presence of the SNP 309 (T/G) in the *MDM2* promoter has been associated with accelerated tumor formation in many types of cancer (Bond and Levine, 2007; Bond et al., 2004). The mechanism of action is thought to be by increasing the binding affinity of the Sp1 TF, leading to the upregulation of *MDM2*. This *MDM2* gene is a negative regulator of p53 (tumor suppressor), and hence, its overexpression leads to suppression of the p53 pathway, resulting in increased tumor formation.

Additionally, germ-line mutations in the *TERT* promoter have been associated with familial melanoma. In the same manner as somatic mutations, germ-line mutations can create new binding motifs for the ETS TFs that lead to the overexpression of telomerase. Moreover, it has been observed that increased expression of the *ELK1* gene (a TF from the ETS family) occurs in female-specific tissues such as placenta and ovary, and therefore, such mutations have been associated with increased ovarian cancer risk in women with these mutations (Horn et al., 2013).

SNPs IN ENHANCERS

It has been observed that a prostate cancer-associated SNP, rs339331, located in chromosome 6q22, occurs in a cell-type-specific enhancer and induces an increase in *HOXB13* binding (Huang et al., 2014). Consequently, the expression of *RFX6* is increased, and as a result, cell growth is induced, potentially leading to prostate cancer.

Another SNP in a superenhancer, that is, a cluster of enhancers located in close genomic proximity (Pott and Lieb, 2015), located in the first intron of *LMO1*, appears to influence neuroblastoma susceptibility due to the differential binding of the GATA TF and direct modulation of *LMO1* expression in *cis* (Oldridge et al., 2015).

SNPs IN ncRNAs AND THEIR BINDING SITES

In a study performed by Chen et al., miRNA binding sites and 3′-UTRs relating to 6000 cancer-associated genes from 32 patients with ovarian cancer were sequenced, with the aim of finding associated germ-line variants (Chen et al., 2015). They found an enrichment of a variant in the 3′-UTR of the pericentriolar material 1 (*PCM1*) gene in patients with ovarian cancer. It is believed that the variant affects miRNA binding, leading to the differential expression of *PCM1* mRNA (Chen et al., 2015).

While cancer-associated polymorphisms are generally found to be related to an increased risk of developing cancer, they can also reduce susceptibility. For example, an SNP in miR-27a has been shown to alter the processing of the pre-miR-27a precursor, leading to expression of its target gene, homeobox A10 (*HOXA10*), and consequently reducing gastric cancer susceptibility (Yang et al., 2014).

GERMLINE-SOMATIC VARIANTS INTERPLAY

The effects of somatic variants may be dependent on the underlying genetic background of an individual. For example, it has been observed that bladder cancer patients with both somatic lesions in the *TERT* promoter and the germinal SNP (rs2853669) showed better survival rates (Rachakonda et al., 2013). The mechanism may be due to the disruption of an ETS2-binding site. This example shows the complex relationship between regulatory variants and cancer susceptibility, oncogenesis, and patient survival (Khurana et al., 2016).

5.2 Asthma

Asthma is a complex syndrome characterized by chronic inflammation in the lower airways resulting in variable airflow obstruction and bronchial hyperresponsiveness. It is a rather common disease, and much genetic study has been made via GWAS. However, replication of initial findings has been complicated by the heterogeneity of the syndrome and the large number of endophenotypes observed (Simon et al., 2012). Further complexity is posed by the presence of epigenetic factors and their interactions with asthma genetics and the environment. However, several genes have been characterized, such as *ORMDL3*, *CD14*, and *IL18*. Moreover, analysis of the noncoding genome has helped us to investigate the mechanisms by which they may be involved in the disease. Several noncoding region SNPs related to asthma are summarized in Table 4.

5.2.1 *ORMDL3*

The orosomucoid 1-like 3 (*ORMDL3*) gene appears to play a key role in allergic inflammation and antiviral responses (Carreras-sureda et al., 2013) and is considered a key asthma candidate gene present in locus 17q21 (Galanter et al., 2008). Several studies have been carried out to investigate the mechanisms by which it may cause asthma. One study identified two main SNPs associated with asthma susceptibility: rs8076131 and rs4065275. Both variants are in a putative *ORMDL3* promoter region, 500bp upstream of the coding region in intron 2, and near the translation start site for exon 2 (Schedel et al., 2015). It is suspected that these genetic variants affect promoter activity, TF binding, and thus changes in *ORMDL3* expression and IL-4/IL13 cytokine levels (Schedel et al., 2015).

5.2.2 *CD14*

The cluster of differentiation 14 (*CD14*) gene encodes for a multifunctional receptor present in macrophages and monocytes and is involved in various innate immune-system processes (Ulevitch and Tobias, 1995). Moreover, it is located on chromosome 5q31.3, a region that has been associated with asthma in genome-wide linkage studies (Bouzigon et al., 2010).

In a recent study by Nabih et al., a specific *CD14* SNP, rs2569190 (G→A), was associated with atopic childhood asthma and with variations of IgE levels in Egyptian children (Nabih et al., 2016). This study suggests an important role for *CD14* SNPs in the pathogenesis of atopic childhood asthma.

TABLE 4 Examples of SNPs Associated With Asthma

SNP	Location	Gene	Molecular Effects	Transcriptional Effects	References
rs8076131	Promoter	*ORMDL3*	Abolishes the ability of USF1/USF2 complex to bind	Increased expression of *ORMDL3*	Schedel et al. (2015)
rs4065275			Changes affinity of SP family of TF		
rs2569190	5q31.3 Chromosome	*CD14*	May alter IgE levels		Nabih et al. (2016)
rs1946518	Promoter	*IL18*	May affect the *CREB* binding site	May decrease the production of IL-18	Ma et al. (2012)
rs187238	Promoter		Changes the *H4TF-1* binding site		Giedraitis et al. (2001)

5.2.3 *Interleukin-18*

Interleukin-18 (IL-18) is a cytokine related to the immune response. Kim et al. found associations between SNPs within the promoter region of *IL-18* and respiratory symptoms in individuals with Baker's asthma. In particular, rs1946518 and rs187238 were shown to be related (Kim et al., 2012). In a study performed by Ma et al., it was proposed that these polymorphisms are associated with increased asthma risk (Ma et al., 2012). Rs1946518 is thought to play a crucial role in asthma development via the disruption of a potential cAMP-responsive element protein binding site, while rs187238 may modify an H4TF-1 nuclear factor binding site (Giedraitis et al., 2001; Haus-Seuffert and Meisterernst, 2000). Additionally, Pawlik et al. (2007) showed that the ability to produce IL-18 by monocytes was higher in individuals with the presence of rs187238 in a Polish population.

6 ANNOTATING THE GENOME WITH NONCODING GENOMIC ELEMENTS

As mentioned, GWAS have helped improve our understanding of complex diseases, and almost 90% of disease-associated variants identified through GWAS are located in noncoding regions of the genome (Edwards et al., 2013). Moreover, we have described several of the myriad ways by which a noncoding variant can contribute to disease. However, given a disease-associated noncoding SNP, it is by no means straightforward to establish how it may affect phenotype. As a first approach, one can use functional genomic element data and look for overlap between genomic elements and the SNP of interest.

Several tools have been developed in recent years for this purpose, using a variety of data sources to annotate the genome with details and positions of functional elements. Although the number of tools is still lower than those for coding regions, it is growing rapidly (Khurana et al., 2016). Most of these tools use data from the ENCODE project (Encyclopedia of DNA Elements, https://www.encodeproject.org/; Bernstein et al., 2012) for annotation. This includes data related to the presence of DNA methylation, TF binding, histone modifications, and more, produced using a wide range of sequencing-based techniques and for many different human cell lines. Other projects include the Roadmap Epigenomics project (Bernstein et al., 2010), which concentrates on DNA methylation and histone marks. Some annotation tools also use data from the MotifMap project, which combines binding motif data with comparative genomic approaches to predict functional elements (http://motifmap.ics.uci.edu/; Daily et al., 2011). Other data sources employed by these tools include the Gene Expression Omnibus (GEO, https://www.ncbi.nlm.nih.gov/geo/; Barrett et al., 2013), containing gene-expression data and information related to human genomic variants, such as dbSNP (https://www.ncbi.nlm.nih.gov/SNP/; Sherry et al., 2001) and ClinVar (https://www.ncbi.nlm.nih.gov/clinvar/; Landrum et al., 2016), which maps variants to disease and the 1000 Genomes Project (http://www.internationalgenome.org/; Siva, 2008). Here, we describe six of the most well-known tools (summarized in Table 5).

TABLE 5 Summary of the Different Tools Described for the Annotation of Regulatory SNPs

Tool	Type of Tool	Key Features	Annotations Sources	Output
RegulomeDB	Web server	Annotates and score variants that overlap with regulatory elements	ENCODE Roadmap Epigenomics Project GEO	Table with ranked scores for all SNPs overlapping with regulatory regions
HaploRegDB	Web server	Determines regulatory variants in haplotype blocks, using annotations of regulatory elements and variants in LD	1000 Genomes Project ENCODE Roadmap Epigenomics Project	Table with found regulatory variants in haplotype blocks and their annotations
FunciSNP	R software package	Assign functionality to unknown variants in noncoding regions	1000 Genomes Project GWAS Catalog ENCODE	Table with the combination of tag SNPs and their assigned annotations
Enlight	Web server	Functional annotation of GWAS SNPs by combining external annotation resources	Several databases from the UCSC Genome Browser (through ANNOVAR functional annotation tool)	Graphic representations of the input SNP and associated regulatory elements
GWAS3D	Web server	Analysis of regulatory variants that potentially disrupt the control of gene regulation by affecting transcription factor binding sites	1000 Genomes Project HapMap dbSNP ENCODE GEO	Visual representation in circle plot, with relations between SNPs and annotations
GREGOR	Software tool developed in Perl	Prioritization and enrichment analysis of regulatory variants that affect transcriptional regulation	1000 Genomes Project ENCODE	Table with enrichment results, with the number of SNPs in LD with regulatory regions

6.1 RegulomeDB

RegulomeDB (http://regulomedb.org/) is an integrated regulatory element database that combines data from multiple resources, including annotations from ENCODE, GEO, and the Roadmap Epigenomics project (Boyle et al., 2012), for the characterization of regulatory elements. The resource includes information on various experimentally validated regulatory regions, including TF binding sites and chromatin state for approximately 100 cell types, and information on expression quantitative trait loci (eQTL). The different data are combined to produce a score based on the number of regulatory elements that a variant overlaps with, which can be interpreted as how likely the SNP is to affect an important regulatory element, and thus can be used for prioritization of, for example, a list of disease-associated SNPs obtained from a GWAS.

6.2 HaploReg

HaploReg (http://compbio.mit.edu/HaploReg) is a bioinformatics tool and database that uses information from different resources to characterize variants (including SNPs and small insertions and deletions, indels) as regulatory, helping the user to determine the functional impact of their query variants (Ward and Kellis, 2012). Instead of analyzing the coordinates of all variants separately, HaploReg combines polymorphisms into haplotype blocks, which can be useful for fine-mapping studies of GWAS results and can improve visualization. The database contains variant annotations from the 1000 Genomes Project, which are used for calculating LD between different SNPs. This can help for predicting the effects of unknown variants that are in LD with annotated ones. In addition, HaploReg provides the user with sequence conservation data from Genomic Evolutionary Rate Profiling (GERP) and SiPhy, SNP functional annotations from dbSNP, gene annotations from RefSeq, and regulatory information from ENCODE and the Roadmap Epigenomics Project (Ward and Kellis, 2015). Starting with a list of variants, HaploReg combines them into haplotype blocks, calculates LD for predicting their effects depending on whether they are within regulatory regions, and produces a summary of the putative regulatory elements that the variants overlap with.

6.3 FunciSNP

FunciSNP is a bioinformatics software package for assigning functionality to variants (SNPs) within genomic regions and associated with complex diseases (Coetzee et al., 2012). Developed in R/Bioconductor, this package integrates information from different sources, including variant information from the 1000 Genomes Project and the GWAS Catalog, and information on functional elements, through sequence-based chromatin maps from ENCODE. From a list of GWAS SNPs and a ChIP-seq peak, FunciSNP can integrate all this information, identify putative functional tag SNPs, and associate them with phenotypes. Unlike the previous two tools, RegulomeDB and HaploReg, FunciSNP has no web portal but requires the user to know basic R for its use. However, this is in some ways an advantage since large data queries can be made more quickly than using web portals and analyses can be saved as scripts and rerun if necessary.

6.4 Enlight

Enlight (http://enlight.wglab.org/) is a web-based tool that annotates GWAS results with functional annotations (Guo et al., 2015). It produces a graphic representation of the input SNPs alongside all the functional elements that it overlaps with, including those that are in LD. The annotations it uses are generated using another bioinformatics tool, ANNOVAR (Wang et al., 2010), which collects functional information to annotate variants and analyze their effects, especially for genes, eQTL, conserved regions, and some predicted regulatory elements, such as TF binding sites. This annotation tool is particularly useful for result interpretation due to its visual output, which can be useful for the identification of putative pathogenic regulatory variants, since it allows the recognition of epigenetic features that overlap with SNPs and other functional annotations in LD with the SNP locations.

6.5 GWAS3D

GWAS3D (http://jjwanglab.org/gwas3d) is a web server for the analysis of putative regulatory disease-associated variants related to the control of gene regulation (TF binding sites), using annotations of conserved regions, chromatin states, and epigenetic modifications, as well as inferred regulatory elements (Li et al., 2013). The different annotations available in this web server are collected from ENCODE and GEO and combined with predicted elements found using ChromHMM (Ernst and Kellis, 2012), a software for the characterization of chromatin states. Moreover, for calculating the binding affinity of TFs, the web server uses variant information from the 1000 Genomes Project, HapMap, and dbSNP. Starting from an input set of SNPs, GWAS3D computes the probability of these variants to affect a regulatory pathway. GWAS3D also offers a visual representation: it can be used to produce circle graphs that relate the input variants to the annotated information and plots with the genomic location of input variants. In comparison to similar tools, this resource integrates a larger collection of genomic features that allows the user to characterize putative regulatory variants related to traits.

6.6 GREGOR

There are cases where the variants found in GWAS are most frequently located in certain regions of the genome. This could indicate that changes in these locations have a greater probability of affecting functional elements. In this case, an evaluation of the variant enrichment could help to prioritize the putative regulatory variants that influence normal transcriptional regulation. With this aim, the genomic regulatory elements and GWAS overlap algorithm (GREGOR) uses a large collection of regulatory information for evaluating the enrichment of variants in genomic regions, in order to fully characterize and determine functionality (Schmidt et al., 2015). This software tool uses information from the 1000 Genomes Project and DNase I hypersensitive sites from ENCODE as indicators of regulatory elements and methylation and TF binding site information. This approach can prioritize the effect of variants that overlap with different regulatory elements of different cell types.

7 rSNP CLASSIFIERS BASED ON MACHINE-LEARNING ALGORITHMS

The previous methods use information on noncoding genomic elements and related data to allow the user to annotate variants and provide clues as to their potential function, helping guide future experiments. Some of them, such as RegulomeDB even provide a score based on the types of genomic elements overlapping a given variant (Boyle et al., 2012). However, these scores are often somewhat arbitrary. There is another type of tool that uses noncoding element data not merely to annotate a SNP, but to build a predictor to determine whether it is likely to be pathogenic. These are known as classifiers and are generally based on machine-learning algorithms, following to some degree the procedures outlined in Fig. 4. Firstly, the developer decides on how to build the classifier, choosing the type of learning model and which genomic features to use for learning (building the classifier). These typically include functional information such as DNA conservation sites, histone modifications, methylation, and TF binding sites, downloaded from different data sources, ENCODE being one of the most recurrent. In addition, the developer must create a pathogenic and nonpathogenic variant dataset. The developer thus creates a matrix of variants and features, labeling each variant as pathogenic or benign. Then comes the training step; in this process, the model is built based on the characteristics of the pathogenic versus benign variants. Once the model has been built, the developer gives the classifier a validation (test) set, containing pathogenic and benign variants that were not used for model construction. The classifier then decides which of the test data variant are likely to be pathogenic and which are likely to be benign. By comparing the results of the classifier for this validation set with their true labels, it is possible to estimate performance, by calculating sensitivity, specificity, and other metrics.

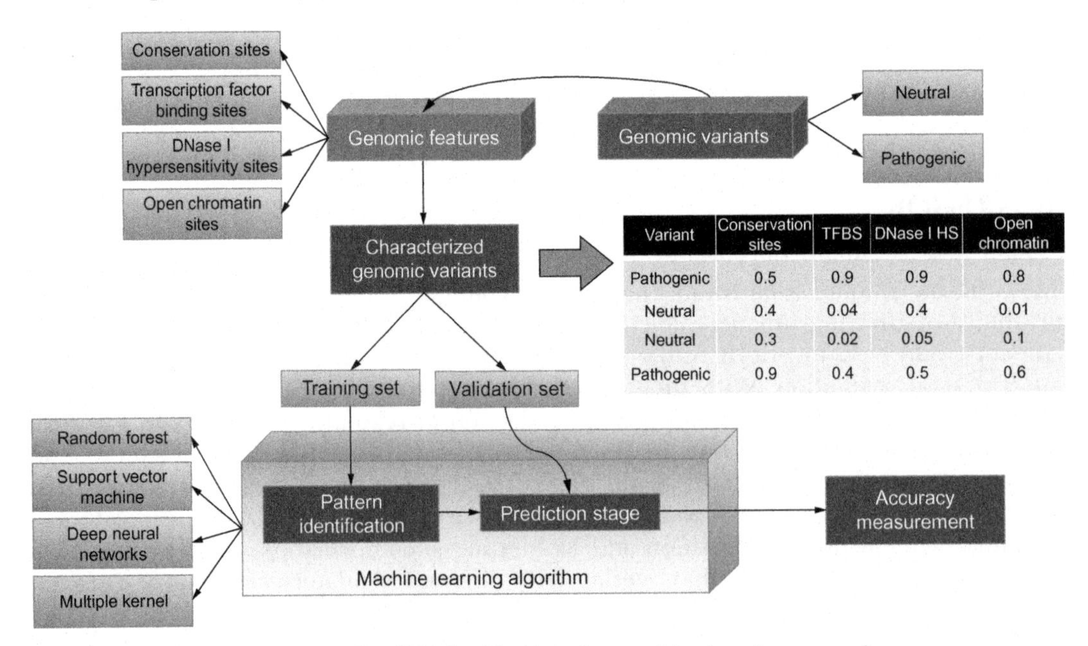

Variant	Conservation sites	TFBS	DNase I HS	Open chromatin
Pathogenic	0.5	0.9	0.9	0.8
Neutral	0.4	0.04	0.4	0.01
Neutral	0.3	0.02	0.05	0.1
Pathogenic	0.9	0.4	0.5	0.6

FIG. 4 General schema for noncoding SNP classifiers based on machine-learning approaches.

These approximations are useful to determine the role of unknown variants and for further characterizing their functional effects. There are numerous types of tools that implement these types of algorithms; we will briefly describe those that are most used to predict the impact of regulatory variants.

7.1 GWAVA

The genome-wide annotation of variants (GWAVA) is a tool that prioritizes both coding and noncoding variants, by combining information from a wide range of annotations (Ritchie et al., 2014). This tool combines gene annotations with regulatory features for identifying putative functional variants: it finds variants that are associated with regulatory elements using a random forest prediction system, which aids to distinguish deleterious variants from benign. GWAVA combines genomic feature annotations, including conservation scores calculated using the GERP method (Cooper et al., 2005) and regulatory elements from ENCODE. To make predictions, it uses variants from the 1000 Genomes Project as neutral/benign and variants from the Human Gene Mutation Database (HGMD) as pathogenic. GWAVA prioritizes variants using a scoring system that allows researchers to determine if their input variants play a role in a disease.

7.2 CADD

The combined annotation-dependent depletion (CADD, http://cadd.gs.washington.edu/; Kircher et al., 2014) method integrates data from a wide-range of genomic annotation sources including ENCODE, conservation data for multiple species close to human, and variant annotation from the Ensembl Variant Effect Predictor resource (McLaren et al., 2016). Using a support vector machine (SVM) algorithm, they calculate a combined score for a given variant, taking into account the different annotation sources. With this approach, the framework can classify input variants (including SNPs and indels) for quantifying and prioritizing their deleteriousness in both coding and noncoding regions. In addition, the CADD website also offers a list of precomputed C scores for variants from different sources, such as the 1000 Genome Project and Illumina BeadChip variants.

7.3 DANN

Since the SVM algorithm used in CADD can only capture linear relationships between different annotations, the Deleterious Annotation of genetic variants using Neural Networks (DANN) was developed to improve the performance of CADD taking into account nonlinear relationships (Quang et al., 2015). DANN uses artificial deep neural networks for predicting nonlinear relationships using the same input data as CADD: annotated alleles and simulated variants.

7.4 FATHMM-MKL

The machine-learning approach FATHMM-MKL (http://fathmm.biocompute.org.uk/; Shihab et al., 2015) uses a multiple kernel (MK) algorithm for predicting the functional impact of coding and regulatory variants. This method uses functional annotations from ENCODE and nucleotide-based sequence conservation measures. During training, the algorithm weighs these annotations according to their significance and creates matrices for performing the MK learning. As in the case of GWAVA, this approach uses variants from the HGMD as pathogenic and variants from the 1000 Genomes Project as controls.

8 FUTURE LINES

We have outlined the importance of accurately determining the links between regulatory regions and their target genes in order to identify the functional effects of variants in the noncoding genome. Besides the techniques described here, many novel and pioneering techniques with this goal are also starting to gain traction. An increasingly important approach is the use of high-throughput chromatin conformation capture technologies such as Hi-C to analyze the physical interactions between distant regions of the genome. Such technologies along with network approaches will help understand the role played by noncoding variants in a variety of illnesses. Additionally, in vivo and in vitro assays focused on analyzing genes associated with disease could help to uncover the functional context of complex diseases, leading to new therapies and preventive strategies.

Unfortunately, even though GWAS have identified thousands of SNPs associated with common diseases, they only detect common variations that account for a small fraction of disease heritability. Therefore, many of the rarer variants underlying phenotypic variation remain unknown.

Next-generation sequencing (NGS) technologies are starting to become popular as they open a new window on heritability by considering the contribution of rare variants. NGS technologies can detect millions of novel rare variants easily and quickly. However, the technology also presents several issues such as potential sequencing errors. Moreover, whole-genome sequencing at deep coverage still remains relatively expensive to perform on a large number of individuals, which may be necessary when considering rare variants (Luo et al., 2011).

Despite such difficulties, sequencing technologies are improving rapidly, and soon, nucleotide and structural variation patterns for an individual will be produced easily and cheaply. Some improvements will come from the use of strategies consisting of examining the less common allelic spectrum, exploiting genetic pleiotropy, fine-mapping loci with biological relevance, and applying biological insight to the discovered associations. These improvements will allow us to unveil the missing heritability unexplained by current GWAS findings. Additionally, the complete genetic mechanisms involved in complex diseases will be deciphered, at both the coding and noncoding levels.

The usage of NGS technologies to uncover disease-related noncoding variants and their putative mechanisms entails challenges associated with data storage and manipulation, quality control, and data analysis and likely involves merging sequencing data with data from transcriptomic, proteomic, phenotypic, and environment studies. Integrating all these levels of complex biomedical information will be vital for the future of human genetics and allow us to connect SNPs in noncoding regions with specific diseases. Although the technological challenges are not insignificant, the effort will be worth it and will lead to improvements in our understanding of disease, potentially opening up new pathways for their effective, personalized treatment.

Acknowledgments

The study was funded by the Institute of Health "Carlos III" of the Ministry of Economy and Competitiveness: RETICS ARADyAL (RD16/0006/0001 RD16/0006/0003 and RD16/0006/0011) and Sara Borrell Program (CD14/00242); Elena Rojano is a researcher from the Plan de Formación de Personal Investigador (FPI) supported by the Andalusian Government. Grants were cofunded by the European Regional Development Fund (SAF2016-78041-C2-1-R; CTS-486). The CIBERER is an initiative from Carlos III.

References

Ahmadiyeh, N., Pomerantz, M.M., Grisanzio, C., Herman, P., Jia, L., Almendro, V., et al., 2010. 8q24 prostate, breast, and colon cancer risk loci show tissue-specific long-range interaction with MYC. Proc. Natl. Acad. Sci. USA 107 (21), 9742–9746. https://doi.org/10.1073/pnas.0910668107.

Barrett, T., Wilhite, S.E., Ledoux, P., Evangelista, C., Kim, I.F., Tomashevsky, M., et al., 2013. NCBI GEO: archive for functional genomics data sets—update. Nucleic Acids Res. 41 (Database issue), D991–D995. https://doi.org/10.1093/nar/gks1193.

Bernstein, B.E., Stamatoyannopoulos, J.A., Costello, J.F., Ren, B., Milosavljevic, A., Meissner, A., et al., 2010. The NIH roadmap epigenomics mapping consortium. Nat. Biotechnol. 28 (10), 1045–1048. https://doi.org/10.1038/nbt1010-1045.

Bernstein, B.E., Birney, E., Dunham, I., Green, E.D., Gunter, C., Snyder, M., 2012. An integrated encyclopedia of DNA elements in the human genome. Nature 489 (7414), 57–74. https://doi.org/10.1038/nature11247.

Boffa, M.B., Maret, D., Hamill, J.D., Bastajian, N., Crainich, P., Jenny, N.S., et al., 2008. Effect of single nucleotide polymorphisms on expression of the gene encoding thrombin-activatable fibrinolysis inhibitor: a functional analysis. Blood 111 (1), 183–189. https://doi.org/10.1182/blood-2007-03-078543.

Bond, G.L., Levine, A.J., 2007. A single nucleotide polymorphism in the p53 pathway interacts with gender, environmental stresses and tumor genetics to influence cancer in humans. Oncogene 26 (9), 1317–1323. https://doi.org/10.1038/sj.onc.1210199.

Bond, G.L., Hu, W., Bond, E.E., Robins, H., Lutzker, S.G., Arva, N.C., et al., 2004. A single nucleotide polymorphism in the MDM2 promoter attenuates the p53 tumor suppressor pathway and accelerates tumor formation in humans. Cell 119 (5), 591–602. https://doi.org/10.1016/j.cell.2004.11.022.

Bouzigon, E., Forabosco, P., Koppelman, G.H., Cookson, W.O.C.M., Dizier, M.-H., Duffy, D.L., et al., 2010. Meta-analysis of 20 genome-wide linkage studies evidenced new regions linked to asthma and atopy. Eur. J. Hum. Genet. 18 (6), 700–706. https://doi.org/10.1038/ejhg.2009.224.

Boyle, A.P., Hong, E.L., Hariharan, M., Cheng, Y., Schaub, M.A., Kasowski, M., et al., 2012. Annotation of functional variation in personal genomes using RegulomeDB. Genome Res. 22 (9), 1790–1797. https://doi.org/10.1101/gr.137323.112.

Brett, D., Pospisil, H., Valcárcel, J., Reich, J., Bork, P., 2002. Alternative splicing and genome complexity. Nat. Genet. 30 (1), 29–30. https://doi.org/10.1038/ng803.

Bush, W.S., Moore, J.H., 2012. Chapter 11: genome-wide association studies. PLoS Comput. Biol. 8 (12), e1002822. https://doi.org/10.1371/journal.pcbi.1002822.

Carreras-sureda, A., Cantero-recasens, G., Rubio-moscardo, F., Kiefer, K., Peinelt, C., Niemeyer, B.A., et al., 2013. ORMDL3 modulates store-operated calcium entry and lymphocyte activation. Hum. Mol. Genet. 22 (3), 519–530. https://doi.org/10.1093/hmg/dds450.

Chang, C.Q., Yesupriya, A., Rowell, J.L., Pimentel, C.B., Clyne, M., Gwinn, M., et al., 2014. A systematic review of cancer GWAS and candidate gene meta-analyses reveals limited overlap but similar effect sizes. Eur. J. Hum. Genet. 22 (3), 402–408. https://doi.org/10.1038/ejhg.2013.161.

Chen, C., Chang, I.-S., Hsiung, C.A., Wasserman, W.W., 2014. On the identification of potential regulatory variants within genome wide association candidate SNP sets. BMC Med. Genet. 7 (1), 1–15. https://doi.org/10.1186/1755-8794-7-34.

Chen, X., Paranjape, T., Stahlhut, C., McVeigh, T., Keane, F., Nallur, S., et al., 2015. Targeted resequencing of the microRNAome and 3′UTRome reveals functional germline DNA variants with altered prevalence in epithelial ovarian cancer. Oncogene 34 (16), 2125–2137. https://doi.org/10.1038/onc.2014.117.

Coetzee, S.G., Rhie, S.K., Berman, B.P., Coetzee, G.A., Noushmehr, H., 2012. FunciSNP: an R/bioconductor tool integrating functional non-coding data sets with genetic association studies to identify candidate regulatory SNPs. Nucleic Acids Res. 40 (18), e139. https://doi.org/10.1093/nar/gks542.

Cooper, T.A., Mattox, W., 1997. The regulation of splice-site selection, and its role in human disease. Am. J. Hum. Genet. 61 (2), 259–266. https://doi.org/10.1086/514856.

Cooper, G.M., Stone, E.A., Asimenos, G., Green, E.D., Batzoglou, S., Sidow, A., 2005. Distribution and intensity of constraint in mammalian genomic sequence. Genome Res. 15 (7), 901–913. https://doi.org/10.1101/gr.3577405.

Daily, K., Patel, V.R., Rigor, P., Xie, X., Baldi, P., 2011. MotifMap: integrative genome-wide maps of regulatory motif sites for model species. BMC Bioinf. 12, 495. https://doi.org/10.1186/1471-2105-12-495.

de Laat, W., Dekker, J., 2012. 3c–Based technologies to study the shape of the genome. Methods 58 (3), 189–191. https://doi.org/10.1016/j.ymeth.2012.11.005.

Durbin, R.M., Altshuler, D.L., Durbin, R.M., Abecasis, G.R., Bentley, D.R., Chakravarti, A., et al., 2010. A map of human genome variation from population-scale sequencing. Nature 467 (7319), 1061–1073. https://doi.org/10.1038/nature09534.

Edwards, S.L., Beesley, J., French, J.D., Dunning, A.M., 2013. Beyond GWASs: illuminating the dark road from association to function. Am. J. Hum. Genet. 93 (5), 779–797. https://doi.org/10.1016/j.ajhg.2013.10.012.

Emahazion, T., Feuk, L., Jobs, M., Sawyer, S.L., Fredman, D., St Clair, D., et al., 2001. SNP association studies in Alzheimer's disease highlight problems for complex disease analysis. Trends Genet. 17 (7), 407–413. https://doi.org/10.1016/S0168-9525(01)02342-3.

Ernst, J., Kellis, M., 2012. ChromHMM: automating chromatin-state discovery and characterization. Nat. Methods 9 (3), 215–216. https://doi.org/10.1038/nmeth.1906.

Ernst, J., Kheradpour, P., Mikkelsen, T., Shoresh, N., Ward, L., Epstein, C., et al., 2011. Mapping and analysis of chromatin state dynamics in nine human cell types. Nature 473 (7345), 1–9. https://doi.org/10.1038/nature09906.

Faham, M., Baharloo, S., Tomitaka, S., DeYoung, J., Freimer, N.B., 2001. Mismatch repair detection (MRD): high-throughput scanning for DNA variations. Hum. Mol. Genet. 10 (16), 1657–1664. https://doi.org/10.1093/hmg/10.16.1657.

Fareed, M., Afzal, M., 2013. Single nucleotide polymorphism in genome-wide association of human population: a tool for broad spectrum service. Egypt. J. Hum. Genet. 14 (2), 123–134. https://doi.org/10.1016/j.ejmhg.2012.08.001.

Feero, W., Guttmacher, A., 2010. Genome wide association studies and assessment of the risk of disease. N. Engl. J. Med. 363 (2), 166–176. https://doi.org/10.1056/NEJMra0905980.

Ferguson, J.A., Steemers, F.J., Walt, D.R., 2000. High-density fiber-optic DNA random microsphere array. Anal. Chem. 72 (22), 5618–5624. https://doi.org/10.1021/ac0008284.

Ferlay, J., Soerjomataram, I., Dikshit, R., Eser, S., Mathers, C., Rebelo, M., et al., 2015. Cancer incidence and mortality worldwide: sources, methods and major patterns in GLOBOCAN 2012. Int. J. Cancer 136 (5), E359–E386. https://doi.org/10.1002/ijc.29210.

Galanter, J., Choudhry, S., Eng, C., Nazario, S., Rodríguez-Santana, J.R., Casal, J., et al., 2008. ORMDL3 gene is associated with asthma in three ethnically diverse populations. Am. J. Respir. Crit. Care Med. 177 (11), 1194–1200. https://doi.org/10.1164/rccm.200711-1644OC.

Giedraitis, V., He, B., Huang, W.-X., Hillert, J., 2001. Cloning and mutation analysis of the human IL-18 promoter: a possible role of polymorphisms in expression regulation. J. Neuroimmunol. 112 (1–2), 146–152. https://doi.org/10.1016/S0165-5728(00)00407-0.

Grisanzio, C., Freedman, M.L., 2010. Chromosome 8q24-associated cancers and MYC. Genes Cancer 1 (6), 555–559. https://doi.org/10.1177/1947601910381380.

Gunderson, K.L., Kuhn, K.M., Steemers, F.J., Ng, P., Murray, S.S., Shen, R., 2006. Whole-genome genotyping of haplotype tag single nucleotide polymorphisms. Pharmacogenomics 7 (4), 641–648. https://doi.org/10.2217/14622416.7.4.641.

Guo, Y., Conti, D.V., Wang, K., 2015. Enlight: web-based integration of GWAS results with biological annotations. Bioinformatics 31 (2), 275–276. https://doi.org/10.1093/bioinformatics/btu639.

Halvorsen, M., Martin, J.S., Broadaway, S., Laederach, A., 2010. Disease-associated mutations that alter the RNA structural ensemble. PLoS Genet. 6 (8), 1–11. https://doi.org/10.1371/journal.pgen.1001074.

Haus-Seuffert, P., Meisterernst, M., 2000. Mechanisms of transcriptional activation of cAMP-responsive element-binding protein CREB. Mol. Cell. Biochem. 212 (1), 5–9. https://doi.org/10.1023/A:1007111818628.

Heidenreich, B., Rachakonda, P.S., Hemminki, K., Kumar, R., 2014. TERT promoter mutations in cancer development. Curr. Opin. Genet. Dev. 24, 30–37. https://doi.org/10.1016/j.gde.2013.11.005.

Henikoff, S., Shilatifard, A., 2016. Histone modification: cause or cog? Trends Genet. 27 (10), 389–396. https://doi.org/10.1016/j.tig.2011.06.006.

Hirschhorn, J.N., Daly, M.J., 2005. Genome-wide association studies for common diseases and complex traits. Nat. Rev. Genet. 6 (2), 95–108. https://doi.org/10.1038/nrg1521.

Horn, S., Figl, A., Rachakonda, P.S., Fischer, C., Sucker, A., Gast, A., et al., 2013. TERT promoter mutations in familial and sporadic melanoma. Science 339 (6122), 959–961. https://doi.org/10.1126/science.1230062.

Huang, Q., Whitington, T., Gao, P., Lindberg, J.F., Yang, Y., Sun, J., et al., 2014. A prostate cancer susceptibility allele at 6q22 increases RFX6 expression by modulating HOXB13 chromatin binding. Nat. Genet. 46 (2), 126–135. https://doi.org/10.1038/ng.2862.

Hughes, J.R., Roberts, N., McGowan, S., Hay, D., Giannoulatou, E., Lynch, M., et al., 2014. Analysis of hundreds of cis-regulatory landscapes at high resolution in a single, high-throughput experiment. Nat. Genet. 46 (2), 205–212. https://doi.org/10.1038/ng.2871.

Jenuwein, T., Allis, C.D., 2001. Translating the histone code. Science 293 (5532), 1074–1080. https://doi.org/10.1126/science.1063127.

Jiang, C., Zhao, Z., 2006. Mutational spectrum in the recent human genome inferred by single nucleotide polymorphisms. Genomics 88 (5), 527–534. https://doi.org/10.1016/j.ygeno.2006.06.003.

Kawase, T., Akatsuka, Y., Torikai, H., Morishima, S., Oka, A., Tsujimura, A., et al., 2007. Alternative splicing due to an intronic SNP in HMSD generates a novel minor histocompatibility antigen. Blood 110 (3), 1055–1063. https://doi.org/10.1182/blood-2007-02-075911.

Kerkel, K., Spadola, A., Yuan, E., Kosek, J., Jiang, L., Hod, E., et al., 2008. Genomic surveys by methylation-sensitive SNP analysis identify sequence-dependent allele-specific DNA methylation. Nat. Genet. 40 (7), 904–908. https://doi.org/10.1038/ng.174.

Khurana, E., Fu, Y., Chakravarty, D., Demichelis, F., Rubin, M.A., Gerstein, M., 2016. Role of non-coding sequence variants in cancer. Nat. Rev. Genet. 17 (2), 93–108. https://doi.org/10.1038/nrg.2015.17.

Kim, S., Misra, A., 2007. SNP genotyping: technologies and biomedical applications. Annu. Rev. Biomed. Eng. 9, 289–320. https://doi.org/10.1146/annurev.bioeng.9.060906.152037.

Kim, S.H., Hur, G.Y., Jin, H.J., Choi, H., Park, H.S., 2012. Effect of interleukin-18 gene polymorphisms on sensitization to wheat flour in bakery workers. J. Korean Med. Sci. 27 (4), 382–387. https://doi.org/10.3346/jkms.2012.27.4.382.

Kircher, M., Witten, D.M., Jain, P., O'Roak, B.J., Cooper, G.M., Shendure, J., et al., 2014. A general framework for estimating the relative pathogenicity of human genetic variants. Nat. Genet. 46 (3), 310–315. https://doi.org/10.1038/ng.2892.

Kouzarides, T., 2007. Chromatin modifications and their function. Cell 128 (4), 693–705. https://doi.org/10.1016/j.cell.2007.02.005.

Kuosmanen, S.M., Viitala, S., Laitinen, T., Peräkylä, M., Pölönen, P., Kansanen, E., et al., 2016. The effects of sequence variation on genome-wide NRF2 binding- new target genes and regulatory SNPs. Nucleic Acids Res. 44 (4), 1760–1775. https://doi.org/10.1093/nar/gkw052.

Kwok, P.-Y., Gu, Z., 1999. Single nucleotide polymorphism libraries: why and how are we building them? Mol. Med. Today 5 (12), 538–543. https://doi.org/10.1016/S1357-4310(99)01601-9.

Landrum, M.J., Lee, J.M., Benson, M., Brown, G., Chao, C., Chitipiralla, S., et al., 2016. ClinVar: public archive of interpretations of clinically relevant variants. Nucleic Acids Res. 44 (D1), D862–D868. https://doi.org/10.1093/nar/gkv1222.

Lee, S., Abecasis, G.R., Boehnke, M., Lin, X., 2014. Rare-variant association analysis: study designs and statistical tests. Am. J. Hum. Genet. 95 (1), 5–23. https://doi.org/10.1016/j.ajhg.2014.06.009.

Li, M.J., Wang, L.Y., Xia, Z., Sham, P.C., Wang, J., 2013. GWAS3D: detecting human regulatory variants by integrative analysis of genome-wide associations, chromosome interactions and histone modifications. Nucleic Acids Res. 41 (Web Server issue), 150–158. https://doi.org/10.1093/nar/gkt456.

Li, M.J., Yan, B., Sham, P.C., Wang, J., 2014. Exploring the function of genetic variants in the non-coding genomic regions: approaches for identifying human regulatory variants affecting gene expression. Brief. Bioinform. 16 (3), 393–412. https://doi.org/10.1093/bib/bbu018.

Lowe, W.L., Reddy, T.E., 2015. Genomic approaches for understanding the genetics of complex disease. Genome Res. 25 (10), 1432–1441. https://doi.org/10.1101/gr.190603.115.

Lu, Z.X., Jiang, P., Xing, Y., 2012. Genetic variation of pre-mRNA alternative splicing in human populations. Wiley Interdiscip. Rev RNA 3 (4), 581–592. https://doi.org/10.1002/wrna.120.

Luo, L., Boerwinkle, E., Xiong, M., 2011. Association studies for next-generation sequencing. Genome Res. 21 (7), 1099–1108. https://doi.org/10.1101/gr.115998.110.

Ma, Y., Zhang, B., Tang, R.-K., Liu, Y., Peng, G.-G., 2012. Interleukin-18 promoter polymorphism and asthma risk: a meta-analysis. Mol. Biol. Rep. 39 (2), 1371–1376. https://doi.org/10.1007/s11033-011-0871-6.

Major, D.A., Sane, D.C., Herrington, D.M., Dahlback, B., Carlsson, M., Svensson, P., et al., 2000. Cardiovascular implications of the factor V Leiden mutation. Am. Heart J. 140 (2), 189–195. https://doi.org/10.1067/mhj.2000.108241.

Mardon, H.J., Sebastio, G., Baralle, F.E., 1987. A role for exon sequences in alternative splicing of the human fibronectin gene. Nucleic Acids Res. 15 (19), 7725–7733.

Mariyama, M., Kishi, K., Nakamura, K., Obata, H., Nishimura, S., 1989. Frequency and types of point mutation at the 12th codon of the c-Ki-ras gene found in pancreatic cancers from Japanese patients. Jpn. J. Cancer Res. 80 (7), 622–626.

Martin, K.C., Ephrussi, A., 2009. mRNA localization: gene expression in the spatial dimension. Cell 136 (4), 719–730. https://doi.org/10.1016/j.cell.2009.01.044.

Maurano, M.T., Humbert, R., Rynes, E., Thurman, R.E., Haugen, E., Wang, H., et al., 2012. Systematic localization of common disease-associated variation in regulatory DNA. Science 337 (6099), 1190–1195. https://doi.org/10.1126/science.1222794.

McLaren, W., Gil, L., Hunt, S.E., Riat, H.S., Ritchie, G.R.S., Thormann, A., et al., 2016. The ensembl variant effect predictor. Genome Biol. 17 (1), 122. https://doi.org/10.1186/s13059-016-0974-4.

McVicker, G., van de Geijn, B., Degner, J.F., Cain, C.E., Banovich, N.E., Raj, A., et al., 2013. Identification of genetic variants that affect histone modifications in human cells. Science 342 (6159), 747–749. https://doi.org/10.1126/science.1242429.

Moser, D., Ekawardhani, S., Kumsta, R., Palmason, H., Bock, C., Athanassiadou, Z., et al., 2009. Functional analysis of a potassium-chloride co-transporter 3 (SLC12A6) promoter polymorphism leading to an additional DNA methylation site. Neuropsychopharmacology 34 (2), 458–467. https://doi.org/10.1038/npp.2008.77.

Nabih, E.S., Kamel, H.F.M., Kamel, T.B., 2016. Association between CD14 polymorphism (−1145G/A) and childhood bronchial asthma. Biochem. Genet. 54 (1), 50–60. https://doi.org/10.1007/s10528-015-9699-4.

Oldridge, D.A., Wood, A.C., Weichert-Leahey, N., Crimmins, I., Sussman, R., Winter, C., et al., 2015. Genetic predisposition to neuroblastoma mediated by a LMO1 super-enhancer polymorphism. Nature 528 (7582), 418–421. https://doi.org/10.1038/nature15540.

Pawlik, A., Kaminski, M., Kuśnierczyk, P., Kurzawski, M., Dziedziejko, V., Adamska, M., et al., 2007. Interleukin-18 promoter polymorphism in patients with atopic asthma. Tissue Antigens 70 (4), 314–318. https://doi.org/10.1111/j.1399-0039.2007.00908.x.

Phillips, C., Lareu, V., Salas, A., Carracedo, A., 2004. Nonbinary single-nucleotide polymorphism markers. Int. Congr. Ser. 1261, 27–29. https://doi.org/10.1016/j.ics.2003.12.008.

Pott, S., Lieb, J.D., 2015. What are super-enhancers? Nat. Genet. 47 (1), 8–12. https://doi.org/10.1038/ng.3167.

Ptashne, M., 2007. On the use of the word "epigenetic". Curr. Biol. 17 (7), 233–236. https://doi.org/10.1016/j.cub.2007.02.030.

Ptashne, M., 2013. Epigenetics: core misconcept. Proc. Natl. Acad. Sci. USA 110 (18), 7101–7103. https://doi.org/10.1073/pnas.1305399110.

Quang, D., Chen, Y., Xie, X., 2015. DANN: a deep learning approach for annotating the pathogenicity of genetic variants. Bioinformatics 31 (5), 761–763. https://doi.org/10.1093/bioinformatics/btu703.

Rachakonda, P.S., Hosen, I., de Verdier, P.J., Fallah, M., Heidenreich, B., Ryk, C., et al., 2013. TERT promoter mutations in bladder cancer affect patient survival and disease recurrence through modification by a common polymorphism. Proc. Natl. Acad. Sci. USA 110 (43), 17426–17431. https://doi.org/10.1073/pnas.1310522110.

Rintisch, C., Heinig, M., Bauerfeind, A., Schafer, S., Mieth, C., Patone, G., et al., 2014. Natural variation of histone modification and its impact on gene expression in the rat genome. Genome Res. 24 (6), 942–953. https://doi.org/10.1101/gr.169029.113.

Ritchie, G.R.S., Dunham, I., Zeggini, E., Flicek, P., 2014. Functional annotation of noncoding sequence variants. Nat. Methods 11 (3), 294–296. https://doi.org/10.1038/nmeth.2832.

Russom, A., Haasl, S., Brookes, A.J., Andersson, H., Stemme, G., 2006. Rapid melting curve analysis on monolayered beads for high-throughput genotyping of single-nucleotide polymorphisms. Anal. Chem. 78 (7), 2220–2225. https://doi.org/10.1021/ac051771u.

Schaub, M.A., Boyle, A.P., Kundaje, A., Batzoglou, S., Snyder, M., 2012. Linking disease associations with regulatory information in the human genome TL-22. Genome Res. 22 (9), 1748–1759. https://doi.org/10.1101/gr.136127.111.

Schedel, M., Michel, S., Gaertner, V.D., Toncheva, A.A., Depner, M., Binia, A., et al., 2015. Polymorphisms related to ORMDL3 are associated with asthma susceptibility, alterations in transcriptional regulation of ORMDL3, and changes in TH2 cytokine levels. J. Allergy Clin. Immunol. 136 (4), 893–903e14. https://doi.org/10.1016/j.jaci.2015.03.014.

Schmidt, S., Barcellos, L.F., DeSombre, K., Rimmler, J.B., Lincoln, R.R., Bucher, P., et al., 2002. Association of polymorphisms in the apolipoprotein E region with susceptibility to and progression of multiple sclerosis. Am. J. Hum. Genet. 70 (3), 708–717. https://doi.org/10.1086/339269.

Schmidt, E.M., Zhang, J., Zhou, W., Chen, J., Mohlke, K.L., Chen, Y.E., Willer, C.J., 2015. GREGOR: evaluating global enrichment of trait-associated variants in epigenomic features using a systematic, data-driven approach. Bioinformatics 31 (16), 2601–2606. https://doi.org/10.1093/bioinformatics/btv201.

Schork, N.J., Fallin, D., Lanchbury, S., 2000. Single nucleotide polymorphisms and the future of genetic epidemiology. Clin. Genet. 58 (4), 250–264. https://doi.org/10.1034/j.1399-0004.2000.580402.x.

Shastry, B.S., 2009. SNPs: impact on gene function and phenotype. In: Komar, A.A. (Ed.), Single Nucleotide Polymorphisms: Methods and Protocols. Humana Press, Totowa, NJ, pp. 3–22. https://doi.org/10.1007/978-1-60327-411-1_1.

Sherry, S.T., Ward, M.H., Kholodov, M., Baker, J., Phan, L., Smigielski, E.M., Sirotkin, K., 2001. dbSNP: the NCBI database of genetic variation. Nucleic Acids Res. 29 (1), 308–311. https://doi.org/10.1093/nar/29.1.308.

Shihab, H.A., Rogers, M.F., Gough, J., Mort, M., Cooper, D.N., Day, I.N.M., et al., 2015. An integrative approach to predicting the functional effects of non-coding and coding sequence variation. Bioinformatics 31 (10), 1536–1543. https://doi.org/10.1093/bioinformatics/btv009.

Shoemaker, R., Deng, J., Wang, W., Zhang, K., 2010. Allele-specific methylation is prevalent and is contributed by CpG-SNPs in the human genome. Genome Res. 20 (7), 883–889. https://doi.org/10.1101/gr.104695.109.

Simon, T., Semsei, A.F., Ungvri, I., Hadadi, E., Virag, V., Nagy, A., Falus, A., 2012. Asthma endophenotypes and polymorphisms in the histamine receptor HRH4 gene. Int. Arch. Allergy Immunol. 159 (2), 109–120. https://doi.org/10.1159/000335919.

Siva, N., 2008. 1000 genomes project. Nat. Biotech. 26 (3), 256. https://doi.org/10.1038/nbt0308-256b.

Stamm, S., Ben-Ari, S., Rafalska, I., Tang, Y., Zhang, Z., Toiber, D., et al., 2005. Function of alternative splicing. Gene 344, 1–20. https://doi.org/10.1016/j.gene.2004.10.022.

Syvänen, A., 2001. Accessing genetic variation: genotyping single nucleotide polymorphisms. Nat Rev Genet 2 (12), 930–942. https://doi.org/10.1038/35103535.

Takai, D., Jones, P.A., 2002. Comprehensive analysis of CpG islands in human chromosomes 21 and 22. Proc. Natl. Acad. Sci. USA 99 (6), 3740–3745. https://doi.org/10.1073/pnas.052410099.

Tazi, J., Bakkour, N., Stamm, S., 2009. Alternative splicing and disease. Biochim. Biophys. Acta Mol. Basis Dis. 1792 (1), 14–26. https://doi.org/10.1016/j.bbadis.2008.09.017.

Tobler, A.R., Short, S., Andersen, M.R., Paner, T.M., Briggs, J.C., Lambert, S.M., et al., 2005. The SNPlex genotyping system: a flexible and scalable platform for SNP genotyping. J. Biomol. Tech. 16 (4), 398–406.

Tost, J., Gut, I.G., 2005. Genotyping single nucleotide polymorphisms by MALDI mass spectrometry in clinical applications. Clin. Biochem. 38 (4), 335–350. https://doi.org/10.1016/j.clinbiochem.2004.12.005.

Tsuge, M., Hamamoto, R., Silva, F.P., Ohnishi, Y., Chayama, K., Kamatani, N., et al., 2005. A variable number of tandem repeats polymorphism in an E2F-1 binding element in the 5′ flanking region of SMYD3 is a risk factor for human cancers. Nat. Genet. 37 (10), 1104–1107. https://doi.org/10.1038/ng1638.

Ulevitch, R.J., Tobias, P.S., 1995. Receptor-dependent mechanisms of cell stimulation by bacterial endotoxin. Annu. Rev. Immunol. 13, 437–457. https://doi.org/10.1146/annurev.iy.13.040195.002253.

Waldispühl, J., Clote, P., 2007. Computing the partition function and sampling for saturated secondary structures of RNA, with respect to the turner energy model. J. Comput. Biol. 14 (2), 190–215. https://doi.org/10.1089/cmb.2006.0012.

Wang, H., Ogawa, M., Wood, J.R., Bartolomei, M.S., Sammel, M.D., Kusanovic, J.P., et al., 2008. Genetic and epigenetic mechanisms combine to control MMP1 expression and its association with preterm premature rupture of membranes. Hum. Mol. Genet. 17 (8), 1087–1096. https://doi.org/10.1093/hmg/ddm381.

Wang, K., Li, M., Hakonarson, H., 2010. ANNOVAR: functional annotation of genetic variants from high-throughput sequencing data. Nucleic Acids Res. 38 (16), e164. https://doi.org/10.1093/nar/gkq603.

Ward, A.J., Cooper, T.A., 2010. The pathobiology of splicing. J. Pathol. 220 (2), 152–163. https://doi.org/10.1002/path.2649.

Ward, L.D., Kellis, M., 2012. HaploReg: a resource for exploring chromatin states, conservation, and regulatory motif alterations within sets of genetically linked variants. Nucleic Acids Res. 40 (D1), D930–D934. https://doi.org/10.1093/nar/gkr917.

Ward, L.D., Kellis, M., 2015. HaploReg v4: systematic mining of putative causal variants, cell types, regulators and target genes for human complex traits and disease. Nucleic Acids Res. 44 (D1), D877–D881. https://doi.org/10.1093/nar/gkv1340.

Westen, A.A., Matai, A.S., Laros, J.F.J., Meiland, H.C., Jasper, M., Leeuw, W.J.F., de Sijen, T., 2009. Genetics Tri-allelic SNP markers enable analysis of mixed and degraded DNA samples. Forensic Sci. Int. 3 (4), 233–241. https://doi.org/10.1016/j.fsigen.2009.02.003.

Wright, J.B., Brown, S.J., Cole, M.D., 2010. Upregulation of c-MYC in cis through a large chromatin loop linked to a cancer risk-associated single-nucleotide polymorphism in colorectal cancer cells. Mol. Cell. Biol. 30 (6), 1411–1420. https://doi.org/10.1128/MCB.01384-09.

Yang, Q., Jie, Z., Ye, S., Li, Z., Han, Z., Wu, J., et al., 2014. Genetic variations in miR-27a gene decrease mature miR-27a level and reduce gastric cancer susceptibility. Oncogene 33 (2), 193–202. https://doi.org/10.1038/onc.2012.569.

Zaina, S., Perez-Luque, E.L., Lund, G., 2010. Genetics talks to epigenetics? The interplay between sequence variants and chromatin structure. Curr. Genomics 11 (5), 359–367. https://doi.org/10.2174/138920210791616662.

Zhang, F., Lupski, J.R., 2015. Non-coding genetic variants in human disease. Hum. Mol. Genet. 24 (R1), R102–R110. https://doi.org/10.1093/hmg/ddv259.

Zhang, Y., Bertolino, A., Fazio, L., Blasi, G., Rampino, A., Romano, R., et al., 2007. Polymorphisms in human dopamine D2 receptor gene affect gene expression, splicing, and neuronal activity during working memory. Proc. Natl. Acad. Sci. USA 104 (51), 20552–20557. https://doi.org/10.1073/pnas.0707106104.

Zhao, N., Han, J.G., Shyu, C.-R., Korkin, D., 2014. Determining effects of non-synonymous SNPs on protein-protein interactions using supervised and semi-supervised learning. PLoS Comput. Biol. 10 (5), e1003592. https://doi.org/10.1371/journal.pcbi.1003592.

Zheng, S.L., Sun, J., Cheng, Y., Li, G., Hsu, F.C., Zhu, Y., et al., 2007. Association between two unlinked loci at 8q24 and prostate cancer risk among European Americans. J. Natl. Cancer Inst. 99 (20), 1525–1533. https://doi.org/10.1093/jnci/djm169.

Zuo, C., Shin, S., Keleş, S., 2015. atSNP: transcription factor binding affinity testing for regulatory SNP detection. Bioinformatics 31 (20), 3353–3355. https://doi.org/10.1093/bioinformatics/btv328.

CHAPTER

8

Synthesis of Magnetic Iron Oxide Nanoparticles

Marcel Wegmann, Melanie Scharr

Furtwangen University, Villingen-Schwenningen, Germany

1 INTRODUCTION

Magnetic nanoparticles are of great interest for different research areas, not only for fundamental scientific interest but also for biomedical applications such as targeted drug delivery systems, magnetic resonance imaging (MRI), magnetic hyperthermia and thermoablation, cellular labeling and separation, or biosensing. For nearly 40 years, the application of small iron oxide nanoparticles (IONPs) has been practiced and intensively developed (Gilchrist et al., 1957). Several types of iron oxides subsist in nature; most common are magnetite (Fe_3O_4), maghemite (γ-Fe_2O_3), and hematite (α-Fe_2O_3) (Cornell et al., 2003). Each of them with unique properties offers capability for specific technical and biomedical applications (Cornell et al., 2003). Hematite, oldest known and widespread iron oxide in rocks and soils, is the most stable iron oxide. As a result, hematite is widely used as catalyst, pigment, gas sensor, and starting material for synthesis of magnetite and maghemite (Klotz et al., 2008; Wu et al., 2010a). Magnetite differs from other iron oxides containing bivalent and trivalent iron. Exhibiting strongest magnetism of any transition metal oxide, it belongs to the most important iron ore and natural resource for the electric industry (Cornelis and Hurlburt, 1977). Maghemite occurs in soils as a weathering product of magnetite or as a heating product of other iron oxides. The crystal structure of the three mentioned iron oxides can be described as close-packed planes of oxygen anions with iron cations in octahedral or tetrahedral interstitial sites (Klotz et al., 2008).

With regard to the magnetic properties, an iron atom has a strong magnetic moment because of four unpaired electrons in its 3d orbitals (Cornell et al., 2003). Usually, when crystals are formed by iron atoms, different magnetic states can arise: first, the paramagnetic state, which is characterized by a zero magnetic moment of the crystal, due to individual randomly aligned atomic magnetic moments. Upon the exposure to an external magnetic field, these moments will align, and the crystal will obtain a small net magnetic moment. If all individual moments are aligned even without an external magnetic field, the crystal will

Precision Medicine
https://doi.org/10.1016/B978-0-12-805364-5.00008-1

act ferromagnetic, whereas the ferrimagnetic crystal is marked by atoms with moments of different strengths, which are arranged antiparallel (Cornell et al., 2003). Magnetic domains exist in a ferromagnetic bulk material, whereas each domain has its own magnetization based on different alignments of atomic magnetic moments within each domain.

The smaller the length scale of the material, the smaller the number of domains, until there is a single domain. The size of IONPs will then be about 15 nm (Wu et al., 2012a). A single domain is said to be superparamagnetic meaning that, there exists a coherent magnetization feature. Superparamagnetic iron oxide nanoparticles (SPIONs) offer a high potential for several biomedical applications, based on their unique chemical, physical, thermal, and mechanical properties (Cornell et al., 2003). Several biomedical applications for IONPs are represented at the end of this chapter.

The magnetic behavior of iron oxides is strongly influenced by size, shape, surface effects, and temperature of the nanoparticles and therefore also depended on the synthesis method used (Wu et al., 2012a; Tronc et al., 2000). Furthermore, aggregation of superparamagnetic nanoparticles is a well-known phenomenon. Thus, wide ranged and suitable synthesis routes for magnetic nanoparticles are available to tune the magnetic properties and to avoid aggregation of bare nanoparticles (Lu et al., 2007). A major challenge for all synthesis routes is the design of magnetic nanoparticles and their magnetic properties, with effective surface coating, providing optimum performance for in vitro and in vivo biological applications (Wu et al., 2012a; Lu et al., 2007). Some of these methods of synthesis are described in the following section.

2 SYNTHESIS OF MAGNETIC IONPs

In view of achieving proper control of particle size, shape, crystallinity, polydispersity, and finally the magnetic properties, many synthesis routes have been developed (Teja and Holm, 2002; Wu et al., 2015). The following methods could be divided into aqueous routes, such as the hydrothermal and solvothermal synthesis, and into nonaqueous routes, such as the thermal decomposition.

2.1 Coprecipitation

The coprecipitation is a simple and most conventional method to obtain IONPs (Fe_3O_4 and γ-Fe_2O_3) from ferric/ferrous salt solution with a 1:2 molar ratio by adding basic solutions at room temperature or at elevated temperature (Massart, 1981). Usually, the reaction proceeds under gas protection (Lee et al., 1996). Size, morphology, and composition of the magnetic particles depend on experimental parameters, such as ferric/ferrous ratio, the salt used (e.g., chlorides, sulfates, or nitrates), the reaction temperature, the pH value, and the ionic strength of the medium (Lee et al., 1996; Wu et al., 2007; Pereira et al., 2012; Blanco-Andujar et al., 2012). In addition, due to fast particle formation rates, the control over particle properties is limited. However, provided that the synthesis conditions are defined, the quality of magnetic particles is completely reproducible. For instance, to receive maghemite particles, magnetite particles are targeted oxidized through dispersing in an acidic medium and finally by adding iron (III) nitrate (Lee et al., 1996).

Problems in terms of aggregation and biocompatibility of IONPs meant that many surfactant and biomolecules have been introduced directly into coprecipitation process. In this case, Salavati-Niasari et al. have reported the synthesis of magnetic nanocrystals via coprecipitation using octanoic acid as surfactant during chemical reaction to improve dispersity (Salavati-Niasari et al., 2012). Moreover, Suh et al. have prepared an in situ synthesis of nonspherical magnetic IONPs in a carboxyl-functionalized polymer matrix. To begin, the iron ions were diffused into polymer particles; next, they formed chelate complexes with the deprotonated carboxyl groups and finally grew up in the polymer particles (Suh et al., 2012). Thus, this method could enable to add several functionalities, for example, biomolecules, after subsequent reactions (Wu et al., 2015).

Overall, the coprecipitation is one of the facile and most conventional methods to synthesize IONPs. Nevertheless, the shortcomings of this method should be taken into account, such as broad particle size distribution of products and the use of a strong base during the reaction process. To overcome the abovementioned problems, different alternative strategies have been developed, such as nonaqueous thermal decomposition.

2.2 Thermal Decomposition

Inspired by the synthesis of high-quality semiconductor nanocrystals and oxides in an aqueous medium through thermal decomposition (Murray et al., 1993; Peng et al., 1998; O'Brien et al., 2001), several similar strategies have been developed for the synthesis of magnetic IONPs with controllable size and morphology. Higher monodispersed and small-sized magnetic nanoparticles could be obtained through thermal decomposition of certain organometallic compounds or coordinated iron precursors in with high-boiling-point, organic solvent, containing stabilizing surfactant (Park et al., 2004; Sun et al., 2004; Redl et al., 2004). Organometallic compound are suitable as follows: metal acetylacetonate ([m(acac)n]) (m = Fe, Mn, Co, Ni, Cr; n = 2 or 3; acac = acetylacetonate) (Wang et al., 2012), metal cupferronate (m_xCup_x) (m = metal ion, Cup = *N*-nitrosophenylhydroxylamine) (Rockenberger et al., 1999), Prussian blue (Fe4 [Fe (CN) 6·14H2O]) (Hu et al., 2012a,b), or carbonyls (Farrell et al., 2003). With regard to surfactants, fatty acids (Jana et al., 2004), oleic acid (Samia et al., 2005), or hexadecylamine (Li et al., 2006) are often added to the reaction process as stabilizers. Accordingly, the nucleation process may be delayed, and adsorption of additives on the growing nanocrystal can be improved. Therefore, favorable small IONPs can arise (Wu et al., 2015). In general, the strategy of thermal decomposition can be divided into hot-injection approaches or conventional reaction strategies. Concerning the hot-injection strategy, the precursors are injected into a hot reaction mixture, whereas in conventional reaction strategies, a reaction mixture is prepared at room temperature and then heated in a closed or open reaction vessel (Sun and Zeng, 2002; Li et al., 2008).

In contrast to coprecipitation, nucleation can be separated from growth, and complex hydrolysis reactions can be avoided, which proved to be huge advantages of the thermal decomposition (Sun and Zeng, 2002; Li et al., 2008). For instance, Lynch et al. (2011) have reported that gas bubbles had a stronger effect on nucleation of IONPs than on their growth.

The thermal decomposition offers the possibility to obtain iron oxide with different shape, such as nanotubes or nanospheres (Amara et al., 2012; Chalasani and Vasudevan, 2011). The magnetic properties are affected by shape and size; hence, these can be tailored by the use of different precursors, additives, and solvents during the preparation process (Shavel and Liz-Marzan, 2009; Demortiere et al., 2001). Still, the process of thermal decomposition involves boiling of the solvents; as a result, the received size and shape are not fully reproducible. To receive monodisperse and reproducible IONPs, Hyeon et al. have reported about an inexpensive synthesis method that uses nontoxic iron chloride that is thermal decomposed in an organic solvent containing iron oleate and surfactant. As a result, in a single reaction, a gram-scale amount of IONPs (in the range of 5–22 nm) was produced without any size-selection processes (Park et al., 2004, 2005). Therefore, the synthetic procedure of Hyeon et al. is highly reproducible and could be applicable to other materials (Wu et al., 2015).

2.3 Hydrothermal and Solvothermal Synthesis

This method is used to obtain a wide range of nanostructured materials and for the synthesis of metal oxides as powders, nanoparticles, and single crystals (Lian et al., 2004; Tavakoli et al., 2007; Dou et al., 2007; Giri et al., 2005; Sorescu et al., 2004; Wang et al., 2004). Until now, the precise reaction mechanism has not been clarified exactly (Lu et al., 2007). In general, the synthesis route includes several wet-chemical techniques, with the objective to crystalize the substance in a sealed container from a high-temperature (in the range of 130–250°C) aqueous or nonaqueous solution under high vapor pressure (0 and 3–4 MPa) (Wu et al., 2008). In contrast to the hydrothermal synthesis, the solvothermal method uses an organic solvent as reaction medium instead of water. Nevertheless, the concept embodied in the hydrothermal synthesis has been derived to the solvothermal process (Wu et al., 2015). Both methods are suitable to obtain highly crystalline IONPs, such as α-Fe_2O_3, γ-Fe_2O_3, or Fe_3O_4 while maintaining good control over their composition (Wu et al., 2015). This synthesis method offers many opportunities for controlling particle size and morphology by modifying the reaction process (Burda et al., 2005; Shaw et al., 1991; Sue et al., 2004). For instance, Hao and Teja have reported of the effects of precursor concentration, temperature, and resistance time on particle size and morphology on the synthesis of IONPs. As a result, particle size and size distribution increased with precursor concentration, but the residence time had a more significant impact on the particle size whereby monodispersed particles were produced by keeping the residence time short (Hao and Teja, 2003). Furthermore, Lester et al. conducted a detailed investigation of the effect of the mixing process inside the reactor. Consequently, the use of a nozzle mixer leads to very low residence times and limits particle growth (in the size range of 6–64 nm) (Lester et al., 2006). Particularly, this process is easy to scale up. Additionally, concerning the hydrothermal synthesis is environmentally friendly and versatile, as no organic solvents or posttreatments are needed. However, postprocessing steps are necessary for surface functionalization, which cannot be accomplished in situ (Teja and Koh, 2009).

2.4 Sol-Gel Reaction and Polyol Method

The sol-gel method belongs to the classical wet-chemical techniques (Wu et al., 2015). In general, a colloidal solution acts as a precursor for an integrated network of either discrete particles or network polymers (Pandey and Mishra, 2011). Concerning the IONPs, typical precursors are iron alkoxides and iron salts that undergo hydrolysis and condensation reactions at room temperature leading to dispersion of oxide particles in a "sol." Finally, heat treatments are needed to acquire a final crystalline state. Afterward, the "sol" is dried or "gelled" by solvent removal (generally water) or by chemical reaction (reaction with acid or base) (Lam et al., 2008). Particularly, a colloidal gel is obtained through the reaction with a basic catalyst, resulting in the agglomeration of colloid particles, whereas a polymeric gel is received with an acid catalyst, causing a polymeric substructure made by the aggregation of subcolloidal particles (Pandey and Mishra, 2011; Lam et al., 2008). The particle size depends on the hydrolysis reaction, solution composition, pH value, and temperature (Tavakoli et al., 2007). The sol-gel process determines the final properties of IONPs (Lemine et al., 2012; Qi et al., 2011). With regard to various shapes and sizes, the polyol method is also suitable to obtain IONPs. In contrast to the sol-gel reaction, resting upon an oxidation reaction, the polyol method is an inversed sol-gel reaction, which uses a reduction reaction (Laurent et al., 2008). Regarding IONPs, an iron precursor is suspended in liquid polyol, then stirred and heated up to the boiling point of polyol. This reaction does not require high pressures compared with the hydrothermal method. However, the polyol solvent determines the morphology, colloidal stability, and magnetic properties of the resulting particles (Cai and Wan, 2007; Caruntu et al., 2007).

The advantages of the sol-gel reaction and the polyol method are on the one hand the simple dispersion in aqueous or polar solvents, due to hydrophilic ligands on the surface of IONPs. On the other hand, the high reaction temperature favors particles with higher crystallinity and magnetization. Yet, these methods are unsuitable owing to relatively high costs of the metal alkoxides (Wu et al., 2015), contaminations from by-products of reactions, and the need for posttreatment of products (Teja and Koh, 2009).

2.5 Synthesis by Microemulsion

The microemulsion is a stable and isotropic dispersion of two nonmixable solutions, in that case oil, water, and surfactant, frequently in combination with a cosurfactant. The surfactant develops a monolayer at the interface between the oil and water layer; thus, the hydrophobic tails of the surface molecules extend into the oil phase and the hydrophilic head groups into the aqueous phase. The aqueous phase contains metal salts or other ingredients, whereas the oil phase composed of different hydrocarbons and olefins (Wongwailikhit and Horwongsakul, 2011). Between two types of microemulsion, it can be distinguished as follows: the direct (oil dispersed in water, o/w) or reversed (water dispersed in oil, w/o) method. Both are suitable for the synthesis of IONPs with tailored shape and size. Nanoparticles could be synthesized by mixing two identical w/o microemulsions and then adding a reducing agent or by bubbling gas like O_2, NH_3, or CO_2 through microemulsion (Pillai et al., 1995). A unique microenvironment is offered by the surfactant-covered water pools for the formation of nanoparticles. Those limit also the growth of nanoparticles (Capek, 2004). Common surfactants for the synthesis of IONPs are bis (2-ethylhexyl) sulfosuccinate (AOT), sodium dodecyl sulfate (SDS), cetyltrimethylammonium bromide (CTAB), and polyvinylpyrrolidone (PVP) (Wongwailikhit and Horwongsakul, 2011; Han et al., 2011; Ladj et al., 2013). The size of the microemulsion droplets depends upon water-to-surfactant ratio, concentration of reactants (especially surfactant), and the nature of surfactants (Darbandi et al., 2012; Okoli et al., 2011, 2012). To prevent aggregation of produced IONPs, several washing processes and further stabilization reagents are needed, which might affect the properties of the particles. Furthermore, due to its disadvantages, there is a small working window for the synthesis and the small yield of particles. In comparison with the coprecipitation or thermal decomposition, this method is rather ineffective and difficult to scale up (Lu et al., 2007).

2.6 Sonochemical Synthesis

This method uses the chemical effect of ultrasound for the production of structures and provides an unusual rout without using high temperatures, high pressures, or long reaction times (Xu et al., 2013). Ultrasound irradiation creates acoustic waves leading to oscillating bubbles, which can accumulate ultrasonic energy while growing in size. The overgrown bubbles can collapse, under right conditions, releasing the concentrated energy and leading to an transient and localized temperature of 5000 K and a pressure of 1000 bar (Suslick, 1990; Bang and Suslick, 2010). Various forms of bare and functionalized IONPs are prepared by the sonication of an aqueous ferro or ferrous salt solution (Mukh-Qasem and Gedanken, 2005). For instance, Theerdhala et al. have reported the synthesis of surface functionalized IONPs by a one-step method through sonochemical synthesis, which could become a promising vehicle for drug delivery (Theerdhala et al., 2010). Furthermore, Zhu et al. functionalized sonochemical synthesized IONPs with hemoglobin as a biosensor to detect H_2O_2 (Zhu et al., 2013). To conclude, this method has several advantages, such as the one-step synthesis and functionalization opportunity, uniformity, and controllable size. Still, this method has several shortcomings, like uncontrollable shape and dispersity (Wu et al., 2015).

2.7 Microwave-Assisted Synthesis

The excitation with electromagnetic radiation causes an intense internal heat, provided by the reorientation of molecules, previously aligning their dipoles within the external field. This creation of intense heat reduce processing time and energy cost, which make those synthesis method more attractive (Wu et al., 2015). Recently, the microwave-assisted synthesis method has been reported through several working groups to obtain IONPs with controllable size and shape (Hu et al., 2007; Wu et al., 2011a; Ai et al., 2010; Qiu et al., 2011). Depending on the experimental conditions, like precursor material or reaction temperature, the received nanoparticles could be slightly different in shape and size as well as in magnetic properties (Sreeja and Joy, 2007; Jiang et al., 2010; Hu et al., 2011). Additionally, surface functionalization, such as dextran (Osborne et al., 2012) or polyacid (Liu et al., 2011) coating, offers the application of biocompatible IONPs. The dextran coating provides water solubility and biocompatibility for in vivo applications. In brief, this synthesis method can be considered as attractive for fabrication of large-scale particles, since no purification steps are needed and particles can be easily dispersed in water (Pascu et al., 2012).

2.8 Biosynthesis by Magnetotactic and Iron Reducing Bacteria

The biosynthesis method for IONPs is a chemical route with reduction/oxidation reactions, due to microbial enzymes (Prathna et al., 2010). Primarily responsible for the traditional biosynthesis are magnetotactic and iron-reducing bacteria, such as *Geobacter metallireducens* or *Magnetospirillum gryphiswaldense* (Vali et al., 2004; Scheffel et al., 2006; Bazylinski and Frankel, 2004). Recently, several groups have reported on new types of bacteria, with the ability to synthesize magnetic IONPs. First, bacterium *Actinobacter* sp. is capable of synthesizing superparamagnetic maghemite nanoparticles of ferric chloride precursors. Unlike magnetotactic and iron-reducing bacteria where the reaction occurs under anaerobic conditions, *Actinobacter* sp. synthesizes under aerobic conditions (Bharde et al., 2008). Furthermore, Sundaram et al. have reported about *Bacillus subtilis* strains isolated from rhizosphere soil that robustly produced Fe_3O_4 (Sundaram et al., 2012). This synthesis method is eco-friendly, even though much more experiments are needed now to clarify the control over size and shape during the synthesis process in living organisms (Wu et al., 2015).

2.9 Other Chemical or Physical Synthesis Routes

In addition to the abovementioned synthesis methods, electrochemical (Pascal et al., 1998; Starowicz et al., 2011; Wang et al., 2011a), flow injection (Salazar-Alvarez et al., 2006), and aerosol/vapor (Huang et al., 2010) methods can be used to synthesize IONPs. The electrochemical methods are characterized by high purity of the obtained products and control over particle size, which depends on the current or the potential applied to the system (Pascal et al., 1998; Starowicz et al., 2011; Wang et al., 2011a). The flow-injection method represents a modified coprecipitation method, whereby the precursor materials are added by injecting with a controllable flow rate. This method distinguishes by high reproducibility, mixing homogeneity, and the possibility of an external control of the process (Salazar-Alvarez et al., 2006). The main aerosol/vapor methods are spray and laser pyrolysis. IONPs are produced by spray pyrolysis through evaporation of ferric salts in a solvent inside a high-temperature atmosphere like the flame spray. The laser pyrolysis differs from the spray method in that the reduction process of the reaction volume is induced by a laser (Huang et al., 2010). Both methods produced small, narrow size, and nonaggregated particles. However, reaction parameters like the evaporation process depend on flame configuration or properties of the starting material (Abid et al., 2013; Costo et al., 2012).

Characteristics as well as advantages and disadvantages of the abovementioned synthesis routes are briefly summarized in Table 1.

To conclude, in terms of simplicity of synthesis, coprecipitation is the preferred route. Relating to proper control of size and morphology, thermal decomposition synthesis route seems to be most developed. Alternatively, through microemulsion, synthesis of monodispersed nanoparticles can be obtained, certainly requiring a lot of solvent. So far, only little is known about the hydro- or solvothermal synthesis, but it enables the synthesis of high-quality nanoparticles. Currently, the particles compounded by coprecipitation and thermal decomposition are best known and could be produced on a large scale. However, the major difficulties in the synthesis of IONPs are the control over size, shape, composition, and size distribution in nanometer range. Thus, in quest of facile and flexible fabrication methods to achieve desirable size and morphology, it is of extreme importance to realize the full potential of these materials in biomedical applications.

TABLE 1 Overview of the Available Synthesis Methods of Magnetic Iron Oxide Nanoparticles

Synthesis Methods	Synthesis Conditions	Reaction temp. (°C)	Reaction Period	Size Distribution	Shape Control	Yield
Coprecipitation	Ambient	20–150	Minutes	Relatively narrow	Not good	High/scalable
Thermal decomposition	Insert atmosphere	100–350	Hours to days	Very narrow	Very good	High/scalable
Hydrothermal/ solvothermal	High pressure	150–220	Hours to days	Very narrow	Very good	High/scalable
Sol-gel	Ambient	25–200	Hours	Narrow	Good	Medium
Microemulsion	Ambient	20–80	Hours	Narrow	Good	Low
Sonolysis/ sonochemical	Ambient	20–50	Minutes	Narrow	Bad	Medium
Microwave-assisted	Ambient	100–200	Minutes	Medium	Good	Medium
Biosynthesis	Ambient	Room temp.	Hours to days	Broad	Bad	Low
Electrochemical	Ambient	Room temp.	Hours- to days	Medium	Medium	Medium
Aerosol/vapor	Insert atmosphere	> 100	Minutes to hours	Relatively narrow	Medium	High/scalable

Modified from Wu, W., Wu, Z., Yu, T., Jiang, C., Kim, W.S., 2015. Recent progress on magnetic iron oxide nanoparticles: synthesis, surface functional strategies and biomedical applications. Sci. Technol. Adv. Mater. 16 (2), 023501.

3 SURFACE FUNCTIONALIZATION OF MAGNETIC NANOPARTICLES

Relevant for biomedical applications is the preparation and storage of nanoparticles in a colloidal form (Wu et al., 2012a). Especially, an enhanced dispersibility in water is important because most biological media are nearly neutral aqueous solutions (Wu et al., 2015). The intrinsic instability over long periods is the main unavoidable problem with IONPs, which manifests in two ways. First, without surface coating, bare particles have a hydrophobic surface with a large surface-area-to-volume ratio. Thus, hydrophobic interactions between these particles lead to aggregation and formation of larger particles to reduce surface energy (Hamley, 2003). Next, surface effects like oxidation processes of bare IONPs, accordingly to high chemical activity, cause loss of magnetism (Tronc et al., 2000). Therefore, magnetic particles are applicable as ferrofluids (Babincova et al., 2001). Ferrofluids are colloidal suspensions of homogenous dispersed magnetic particles into suitable solvent, which exhibit magnetization in most intense magnetic fields (Babincova et al., 2001; Wang et al., 2001).

Hence, surface coating of bare IONPs is necessary to achieve dispersibility and stabilization against damage during or after biomedical application (Wang et al., 2001; Bonnemain, 1998). Finally, coatings are suitable for further functionalization by the attachment of bioactive molecules such as targeted drug delivery systems (Berry and Curtis, 2003).

3.1 Fabricating Types of Magnetic Composite Nanomaterials

Four fabricating types of iron oxide-based materials such as core-shell structure, matrix-dispersed, Janus, and shell-core-shell structures have been evolved with regard to their application. Fig. 1 shows the typical morphologies of magnetic composite nanomaterials, which are described in detail below.

3.1.1 *Core-Shell Structure*

The iron oxide core is encapsulated in an inorganic or organic coating, which serves for particle stabilization and biocompatibility. The iron oxide particle could be located not exactly at the center of the coating material rather peripherally and is also known as York structure. To the contrary, the inverse core-shell structure is characterized by an iron oxide coating of nonmagnetic functional material. In addition, the combination of one or more functional materials and further coating of already functionalized surface is feasible (Hu et al., 2013; Luo et al., 2012a; Wei et al., 2011).

3.1.2 *Matrix-Dispersed Structure*

To prevent IONPs from aggregation into larger particles, those are dispersed in a matrix. IONPs can be established as amorphous matrix, as mesoscale particles, or as three-dimensional superstructures (Behrens, 2011).

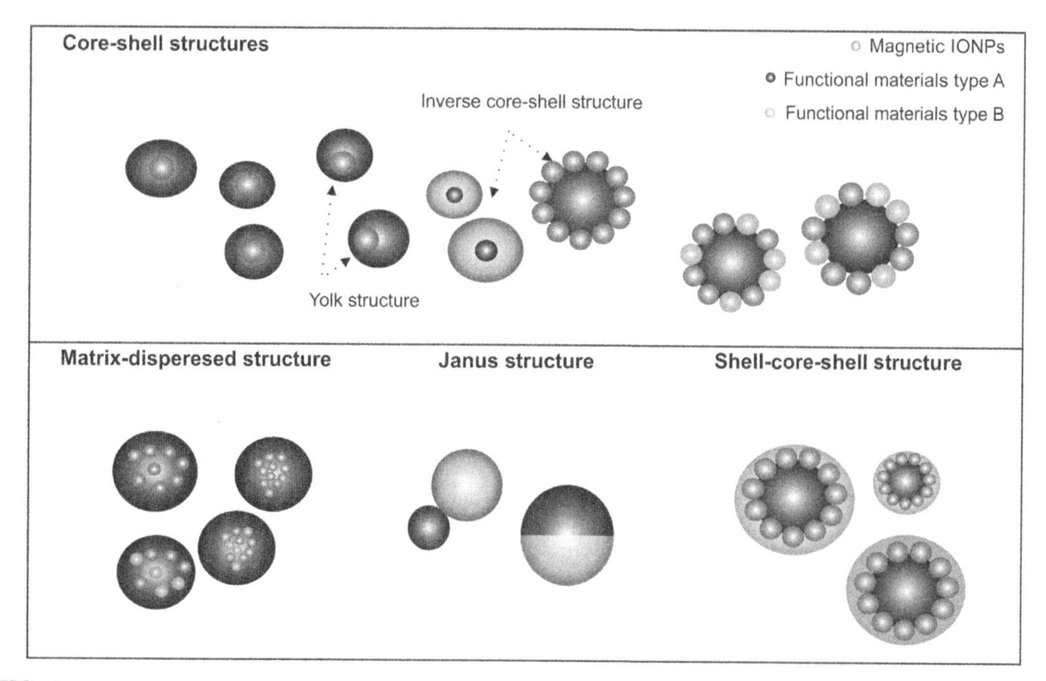

FIG. 1 Fabricating types of magnetic composite nanomaterials. Gray spheres represent iron oxide nanoparticles, whereby nonmagnetic composite materials are represented in other colors. *Modified from Wu, W., Wu, Z., Yu, T., Jiang, C., Kim, W.S., 2015. Recent progress on magnetic iron oxide nanoparticles: synthesis, surface functional strategies and biomedical applications. Sci. Technol. Adv. Mater. 16 (2) 023501.*

3.1.3 *Janus Structure*

This structure is characterized by one magnetic iron oxide side and one functional material side. For example, magnetic γ-Fe_2O_3||SiO_2 Janus particles have synthesized by Zhao and Gao through flame synthesis, which are highly uniform, dispersible in aqueous solvents, and suitable for further manipulation (Zhao and Gao, 2009). Furthermore, Sun et al. have reported about the synthesis of dumbbell-like, small-sized Au-Fe_3O_4 and Pt-Fe_3O_4 nanoparticles that are used in target-specific platin delivery (Yu et al., 2005; Wang et al., 2009a; Xu et al., 2009).

3.1.4 *Shell-Core-Shell Structure*

The last possible fabricating type is the location of an iron oxide particle between two functional layers. One potential "sandwich" construction is a layer of magnetic IONPs between a luminescent and a biocompatible polymer layer that allows further manipulation (Sperling and Parak, 2010).

3.2 Organic Materials

In situ coating, where the coating materials are added during synthesis of IONPs, and post-synthesis coatings are the mainly common routes for surface coatings of IONPs (Tsai et al., 2010; Kloust et al., 2013; Li et al., 2013a). In this connection, following organic materials as an example, dextran, starch, PEG (polyethylene glycol), or PLA (poly D, L-lactide) is available.

3.2.1 *Small Molecules and Surfactants*

Different bioactive molecules can be attached through functionalizing iron oxide surface with special groups like –OH, –COOH, –NH_2, or –SH. In the range of small molecules, silane is often used to modify or to equip bare IONPs with functionalized end groups. The most common silane to anchor –NH_2 and –SH, respectively, are 3-aminopropyltriethyloxysilane (APTES), *p*-aminophenyl-trimethoxysilane (APTS), and mercaptopropyltriethoxysilane (MPTES). For example, Shen et al. have reported that APTES-coated Fe_3O_4 particles enable particle cytocompatibility and hemocompatibility due to acetylation of amine groups on the surface (Shen et al., 2012). Moreover, Wei et al. have showed that APTES is beneficial for preservation of Fe_3O_4 particles morphology (Wei et al., 2007). Advantageous at this occasion is generally that the magnetic properties are not affected by silane modification (Shen et al., 2012).

Other possibilities for surface coating with small molecules are oleic acid and oleylamine to obtain oil-soluble monodisperse nanoparticles. These consist of a C18 tail with cis-double-bond in the middle, forming a kink that might be necessary for an effective stabilization. Especially, oleic acid is widely used to establish a dense protective monolayer upon IONPs, providing highly uniform particles (Wu et al., 2008). Additionally, oleic acid coating has no appreciable effect upon magnetic properties of IONPs (Barbeta et al., 2010). To transfer hydrophobic coated IONPs into an aqueous phase, the coating has to be replaced by a hydrophilic coating (Wu et al., 2015). This could be achieved directly during synthesis or indirectly by a ligand-exchange method. Water-soluble IONPs can be synthesized directly by using small molecules like amino acid, citric acid, and vitamin in the reaction process (Lartigue et al., 2009; Chalasani and Vasudevan, 2012; Xiao et al., 2011). For instance, Majeed et al. have synthesized monodisperse and water-soluble magnetic IONPs through coprecipitation by using the water-soluble polymer ligand dodecanethiol-polymethacrylic acid (DDT-PMAA).

These synthesized particles were conjugated with the anticancer drug doxorubicin (DOX) as a drug delivery system model. Thus, the ligand allowed an effective control over the size of IONPs during synthesis, leading to longtime stability against aggregation and oxidation, improved water solubility, increased biocompatibility, and enabled a multifunctional surface (Majeed et al., 2013).

To conclude, it should be mentioned that the polarity can be changed by a ligand-exchange procedure, resulting in a change of the hydrophobic layer to being hydrophilic. This will be achieved by adding an excess of ligands to the nanoparticle solution. Consequently, the original ligand on the surface of the nanoparticle will be displaced (Wang et al., 2009b; Yang et al., 2013; Song et al., 2012). As an example, Dong et al. have reported about a sequential surface functionalization and phase transfer method of colloidal IONPs to obtain stabilized nanoparticles in various polar media without aggregation. Therefore, nitrosonium tetrafluoroborate ($NOBF_4$) was used to displace the organic ligand.

Due to the low binding affinity of the BF_4^- anions, secondary surface modifications were possible (Dong et al., 2010). Another commonly used possibility to modify the surfaces of nanoparticles are polymers. Polymers are generally advantageous for functionalization regarding the biological fate, like pharmacokinetics and biodistribution, of IONPs (Yang et al., 2010).

3.2.2 *Natural and Synthetic Polymer Coating*

A wide range of natural and synthetic biodegradable polymers are under evaluation as functionalizing materials, such as polyaspartate (Ray et al., 2012), polysaccharides (Chen et al., 2009; Chang et al., 2011; Dias et al., 2011), gelatine (Gaihre et al., 2008, 2009; Mahmoudi et al., 2011), starch (Dung et al., 2009; Alidokht et al., 2011; Lu et al., 2013), alginate (Hernández and Mijangos, 2009; Wu et al., 2010b; Kim et al., 2011a), polyacrylic acid (PAA) (Sondjaja et al., 2009; Moscoso-Londoño et al., 2012; Kang et al., 2012), PEG (Mukhopadhyay et al., 2011; Cole et al., 2011), PLA (Prashant et al., 2010; Tan et al., 2011; Zheng et al., 2009), chitosan (Wang et al., 2013a; Bhattacharya et al., 2011a), and polymethylmethacrylate (PMMA) (Rovers et al., 2010; Pimpha et al., 2012; Lan et al., 2011). The common in situ synthesis routes are mini-/microemulsion polymerization and the sol-gel process (Oh and Park, 2011; Zhou et al., 2012a), obtaining conventionally a core-shell or matrix-dispersed structure (El-Sherif et al., 2010). The control of colloidal stability and thickness of the obtained structures are still problematic. Therefore, postsynthesis routes are preferred, like one-pot route, self-assembly, or heterogeneous polymerization (Pimpha et al., 2012). Especially, the one-pot route is suitable to receive simple polymer-coated composite nanomaterials (Shkilnyy et al., 2010). Other functional techniques are physical adsorption, anchoring of functional groups, covalent bonding, and cross-linking that is based on alkyl chain or carboxylic acid functionalized thiol and hydrogen bonding (Lu et al., 2010). Furthermore, several heterogeneous polymerizations with water-soluble monomers show the successful preparation of well-defined core-shell or matrix-dispersed polymer-coated nanoparticles for biomedical applications (Amara and Margel, 2012; Douadi-Masrouki et al., 2010).

The usages of polymer coatings enabled an enhanced stability of IONPs and further an extended application field through varying multiple functional groups. For instance, conjugated polymers with a delocalized electronic structure are enabled by coupling with optoelectronic segments, imaging, diagnosis, and therapy (Zhu et al., 2012a). In addition, Wang et al. have reported about an organic/inorganic hybrid nanoparticle, which are magnetic IONPs modified with the fluorescent conjugated polyelectrolyte (BtPFN). The nanoparticles showed a response toward light excitation and external magnetic field. Therefore, these particles are suitable for cell imaging or can be used as multicolor probes to detect interactions of particles within living cells (Sun et al., 2010). Another possibility that is frequently used are IONPs functionalized with light-sensitive, pH-sensitive, or temperature-sensitive polymers to respond upon an environmental stimulus (Fan et al., 2013; Zhang et al., 2008). These could be used as drug delivery systems, in the framework of MRI or as biosensors (Leal et al., 2012; Santra et al., 2009; Chan and Gu, 2013). To conclude, IONPs with amphiphilic block copolymers, which incorporate a high number of more functional groups, are a field of great interest (Zhu and Hayward, 2008). Those are synthesized by a self-assembly method to obtain a designed stable complex of IONPs for film platforms, drug carriers, or trackers for MRI (Truby et al., 2013). To achieve higher biocompatibility, biomolecule functionalized magnetic IONPs are of great interest and will be described in the following.

3.2.3 *Functionalized Biomolecules as Targeting Ligands*

The functionalized biomolecules include enzymes, antibodies, proteins, biotin, bovine/human serum albumin, avidin, and polypeptides (Samanta et al., 2008; Iwaki et al., 2012; Okuda et al., 2012; Marcelo et al., 2013; Xie et al., 2010). For example, Magro et al. have reported about the synthesis and application of maghemite nanoparticles coated with avidin. They used the prepared nanoparticles for the large-scale and magnetic purification of recombinant biotinylated human sarco-/endoplasmic reticulum Ca^{2+}-ATPase (hSERCA-2a), expressed by *Saccharomyces cerevisiae*. As a result, 70% pure hSECRA-2a was obtained from 4L of yeast culture with a purification yield of 64% (Magro et al., 2012). The biomolecule modified IONPs can be used not only for purification but also for the detection of microbes. For instance, Bhattacharya et al. (2011b) used antibody-labeled Au-Fe_3O_4 nanocomposites against *Staphylococcus aureus* at ultralow concentrations. In conclusion, biomolecule functionalized nanoparticles can be an effective and successful achievement for the separation of different biologic species, such as proteins or DNA.

3.3 Inorganic Materials

Inorganic materials offer several properties, such as high electron density, strong optical absorption (e.g., Au and Ag), photoluminescence in the form of fluorescence (e.g., semiconductor quantum dots of CdSe or CdTe) or phosphorescence (doped oxide materials Y_2O_3), or magnetic moments (e.g., manganese or cobalt oxide nanoparticles) (Zedan et al., 2013; Wang et al., 2013b; Gowd et al., 2013; Ye et al., 2012). They are useful in providing stability to the nanoparticles in solution and valuable in binding various biological ligands to the iron oxide surface, such as described silica coatings, metal, or metal oxide coatings (Zhao et al., 2013; He et al., 2013; Wang et al., 2013c).

3.3.1 *Silica Coatings*

Silica coatings have different advantages, such as an enhanced dispersion in solutions due to the preservation of the magnetic dipolar attraction between magnetic IONPs. Next, the silica-coated particles provide an increased stability and resistance to acidic environment.

Lastly, the abundant silanol groups could be easily activated to bind various functional groups on the surface. Three synthesis methods exist to fabricate silica-coated nanoparticles. The first well-known method is the classical Stöber process (Stöber et al., 1968). Therefore, sol-gel precursors are hydrolyzed and condensed to obtain SiO_2-coated nanoparticles. The most common silanes, which easily bind on the surface of IONPs due to OH groups, are tetraethoxysilane (TEOS), vinyltriethoxysilane (VTEOS), and octadecyltrimethoxysilane (OTMS) (Zhu et al., 2010a; Shao et al., 2013; Cao et al., 2013). For example, with this method obtained, silica-coated particles are used as contrast agent in MRI or as drug carriers (Xuan et al., 2012). For a standardized application, experiment parameters, such as the amount of silane used, are important to obtain an adequate silica shell thickness, which allows efficient modifications with targeting ligands or a high response as contrast agent (Pinho et al., 2010).

The second method is based upon microemulsion synthesis. This method is suitable to obtain uniform silica-coated nanoparticles with a controllable thickness in the nanometer range. This method requires much coating regulations to separate core-shell nanoparticles from the large amount of surfactants associated with the microemulsion system (Park et al., 2010). For instance, Ding et al. have reported about several coating regulations for silica-coated Fe_3O_4 nanoparticles obtained by the reverse microemulsion method. As an example, they could present in detail the influence of the aqueous domain on coating thickness. Hence, ultrathin silica shells were obtained by a small aqueous domain, whereas the larger aqueous domain favors thicker shells (Ding et al., 2012). The last method is the aerosol pyrolysis of a precursor mixture composed of silicon alkoxides and metal compounds (Li et al., 2013b). Basak et al. (2011) have synthesized silica-coated γ-Fe_2O_3 particles and have confirmed the dependence on proper precursor materials to obtain efficient silica-coated nanoparticles.

3.3.2 *Carbon Coatings*

Carbon coatings are of great interest due to several advantages. Those offer good chemical and thermal stability, an intrinsic high electric conductivity, and an effective oxidation barrier and prevent the magnetic core material from corrosion. In addition, better dispersibility and stability is enabled by hydrophilic carbon coatings (Bae et al., 2012). Core-shell nanostructures are obtained through different synthesis approaches, for example, the three-step process, in which magnetic IONPs are prepared as seeds by various methods, then polymer-coated through polymerization process, and finally carbonated by annealing treatment. For instance, Lei et al. (2013) have demonstrated the synthesis of dopamine-coated α-Fe_2O_3 followed by a carbonization process, which showed good electrochemical performance as superior lithium ion anodes in batteries. Furthermore, much attention has been paid to the synthesis of Fe_3O_4/graphene hybrid materials as targeted drug delivery systems and MRI agents (Fan et al., 2013; Li et al., 2011a; Cong et al., 2010). Therefore, Chen et al. (2011) have demonstrated that those hybrid materials with an aminodextran coating had no effect on cellular viability and proliferation in comparison with bare IONPs and as a result are suitable for cellular MRI. A further possibility to modify IONPs is the functionalization with metallic or metallic oxides/sulfides components.

3.3.3 *Metallic and Metallic Oxides/Sulphides (Surface Functionalization of IONPs)*

Binary properties are offered by monodispersed iron oxide/metal nanostructures, such as core-shell, core-satellites, and dumbbell structures. With different surface functionalizations, like charges, functional groups or moieties, the stability and biocompatibility can be improved (Arsianti et al., 2011; Yallapu et al., 2011; Guo et al., 2009). The sequential growth of metallic components onto the surface of the IONPs is the most efficient and facile method. The appropriate methods to obtain core-shell, core-satellite, or dumbbell structures are the microemulsion or thermal decomposition method (Kirui et al., 2010; Chiang and Chen, 2009). However, the direct coating of IONPs with metal by thermal decomposition process is quite difficult, owing to the dissimilar nature of the two surfaces and lattice (Gu et al., 2004). The ideal coatings, due to their low reactivity with the IONPs, are gold and silver. In addition, the use of surfactants and additives to the synthesis process can modify stability and surface properties of the particles. Oleylamine is a common capping, stabilizing, and reductant agent (Kim et al., 2011b). Xu and Sun have reported about the synthesis of oleylamine-capped 10 nm Fe_3O_4 nanoparticles by thermal decomposition of iron (III) oleate. Afterward, those capped particles were used to obtain Fe_3O_4-Au and Fe_3O_4-Au-Ag particles by deposition Au and Ag on the surface of the Fe_3O_4 particles. Modifying the shell thickness by the amount of Au and Ag, the plasmonic property of the core-shell structure could be tuned to either redshifted or blueshifted (Xu and Sun, 2007). Multistep methods including the seed-mediated and emulsion methods are also suitable to obtain core-shell, aggregated, or hybrid metal/IONPs. The most known structure is the monolayer-capped core-shell Fe_3O_4-Au structure, which can be obtained by using Fe_3O_4 nanoparticles as the seed and afterward reducing Au ions. Those particles exhibit controllable surface properties (Wang et al., 2005).

Another multistep method is the layer-by-layer self-assembly to prepare multilayers or hybrid iron oxide/metal structures (Chiang and Chen, 2009; Zhai et al., 2009). Thereby, nanoparticles or discrete structures spontaneously organize into ordered structures by molecular interactions (Grzelczak et al., 2010). The chemical conjugation of organic surfactants with aliphatic/hydrophilic tail groups and/or hydrocarbon solvents is responsible for the formation of iron oxide/metal particles (Chiang and Chen, 2009; Ma et al., 2009). For instance, hybrid Fe_3O_4-Au nanorods were obtained by a layer-by-layer synthesis method, in which Au nanoparticles annealed onto the surface of FeOOH nanorods due to the strong electrostatic attraction between metal ions and polyelectrolyte-modified FeOOH nanorods (Ou et al., 2010). Furthermore, Truby et al. have reported about the synthesis of hybrid plasmonic-superparamagnetic particles (gold nanorods/SPIONs) by an aqueous-based self-assembly process. In detail, the Au nanorods were functionalized by carboxyl-bearing surface ligands, and afterward, the two components were mixed. Due to chemisorption between the carboxyl groups and the IONPs, the Au nanorods were equipped with SPIONs (Truby et al., 2013).

The abovementioned structures are based upon molecular or charged links between the IONPs and the metal component, whereby the dumbbell iron oxide/metal structures are premised on electron transfer across the nanometer contact at the interface of IONPs and the metal nanoparticles. This impact offers new properties, which are not present in each individual component (Wang et al., 2009c). For example, Sun et al. prepared dumbbell iron oxide/metal particles by controlling the nucleation and growth of only one Fe_3O_4 on each Au, Pt, or Pd seeding nanoparticle (Wang et al., 2009c; Sun et al., 2012a). Moreover, they investigated the formation mechanism that might be based on the "tug-of-war mechanism." The overgrowth of Au onto the Fe_3O_4-Au nanoparticles resulted in ternary nanostructures. Sun et al. would take into consideration that such processes result in an unbalanced stress across the interface. As a result, the added Au particles will extract the actual existing Au particles from the conjugation with Fe_3O_4 (Wang et al., 2009d). In conclusion, total synthesis frameworks for hybrid nanoparticle architectures have developed by Buck et al. to obtain complex hybrid molecules. These include M-Pt-Fe_3O_4 (M = Au, Ag, Ni, Pd) heterotrimers and M_xS-Au-Pt-Fe_3O_4 (M = Pb, Cu) heterotetramers (Buck et al., 2011).

Nevertheless, the limited chemical stability of the metal-coated iron oxide core-shell, aggregated, or dumbbell-shaped particles must be paid attention. Therefore, several multilayer iron oxide particle/metal components have been developed to enhance the protection of the core from oxidation and corrosion and to exhibit biocompatibility. In general, copolymers and branched polymers are used as stabilizers for the composites (Goon et al., 2009; Lim et al., 2009; Hsiao et al., 2010). For promoting colloidal stability in elevated ionic strength media, Lim et al. have reported about three different macromolecules (Pluronic F127, cationic polyelectrolyte polydiallyldimethylammonium chloride (PDDA), and PEG) to coat the performed core-shell iron oxide/Au particles. As a result, the combination of Pluronic F127 and PDDA yielded longer stable dispersion than the single coating with PEG (Yeap et al., 2012). Another effective route for promoting stability and protection of the composites is encapsulation. In addition, new structures, like the rattle structure could be obtained (Zhou et al., 2010). A universal material for encapsulation is SiO_2. For example, Fe_3O_4-Au hybrid nanocrystals were encapsulated in silica nanospheres, whereby after the nucleation of Au at the Fe_3O_4 surface, the Fe_3O_4 was dissolved through reduction. The maintained particles had a rattle structure that consisted of a porous silica nanoshell and Au nanocrystals (Kyung et al., 2010).

Metallic oxides/sulfides offer unique chemical and physical properties to protect or functionalize IONPs. The most common compounds, which are used to functionalize IONPs are oxide and sulfide semiconductors, like TiO_2 (Sun et al., 2012b, 2013; Wu et al., 2011b), ZnO (Wu et al., 2012b), SnO_2 (Wu et al., 2011c; Zhang et al., 2013), WO_3 (Xi et al., 2011), Cu_2O (Li et al., 2011b), CdS (Joseph et al., 2012; Zhou et al., 2012b; Shi et al., 2012), ZnS (Yu et al., 2009; Liu et al., 2013), PbS (Zhou et al., 2011), or Bi_2S_3 (Luo et al., 2012b). Several synthesis processes are available to obtain metallic oxides/sulfides functionalized IONPs. For example, Lee et al. have reported about a sol-gel reaction process, in which tantalum (V) ethoxide reacts in a microemulsion containing Fe_3O_4 particles, resulting in biocompatible and multifunctional Fe_3O_4/TaO_x core-shell nanoparticles. In addition, the functionalized particles exhibited a prolonged circulation time and could be observed by computer tomography (CT) and MRI (Lee et al., 2012). Furthermore, superparamagnetic and fluorescent Fe_3O_4/ZnS hollow nanospheres, with diameters smaller than 100 nm, were synthesized by Wu et al. using a simple synthesis strategy by combining the corrosion and Ostwald ripening process. The nontoxic nanospheres offered a porous shell and a good magnetic resonance and fluorescence (Wang et al., 2009e). Indeed, with semiconductor, coated IONPs are commonly used to obtain a bifunctional composite nanoparticle with different characteristics (Ou et al., 2010). The combination of such two different magnetic materials will generate new magnetic properties with new potential applications. For instance, Manna et al. have showed that the magnetization of core-shell nanoparticles composed of a ferrimagnetic Fe_3O_4 core-ferrimagnetic γ-Mn_2O_3 shell is greater than of bare IONPs. In addition, Liu et al. have synthesized multifunctional magnetic core-shell heteronanoarchitectures, which consisted of Fe_3O_4/NiO and Fe_3O_4/Co_3O_4 by an in situ solvothermal-coating/decomposition approach. Those particles exhibited an excellent magnetism and a large surface exposure, stable recyclability, and controllable shell thickness. Metal ions on the shell surface show a high affinity to biomolecules. Therefore, the

Fe_3O_4/NiO particles can be used to separate, for example, His-tagged proteins from cell lysates or to enrich peptides from complex samples as a purification step for mass spectrometry (Liu et al., 2012a).

In conclusion, magnetic IONPs that exhibit long blood retention time, biodegradability, and low toxicity are the preferred primary nanomaterials for biomedical applications. Nevertheless, a balancing act has to be performed between the final size, the nature of magnetically responsive components, and the composites including core and coatings. Those are important criteria to take into account concerning the biocompatibility and toxicity, more specifically the biomedical application. In the following section, several examples for biomedical applications of magnetic IONPs will be presented.

4 BIOMEDICAL APPLICATIONS

Due to their coating and functionalization and hence their broad variety of properties, magnetic IONPs are applied in several fields. They have especially shown to be useful in biomedicine. For this reason, the following chapter is dedicated to their importance for biomedical applications. The main focus will be put on the engagement of IONPs in either in vitro or in vivo applications. Among the in vitro applications, IONPs are used for building biosensors, improving bioseparation, detoxifying biological fluids, enabling magnetofection (MF), or repairing tissues. In contrast, the fields of chemotherapy and drug delivery, MRI and thermoablation rank among the possible in vivo applications (Wu et al., 2015; Gupta and Gupta, 2005). There are certain properties that increase the efficiency of IONPs when employed, regardless of the exact kind of application. Those features that have shown to be the most important are as follows: long blood retention time, biodegradability, low toxicity, and a core-shell-type functionalization (Wu et al., 2015; Mahmoudi et al., 2011). First, the benefits of IONPs employed in vitro will be depicted in the following paragraphs. Second, possible in vivo engagements of IONPs will be outlined.

4.1 In Vitro Applications

4.1.1 Biosensors

Biosensors are generally used for the detection of biomolecules like proteins, mRNA, pathogens, cells, and enzymatic activity (Perez et al., 2004; Grimm et al., 2004). Moreover, slight changes in MRI signal or molecular interactions can be identified using biosensors. As can be seen in Table 2, there are several approaches for designing biosensors and diagnostic platforms. Especially, the optical-magnetic bead-based approach containing modified IONPs has crucial advantage over other approaches. This usage promises an excellent optical performance. The explanation for this phenomenon can be given with regard to the strong signals measured by common methods such as localized surface plasmon resonance (LSPR), surface-enhanced Raman scattering (SERS), or fluorescence signal measuring (Wang et al., 2011b; Depalo et al., 2011). To optimize the detection rate of IONP-containing magnetic bead-based biosensors and, hence, the overall efficiency of these biosensors, they can be coated with either gold or silver. Further, quantum dots can be used, or fluorescent molecules can

TABLE 2 Magnetic Bead-Based Biosensors Sorted by Different Key Features

Functional Material Groups	Materials	Groups		
	Polymers, nanotubes, metallic NPs	Hydroxyl, amine, phosphate, etc.		
Detected Signals	**Optical**	**Electrochemical**	**Magnetic**	**Mechanical**
	Fluorescence, LSPR, SERS	Potentiometric, amperometric, voltammetric	Magnetoresistance, magnetoimpedance, Hall effect	Suspended force
Targeted Receptors	**Biomolecules**	**Aptamer**	**Molecules**	
	Enzymes, proteins, antibodies, lectins	ssDNA, RNA cells,	Polymers, drugs, dye, putrescine	

This table provides an overview but does not claim to be complete. Abbreviations: *LSPR*, localized surface plasmon resonance; *NPs*, nanoparticles; *RNA*, ribonucleic acid; *SERS*, surface-enhanced Raman scattering; *ssDNA*, single strand deoxyribonucleic acid.
Based on Wu, W., Wu, Z., Yu, T., Jiang, C., Kim, W.S., 2015. Recent progress on magnetic iron oxide nanoparticles: synthesis, surface functional strategies and biomedical applications. Sci. Technol. Adv. Mater. 16 (2) 023501.

be attached to them. Moreover, one-dimensional nanostructures are applied as functional components for electrochemical magnetic bead-based biosensors (Li et al., 2009; Pérez-López and Merkoçi, 2011).

In clinical diagnosis, they are applied as immunosensors to determine the concentration of glucose, lactate, cholesterol, urea, creatine, and creatinine (Stanley et al., 2012; Nouira et al., 2011; Teymourian et al., 2012; Sun et al., 2012c). These modern biosensors have been shown to have higher sensitivity and specificity levels and a quicker respond time (Yang et al., 2009). They are constantly improved by aiming at the further increase of sensitivity and specificity (Li et al., 2012; Bellan et al., 2011). For instance, this could be achieved attaching antibodies against certain epitopes of β-human chorionic gonadotropin (βhCG) to the surface of an IONP-enhanced grating-coupled surface plasmon resonance (GC-SPR) sensor (Wang et al., 2011b). Considering current demands concerning biosensors, the trend is moving toward the development of multiplexed magnetic bead-based biosensors for multidetection (Arrabito and Pignataro, 2012). The advantages and disadvantages of biosensors for clinical use have still to be outweighed and evaluated since there are several challenges and limitations that have to be overcome and the only reliable application at present are magnetic bead-based biosensors detecting glucose (Monošík et al., 2012). In addition to this mode of engagement, magnetic nanoparticles can also be applied as biosensors in the framework of bioseparation.

4.1.2 Bioseparation

SPIONs find further use in the separation of biological components, for example, in vitro DNA, antibodies, proteins, genes, enzymes, cells, viruses, and bacteria (Shao et al., 2009; Chen et al., 2010; Du et al., 2010; Huidan et al., 2010). They are especially adapted for this task due to their superparamagnetic properties that allow control over NPs by an external magnetic field. Thus, their place of residence can be externally determined. For this purpose, a common magnet can be used making this method very cost-effective and faster when compared with column affinity chromatography (Wu et al., 2009). Efficiency can further be increased by adding functional end groups (Latham and Williams, 2008). All these advantages can only be taken for granted, when certain parameters are under constant monitoring. These include chemical composition, particle size, size distribution, stability of magnetic properties, morphology, adsorption properties, and low toxicity (Wu et al., 2015). In this course, it is very important to ensure a core-shell-type design as mentioned above in addition to a silica coating (Wu et al., 2015). This combination has been reported to have positive effects like enhanced removal capability of these nanoparticles from different fluids and can therefore be reused in further experiments. The adsorption rate to the biological component of interest could also be increased which in this case was the His-tagged protein from *Escherichia coli* lysate (Shao et al., 2012). These findings emphasize the relevance and necessity of further improvements in the field of bioseparation using magnetic nanoparticles. Moreover, there is an approach that takes advantage of magnetic nanoparticles relating to some extent to the principles of bioseparation. However, due to its unique features, the so-called detoxification of biological fluids will be separately discussed in the following paragraphs.

4.1.3 Detoxification of Biological Fluids

This method describes the process of isolating living cells from biological fluids containing toxic substances. Detoxification is implemented by using cell surface antigens for cell nanoparticle binding. It has been shown that positive selection of tumor cells from cell suspension could be achieved by coupling an antibody against tumor surface antigens, namely, epithelial specific antigen (ESA) to magnetic beads leaving erythrocytes and leukocytes behind, while retracting tumor cells bound to the antibody (Kemmner et al., 1992). The size of IONPs varied between 50nm and a few microns. In some cases, the nanoparticle matrix consists of silica or polystyrene (Gupta and Gupta, 2005). The extracted tumor cells were then purified on a magnet rack. The purity, recovery rate, and condition of the isolated cells depended on the number of washing steps and the composition of used buffers and specifications of the beads (Kronick and Gilpin, 1986). Further, it has been reported about the employment of magnetic nanoparticles for hemosorption when injected into extracorporeal systems. In comparison with other approaches serving the same purpose (hemoperfusion, hemodialysis, and plasmapheresis), it is simpler and has higher efficacy rates and lower expenses. Magnetic beads do not destroy blood cells and can easily be removed by using a high-gradient magnetic separator where the purified blood is returned to the organism afterward. This has been shown in an in vivo model for the removal of barbiturates (Kutushov et al., 1997). More recent research shows that it is possible to perform biohazard detoxification of biological, chemical, or radioactive substances from humans using magnetic PEGylated PLGA/PLA nanospheres with a size range of 100–500nm. They were coupled to receptor antigens that had high binding affinities toward the toxin of interest and were injected intravascularly. After the capturing of blood-borne toxins, removal was performed by a handheld magnetic filter unit inserted in a suitable artery or vein. The clean blood is returned to the body after purification of the blood with the magnetic filter unit (Rosengart, 2007).

4.1.4 Magnetofection

Another attempt to apply magnetic IONPs is the so-called magnetofection (MF) approach. Key factors enabling this method are IONPs that are coupled to vector DNA and guided by the influence of an external magnetic field. By this means, DNA can be transfected into cells of interest. One possibility to enable enhanced binding capabilities of the negatively charged DNA to magnetic IONP beads is the coating IONPs with a positively charged material such as polyethylenimine. The efficiency of the vectors has hence shown to increase up to several thousand times (Scherer et al., 2002). The above depicted engagement of IONPs in MF has shown to be universally applicable to viral and nonviral vectors. This is mostly because it is very rapid and simple. Furthermore, it is a very attractive approach since it yields saturation level transfection at low-dose in vitro (Krotz et al., 2003). Fernandes and Chari (2016) have demonstrated an approach delivering DNA minicircles (mcDNA) to neural stem cells (NSCs) by means of MF. DNA minicircles are small DNA vectors encoding essential gene expression components but devoid of a bacterial backbone, thereby reducing construct size versus conventional plasmids. This could be shown to be very beneficial for the use of genetically engineered NSC transplant populations in regenerative neurology. The aim was to improve the release of biomolecules in ex vivo gene therapy. It could be demonstrated that MF of DNA minicircles is very safe and provided for sustained gene expression for up to 4 weeks. It is described to have high potential as clinically translatable genetic modification strategy for cell therapy (Fernandes and Chari, 2016). The last in vitro application for magnetic nanoparticles to be presented in this chapter will be tissue repair.

4.1.5 Tissue Repair

Tissue repair is accomplished by choosing either of the following two methods: welding and soldering. Welding describes the process of opposing two tissue surfaces and heating the tissues sufficiently to join them. As for soldering, protein or synthetic polymer-coated nanoparticles are placed between two tissue surfaces to facilitate joining of the two tissues. The minimum temperature needed to implement enable tissue joining is about 50°C (Lobel et al., 2000). In the case of magnetic nanoparticles, it is taken advantage of their strong light-absorbing properties. This is due to the fact that heat generation and subsequent joining of the tissues of interest is initiated by lasers. The nanoparticles that are employed for this method are coated between the surfaces that should be connected. For the application of gold- or silica-coated IONPs, the successful minimization of tissue damage has been reported (Xu et al., 2004; Sokolov et al., 2003). Importantly, this could be taken advantage for stem cell targeting in vivo. Since current limitations of stem-cell-mediated replacement or reparation of damaged cells make it difficult to establish a stem-cell-based therapy, new and improved methods are needed (Gupta and Gupta, 2005). Coupling SPIONs to the desired cells could make it possible to guide stem cells more precisely to their target in vivo. In addition, growth factors and other proteins having a positive effect on tissue development could be used as supplementary substances (Sensenig et al., 2012). A successful application of this method would promise better prognosis for patients suffering from degenerative diseases like Alzheimer's or Parkinson's disease. To make this method feasible, the ability to target and activate these stem cells via magnetic particle technology would be used (Bulte et al., 2001). Since we took a closer look to some feasible in vitro applications, the aim of the following paragraphs will be to complement these by some important in vivo applications of which the first topic to be discussed about is the employment of magnetic nanoparticles as magnetic vectors.

4.2 In Vivo Applications

4.2.1 *Chemotherapy and Drug Delivery*

In this framework, magnetic nanoparticles are often used for targeting purposes in targeted drug delivery. Because of their magnetic properties, guiding of these particles by an external magnetic field is implementable as it has been described in the chapter "bioseparation." Hereby, only the tissue of interest is being targeted, leaving out healthy tissue. This approach has shown to be very beneficial in the treatment of cancer, where conventional chemotherapeutics cannot be targeted as precise (Marszałł, 2011). This often results in the destruction of tumor cells and healthy cells, which contributes to the adverse effects of cytotoxic agents. Using magnetic nanoparticles, this hurdle has partially been overcome (Cao et al., 2008). One of the key factors for this is the efficient intracellular delivery of cytotoxic agents. The overall structure of this new composite consists of the magnetic nanoparticle as the core and biocompatible polymers as the shell. Within the shell of these therapeutic particles, the actual cytotoxic agent or any other kind of drug is embedded in the polymer matrix allowing for the drug to be more easily released at the target. Considering the particle size, which ranges from 10 to 100 nm, they are rather designated for intravenous administration (Laurent et al., 2008; Amstad et al., 2011). Furthermore, their behavior in body fluids is tunable in small increments by changing key properties. These include charge, hydrophobicity, and stability to mention a few (Banerjee and Chen, 2010).

This contributes to the fact that this so-called active-targeting strategy has been used in different tumor-related contexts, yielding promising results. For instance, IONPs have been shown to be the most promising alternative in terms of gene carriers (Hasenpusch et al., 2012) and drug carriers (Banerjee and Chen, 2010; Zhu et al., 2010b; Talelli et al., 2009). The first was the case for Qiu et al. whose aim was to transfect cells with mcDNA and establishes a noninvasive monitoring method. They were able to get a strong MRI signal using a special gene delivery system. It consisted of SPION nanocrystals in the core and a shell of biodegradable stearic acid-modified low-molecular-weight polyethyleneimine (Stearic-LWPEI) formed by self-assembly. This made it possible to track the genes on their way to the target site. On one hand, this was partially due to beneficial properties of the magnetic nanocrystals like ultrasensitive imaging capacity or a narrow size distribution. On the other hand, high binding affinity of the Stearic-LWPEI system provided for protection of mcDNA from enzymatic degradation and provided a successful MF in the first place (Wan et al., 2013). Concerning the release of cytotoxic agents, instancing DOX, different mechanisms have been described in which release upon enzymatic activity was mentioned. Moreover, changes in physiological condition like enhanced pH sensitivity due to functional groups on the particle surface were described (Kim et al., 2009). Further desired properties include thermolability or photomagnetic sensitivity (Zhu et al., 2012b). Moreover, diseases originating in the brain, namely, Alzheimer's disease or brain tumors, could be addressed more easily harnessing the features of nanoparticles. For example, they are capable of crossing the blood-brain barrier, a feature ranging among the most important improvements of these targeted drug delivery systems (Patel et al., 2010). The future tendencies for a patient's treatment with drug delivery systems will be their applicability in both diagnostic and therapeutic issues. As this is becoming more and more important, a special term for this combinatorial approach was coined: theranostics.

To fill this emerging niche, a lot of research groups are currently working on different approaches with the aim of creating even more effective therapeutic systems. One of these was the attempt to combine SPIONs with DOX embedded in a matrix of PEG and folic acid (FA) (Kaaki et al., 2012). This therapeutic setting was used to enable bimodal cancer imaging by MRI and fluorescence in addition to bimodal cancer therapy using DOX and hyperthermia. The latter therapeutic approach will be discussed in an own chapter further below. Even though there is a good yield of promising results, therapeutic systems consisting of magnetic nanoparticles in combination with the drug are of interest. However, crucial limitations are persisted especially to clinical trials. This relates especially to clinical trials, where these particles have shown their biocompatibility on the one hand, but on the other hand, they have also to be held responsible for immune responses and toxicity (Wu et al., 2015).

4.2.2 *Magnetic Resonance Imaging*

As previously mentioned, magnetic nanoparticles are also applicable as magnetic contrast agents in MRI. It is particularly the SPIONs that are made use of in this application. Independent of the desired weighing of the image for the MRI graph, SPIONs are capable of enhancing the overall quality of the image. Among the most frequent field of employment, oncology-related imaging is to be listed (Peng et al., 2008; Yue-Jian et al., 2010; Chow et al., 2010). Active targeting strategies do only partly rely on an external magnetic field. Nevertheless, cells can easily be tracked and labeled using this method. This could be important in the course of the analysis of biological processes or the monitoring of cell therapies (Tsai et al., 2010; Wu et al., 2006; Asanuma et al., 2010). Magnetic contrast agents have some advantages over conventional ones in terms of delay until imaging, precision of tissue differentiation, and easier detection of artifacts. The first aspect has to be understood as the time of contrast agent injection to the time point when first images may be recorded. This shortage of delay is beneficial in many regards. First, neither the patient nor the examiner are ought to wait a long time until examination can take place. Second, magnetic agents would not have to last a long time in the patient's body, sparing him from side effects due to long time to agent excretion since most of the conventional MRI contrast agents have shown to be very toxic (Kanda et al., 2016; Semelka et al., 2016). Concerning the second advantage, magnetic MRI contrast agents have over conventional agents; it has been reported about more pronounced differentiation capabilities of different tissue types. This was the case for liver and blood vessel tissue leading to easier determination of liver lesions. The third advantage is mostly related to artifacts emerging due to aortic pulsation that could be recognized more pronounced preventing from hasty diagnostic conclusions to be drawn (Bulte and Kraitchman, 2004).

4.2.3 *Magnetic Hyperthermia or Thermoablation*

Another promising field of application especially for SPIONs is their employment as hyperthermia or thermoablation agents in the framework of cancer treatment (Pankhurst et al., 2009; Roca et al., 2009). Therefore, it is important to know the difference between hyperthermia and thermoablation. The terms hyperthermia and thermoablation refer to the artificial overheating of the body. For hyperthermia, this is the case for either local (37°C), regional (42–47°C), or systemic (40–42°C) overheating for 3–4 h, 30–60, or 60–90 min, respectively, whereas thermoablation describes the process of regional overheating with a minimum of 50°C for 4–6 min (van der Zee et al., 2008). There are several techniques to implement hyperthermia in tumor therapy including local external, intraluminal, or interstitial approaches. Furthermore, hyperthermia can be induced by perfusion of heated fluids through the body region of interest. Another possibility is the introduction of energy to the whole body while energy losses are minimized (van der Zee et al., 2008).

As regards thermotolerance of normal tissue, it accepts overheating up to 43°C for a period of 5 min. Hence, this heat-induced stress allows more effective application of chemotherapy and radiotherapy due to higher cell sensitivity (Hildebrandt et al., 2002). It has been demonstrated that the preferential death of tumor cells can be achieved with temperatures ranging from 41°C to 47°C (Thomas et al., 2009). After injection of SPIONs with a size of 2–20 nm and their active magnetic guiding to the tissue of interest, these particles agglomerate either inside cell organelles like mitochondria if positively charged and disturb tumor cell membrane integrity if negatively charged (Mahmoudi et al., 2011). This mechanism is also depicted in Fig. 2. Heating of SPIONs is achieved using an alternating magnetic field (AMF). It is hypothesized that the actual reason for tumor cell

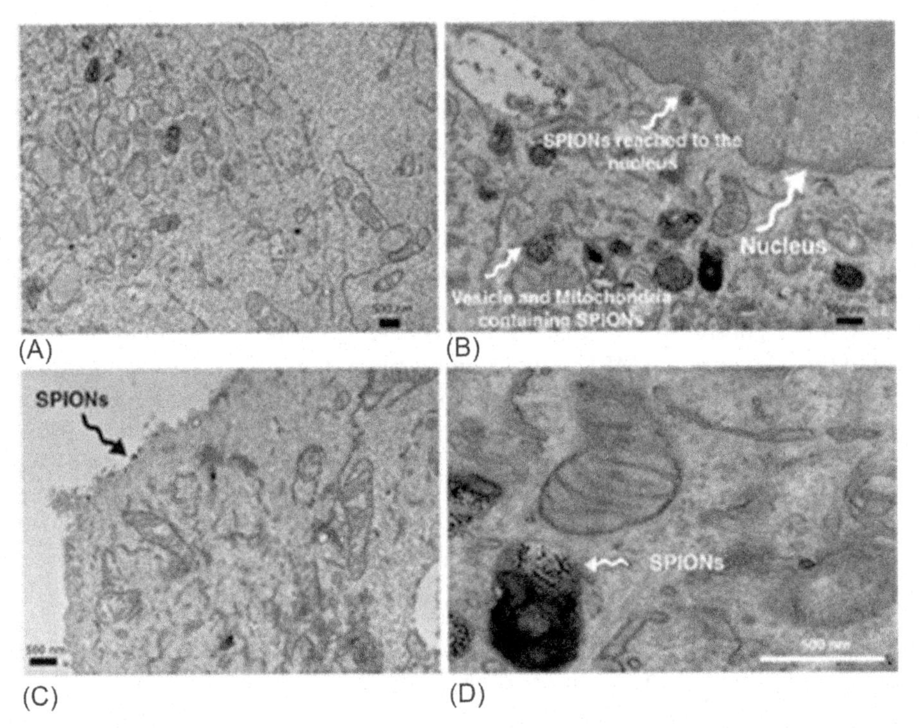

FIG. 2 TEM images of HeLa cells: (A) control and exposed to (B) positively or (C) negatively charged SPIONs and (D) the internalization magnetite nanoparticles inside of mitochondria. *Reprinted with permission from Mahmoudi, M., Sant, S., Wang, B., Laurent, S., Sen, T., 2011. Superparamagnetic iron oxide nanoparticles (SPIONs): development, surface modification and applications in chemotherapy. Adv. Drug Deliv. Rev. 63, 24–46. doi:10.1016/j.addr.2010.05.006 (unpublished work by M. Mahmoudi).*

death can be explained by the frequency of the AMF that make magnetic nanoparticles release heat as soon as particle relaxation is allowed when the AMF is switched off. This is supposed to cause necrosis of the tissue (Ivkov et al., 2005). Another theory states that due to alignment of internalized and aggregated SPIONs along an external magnetic field, magnetic moments and thus field friction are generated due to coherent movements of SPIONs. Hence, changes in cytoskeleton structure and subsequently tumor cell death are caused (Hapuarachchige et al., 2016). To quantify the extent to which SPIONs are capable of producing heat, a parameter called specific adsorption rate (SAR) has been defined. The SAR is proportional to the rate of the temperature increase in a certain tissue. It has to be considered that this value heavily depends on the geometry and the location of the body part that is exposed to the AMF, the energy of the applied radio frequency (RF) (Virtanen et al., 2006). Another important parameter is the particle size that has been demonstrated to correlate also with SAR inasmuch as SAR values increased with growing particle size (Gonzales-Weimuller et al., 2009). Positive effects on SAR and subsequently on increased heating rates in magnetic IONPs could be observed under conditions where the ferrofluidic solution was monodisperse, anisotropy was enabled, the electric field strength was high, and surface functionalization and coating of the particles were appropriately designed and implemented. This is the case for a decrease in coating thickness and the inorganic nature of the coating, which applies, for example, for a gold coating (Gonzales-Weimuller et al., 2009; Liu et al., 2012b; Mohammad et al., 2010). The range of feasible applications for magnetic IONPs in hyperthermia or thermoablation therapy is very diversified. For example, Kalber et al. have recently reported about a tumor-targeting approach using mesenchymal stem cells (MSC) as carriers for magnetic IONPs in the framework of hyperthermia therapy. AMF was used to heat IONPs upon arrival at the target site. Among the six different coatings that were tested, SPIONs with carboxydextran coating yielded the most promising results. The experimental setting intended for an immunosuppressed mouse model infected with OVCAR-3 cells. It could be shown that even though a heat increase could be achieved in IONPs upon exposure to AMF, there were almost no differences in tumor volume. Thus, a model of immunocompetent mice from another research group served as control. The results have demonstrated that AMF therapy was more successful in mice whose immune system was intact implying that it must act as an initiator of apoptosis upon heating of IONPs (Kalber et al., 2016). Other interesting results regarding tumor-targeting strategies using IONPs have recently been published (Haghniaz et al., 2016; Pala et al., 2014). Besides these studies, there also is a product readily available on the market that is already approved for the treatment of brain tumors. This product is called NanoTherm and is produced by MagForce, Berlin. As for other fields of application like prostate CA and esophagus CA, it still has to undergo clinical trials (Jordan et al., 2007). Besides all these promising data, there are some hurdles to be overcome, for example, the controllability of size and shape when manufacturing SPIONs.

Moreover, the biocompatibility of SPIONs is rather low, and their employment in fine tissues is questionable and yet to be tested thoroughly (Wu et al., 2015). In contrast, recent studies have shown that even though considerably high concentrations of injected SPIONs remain at the injection site for at least 3 weeks, there was no proof of toxicity for the organism as long as concentrations were within the desired dose range (Jarockyte et al., 2016).

5 CONCLUSION REMARKS AND FURTHER DIRECTIONS

Firstly, the effort of researchers developing and improving synthesis methods for IONPs could be pointed out. Among the existing synthesis routes, the coprecipitation and the thermal decomposition method could ensure the highest level of monodispersity, including control of size and morphology. Concerning the other synthesis methods, there are still certain flaws to be patched. Secondly, with regard to surface functionalization, yet another source for the amelioration of IONPs can be identified. Depending on the coating and the experimental setting and how those were used, key parameters allowing for the outcome of the experiment to be influenced in a positive way could be established. These features include long blood retention time, biodegradability and low toxicity to name the most important. These establishments have enabled further possibilities for biomedical applications. These prerequisites have inevitably led to the versatile application of magnetic IONPs. It has become clear that these particles do not only offer an alternative for already established biomedical applications but also have outperformed some of them. This is due to the easy synthesis on one hand and the favorable alteration of nanoparticle properties on the other hand. Since this field of research is still young, it should only be a matter of time until new areas are opened up leading to new promising approaches. Concerning the previously mentioned aspects, it has clearly emerged that IONPs have more and more become an essential component of biomedicine. However, there is a problem with standardized survey methods for toxicity determination. Even though over 1700 studies have been published with regard to nanotoxicity, IONP toxicity is still being controversially discussed. This is mostly owed to the fact that there are an unmanageable variety of studies that cannot be easily compared among themselves (Wu et al., 2015; Valdiglesias et al., 2016). Further, there are a very high number of factors by which toxicity is influenced, such as cell and tissue type, time of exposition, or way of administration, just to name a few, and this only applies to the in vitro results. Concerning the in vivo studies, it has to be mentioned that there are almost no useful data to be consulted regarding this question. The same applies to epidemiological studies about magnetic IONPs. The number of new developed and examined nanoparticles increases more rapidly than the number of survey methods to evaluate their potential toxicity (Valdiglesias et al., 2016).

References

Abid, A.D., Kanematsu, M., Young, T.M., Kennedy, I.M., 2013. Arsenic removal from water using flame-synthesized iron oxide nanoparticles with variable oxidation states. Aerosp. Sci. Technol. 47, 169–176.

Ai, Z., Deng, K., Wan, Q., Zhang, L., Lee, S., 2010. Facile microwave-assisted synthesis and magnetic and gas sensing properties of Fe_3O_4 nanoroses. J. Phys. Chem. C 114, 6237–6242.

Alidokht, L., Khataee, A., Reyhanitabar, A., Oustan, S., 2011. Reductive removal of Cr (VI) by starch-stabilized Fe0 nanoparticles in aqueous solution. Desalination 270, 105–110.

Amara, D., Margel, S., 2012. Synthesis and characterization of superparamagnetic core–shell micrometresized particles of narrow size distribution by a swelling process. J. Mater. Chem. 22, 9268.

Amara, D., Grinblat, J., Margel, S., 2012. Solventless thermal decomposition of ferrocene as a new approach for one-step synthesis of magnetite nanocubes and nanospheres. J. Mater. Chem. 22, 2188–2195.

Amstad, E., Textor, M., Reimhult, E., 2011. Stabilization and functionalization of iron oxide nanoparticles for biomedical applications. Nanoscale 3 (7), 2819–2843.

Arrabito, G., Pignataro, B., 2012. Solution processed micro- and nano-bioarrays for multiplexed biosensing. Anal. Chem. 84 (13), 5450–5462.

Arsianti, M., Lim, M., Lou, S.N., Goon, I.Y., Marquis, C.P., Amal, R., 2011. Bi-functional gold-coated magnetite composites with improved biocompatibility. J. Colloid Interface Sci. 354 (2), 536–545.

Asanuma, T., Ono, M., Kubota, K., Hirose, A., Hayashi, Y., Saibara, T., et al., 2010. Super paramagnetic Iron Oxide MRI shows defective Kupffer cell uptake function in non-alcoholic fatty liver disease. Gut 59 (2), 258–266.

Babincova, M., et al., 2001. High-gradient magnetic capture of ferrofluids: implications for drug targeting and tumor immobilization. Z. Naturforsch. C 56, 909–911.

Bae, H., Ahmad, T., Rhee, I., Chang, Y., Jin, S.U., Hong, S., 2012. Carbon-coated iron oxide nanoparticles as contrast agents in magnetic resonance imaging. Nanoscale Res. Lett 7 (1), 44–48.

Banerjee, S.S., Chen, D.-H., 2010. Grafting of 2-hydroxypropyl-β-cyclodextrin on gum arabic-modified iron oxide nanoparticles as a magnetic carrier for targeted delivery of hydrophobic anticancer drug. Int. J. Appl. Ceram. Technol. 7 (1), 111–118.

Bang, J.H., Suslick, K.S., 2010. Applications of ultrasound to the synthesis of nanostructured materials. Adv. Mater. 22 (10), 1039–1059.

Barbeta, V.B., Jardim, R.F., Kiyohara, P.K., Effenberger, F.B., Rossi, L.M., 2010. Magnetic properties of Fe_3O_4 nanoparticles coated with oleic and dodecanoic acids. J. Appl. Phys. 107 (7), 073913.

Basak, S., Tiwari, V., Fan, J., Achilefu, S., Sethi, V., Biswas, P., 2011. Single step aerosol synthesis of nanocomposites by aerosol routes: γ- Fe_2O_3/SiO_2 and their functionalization. J. Mater. Res. 26 (10), 1225–1233.

Bazylinski, D.A., Frankel, R.B., 2004. Magnetosome formation in prokaryotes. Nat. Rev. Microbiol. 2, 217–230.

Behrens, S., 2011. Preparation of functional magnetic nanocomposites and hybrid materials: recent progress and future directions. Nanoscale 3, 877–892.

Bellan, L.M., Wu, D., Langer, R.S., 2011. Current trends in nanobiosensor technology. Wiley Interdiscip. Rev. Nanomed. Nanobiotechnol. 3 (3), 229–246.

Berry, C.C., Curtis, A.S.G., 2003. Functionalisation of magnetic nanoparticles for applications in biomedicine. J. Phys. D. Appl. Phys. 36, 198–206.

Bharde, A., Parikh, R.Y., Baidakova, M., Jouen, S., Hannoyer, B., Enoki, T., Prasad, B.L.V., Shouche, Y.S., Ogale, S., Sastry, M., 2008. Bacteria-mediated precursordependent biosynthesis of superparamagnetic iron oxide and iron sulfide nanoparticles. Langmuir 24 (11), 5787–5794.

Bhattacharya, D., Das, M., Mishra, D., Banerjee, I., Sahu, S.K., Maiti, T.K., Pramanik, P., 2011a. Folate receptor targeted, carboxymethyl chitosan functionalized iron oxide nanoparticles: a novel ultradispersed nanoconjugates for bimodal imaging. Nanoscale 3 (4), 1653–1662.

Bhattacharya, D., Chakraborty, S.P., Pramanik, A., Baksi, A., Roy, S., Maiti, T.K., Ghosh, S.K., Pramanik, P., 2011b. Detection of total count of Staphylococcus aureus using anti-toxin antibody labelled gold magnetite nanocomposites: a novel tool for capture, detection and bacterial separation. J. Mater. Chem. 21, 17273–17282.

Blanco-Andujar, C., Ortega, D., Pankhurst, Q.A., Thanh, N.T.K., 2012. Elucidating the morphological and structural evolution of iron oxide nanoparticles formed by sodium carbonate in aqueous medium. J. Mater. Chem. 22, 12498–12506.

Bonnemain, B., 1998. Superparamagnetic agents in magnetic resonance imaging: physiochemical characteristics and clinical applications—a review. J. Drug Target. 6, 167–174.

Buck, M.R., Bondi, J.F., Schaak, R.E., 2011. A total-synthesis framework for the construction of high-order colloidal hybrid nanoparticles. Nat. Chem. 4 (1), 37–44.

Bulte, J.W.M., Kraitchman, D.L., 2004. Iron oxide MR contrast agents for molecular and cellular imaging. NMR Biomed. 17 (7), 484–499.

Bulte, J.W., Douglas, T., Witwer, B., Zhang, S.C., Strable, E., Lewis, B.K., et al., 2001. Magnetodendrimers allow endosomal magnetic labeling and in vivo tracking of stem cells. Nat. Biotechnol. 19 (12), 1141–1147.

Burda, C., Chen, X., Narayanan, R., El-Sayed, M.A., 2005. Chemistry and properties of nanocrystals of different shapes. Chem. Rev. 105 (4), 1025–1102.

Cai, W., Wan, J., 2007. Facile synthesis of superparamagnetic magnetite nanoparticles in liquid polyols. J. Colloid Interface Sci. 305 (2), 366–370.

Cao, S.-W., Zhu, Y.-J., Ma, M.-Y., Li, L., Zhang, L., 2008. Hierarchically nanostructured magnetic hollow spheres of Fe3O4 and γ-Fe2O3: preparation and potential application in drug delivery. J. Phys. Chem. C 112 (6), 1851–1856.

Cao, Z., Yang, L., Ye, Q., Cui, Q., Qi, D., Ziener, U., 2013. Transition-metal salt-containing silica nanocapsules elaborated via salt-induced interfacial deposition in inverse miniemulsions as precursor to functional hollow silica particles. Langmuir 29 (22), 6509–6518.

Capek, I., 2004. Preparation of metal nanoparticles in water-in-oil (w/o) microemulsions. Adv. Colloid Interf. Sci. 110 (1–2), 49–74.

Caruntu, D., Caruntu, G., O'Connor, C.J., 2007. Magnetic properties of variable-sized Fe_3O_4 nanoparticles synthesized from non-aqueous homogeneous solutions of polyols. J. Phys. D. Appl. Phys. 40 (19), 5801.

Chalasani, R., Vasudevan, S., 2011. Form, content, and magnetism in iron oxide nanocrystals. J. Phys. Chem. C 115 (37), 18088–18093.

Chalasani, R., Vasudevan, S., 2012. Cyclodextrin functionalized magnetic iron oxide nanocrystals: a hostcarrier for magnetic separation of non-polar molecules and arsenic from aqueous media. J. Mater. Chem. 22, 14925–14931.

Chan, T., Gu, F., 2013. Development of acolorimetric, superparamagnetic biosensor for the capture and detection of biomolecules. Biosens. Bioelectron. 42, 12–16.

Chang, P.R., Yu, J., Ma, X., Anderson, D.P., 2011. Polysaccharides as stabilizers for the synthesis of magnetic nanoparticles. Carbohydr. Polym. 83 (2), 640–644.

Chen, H., Deng, C., Li, Y., Dai, Y., Yang, P., Zhang, X., 2009. A facile synthesis approach to C8-functionalized magnetic carbonaceous polysaccharide microspheres for the highly efficient and rapid enrichment of peptides and direct MALDI-TOF-MS analysis. Adv. Mater. 21, 2200–2205.

Chen, F., Shi, R., Xue, Y., Chen, L., Wan, Q.-H., 2010. Templated synthesis of monodisperse mesoporous maghemite/silica microspheres for magnetic separation of genomic DNA. J. Magn. Magn. Mater. 322 (16), 2439–2445.

Chen, W.H., Yi, P.W., Zhang, Y., Zhang, L.M., Deng, Z.W., Zhang, Z.J., 2011. Composites of aminodextran-coated Fe_3O_4 nanoparticles and graphene oxide for cellular magnetic resonance imaging. ACS Appl. Mater. Interfaces 3 (10), 4085–4091.

Chiang, I.C., Chen, D.H., 2009. Structural characterization and self-assembly into superlattices of iron oxide-gold core–shell nanoparticles synthesized via a hightemperature organometallic route. Nanotechnology 20 (1), 015602.

Chow, A.M., Chan, K.W.Y., Cheung, J.S., Wu, E.X., 2010. Enhancement of gas-filled microbubble R2* by iron oxide nanoparticles for MRI. Magn. Reson. Med. 63 (1), 224–229.

Cole, A.J., David, A.E., Wang, J., Galbán, C.J., Yang, V.C., 2011. Magnetic brain tumor targeting and biodistribution of long-circulating PEG-modified, cross-linked starch-coated iron oxide nanoparticles. Biomaterials 32 (26), 6291–6301.

Cong, H.P., He, J.J., Lu, Y., Yu, S.H., 2010. Water-soluble magnetic-functionalized reduced graphene oxide sheets: in situ synthesis and magnetic resonance imaging applications. Small 6 (2), 169–173.

Cornelis, K., Hurlburt, C.S., 1977. Manual of Mineralogy, 19th ed. Wiley, New York.

Cornell, R.M., Schwertmann, U., 2003. In: Weinheim (Ed.), The Iron Oxides: Structure, Properties, Reactions, Occurrences and Uses, second ed. Wiley-VCH.

Costo, R., Bello, V., Robic, C., Port, M., Marco, J.F., Puerto Morales, M., Veintemillas-Verdaguer, S., 2012. Ultrasmall iron oxide nanoparticles for biomedical applications: improving the colloidal and magnetic properties. Langmuir 28 (1), 178–185. http://pubs.acs.org/doi/ipdf/10.1021/la203428z.

Darbandi, M., Stromberg, F., Landers, J., Reckers, N., Sanyal, B., Keune, W., Wende, H., 2012. Nanoscale size effect on surface spin canting in iron oxide nanoparticles synthesized by the microemulsion method. J. Phys. D. Appl. Phys. 45 (19), 195001.

Demortiere, A., Panissod, P., Pichon, B.P., Pourroy, G., Guillon, D., Donnio, B., Begin-Colin, S., 2001. Size-dependent properties of magnetic iron oxide nanocrystals. Nanoscale 3 (1), 225–232.

Depalo, N., Carrieri, P., Comparelli, R., Striccoli, M., Agostiano, A., Bertinetti, L., et al., 2011. Biofunctionalization of anisotropic nanocrystalline semiconductor–magnetic heterostructures. Langmuir 27 (11), 6962–6970.

Dias, A., Hussain, A., Marcos, A., Roque, A., 2011. A biotechnological perspective on the application of iron oxide magnetic colloids modified with polysaccharides. Biotechnol. Adv. 29 (1), 142–155. http://www.sciencedirect.com/science/article/pii/S0734975010001382?via%3Dihub.

Ding, H.L., Zhang, Y.X., Wang, S., Xu, J.M., Xu, S.C., Li, G.H., 2012. Fe_3O_4@SiO_2 core/shell nanoparticles: the silica coating regulations with a single core for different core sizes and shell thicknesses. Chem. Mater. 24 (23), 4572–4580. http://pubs.acs.org/doi/abs/10.1021/cm302828d.

Dong, A., Ye, X., Chen, J., Kang, Y., Gordon, T., Kikkawa, J.M., Murray, C.B., 2010. A generalized ligand-exchange strategy enabling sequential surface functionalization of colloidal nanocrystals. J. Am. Chem. Soc. 133 (4), 998–1006.

Dou, Q.S., Zhang, H., Wu, J.B., Yang, D.R., 2007. Synthesis and characterization of Fe_2O_3 and FeOOH nanostructures prepared by ethylene glycol assisted hydrothermal process. J. Inorg. Mater. 22 (2), 213–218.

Douadi-Masrouki, S., Frka-Petesic, B., Save, M., Charleux, B., Cabuil, V., Sandre, O., 2010. Incorporation of magnetic nanoparticles into lamellar polystyrene-b-poly(n-butyl methacrylate) diblock copolymer films: influence of the chain end-groups on nanostructuration. Polymer 51 (21), 4673–4685.

Du, C.L., Du, W., Wang, B., Feng, W.Y., Wang, Z., Zhao, Y.L., 2010. Bioseparation and bioassay based on iron oxide nanomaterials properties. Chin. J. Anal. Chem. 38, 902.

Dung, T., Danh, T., Hoa, L., Chien, D., Duc, N., 2009. Structural and magnetic properties of starch-coated magnetite nanoparticles. J. Exp. Nanosci. 4 (3), 259–267.

El-Sherif, H., El-Masry, M., Emira, H.S., 2010. Magnetic polymer composite particles via in situ inverse mini-emulsion polymerization process. J. Macromol. Sci. A 47 (11), 1096–1103.

Fan, X., Jiao, G., Zhao, W., Jin, P., Li, X., 2013. Magnetic Fe_3O_4-graphene composites as targeted drug nanocarriers for pH-activated release. Nanoscale 5 (3), 1143–1152.

Farrell, D., Majetich, S.A., Wilcoxon, J.P., 2003. Preparation and characterization of Monodisperse Fe nanoparticles. J. Phys. Chem. B 107 (40), 11022–11030.

Fernandes, A.R., Chari, D.M., 2016. Part I: minicircle vector technology limits DNA size restrictions on ex vivo gene delivery using nanoparticle vectors: overcoming a translational barrier in neural stem cell therapy. J. Control. Release 238, 289–299.

Gaihre, B., Aryal, S., Khil, M.S., Kim, H.Y., 2008. Encapsulation of Fe_3O_4 in gelatin nanoparticles: effect of different parameters on size and stability of the colloidal dispersion. J. Microencapsul. 25 (1), 21–30.

Gaihre, B., Khil, M.S., Lee, D.R., Kim, H.Y., 2009. Gelatincoated magnetic iron oxide nanoparticles as carrier system: drug loading and in vitro drug release study. Int. J. Pharm. 365, 180–189.

Gilchrist, R.K., Medal, R., Shorey, W.D., Hanselman, R.C., Parrot, J.C., Taylor, C.B., 1957. Selective inductive heating of lymph nodes. Ann. Surg. 146 (4), 596–606.

Giri, S., Samanta, S., Maji, S., Ganguli, S., Bhaumik, A., 2005. Magnetic properties of α-Fe_2O_3 nanoparticle synthesized by a new hydrothermal method. J. Magn. Magn. Mater. 285 (1–2), 296–302.

Gonzales-Weimuller, M., Zeisberger, M., Krishnan, K.M., 2009. Size-dependant heating rates of iron oxide nanoparticles for magnetic fluid hyperthermia. J. Magn. Magn. Mater. 321 (13), 1947–1950.

Goon, I.Y., Lai, L.M.H., Lim, M., Munroe, P., Gooding, J.J., Amal, R., 2009. Fabrication and dispersion of gold-shellprotected magnetite nanoparticles: systematic control using polyethyleneimine. Chem. Mater. 21 (4), 673–681.

Gowd, G.S., Patra, M.K., Mathew, M., Shukla, A., Songara, S., Vadera, S.R., Kumar, N., 2013. Synthesis of $_{Fe3O4}$@Y_2O_3: Eu^{3+} core–shell multifunctional nanoparticles and their magnetic and luminescence properties. Opt. Mater 35 (9), 1685–1692. https://www.sciencedirect.com/science/article/pii/S0925346713002267.

Grimm, J., Perez, J.M., Josephson, L., Weissleder, R., 2004. Novel nanosensors for rapid analysis of telomerase activity. Cancer Res. 64 (2), 639–643.

Grzelczak, M., Vermant, J., Furst, E.M., Liz-Marzan, L.M., 2010. Directed self-assembly of nanoparticles. ACS Nano 4 (7), 3591–3605. http://pubs.acs.org/doi/abs/10.1021/nn100869j.

Gu, H.W., Zheng, R.K., Zhang, X.X., Xu, B., 2004. Facile onepot synthesis of bifunctional heterodimers of nanoparticles: a conjugate of quantum dot and magnetic nanoparticles. J. Am. Chem. Soc. 126 (18), 5664–5665. http://pubs.acs.org/doi/abs/10.1021/ja0496423.

Guo, S., Dong, S., Wang, E., 2009. A general route to construct diverse multifunctional Fe_3O_4/metal hybrid nanostructures. Chem. Eur. J. 15 (10), 2416–2424.

Gupta, A.K., Gupta, M., 2005. Synthesis and surface engineering of iron oxide nanoparticles for biomedical applications. Biomaterials 26 (18), 3995–4021.

Haghniaz, R., Umrani, R.D., Paknikar, K.M., 2016. Hyperthermia mediated by dextran-coated La(0.7)Sr(0.3)MnO(3) nanoparticles: in vivo studies. Int. J. Nanomedicine 11, 1779–1791.

Hamley, I.W., 2003. Nanotechnology with soft materials. Angew. Chem. Int. Ed. 42 (15), 1692–1712.

Han, L.H., Liu, H., Wei, Y., 2011. In situ synthesis of hematite nanoparticles using a low-temperature microemulsion method. Powder Technol. 207 (1–3), 42–46.

Hao, Y.L., Teja, A.S., 2003. Continuous hydrothermal crystallization of α-Fe_2O_3 and Co_3O_4 nanoparticles. J. Mater. Res. 18 (2), 415–422.

Hapuarachchige, S., Kato, Y., Ngen, E.J., Smith, B., Delannoy, M., Artemov, D., 2016. Non-temperature induced effects of magnetized iron oxide nanoparticles in alternating magnetic field in cancer cells. PLoS One 11 (5), e0156294.

Hasenpusch, G., Geiger, J., Wagner, K., Mykhaylyk, O., Wiekhorst, F., Trahms, L., et al., 2012. Magnetized aerosols comprising superparamagnetic iron oxide nanoparticles improve targeted drug and gene delivery to the lung. Pharm. Res. 29 (5), 1308–1318.

He, X., Tan, L., Chen, D., Wu, X., Ren, X., Zhang, Y., Meng, X., Tang, F., 2013. Fe_3O_4-Au@mesoporous SiO_2 microsphere: an ideal artificial enzymatic cascade system. Chem. Commun. 49, 4643–4645.

Hernández, R., Mijangos, C., 2009. In situ synthesis of magnetic iron oxide nanoparticles in thermally responsive alginate-poly (N-isopropylacrylamide) semi-interpenetrating polymer networks. Macromol. Rapid Commun. 30 (3), 176–181.

Hildebrandt, B., Wust, P., Ahlers, O., Dieing, A., Sreenivasa, G., Kerner, T., et al., 2002. The cellular and molecular basis of hyperthermia. Crit. Rev. Oncol. Hematol. 43 (1), 33–56.

Hsiao, S.C., Ou, J.L., Sung, Y., Chang, C.P., Ger, M.D., 2010. Preparation of sulfate-and carboxyl-functionalized magnetite/polystyrene spheres for further deposition of gold nanoparticles. Colloid Polym. Sci. 288 (7), 787–794.

Hu, X., Yu, J.C., Gong, J., Li, Q., Li, G., 2007. α-Fe_2O_3 nanorings prepared by a microwave-assisted hydrothermal process and their sensing properties. Adv. Mater. 19 (17), 2324–2329.

Hu, L., Percheron, A., Chaumont, D., Brachais, C.H., 2011. Microwave-assisted one-step hydrothermal synthesis of pure iron oxide nanoparticles: magnetite, maghemite and hematite. J. Sol-Gel Sci. Technol. 60, 198.

Hu, M., Jiang, J.S., Bu, F.X., Cheng, X.L., Lin, C.C., Zeng, Y., 2012a. Hierarchical magnetic iron (III) oxides prepared by solid-state thermal decomposition of coordination polymers. RSC Adv. 2, 4782.

Hu, M., Jiang, J.S., Zeng, Y., 2012b. Prussian blue microcrystals prepared by selective etching and their conversion to mesoporous magnetic iron(III) oxides. Chem. Commun. 46, 1133–1135.

Hu, Q., Lee, S.C., Baek, Y.J., Lee, H.H., Kang, C.J., Kim, H.M., Kim, K.B., Yoon, T.S., 2013. Non-volatile nano-floating gate memory with Pt- Fe_2O_3 composite nanoparticles and indium gallium zinc oxide channel. J. Nanopart. Res. 15, 1435.

Huang, H.Y., Shieh, Y.T., Shih, C.M., Twu, Y.K., 2010. Magnetic chitosan/iron (II, III) oxide nanoparticles prepared by spray-drying. Carbohydr. Polym. 81, 906–910.

Huidan, W., Hao, Q., Zhijun, Y., Wenhu, Y., 2010. An effective magnetic separation technique for antibody based on thiophilic paramagnetic polymer beads. Acta Polym. Sin. 1 (10), 1238–1244.

Ivkov, R., DeNardo, S.J., Daum, W., Foreman, A.R., Goldstein, R.C., Nemkov, V.S., et al., 2005. Application of high amplitude alternating magnetic fields for heat induction of nanoparticles localized in cancer. Clin. Cancer Res. 11 (19 Pt 2), 7093s–7103s.

Iwaki, Y., Kawasaki, H., Arakawa, R., 2012. Human serum albumin-modified Fe_3O_4magnetic nanoparticles for affinity- SALDI-MS of small-molecule drugs in biological liquids. Anal. Sci. 28 (9), 893–900.

Jana, N.R., Chen, Y., Peng, X., 2004. Size- and shape-controlled magnetic (Cr, Mn, Fe, Co, Ni) oxide nanocrystals via a simple and general approach. Chem. Mater. 16 (20), 3931–3935.

Jarockyte, G., Daugelaite, E., Stasys, M., Statkute, U., Poderys, V., Tseng, T.-C., et al., 2016. Accumulation and toxicity of superparamagnetic iron oxide nanoparticles in cells and experimental animals. Int. J. Mol. Sci. 17 (8), 1193.

Jiang, F.Y., Wang, C.M., Fu, Y., Liu, R.C., 2010. Synthesis of iron oxide nanocubes via microwave-assisted solvolthermal method. J. Alloys Compd. 503 (2), L31–L33.

Jordan, A., Maier-Hauff, K., Wust, P., Rau, B., Johannsen, M., 2007. Thermotherapy using magnetic nanoparticles. Onkologe 13 (10), 894–902.

Joseph, J., Nishad, K.K., Sharma, M., Gupta, D.K., Singh, R.R., Pandey, R.K., 2012. Fe_3O_4 and CdS based bifunctional core–shell nanostructure. Mater. Res. Bull. 47 (6), 1471–1477.

Kaaki, K., Hervé-Aubert, K., Chiper, M., Shkilnyy, A., Soucé, M., Benoit, R., et al., 2012. Magnetic nanocarriers of doxorubicin coated with poly(ethylene glycol) and folic acid: relation between coating structure, surface properties, colloidal stability, and cancer cell targeting. Langmuir 28 (2), 1496–1505.

Kalber, T.L., Ordidge, K.L., Southern, P., Loebinger, M.R., Kyrtatos, P.G., Pankhurst, Q.A., et al., 2016. Hyperthermia treatment of tumors by mesenchymal stem cell-delivered superparamagnetic iron oxide nanoparticles. Int. J. Nanomedicine 11, 1973–1983.

Kanda, T., Nakai, Y., Oba, H., Toyoda, K., Kitajima, K., Furui, S., 2016. Gadolinium deposition in the brain. Magn. Reson. Imaging 34 (10), 1346–1350.

Kang, X.J., Dai, Y.L., Ma, P.A., Yang, D.M., Li, C.X., Hou, Z.Y., Cheng, Z.Y., Lin, J., 2012. Poly (acrylic acid)-Modified Fe_3O_4 microspheres for magnetic-targeted and pH-triggered anticancer drug delivery. Chem. Eur. J. 18 (49), 15676–15682.

Kemmner, W., Moldenhauer, G., Schlag, P., Brossmer, R., 1992. Separation of tumor cells from a suspension of dissociated human colorectal carcinoma tissue by means of monoclonal antibody-coated magnetic beads. J. Immunol. Methods 147 (2), 197–200.

Kim, S., Kim, J.-H., Jeon, O., Kwon, I.C., Park, K., 2009. Engineered polymers for advanced drug delivery. Eur. J. Pharm. Biopharm. 71 (3), 420–430.

Kim, J., Arifin, D.R., Muja, N., Kim, T., Gilad, A.A., Kim, H., Arepally, A., Hyeon, T., Bulte, J.W., 2011a. Multifunctional capsule-in-capsules for immunoprotection and trimodal imaging. Angew. Chem. 123 (10), 2237–2365.

Kim, D., Yu, M.K., Lee, T.S., Park, J.J., Jeong, Y.Y., Jon, S., 2011b. Amphiphilic polymer-coated hybrid nanoparticles as CT/MRI dual contrast agents. Nanotechnology 22 (15), 155101.

Kirui, D.K., Rey, D.A., Batt, C.A., 2010. Gold hybrid nanoparticles for targeted phototherapy and cancer imaging. Nanotechnology 21 (10), 105105.

Klotz, S., Steinle-Neumann, G., Strassle, T., Philippe, J., Hansen, T., Wenzel, M.J., 2008. Magnetism and the Verwey transition in Fe_3O_4 under pressure. Phys. Rev. B 77 (1), 012411.

Kloust, H., et al., 2013. In situ functionalization and PEO coating of iron oxide nanocrystals using seeded emulsion polymerization. Langmuir 29 (15), 4915–4921.

Kronick, P., Gilpin, R.W., 1986. Use of superparamagnetic particles for isolation of cells. J. Biochem. Biophys. Methods 12 (1), 73–80.

Krotz, F., de Wit, C., Sohn, H.Y., Zahler, S., Gloe, T., Pohl, U., et al., 2003. Magnetofection—a highly efficient tool for antisense oligonucleotide delivery in vitro and in vivo. Mol. Ther. 7 (5 Pt 1), 700–710.

Kutushov, M.V., Kuznetsov, A.A., Filippov, V.I., Kuznetsov, O.A., 1997. New method of biological fluid detoxification based on magnetic adsorbents. In: Häfeli, U., Schütt, W., Teller, J., Zborowski, M. (Eds.), Scientific and Clinical Applications of Magnetic Carriers. Springer US, Boston, MA, pp. 391–397.

Kyung, M.Y., Jongmin, S., In, S.L., 2010. Reductive dissolution of Fe_3O_4 facilitated by the Au domain of an Fe_3O_4/Au hybrid nanocrystal: formation of a nanorattle structure composed of a hollow porous silica nanoshell and entrapped Au nanocrystal. Chem. Commun. 46 (1), 64–66.

Ladj, R., Bitar, A., Eissa, M., Mugnier, Y., Le Dantec, R., Fessi, H., Elaissari, A., 2013. Individual inorganic nanoparticles: preparation, functionalization and in vitro biomedical diagnostic applications. J. Mater. Chem. B 1 (10), 1381–1396.

Lam, U.T., Mammucari, R., Suzuki, K., Foster, N.R., 2008. Processing of iron oxide nanoparticles by supercritical fluids. Ind. Eng. Chem. Res. 47 (3), 599–614.

Lan, F., Liu, K.X., Jiang, W., Zeng, X.B., Wu, Y., Gu, Z.W., 2011. Facile synthesis of monodisperse superparamagnetic Fe_3O_4/PMMA composite nanospheres with high magnetization. Nanotechnology 22 (22), 225604.

Lartigue, L., et al., 2009. Water-soluble rhamnose-coated Fe_3O_4 nanoparticles. Org. Lett. 11 (14), 2992–2995. http://pubs.acs.org/doi/pdf/10.1021/ol900949y.

Latham, A.H., Williams, M.E., 2008. Controlling transport and chemical functionality of magnetic nanoparticles. Acc. Chem. Res. 41 (3), 411–420.

Laurent, S., Forge, D., Port, M., Roch, A., Robic, C., Elst, L.V., Muller, R.N., 2008. Magnetic iron oxide nanoparticles: synthesis, stabilization, vectorization, physicochemical characterizations, and biological applications. Chem. Rev. 108 (6), 2064–2110.

Leal, M.P., Torti, A., Riedinger, A., La Fleur, R., Petti, D., Cingolani, R., Bertacco, R., Pellegrino, T., 2012. Controlled release of doxorubicin loaded within magnetic thermoresponsive nanocarriers under magnetic and thermal actuation in a microfluidic channel. ACS Nano 6 (12), 10535–10545.

Lee, J., Isobe, T., Senna, M., 1996. Magnetic properties of ultrafine magnetite particles and their slurries prepared via in-situ precipitation. Colloids Surf. A Physicochem. Eng. Asp. 109, 121–127.

Lee, N., et al., 2012. Multifunctional Fe_3O_4/TaOx core/shell nanoparticles for simultaneous magnetic resonance imaging and X-ray computed tomography. J. Am. Chem. Soc. 134 (25), 10309–10312.

Lei, C., Han, F., Li, D., Li, W.C., Sun, Q., Zhang, X.Q., Lu, A.H., 2013. Dopamine as the coating agent and carbon precursor for the fabrication of N-doped carbon coated Fe_3O_4 composites as superior lithium ion anodes. Nanoscale 5 (3), 1168–1175.

Lemine, O.M., Omri, K., Zhang, B., El Mir, L., Sajieddine, M., Alyamani, A., Bououdina, M., 2012. Sol–gel synthesis of 8 nm magnetite (Fe_3O_4) nanoparticles and their magnetic properties. Superlattice. Microst. 52 (4), 793–799.

Lester, E., Blood, P., Denyer, J., Giddings, D., Azzopardi, B., Poliakoff, M., 2006. Reaction engineering: the supercritical water hydrothermal synthesis of nano-particles. J. Supercrit. Fluids 37 (2), 209–214.

Li, Y., Afzaal, M., O'Brien, P., 2006. The synthesis of amine-capped magnetic (Fe, Mn, Co, Ni) oxide nanocrystals and their surface modification for aqueous dispersibility. J. Mater. Chem. 16, 2175–2180.

Li, Z., Tan, B., Allix, M., Cooper, A.I., Rosseinsky, M., 2008. Direct coprecipitation route to monodisperse dualfunctionalized magnetic iron oxide nanocrystals without size selection. Small 4 (9), 231–239.

Li, J., Zhu, W., Wang, H., 2009. Novel magnetic single-walled carbon nanotubes/methylene blue composite amperometric biosensor for DNA determination. Anal. Lett. 42 (2), 366–380.

Li, X.Y., Huang, X.L., Liu, D.P., Wang, X., Song, S.Y., Zhou, L., Zhang, H.J., 2011a. Synthesis of 3D hierarchical Fe_3O_4/graphene composites with high lithium storage capacity and for controlled drug delivery. J. Phys. Chem. C 115 (5), 1256–1266.

Li, S.K., Huang, F.Z., Wang, Y., Shen, Y.H., Qiu, L.G., Xie, A.J., Xu, S.J., 2011b. Magnetic Fe_3O_4@C@Cu_2O composites with bean-like core/shell nanostructures: synthesis, properties and application in recyclable photocatalytic degradation of dye pollutants. J. Mater. Chem. 21, 7459–7466.

Li, J., Li, S., Yang, C.F., 2012. Electrochemical biosensors for cancer biomarker detection. Electroanalysis 24 (12), 2213–2229.

Li, L., Jiang, W., Luo, K., Song, H., Lan, F., Wu, Y., Gu, Z., 2013a. Superparamagnetic iron oxide nanoparticles as MRI contrast agents for non-invasive stem cell labeling and tracking. Theranostics 3 (8), 595–615.

Li, Y., Hu, Y., Jiang, H., Li, C., 2013b. Double-faced γ- Fe_2O_3||SiO_2 nanohybrids: flame synthesis, in situ selective modification and highly interfacial activity. Nanoscale 5 (12), 5360–5367. https://www.ncbi.nlm.nih.gov/pubmed/23649103.

Lian, S.Y., Wang, E., Kang, Z.H., Bai, Y.P., Gao, L., Jiang, M., Hu, C.W., Xu, L., 2004. Synthesis of magnetite nanorods and porous hematite nanorods. Solid State Commun. 129 (8), 485–490.

Lim, J.K., Majetich, S.A., Tilton, R.D., 2009. Stabilization of superparamagnetic iron oxide core-gold shell nanoparticles in high ionic strength media. Langmuir 25 (23), 13384–13393.

Liu, S., Lu, F., Jia, X., Cheng, F., Jiang, L.P., Zhu, J.J., 2011. Microwave-assisted synthesis of a biocompatible polyacid conjugated Fe_3O_4 superparamagnetic hybrid. CrystEngComm 13 (7), 2425–2429.

Liu, Z., Li, M., Pu, F., Ren, J., Yang, X., Qu, X., 2012a. Hierarchical magnetic core–shell nanoarchitectures: nonlinker reagent synthetic route and applications in a biomolecule separation system. J. Mater. Chem. 22, 2935–2942.

Liu, X.L., Fan, H.M., Yi, J.B., Yang, Y., Choo, E.S.G., Xue, J.M., et al., 2012b. Optimization of surface coating on Fe3O4 nanoparticles for high performance magnetic hyperthermia agents. J. Mater. Chem. 22 (17), 8235–8244.

Liu, L., Xiao, L., Zhu, H.Y., Shi, X.W., 2013. Studies on interaction and illumination damage of CS-Fe_3O_4@ZnS: Mn to bovine serum albumin. J. Nanopart. Res. 15, 1394–1405.

Lobel, B., Eyal, O., Kariv, N., Katzir, A., 2000. Temperature controlled CO(2) laser welding of soft tissues: urinary bladder welding in different animal models (rats, rabbits, and cats). Lasers Surg. Med. 26 (1), 4–12.

Lu, A.H., Salabas, E.L., Schüth, F., 2007. Magnetic nanoparticles: synthesis, protection, functionalization, and application. Angew. Chem. Int. Ed. 46 (8), 1222–1244.

Lu, C.C., Quan, Z.S., Sur, J.C., Kim, S.H., Lee, C.H., Chai, K.Y., 2010. One-pot fabrication of carboxyl-functionalized biocompatible magnetic nanocrystals for conjugation with targeting agents. New J. Chem. 34 (9), 2040–2046.

Lu, W., Shen, Y., Xie, A., Zhang, W., 2013. Preparation and protein immobilization of magnetic dialdehyde starch nanoparticles. J. Phys. Chem. B 117 (14), 3720–3725.

Luo, Y., et al., 2012a. Seed-assisted synthesis of highly ordered TiO2@α- Fe_2O_3 core/shell arrays on carbon textiles for lithium-ion battery applications. Energy Environ. Sci. 5, 6559–6566.

Luo, S.R., Chai, F., Zhang, L.Y., Wang, C.G., Li, L., Liu, X.C., Su, Z.M., 2012b. Facile and fast synthesis of urchin-shaped Fe_3O_4@Bi2S3 core–shell hierarchical structures and their magnetically recyclable photocatalytic activity. J. Mater. Chem. 22 (11), 4832–4836.

Lynch, J., Zhuang, J., Wang, T., LaMontagne, D., Wu, H., Cao, Y.C., 2011. Gas-bubble effects on the formation of colloidal iron oxide nanocrystals. J. Am. Chem. Soc. 133 (32), 12664–12674.

Ma, L.L., et al., 2009. Small multifunctional nanoclusters (nanoroses) for targeted cellular imaging and therapy. ACS Nano 3 (9), 2686–2696. http://pubs.acs.org/doi/abs/10.1021/nn900440e.

Magro, M., Faralli, A., Baratella, D., Bertipaglia, I., Giannetti, S., Salviulo, G., Zboril, R., Vianello, F., 2012. Avidin functionalized maghemite nanoparticles and their application for recombinant human biotinyl-SERCA purification. Langmuir 28 (43), 15392–15401.

Mahmoudi, M., Sant, S., Wang, B., Laurent, S., Sen, T., 2011. Superparamagnetic iron oxide nanoparticles (SPIONs): development, surface modification and applications in chemotherapy. Adv. Drug Deliv. Rev. 63 (1-2), 24–46.

Majeed, M.I., Lu, Q., Yan, W., Li, Z., Hussain, I., Tahir, M.N., Tremel, W., Tan, B., 2013. Highly water-soluble magnetic iron oxide (Fe_3O_4) nanoparticles for drug delivery: enhanced in vitro therapeutic efficacy of doxorubicin and MION conjugates. J. Mater. Chem. B 1 (22), 2874–2884.

Marcelo, G., Muñoz-Bonilla, A., Rodríguez-Hernández, J., Fernández-García, M., 2013. Hybrid materials achieved by polypeptide grafted magnetite nanoparticles through a dopamine biomimetic surface anchored initiator. Polym. Chem. 4, 558–567.

Marszałł, M.P., 2011. Application of magnetic nanoparticles in pharmaceutical sciences. Pharm. Res. 28 (3), 480–483.

Massart, R., 1981. Preparation of aqueous magnetic liquids in alkaline and acidic media. IEEE Trans. Magn. 17 (2), 1247–1248.

Mohammad, F., Balaji, G., Weber, A., Uppu, R.M., Kumar, C.S.S.R., 2010. Influence of gold nanoshell on hyperthermia of superparamagnetic iron oxide nanoparticles. J. Phys. Chem. C 114 (45), 19194–19201.

Monošík, R., Stred'anský, M., Šturdík, E., 2012. Application of electrochemical biosensors in clinical diagnosis. J. Clin. Lab. Anal. 26 (1), 22–34.
Moscoso-Londoño, O., Gonzalez, J., Muraca, D., Hoppe, C., Alvarez, V., López-Quintela, A., Socolovsky, L.M., Pirota, K.R., 2012. Structural and magnetic behavior of ferrogels obtained by freezing thawing of polyvinyl alcohol/poly (acrylic acid)(PAA)-coated iron oxide nanoparticles. Eur. Polym. J. 49, 279–289.
Mukhopadhyay, A., Joshi, N., Chattopadhyay, K., De, G., 2011. A facile synthesis of PEG-coated magnetite (Fe_3O_4) nanoparticles and their prevention of the reduction of cytochrome; C. ACS Appl. Mater. Interfaces 4 (1), 142–149.
Mukh-Qasem, R.A., Gedanken, A., 2005. Sonochemical synthesis of stable hydrosol of Fe_3O_4 nanoparticle. J. Colloid Interface Sci. 284 (2), 489–494.
Murray, C.B., Norris, D.J., Bawendi, M.G., 1993. Synthesis and characterization of nearly monodisperse CdE (E = S, Se, Te) semiconductor nanocrystallites. J. Am. Chem. Soc. 115 (19), 8706–8715.
Nouira, W., Maaref, A., Siadat, M., Errachid, A., Jaffrezic-Renault, N., 2011. Conductometric biosensors based on layer-by-layer coated paramagnetic nanoparticles for urea detection. Sens. Lett. 9 (6), 2272–2274.
O'Brien, S., Brus, L., Murray, C.B., 2001. Synthesis of monodisperse nanoparticles of barium titanate: toward a generalized strategy of oxide nanoparticle synthesis. J. Am. Chem. Soc. 123 (48), 12085–12086.
Oh, J.K., Park, J.M., 2011. Iron oxide-based superparamagnetic polymeric nanomaterials: design, preparation, and biomedical application. Prog. Polym. Sci. 36 (1), 168–189.
Okoli, C., Boutonnet, M., Mariey, L., Järås, S., Rajarao, G., 2011. Application of magnetic iron oxide nanoparticles prepared from microemulsions for protein purification. J. Chem. Technol. Biotechnol. 86 (11), 1386–1393.
Okoli, C., Sanchez-Dominguez, M., Boutonnet, M., Järås, S., Civera, C., Solans, C., Kuttuva, G.R., 2012. Comparison and functionalization study of microemulsion-prepared magnetic iron oxide nanoparticles. Langmuir 28 (22), 8479–8485.
Okuda, M., Eloi, J.C., Jones, S.E.W., Sarua, A., Richardson, R.M., Schwarzacher, W., 2012. Fe_3O_4 nanoparticles: protein-mediated crystalline magnetic superstructures. Nanotechnology 23 (41), 415601.
Osborne, E.A., Atkins, T.M., Gilbert, D.A., Kauzlarich, S.M., Liu, K., Louie, A.Y., 2012. Rapid microwave-assisted synthesis of dextran-coated iron oxide nanoparticles for magnetic resonance imaging. Nanotechnology 23 (21), 215602.
Ou, G.F., Zhu, H.L., Zhu, E.Z., Gao, L.H., Chen, J.J., 2010. Fe_3O_4-Au and Fe_2O_3-Au hybrid nanorods: layer-by-layer assembly synthesis and their magnetic and optical properties. Nanoscale Res. Lett. 5 (11), 1755–1761.
Pala, K., Serwotka, A., Jeleń, F., Jakimowicz, P., Otlewski, J., 2014. Tumor-specific hyperthermia with aptamer-tagged superparamagnetic nanoparticles. Int. J. Nanomedicine 9, 67–76.
Pandey, S., Mishra, S., 2011. Sol–gel derived organic–inorganic hybrid materials: synthesis, characterizations and applications. J. Sol-Gel Sci. Technol. 59 (1), 73–94.
Pankhurst, Q.A., Thanh, N.T.K., Jones, S.K., Dobson, J., 2009. Progress in applications of magnetic nanoparticles in biomedicine. J. Phys. D. Appl. Phys. 42 (22), 224001.
Park, J., An, K., Hwang, Y., Park, J.G., Noh, H.J., Kim, J.-Y., Park, J.H., Hwang, N.M., Hyeon, T., 2004. Ultra-large-scale syntheses of monodisperse nanocrystals. Nat. Mater. 3 (12), 891–895.
Park, J., et al., 2005. One-nanometer-scale size-controlled synthesis of monodisperse magnetic iron oxide nanoparticles. Angew. Chem. Int. Ed. 44 (19), 2872–2877.
Park, J.N., Zhang, P., Hu, Y.S., McFarland, E.W., 2010. Synthesis and characterization of sintering-resistant silicaencapsulated Fe_3O_4 magnetic nanoparticles active for oxidation and chemical looping combustion. Nanotechnology 21 (22), 225708.
Pascal, C., Pascal, J.L., Favier, F., Elidrissi Moubtassim, M.L., Payen, C., 1998. Electrochemical synthesis for the control of γ- Fe_2O_3nanoparticle size: morphology, microstructure, and magnetic behaviour. Chem. Mater. 11, 1141–1147.
Pascu, O., Carenza, E., Gich, M., Estrade, S., Peiro, F., Herranz, G., Roig, A., 2012. Surface reactivity of iron oxide nanoparticles by microwave-assisted synthesis; comparison with the thermal decomposition route. J. Phys. Chem. C 116 (7), 2425–2429.
Patel, M.P., Patel, R.R., Patel, J.K., 2010. Chitosan mediated targeted drug delivery system: a review. J. Pharm. Pharm. Sci. 13 (4), 536–557.
Peng, X., Wickham, J., Alivisatos, A.P., 1998. Kinetics of II-VI and III-V colloidal semiconductor nanocrystal growth: "focusing" of size distributions. J. Am. Chem. Soc. 120 (21), 5343–5344.
Peng, X.-H., Qian, X., Mao, H., Wang, A.Y., Chen, Z., Nie, S., et al., 2008. Targeted magnetic iron oxide nanoparticles for tumor imaging and therapy. Int. J. Nanomedicine 3 (3), 311–321.
Pereira, C., et al., 2012. Superparamagnetic MFe_2O_4 (M = Fe, Co, Mn) nanoparticles: tuning the particle size and magnetic properties through a novel one-step co-precipitation route. Chem. Mater. 24 (8), 1496–1504.

Perez, J.M., Josephson, L., Weissleder, R., 2004. Use of magnetic nanoparticles as nanosensors to probe for molecular interactions. ChemBioChem 5 (3), 261–264.

Pérez-López, B., Merkoçi, A., 2011. Magnetic nanoparticles modified with carbon nanotubes for electrocatalytic magnetoswitchable biosensing applications. Adv. Funct. Mater. 21 (2), 255–260.

Pillai, V., Kumar, P., Hou, M.J., Ayyub, P., Shah, D.O., 1995. Preparation of nanoparticles of silver halides, superconductors and magnetic materials using water-in-oil microemulsions as nano-reactors. Adv. Colloid Interf. Sci. 55, 241–269.

Pimpha, N., Chaleawlert-umpon, S., Sunintaboon, P., 2012. Core/shell polymethyl methacrylate/polyethyleneimine particles incorporating large amounts of iron oxide nanoparticles prepared by emulsifier-free emulsion polymerization. Polymer 53, 2015–2022.

Pinho, S.L.C., et al., 2010. Fine tuning of the relaxometry of γ- Fe_2O_3@SiO_2 nanoparticles by tweaking the silica coating thickness. ACS Nano 4 (9), 5339–5349. http://pubs.acs.org/doi/abs/10.1021/nn101129r.

Prashant, C., Dipak, M., Yang, C.T., Chuang, K.H., Jun, D., Feng, S.S., 2010. Superparamagnetic iron oxide-loaded poly (lactic acid)-d-α-tocopherol polyethylene glycol 1000 succinate copolymer nanoparticles as MRI contrast agent. Biomaterials 31 (21), 5588–5597.

Prathna, T.C., Mathew, L., Chandrasekaran, N., Raichur, A.M., Mukherjee, A., 2010. Biomimetic synthesis of nanoparticles: science, technology & applicability. In: Mukherjee, A. (Ed.), Biomimetics Learning from Nature. InTech. https://doi.org/10.5772/8776. Available from: https://www.intechopen.com/books/biomimetics-learning-from-nature/biomimetic-synthesis-of-nanoparticles-science-technology-amp-applicability.

Qi, H., Yan, B., Lu, W., Li, C., Yang, Y., 2011. A non-alkoxide sol-gel method for the preparation of magnetite (Fe_3O_4) nanoparticles. Curr. Nanosci. 7 (3), 381–388.

Qiu, G., Huang, H., Genuino, H., Opembe, N., Stafford, L., Dharmarathna, S., Suib, S.L., 2011. Microwave-assisted hydrothermal synthesis of nanosized α-Fe_2O_3 for catalysts and adsorbents. J. Phys. Chem. C 115, 19626–19631.

Ray, J.R., Lee, B., Baltrusaitis, J., Jun, Y.S., 2012. Formation of iron (III)(hydr) oxides on polyaspartate-and alginate-coated substrates: effects of coating hydrophilicity and functional group. Environ. Sci. Technol. 46 (24), 13167–13175. http://pubs.acs.org/doi/abs/10.1021/es302124g.

Redl, F.X., Black, C.T., Papaefthymiou, G.C., Sandstrom, R.L., Yin, M., Zeng, H., Murray, C.B., 2004. Magnetic, electronic, and structural characterization of nonstoichiometric iron oxides at the nanoscale. J. Am. Chem. Soc. 126 (44), 14583–14599.

Roca, A.G., Costo, R., Rebolledo, A.F., Veintemillas-Verdaguer, S., Tartaj, P., González-Carreño, T., et al., 2009. Progress in the preparation of magnetic nanoparticles for applications in biomedicine. J. Phys. D. Appl. Phys. 42 (22), 224002.

Rockenberger, J., Scher, E.C., Alivisatos, A.P., 1999. A new nonhydrolytic single-precursor approach to surfactant-capped nanocrystals of transition metaloxides. J. Am. Chem. Soc. 121 (49), 11595–11596.

Rosengart, A.J., 2007. In: Biohazard detoxification method utilizing magnetic particles. Medicine and Medical Research Toxicology Chemical, Biological and Radiological Warfare. ADA469264.

Rovers, S.A., Dietz, C.H., vd Poel, L.A., Hoogenboom, R., Kemmere, M.F., Keurentjes, J.T., 2010. Influence of distribution on the heating of superparamagnetic iron oxide nanoparticles in poly (methyl methacrylate) in an alternating magnetic field. J. Phys. Chem. C 114 (18), 8115–8678.

Salavati-Niasari, M., Mahmoudi, T., Amiri, O., 2012. Easy synthesis of magnetite nanocrystals via co-precipitation method. J. Clust. Sci. 23 (2), 597–602.

Salazar-Alvarez, G., Muhammed, M., Zagorodni, A.A., 2006. Novel flow injection synthesis of iron oxide nanoparticles with narrow size distribution. Chem. Eng. Sci. 61 (14), 4625–4633.

Samanta, B., Yan, H., Fischer, N.O., Shi, J., Jerry, D.J., Rotello, V.M., 2008. Protein-passivated Fe_3O_4 nanoparticles: low toxicity and rapid heating for thermal therapy. J. Mater. Chem. 18 (11), 1204–1208.

Samia, A.C.S., Hyzer, K., Schlueter, J.A., Qin, C.J., Jiang, J.S., Bader, S.D., Lin, X.M., 2005. Ligand effect on the growth and the digestion of Co nanocrystals. J. Am. Chem. Soc. 127 (12), 4126–4127.

Santra, S., Kaittanis, C., Grimm, J., Perez, J.M., 2009. Drug/dyeloaded, multifunctional iron oxide nanoparticles for combined targeted cancer therapy and dual optical/magnetic resonance imaging. Small 5 (16), 1862–1868.

Scheffel, A., Gruska, M., Faivre, D., Linaroudis, A., Plitzko, J.M., Schuler, D., 2006. An acidic protein aligns magnetosomes along a filamentous structure in magnetotactic bacteria. Nature 440, 110–114.

Scherer, F., Anton, M., Schillinger, U., Henke, J., Bergemann, C., Kruger, A., et al., 2002. Magnetofection: enhancing and targeting gene delivery by magnetic force in vitro and in vivo. Gene Ther. 9 (2), 102–109.

Semelka, R.C., Ramalho, J., Vakharia, A., Al Obaidy, M., Burke, L.M., Jay, M., et al., 2016. Gadolinium deposition disease: initial description of a disease that has been around for a while. Magn. Reson. Imaging 34 (10), 1383–1390.

Sensenig, R., Sapir, Y., MacDonald, C., Cohen, S., Polyak, B., 2012. Magnetic nanoparticle-based approaches to locally target therapy and enhance tissue regeneration in vivo. Nanomedicine (Lond.) 7 (9), 1425–1442.

Shao, D., Xu, K., Song, X., Hu, J., Yang, W., Wang, C., 2009. Effective adsorption and separation of lysozyme with PAA-modified Fe3O4@silica core/shell microspheres. J. Colloid Interface Sci. 336 (2), 526–532.

Shao, M., Ning, F., Zhao, J., Wei, M., Evans, D.G., Duan, X., 2012. Preparation of Fe3O4@SiO2@layered double hydroxide core–shell microspheres for magnetic separation of proteins. J. Am. Chem. Soc. 134 (2), 1071–1077.

Shao, J., Xie, X., Xi, Y., Liu, X., Yang, Y., 2013. Characterization of Fe_3O_4/SiO_2 composite core–shell nanoparticles synthesized in isopropanol medium. Glas. Phys. Chem. 39 (3), 329–335.

Shavel, A., Liz-Marzan, L.M., 2009. Shape control of iron oxide nanoparticles. Phys. Chem. Chem. Phys. 11 (19), 3762–3766.

Shaw, R.W., Brill, T.B., Clifford, A.A., Eckert, C.A., Franck, E.U., 1991. Supercritical water: a medium for chemistry. Chem. Eng. News 69 (51), 26–39.

Shen, M., Cai, H., Wang, X., Cao, X., Li, K., Wang, S., Guo, R., Zheng, L., Zhang, G., Shi, X., 2012. Facile one-pot preparation, surface functionalization, and toxicity assay of APTS-coated iron oxide nanoparticles. Nanotechnology 23 (10), 105601.

Shi, Y., Li, H.Y., Wang, L., Shen, W., Chen, H.Z., 2012. Novel α-Fe_2O_3/CdS cornlike nanorods with enhanced photocatalytic performance. ACS Appl. Mater. Interfaces 4 (9), 4800–4806. http://pubs.acs.org/doi/abs/10.1021/am3011516.

Shkilnyy, A., Munnier, E., Hervé, K., Soucé, M., Benoit, R., Cohen-Jonathan, S., Limelette, P., Saboungi, M.L., Dubois, P., Chourpa, I., 2010. Synthesis and evaluation of novel biocompatible super-paramagnetic iron oxide nanoparticles as magnetic anticancer drug carrier and fluorescence active label. J. Phys. Chem. C 114 (13), 5850–5858. http://pubs.acs.org/doi/abs/10.1021/jp9112188.

Sokolov, K., Follen, M., Aaron, J., Pavlova, I., Malpica, A., Lotan, R., et al., 2003. Real-time vital optical imaging of precancer using anti-epidermal growth factor receptor antibodies conjugated to gold nanoparticles. Cancer Res. 63 (9), 1999–2004.

Sondjaja, R., Alan Hatton, T., Tam, M.K., 2009. Clustering of magnetic nanoparticles using a double hydrophilic block copolymer, poly (ethylene oxide)-b-poly (acrylic acid). J. Magn. Magn. Mater. 321 (16), 2393–2508.

Song, M., Zhang, Y., Hu, S., Song, L., Dong, J., Chen, Z., Gu, N., 2012. Influence of morphology and surface exchange reaction on magnetic properties of monodisperse magnetite nanoparticles. Colloids Surf. A Physicochem. Eng. Asp. 408, 114–121.

Sorescu, M., Diamandescu, L., Tarabasanu-Mihaila, D., 2004. α-Fe_2O_3–In_2O_3 mixed oxide nanoparticles synthesized under hydrothermal supercritical conditions. J. Phys. Chem. Solids 65 (10), 1719–1725.

Sperling, R.A., Parak, W.J., 2010. Surface modification, functionalization and bioconjugation of colloidal inorganic nanoparticles. Phil. Trans. R. Soc. A 368 (1915), 1333–1383.

Sreeja, V., Joy, P.A., 2007. Microwave–hydrothermal synthesis of γ-Fe_2O_3 nanoparticles and their magnetic properties. Mater. Res. Bull. 42, 1570–1576.

Stanley, S.A., Gagner, J.E., Damanpour, S., Yoshida, M., Dordick, J.S., Friedman, J.M., 2012. Radio-wave heating of iron oxide nanoparticles can regulate plasma glucose in mice. Science 336 (6081), 604–608.

Starowicz,M.,Starowicz,P.,Żukrowski,J.,Przewoźnik,J.,Lemański,A.,Kapusta,C.,Banaś,J.,2011.Electrochemicalsynthesis of magnetic iron oxide nanoparticles with controlled size. J. Nanopart. Res. 13 (12), 7167–7176.

Stöber, W., Fink, A., Bohn, E., 1968. Controlled growth of monodisperse silica spheres in the micron size range. J. Colloid Interface Sci. 26, 62–69.

Sue, K., Kimura, K., Arai, K., 2004. Hydrothermal synthesis of ZnO nanocrystals using microreactor. Mater. Lett. 58 (25), 3229–3231.

Suh, S.K., Yuet, K., Hwang, D.K., Bong, K.W., Doyle, P.S., Hatton, T.A., 2012. Synthesis of nonspherical superparamagnetic particles: in situ co-precipitation of magnetic nanoparticles in microgels prepared by stop-flow lithography. J. Am. Chem. Soc. 134 (17), 7337–7743.

Sun, S., Zeng, H., 2002. Size-controlled synthesis of magnetite nanoparticles. J. Am. Chem. Soc. 124 (28), 8204–8205.

Sun, S., Zeng, H., Robinson, D.B., Raoux, S., Rice, P.M., Wang, S.X., Li, G., 2004. Monodisperse MFe2O4 (M = Fe, Co, Mn) nanoparticles. J. Am. Chem. Soc. 126 (1), 273–279.

Sun, B., Sun, M.J., Gu, Z., Shen, Q.D., Jiang, S.J., Xu, Y., Wang, Y., 2010. Conjugated polymer fluorescence probe for intracellular imaging of magnetic nanoparticles. Macromolecules 43 (24), 10348–10354. http://pubs.acs.org/doi/abs/10.1021/ma101680g.

Sun, X., Guo, S., Liu, Y., Sun, S., 2012a. Dumbbell-like PtPd– Fe_3O_4 nanoparticles for enhanced electrochemical detection of H_2O_2. Nano Lett. 12 (9), 4859–4863.

Sun, L.L., Wu, W., Zhang, S.F., Zhou, J., Cai, G.X., Ren, F., Xiao, X.H., Dai, Z.G., Jiang, C.Z., 2012b. Novel doping for synthesis monodispersed TiO_2 grains filled into spindle-like hematite bi-component nanoparticles by ion implantation. AIP Adv. 2, 032179.

Sun, X., Li, Q., Wang, X., Du, S., 2012c. Amperometric immunosensor based on gold nanoparticles/Fe3O4-FCNTs-CS composite film functionalized interface for carbofuran detection. Anal. Lett. 45 (12), 1604–1616.

Sun, L.L., Wu, W., Zhang, S.F., Liu, Y.C., Xiao, X.H., Ren, F., Cai, G.X., Jiang, C.Z., 2013. Spindle-like alpha- Fe_2O_3 embedded with TiO_2 nanocrystalline: ion implantation preparation and enhanced magnetic properties. J. Nanosci. Nanotechnol. 13 (8), 5428–5433.

Sundaram, P.A., Augustine, R., Kannan, M., 2012. Extracellular biosynthesis of iron oxide nanoparticles by Bacillus subtilis strains isolated from rhizosphere soil. Biotechnol. Bioprocess Eng. 17 (4), 835–840.

Suslick, K.S., 1990. Sonochemistry. Science 247, 1439–1445.

Talelli, M., Rijcken, C.J.F., Lammers, T., Seevinck, P.R., Storm, G., van Nostrum, C.F., et al., 2009. Superparamagnetic iron oxide nanoparticles encapsulated in biodegradable thermosensitive polymeric micelles: toward a targeted nanomedicine suitable for image-guided drug delivery. Langmuir 25 (4), 2060–2067.

Tan, Y.F., Chandrasekharan, P., Maity, D., Yong, C.X., Chuang, K.H., Zhao, Y., Wang, S., Ding, J., Feng, S.S., 2011. Multimodal tumor imaging by iron oxides and quantum dots formulated in poly (lactic acid)-d-alpha-tocopheryl polyethylene glycol 1000 succinate nanoparticles. Biomaterials 32 (11), 2969–2978.

Tavakoli, A., Sohrabi, M., Kargari, A., 2007. A review of methods for synthesis of nanostructured metals with emphasis on iron compounds. Chem. Pap. 61 (3), 151–170.

Teja, A.S., Holm, L.J., 2002. Production of magnetic nanoparticles using supercritical fluids. In: Sun, Y.-P. (Ed.), Supercritical Fluid Technology in Materials Science and Engineering: Synthesis, Properties, and Applications. CRC Press, Boca Raton, FL, pp. 327–347.

Teja, A.S., Koh, P.Y., 2009. Synthesis, properties, and applications of magnetic iron oxide nanoparticles. Prog. Cryst. Growth Charact. Mater. 55 (1–2), 22–45.

Teymourian, H., Salimi, A., Hallaj, R., 2012. Low potential detection of NADH based on Fe_3O_4 nanoparticles/multiwalled carbon nanotubes composite: fabrication of integrated dehydrogenase-based lactate biosensor. Biosens. Bioelectron. 33 (1), 60–68.

Theerdhala, S., Bahadur, D., Vitta, S., Perkas, N., Zhong, Z.Y., Gedanken, A., 2010. Sonochemical stabilization of ultrafine colloidal biocompatible magnetite nanoparticles using amino acid, L-arginine, for possible bio applications. Ultrason. Sonochem. 17 (4), 730–737.

Thomas, L.A., Dekker, L., Kallumadil, M., Southern, P., Wilson, M., Nair, S.P., et al., 2009. Carboxylic acid-stabilised iron oxide nanoparticles for use in magnetic hyperthermia. J. Mater. Chem. 19 (36), 6529–6535.

Tronc, E., Ezzir, A., Cherkaoui, R., Chaneac, C., Nogues, M., Kachkachi, H., Fiorani, D., Testa, A.M., Greneche, J.M., Jolivet, J.P., 2000. Surface-related properties of γ-Fe_2O_3 nanoparticles. J. Magn. Magn. Mater. 221 (1–2), 63–79.

Truby, R.L., Emelianov, S., Homan, K.A., 2013. Ligand mediated self-assembly of hybrid plasmonic and superparamagnetic nanostructures. Langmuir 29 (8), 2465–2470.

Tsai, Z.T., Wang, J.F., Kuo, H.Y., Shen, C.R., Wang, J.J., Yen, T.C., 2010. In situ preparation of high relaxivity iron oxide nanoparticles by coating with chitosan: a potential MRI contrast agent useful for cell tracking. J. Magn. Magn. Mater. 322, 208–213.

Valdiglesias, V., Fernandez-Bertolez, N., Kilic, G., Costa, C., Costa, S., Fraga, S., et al., 2016. Are iron oxide nanoparticles safe? Current knowledge and future perspectives. J. Trace Elem. Med. Biol. 38, 53–63.

Vali, H., Weiss, B., Li, Y.L., Sears, S.K., Kim, S.S., Kirschvink, J.L., Zhang, L., 2004. Formation of tabular single-domain magnetite induced by Geobacter metallireducens GS-15. Proc. Natl. Acad. Sci. 101 (46), 16121–16126.

van der Zee, J., Vujaskovic, Z., Kondo, M., Sugahara, T., 2008. The Kadota fund international forum 2004—clinical group consensus. Int. J. Hyperth. 24 (2), 111–122.

Virtanen, H., Keshvari, J., Lappalainen, R., 2006. Interaction of radio frequency electromagnetic fields and passive metallic implants—a brief review. Bioelectromagnetics 27 (6), 431–439.

Wan, Q., Xie, L., Gao, L., Wang, Z., Nan, X., Lei, H., et al., 2013. Self-assembled magnetic theranostic nanoparticles for highly sensitive MRI of minicircle DNA delivery. Nanoscale 5 (2), 744–752.

Wang, Y.X., Hussain, S.M., Krestin, G.P., 2001. Superparamagnetic iron oxide contrast agents: physicochemical characteristics and applications in MR imaging. Eur. Radiol. 11, 2319–2331.

Wang, X., Chen, X.Y., Gao, L.S., Zheng, H.G., Ji, M.R., Tang, C.M., Shen, T., Zhang, Z.D., 2004. Synthesis of β-FeOOH and α-Fe_2O_3 nanorods and electrochemical properties of β-FeOOH. J. Mater. Chem. 14 (5), 905–907.

Wang, L.Y., Luo, J., Maye, M.M., Fan, Q., Qiang, R.D., Engelhard, M.H., Wang, C.M., Lin, Y.H., Zhong, C.J., 2005. Iron oxide-gold core–shell nanoparticles and thin film assembly. J. Mater. Chem. 15, 1821–1832.

Wang, C., Daimon, H., Sun, S.H., 2009a. Dumbbell-like Pt- Fe_3O_4 nanoparticles and their enhanced catalysis for oxygen reduction reaction. Nano Lett. 9 (4), 1493–1496.

Wang, L., Neoh, K., Kang, E., Shuter, B., Wang, S.C., 2009b. Superparamagnetic hyperbranched polyglycerol-grafted Fe_3O_4 nanoparticles as a novel magnetic resonance imaging contrast agent: an in vitro assessment. Adv. Funct. Mater. 19 (16), 2615–2622.

Wang, C., Xu, C.J., Zeng, H., Sun, S.H., 2009c. Recent progress in syntheses and applications of dumbbell-like nanoparticles. Adv. Mater. 21 (30), 3045–3052.

Wang, C., Wei, Y.J., Jiang, H.Y., Sun, S.H., 2009d. Tug-of-war in nanoparticles: competitive growth of Au on Au- Fe_3O_4 nanoparticles. Nano Lett. 9, 4544–4547.

Wang, Z., Wu, L., Chen, M., Zhou, S., 2009e. Facile synthesis of superparamagnetic fluorescent Fe_3O_4/ZnS hollow nanospheres. J. Am. Chem. Soc. 131 (32), 11276–11277.

Wang, G., Ling, Y., Wheeler, D.A., George, K.E.N., Horsley, K., Heske, C., Zhang, J.Z., Li, Y., 2011a. Facile synthesis of highly photoactive α-Fe_2O_3-based films for water oxidation. Nano Lett. 11 (8), 3503–3509.

Wang, Y., Dostalek, J., Knoll, W., 2011b. Magnetic nanoparticle-enhanced biosensor based on grating-coupled surface plasmon resonance. Anal. Chem. 83 (16), 6202–6207.

Wang, Y., Zhu, Z., Xu, F., Wei, X., 2012. One-pot reaction to synthesize superparamagnetic iron oxide nanoparticles by adding phenol as reducing agent and stabilizer. J. Nanopart. Res. 14 (1), 755.

Wang, Q., Zhang, J., Wang, A., 2013a. Spray-dried magnetic chitosan/Fe_3O_4/halloysite nanotubes/ofloxacin microspheres for sustained release of ofloxacin. RSC Adv. 3 (45), 23423–23431.

Wang, G., Jin, L., Dong, Y., Niu, L., Liu, Y., Ren, F., Su, X., 2013b. Multifunctional Fe_3O_4-CdTe@ SiO_2@carboxymethyl chitosan drug nanocarriers: synergistic effect towards magnetic targeted drug delivery and cell imaging. New J. Chem. 38, 700–708.

Wang, D.W., et al., 2013c. Folate-conjugated Fe_3O_4@SiO_2@gold nanorods@ mesoporous SiO_2 hybrid nanomaterial: a theranostic agent for magnetic resonance imaging and photothermal therapy. J. Mater. Chem. B 1, 2934–2942.

Wei, W., Quanguo, H., Hong, C., 2007. ICBBE 2007: The 1st Int. Conf. on Bioinformatics and Biomedical Engineering, Wuhan. IEEE Xplore, pp. 76–79.

Wei, Z., Zhou, Z., Yang, M., Lin, C., Zhao, Z., Huang, D., Chen, Z., Gao, J., 2011. Multifunctional Ag@Fe_2O_3 yolkshell nanoparticles for simultaneous capture, kill, and removal of pathogen. J. Mater. Chem. 21, 16344–16348.

Wongwailikhit, K., Horwongsakul, S., 2011. The preparation of iron (III) oxide nanoparticles using w/o microemulsion. Mater. Lett. 65 (17–18), 2820–2822.

Wu, Y.L., Ye, Q., Foley, L.M., Hitchens, T.K., Sato, K., Williams, J.B., et al., 2006. In situ labeling of immune cells with iron oxide particles: an approach to detect organ rejection by cellular MRI. Proc. Natl. Acad. Sci. USA 103 (6), 1852–1857.

Wu, W., He, Q.G., Hu, R., Huang, J.K., Chen, H., 2007. Preparation and characterization of magnetite Fe_3O_4 nanopowders. Rare Metal Mat. Eng. 36, 238.

Wu, W., He, Q.G., Jiang, C.Z., 2008. Magnetic iron oxide nanoparticles: synthesis and surface functionalization strategies. Nanoscale Res. Lett. 3 (11), 397–415.

Wu, W., Xiao, X., Zhang, S., Li, H., Zhou, X., Jiang, C., 2009. One-pot reaction and subsequent annealing to synthesis hollow spherical magnetite and maghemite nanocages. Nanoscale Res. Lett. 4 (8), 926–931.

Wu, W., Xiao, X.H., Zhang, S.F., Zhou, J.A., Fan, L.X., Ren, F., Jiang, C.Z., 2010a. Large-scale and controlled synthesis of iron oxide magnetic short nanotubes: shape evolution, growth mechanism, and magnetic properties. J. Phys. Chem. C 114 (39), 16092–16103.

Wu, D., Zhao, J., Zhang, L., Wu, Q., Yang, Y., 2010b. Lanthanum adsorption using iron oxide loaded calcium alginate beads. Hydrometallurgy 101 (1–2), 76–83.

Wu, L.H., Yao, H.B., Hu, B., Yu, S.H., 2011a. Unique lamellar sodium/potassium iron oxide nanosheets: facile microwave-assisted synthesis and magnetic and electrochemical properties. Chem. Mater. 23 (17), 3946–3952.

Wu, W., Xiao, X.H., Zhang, S.F., Ren, F., Jiang, C.Z., 2011b. Facile method to synthesize magnetic iron oxides/TiO_2 hybrid nanoparticles and their photodegradation application of methylene blue. Nanoscale Res. Lett. 6 (1), 533–547.

Wu, W., Zhang, S.F., Ren, F., Xiao, X.H., Zhou, J., Jiang, C.Z., 2011c. Controlled synthesis of magnetic iron oxides@ SnO_2 quasi-hollow core–shell heterostructures: formation mechanism, and enhanced photocatalytic activity. Nanoscale 3 (11), 4676–4684.

Wu, W., Xiao, X.H., Ren, F., Zhang, S.F., Jiang, C.Z., 2012a. A comparative study of the magnetic behavior of single and tubular clustered magnetite nanoparticles. J. Low Temp. Phys. 168 (5), 306–313.

Wu, W., Zhang, S.F., Xiao, X.H., Zhou, J., Ren, F., Sun, L.L., Jiang, C.Z., 2012b. Controllable synthesis, magnetic properties, and enhanced photocatalytic activity of spindlelike mesoporous α- Fe_2O_3/ZnO core–shell heterostructures. ACS Appl. Mater. Interfaces 4 (7), 3602–3609.

Wu, W., Wu, Z., Yu, T., Jiang, C., Kim, W.S., 2015. Recent progress on magnetic iron oxide nanoparticles: synthesis, surface functional strategies and biomedical applications. Sci. Technol. Adv. Mater. 16 (2), 023501.

Xi, G.C., Yue, B., Cao, J.Y., Ye, J.H., 2011. Fe_3O_4/WO_3 hierarchical core–shell structure: high-performance and recyclable visible-light photocatalysis. Chem. Eur. J. 17 (18), 5145–5154.

Xiao, L., et al., 2011. Water-soluble superparamagnetic magnetite nanoparticles with biocompatible coating for enhanced magnetic resonance imaging. ACS Nano 5 (8), 6315–6324. http://pubs.acs.org/doi/abs/10.1021/nn201348s.

Xie, J., Wang, J., Niu, G., Huang, J., Chen, K., Li, X., Chen, X., 2010. Human serum albumin coated iron oxide nanoparticles for efficient cell labeling. Chem. Commun. 46 (3), 433–435.

Xu, Y.H.Z., Sun, S., 2007. Magnetic core/shell Fe_3O_4/Au and Fe_3O_4/Au/Ag nanoparticles with tunable plasmonic properties. J. Am. Chem. Soc. 129, 8698–8699.

Xu, H.H., Smith, D.T., Simon, C.G., 2004. Strong and bioactive composites containing nano-silica-fused whiskers for bone repair. Biomaterials 25 (19), 4615–4626.

Xu, C.J., Wang, B.D., Sun, S.H., 2009. Dumbbell-like Au- Fe_3O_4 nanoparticles for target-specific platin delivery. J. Am. Chem. Soc. 131 (2), 4216–4217.

Xu, H., Zeiger, B.W., Suslick, K.S., 2013. Sonochemical synthesis of nanomaterials. Chem. Soc. Rev. 42, 2555–2567.

Xuan, S.H., et al., 2012. Photocytotoxicity and magnetic relaxivity responses of dual-porous γ-Fe_2O_3@meso-SiO_2 microspheres. ACS Appl. Mater. Interfaces 4 (4), 2033–2040. https://www.ncbi.nlm.nih.gov/pubmed/22409402.

Yallapu, M.M., Othman, S.F., Curtis, E.T., Gupta, B.K., Jaggi, M., Chauhan, S.C., 2011. Multi-functional magnetic nanoparticles for magnetic resonance imaging and cancer therapy. Biomaterials 32 (7), 1890–1905.

Yang, L., Ren, X., Tang, F., Zhang, L., 2009. A practical glucose biosensor based on Fe3O4 nanoparticles and chitosan/nafion composite film. Biosens. Bioelectron. 25 (4), 889–895.

Yang, X.Q., Grailer, J.J., Rowland, I.J., Javadi, A., Hurley, S.A., Steeber, D.A., Gong, S.Q., 2010. Multifunctional SPIO/DOX-loaded wormlike polymer vesicles for cancer therapy and MR imaging. Biomaterials 31 (34), 9065–9073.

Yang, D., Ma, J.Z., Gao, M., Peng, M., Luo, Y., Hui, W., Chen, C., Wang, Z., Cui, Y., 2013. Suppression of composite nanoparticle aggregation through steric stabilization and ligand exchange for colorimetric protein detection. RSC Adv. 3 (25), 9681–9686.

Ye, Y., Kuai, L., Geng, B., 2012. A template-free route to a Fe_3O_4-Co_3O_4 yolk-shell nanostructure as a noble-metal free electrocatalyst for ORR in alkaline media. J. Mater. Chem. 22 (36), 19132–19138.

Yeap, S.P., Toh, P.Y., Ahmad, A.L., Low, S.C., Majetich, S.A., Lim, J., 2012. Colloidal stability and magnetophoresis of goldcoated iron oxide nanorods in biological media. J. Phys. Chem. C 116 (42), 22561–22569.

Yu, H., Chen, M., Rice, P.M., Wang, S.X., White, R.L., Sun, S.H., 2005. Dumbbell-like bifunctional Au-Fe_3O_4 nanoparticles. Nano Lett. 5 (2), 379–382.

Yu, X., Wan, J., Shan, Y., Chen, K., Han, X., 2009. A facile approach to fabrication of bifunctional magnetic-optical Fe_3O_4@ ZnS microspheres. Chem. Mater. 21 (20), 4892–4898. http://pubs.acs.org/doi/abs/10.1021/cm902667b.

Yue-Jian, C., Juan, T., Fei, X., Jia-Bi, Z., Ning, G., Yi-Hua, Z., et al., 2010. Synthesis, self-assembly, and characterization of PEG-coated iron oxide nanoparticles as potential MRI contrast agent. Drug Dev. Ind. Pharm. 36 (10), 1235–1244.

Zedan, A.F., Abdelsayed, V., Mohamed, M.B., El-Shall, M.S., 2013. Rapid synthesis of magnetic/luminescent (Fe_3O_4/CdSe) nanocomposites by microwave irradiation. J. Nanopart. Res. 15, 1312–1319.

Zhai, Y.M., Zhai, J.F., Wang, Y.L., Guo, S.J., Ren Wand Dong, S.J., 2009. Fabrication of iron oxide core/gold shell submicrometer spheres with nanoscale surface roughness for efficient surface-enhanced raman scattering. J. Phys. Chem. C 113 (17), 7009–7014. http://pubs.acs.org/doi/abs/10.1021/jp810561q.

Zhang, S.M., Zhang, L.N., He, B.F., Wu, Z.S., 2008. Preparation and characterization of thermosensitive PNIPAA-coated iron oxide nanoparticles. Nanotechnology 19 (32), 325608.

Zhang, S.F., Ren, F., Wu, W., Zhou, J., Xiao, X.H., Sun, L.L., Liu, Y., Jiang, C.Z., 2013. Controllable synthesis of recyclable core–shell γ- Fe_2O_3@SnO_2 hollow nanoparticles with enhanced photocatalytic and gas sensing properties. Phys. Chem. Chem. Phys. 15 (21), 8228–8236.

Zhao, N., Gao, M.Y., 2009. Magnetic janus particles prepared by a flame synthetic approach: synthesis, characterizations and properties. Adv. Mater. 21, 184–187.

Zhao, Y., Zhang, W., Lin, Y., Du, D., 2013. The vital function of Fe_3O_4@Au nanocomposites for hydrolase biosensor design and its application in detection of methyl parathion. Nanoscale 5 (3), 1121–1126.
Zheng, X., Zhou, S., Xiao, Y., Yu, X., Li, X., Wu, P., 2009. Shape memory effect of poly (d, l-lactide)/Fe_3O_4 nanocomposites by inductive heating of magnetite particles. Colloids Surf. B 71 (1), 67–72.
Zhou, L., Gao, C., Xu, W.J., 2010. Robust Fe_3O_4/SiO_2-Pt/Au/Pd magnetic nanocatalysts with multifunctional hyperbranched polyglycerol amplifiers. Langmuir 26 (13), 11217–11225.
Zhou, W., Chen, Y., Wang, X., Guo, Z., Hu, Y., 2011. Synthesis of Fe_3O_4@PbS hybrid nanoparticles through the combination of surface-initiated atom transfer radical polymerization and acidolysis by H2S. J. Nanosci. Nanotechnol. 11 (1), 98–105. https://www.ncbi.nlm.nih.gov/pubmed/21446412.
Zhou, L., He, B., Zhang, F., 2012a. Facile one-pot synthesis of iron oxide nanoparticles cross-linked magnetic poly(vinylalcohol) gel beads for drug delivery. ACS Appl. Mater. Interfaces 4 (1), 192–199.
Zhou, S., Chen, Q.W., Hu, X.Y., Zhao, T.Y., 2012b. Bifunctional luminescent superparamagnetic nanocomposites of CdSe/CdS-Fe_3O_4 synthesized via a facile method. J. Mater. Chem. 22, 8263–8270.
Zhu, J., Hayward, R.C., 2008. Spontaneous generation of amphiphilic block copolymer micelles with multiple morphologies through interfacial Instabilities. J. Am. Chem. Soc. 130 (23), 7496–7502.
Zhu, Y., Ikoma, T., Hanagata, N., Kaskel, S., 2010a. Rattle-type Fe_3O_4@SiO_2 hollow mesoporous spheres as carriers for drug delivery. Small 6, 471–478.
Zhu, Y., Fang, Y., Kaskel, S., 2010b. Folate-conjugated Fe3O4@SiO2 hollow mesoporous spheres for targeted anticancer drug delivery. J. Phys. Chem. C 114 (39), 16382–16388.
Zhu, C.L., Liu, L.B., Yang, Q., Lv, F.T., Wang, S., 2012a. Water-soluble conjugated polymers for imaging, diagnosis, and therapy. Chem. Rev. 112 (8), 4687–4735. http://pubs.acs.org/doi/abs/10.1021/cr200263w.
Zhu, X.-M., Yuan, J., Leung, K.C.-F., Lee, S.-F., Sham, K.W.Y., Cheng, C.H.K., et al., 2012b. Hollow superparamagnetic iron oxide nanoshells as a hydrophobic anticancer drug carrier: intracelluar pH-dependent drug release and enhanced cytotoxicity. Nanoscale 4 (18), 5744–5754.
Zhu, S., Guo, J., Dong, J., Cui, Z., Lu, T., Zhu, C., Zhang, D., Ma, J., 2013. Sonochemical fabrication of Fe_3O_4 nanoparticles on reduced graphene oxide for biosensors. Ultrason. Sonochem. 20 (3), 872–880.

CHAPTER

9

Magnetic Particle Imaging

Anna Bakenecker, Mandy Ahlborg, Christina Debbeler, Christian Kaethner, Kerstin Lüdtke-Buzug

University of Luebeck, Luebeck, Germany

1 INTRODUCTION

Several medical imaging techniques, such as computed tomography (CT), magnetic resonance imaging (MRI), or positron emission tomography (PET), are applied in clinical routine. CT utilizes the attenuation of X-rays depending on the material density to image anatomical information. Thus, it is often used when imaging dense material such as bone structures. MRI measures the tissue-dependent proton density. External magnetic fields are applied, which excite the nuclear spin of hydrogen atoms in the tissue. The relaxation time of the spins after such an excitation is tissue-dependent. Thus, high-contrast images can be obtained from soft tissue, containing mostly fat and water. The physical principles of both, CT and MRI, are based on a direct interaction with tissue material. In contrast, PET is a tracer-based imaging modality. Radioactive markers are attached to functional molecules; hence, PET is a suitable imaging technique for the investigation of metabolic processes.

Magnetic particle imaging (MPI) is a novel imaging technique, which is not yet in clinical use but is highly promising for medical applications. MPI was invented by Bernhard Gleich and Jürgen Weizenecker at the Philips Research Laboratories, Hamburg, and was first published in 2005 (Gleich and Weizenecker, 2005). No harmful radiation, such as X-rays in CT or gamma rays caused by radioactive tracers in PET, is needed. Similar to PET, MPI images the spatial and temporal distribution of a tracer, but instead of using a radioactive material, MPI uses magnetic nanoparticles (MNPs), which are biocompatible, well tolerated and offer the potential to be quantitatively imaged. Promising experiments showed the potential of a spatial resolution in the submillimeter range and that a high sensitivity of MNP detection can be reached. Furthermore, a fast acquisition time enables real-time and interventional imaging.

To explain the basic physical principle of MPI, we consider an MNP having a magnetic moment imaginable as a compass needle. An externally applied alternating magnetic field affects the magnetic moment to flip, such as a compass needle that aligns with an external magnetic field. This externally applied alternating magnetic field can be described mathematically by its strength, its direction, and its excitation frequency and phase. The magnetic

Precision Medicine
https://doi.org/10.1016/B978-0-12-805364-5.00009-3

moment of the MNP then flips with the same frequency as the excitation frequency. However, when the magnetic moments of the particles are completely aligned to the externally applied magnetic field, the particles are saturated, and the magnetization reaches a maximum—the saturation magnetization. The maximum deflection of a compass needle is reached, when it is completely aligned with the external magnetic field. The change of the magnetic moments induces a characteristic particle signal, which is detectable and the key to MPI. Due to the saturation magnetization, the temporal evolution of the magnetization has a plateau at its maximum and minimum. To mathematically describe the induced particle signal, which is the derivation of the magnetization, not only the excitation frequency but also higher harmonics are needed. The induced signal, often shown in a frequency spectrum, is a fingerprint of the MNPs and can be used to determine the spatial and temporal distribution of MNP concentrations. Commonly, superparamagnetic iron oxide nanoparticles (SPIONs) are used as a tracer material, which are usually coated for biocompatibility reasons and to prevent agglomeration.

Based on this imaging principle, a variety of different scanner prototypes have been realized, mostly for preclinical use. Since several medical application scenarios are of potential interest, current investigations are considering the upscaling toward human-sized scanners and improving the sensitivity and the spatial and temporal resolution. Further, the development of MPI-specific MNPs and image reconstruction methods are part of ongoing research.

This chapter is structured as follows. In Section 2, the *basic principle of MPI* will be explained. This includes the signal generation and spatial encoding of the MNP signal. Further, different possibilities of the volume sampling are introduced.

In Section 3, the *development of MPI tracers* is explained. The most common synthesis routes of SPIONs are briefly described. There are several analysis tools available to characterize the synthesized particles. A selection of these techniques will be presented with focus on the evaluation of physical characteristics necessary for MPI. Theoretical approaches will be introduced that are mainly used to identify optimal MPI tracers, and dedicated devices for MNP analysis, so-called magnetic particle spectrometers, are presented.

Section 4 will provide a selection of different *scanner designs*. The development and ongoing research on the hardware construction focus on improving the sensitivity, time resolution, and upscaling of MPI.

The *image reconstruction* in MPI can, in general, be performed in two different ways in MPI. Both possibilities are described and discussed in Section 5. Further, specific challenges for image reconstruction are presented.

Finally, different *innovative applications of MPI* will be introduced in Section 6. An outlook and future prospects of this novel imaging technique will be addressed.

2 BASIC IMAGING PRINCIPLES

The basic imaging principles of MPI rely on the interaction of the MNPs with specific magnetic fields that are externally applied. Because of this interaction, a characteristic particle signal can be detected and later reconstructed to an image representation of the particle suspension (see Section 5).

In Section 2.1, an overview of the generation and the encoding principles of MPI will be given. The basic spatial encoding principles will be covered in Section 2.2. Due to simplicity, both, the principles of the signal and the spatial encoding, will be described for the 1D case. Since MPI is an imaging method that is intrinsically 3D, it is of interest to discuss different possibilities for a volume sampling (see Section 2.3). In the last part of this section (see Section 2.4), current limitations of MPI in terms of its clinical application possibilities will be summarized.

2.1 Signal Generation and Encoding

The generation of a detectable signal in MPI is based on the interaction of a fluidal solution containing MNPs with magnetic fields (Gleich and Weizenecker, 2005). These fields, named drive fields, $H^D(t)$, feature a high homogeneity and an oscillating magnetic field strength and thus allow for a change of the particles' magnetization direction (Gleich and Weizenecker, 2005; Graeser et al., 2015a). In a simplified setting, the magnetization behavior of the particles can be described by the Langevin theory of paramagnetism (see Section 3.2.2). In terms of the applied frequency for the drive fields, it is often chosen around a base frequency of $f^B = 25\,\text{kHz}$ (Erbe et al., 2013; Gleich and Weizenecker, 2005; Sattel et al., 2009), although the signal generation is not limited to this frequency, and other frequencies can be chosen as well (Borgert et al., 2013; Goodwill et al., 2009). The mathematical description of the drive field can be expressed as $H_i^D(t) = A_i \sin(2\pi f_i t + \phi_i)$ where $i \in \{x, y, z\}$ indicates the direction. Here, A_i represents the field amplitude of the oscillating magnetic field in Tesla, f_i is the applied frequency, and ϕ_i describes a phase shift of the field.

Since the MNPs feature a nonlinear magnetization curve (see Section 3.2.2), the sinusoidal progression of the applied drive field $H^D(t)$ is modulated by its interaction with the particles. According to Faraday's law, this modulation of the magnetization causes an induction of a characteristic voltage $u(t)$ in a receiving electromagnetic coil. Such a voltage signal is often referred to as the particle signal and resembles a unique fingerprint of the particle sample, because it contains higher harmonics of the base frequency f^B that specifically depend on the underlying particle solution (Buzug et al., 2012; Gleich and Weizenecker, 2005; Knopp and Buzug, 2012).

An example illustration of the signal encoding principle in MPI is given by Fig. 1. In this context, it is important to keep in mind that the illustration only covers a particle excitation using one drive field, $H^D(t)$, that is, a 1D imaging approach.

2.2 Spatial Encoding

By the application of a drive field such as described in Section 2.1, the magnetization direction of the MNPs changes according to the drive field, which yields that all MNPs contribute to a detectable particle signal. To spatially distinguish between the individual particle signals, only MNPs in a defined area should be allowed to contribute to the signal at a given time. This can be realized by the use of an additional magnetic field featuring a high magnetic field strength and spatial inhomogeneity, that is, a strong magnetic gradient field. Such a field, referred to as a selection field, $H^S(i)$, allows for a saturation of all MNPs except those in a specific area; see Fig. 1. This means that the magnetization of the MNPs by the selection field $H^S(i)$ outside this area is high enough to neglect the effect of the temporally variable drive

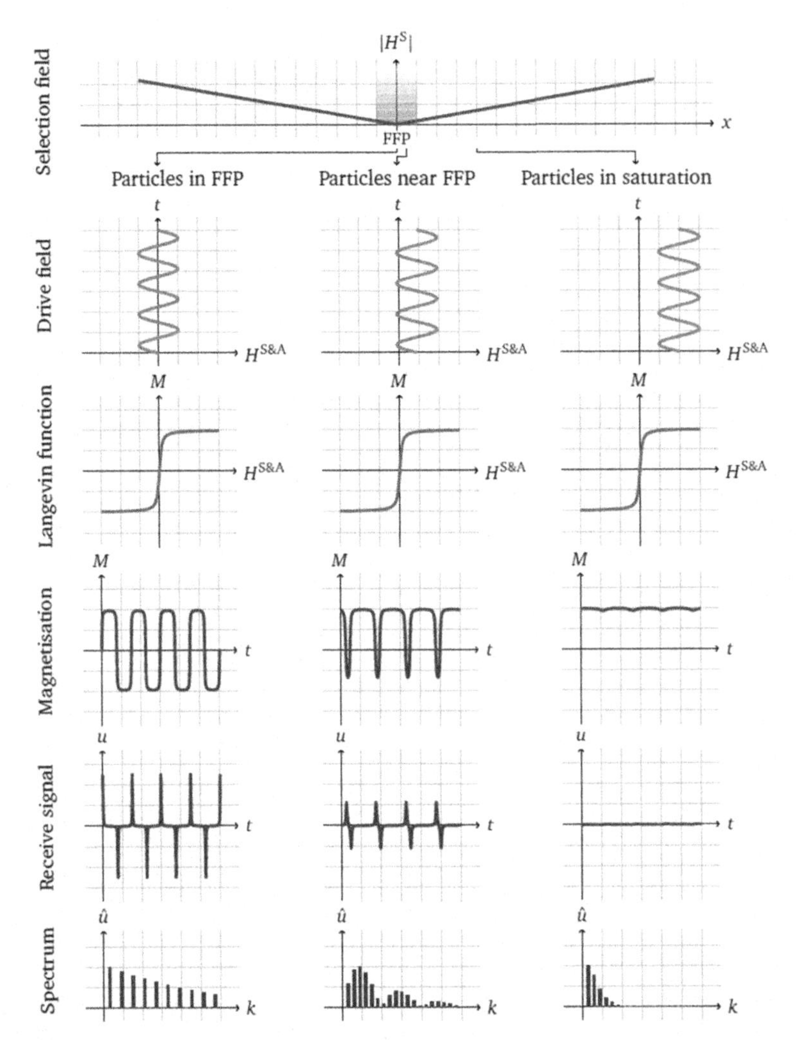

FIG. 1 The signal encoding principle in MPI is based on the excitation of MNPs by an oscillating magnetic field called drive field $H^D(t)$. Based on a modulation of the drive field by the particles' nonlinear magnetization curve (modeled by the Langevin function), a characteristic receive signal $u(t)$ can be induced. As shown by the frequency spectrum $\hat{u}(t)$ of such a voltage signal, the modulation of $H^D(t)$ causes the generation of higher harmonics of the applied frequency. The principle for spatial encoding in MPI is based on the application of a selection field, that is, magnetic gradient field or superimposed to the drive field, $H^D(t)$. Based on a gradient field featuring an FFP, all MNPs outside the area of the FFP or in a close vicinity to it remain in saturation, and thus do not contribute to the particle signal.

field, $H^D(t)$. The mathematical description of such a field is given by $H_i^S(i) = G_i i$ with G_i being the magnetic gradient strength and $i \in \{x, y, z\}$ the spatial direction. In an ideal setting, the field strength of the selection field $H^S(i)$ is linearly increasing in space.

The specific area that the selection field forms to spatially encode the particle signals can be generated as a field-free point (FFP) (Sattel et al., 2009; Borgert et al., 2013; Goodwill et al., 2012c; Vogel et al., 2014a; Weizenecker et al., 2009) or a field-free line (FFL) (Bente et al., 2015;

Goodwill et al., 2012a; Knopp et al., 2010b; Weizenecker et al., 2008). In both settings, the magnetic field strength of the selection field $H^S(i)$ is zero in the FFP or the FFL, respectively. The main difference of both encoding schemes is the shape of the field-free area and thus the number of MNPs that contribute to the received signal. While the FFP offers a spatial encoding in a defined hot spot, the FFL allows for a spatial encoding along a line, which yields among an increased sensitivity also an increased signal-to-noise ratio (SNR) compared with the FFP approach (Weizenecker et al., 2008). In terms of their basic imaging principle, both approaches are nearly identical and only differ in the applied imaging sequence (Knopp et al., 2009a; Weizenecker et al., 2008). The use of an FFL can also be seen in analogy to the imaging principles of computed tomography (CT). This is particularly interesting when it comes to the reconstruction of the received particle signals (see Section 5) and the possibility of applying sophisticated CT algorithms (Knopp et al., 2011; Konkle et al., 2013a; Buzug, 2008; Kalender, 2011). The spatial encoding principle in MPI by use of an FFP for a 1D scenario is illustrated in Fig. 1.

2.3 Volume Sampling

Up to here, the described basic principles of MPI allow for a signal and a spatial encoding at a defined spatiotemporal position in the measuring field. Since it is of interest to not only cover one position but also an entire area or volume given by a defined field of view (FOV), it is necessary to change the position of the FFP or the FFL relative to the imaged object containing the MNPs. To achieve this, there are several ways that differ in time consumption and complexity to realize a volume sampling.

An approach that seems natural to use for a sampling of the entire FOV is a stepwise movement of the object while not changing the position of the FFP or FFL (Gleich and Weizenecker, 2005; Goodwill et al., 2009). In addition to being very time-consuming, the spatial resolution of this technique strongly depends on the precision of the object movement. In the same context, the object can remain at its position, and the FFP or FFL can be moved stepwise to different positions by adapting the selection field, $H^S(i)$. Despite their disadvantages, both techniques allow the construction of a very simple imaging device and served as a basis for the first acquired images in MPI (Gleich and Weizenecker, 2005).

A more sophisticated approach that significantly shortens the acquisition time is the continuous variation of the magnetic field strength of the drive field $H^D(t)$ with an increased amplitude as proposed in Gleich and Weizenecker (2005). Thus, in a 1D setting, the FFP or FFL can be moved back and forth along a trajectory given by a straight line, whereas the amplitude A_i directly correlates with the displacement of the FFP or FFL with respect to the gradient strength, G_i. It follows that the area covered in such a 1D approach is given by $[-A_i/G_i, A_i/G_i]$. The theoretical basics of this concept are easily transferable to a multidimensional imaging approach (Gleich et al., 2008; Weizenecker et al., 2009). Instead of moving the FFP or FFL along a line, the FOV can be covered by a sequential movement (see Fig. 2A) or more complex data acquisition paths such as a Lissajous trajectory as shown in Fig. 2B. More details about possible trajectories in MPI can be found, for example, in Knopp et al. (2009a) and Szwargulski et al. (2015a). In this context, it is important to mention that such an approach using multiple drive fields causes the generation of mixing frequencies in the frequency spectrum as described for the 1D case in Section 2.1 (Rahmer et al., 2012).

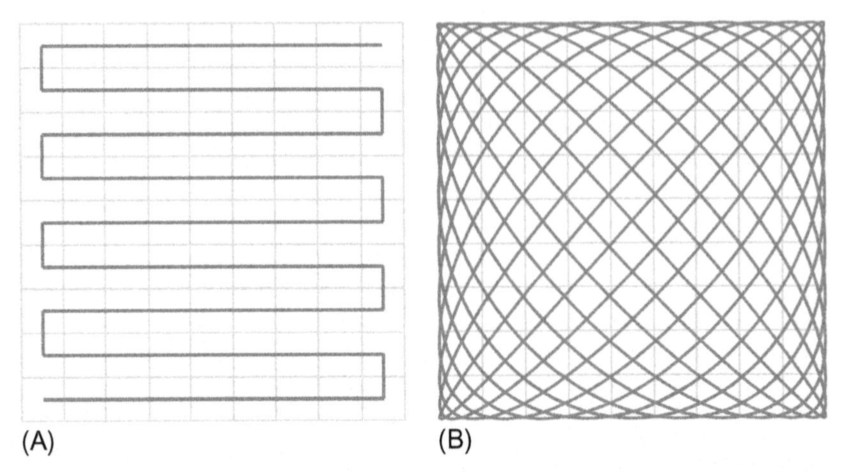

FIG. 2 By use of a continuous data acquisition, different trajectories can be applied to cover the FOV. For a 2D or even 3D imaging approach, a sequential movement (A) or a Lissajous trajectory (B) can be used.

An advantage of using the drive field this way is that the temporal resolution can be significantly increased compared with the approach based on a stepwise displacement (Gleich et al., 2008). In terms of potential application scenarios (see Section 6), it becomes even possible to acquire data in real time (Weizenecker et al., 2009). However, the use of a continuous data acquisition path also comes with the price of a significantly higher complexity with respect to the scanner construction (see Section 4) and, thus, with much higher costs. In addition to this, the size of the FOV that can be covered by the FFP or FFL this way is limited. Some reasons for this are an increased electric power consumption and potential medical restrictions, which will be covered in Section 2.4.

A promising approach to overcome such limitations and to further increase the size of the FOV is the use of additional homogenous magnetic fields (Rahmer et al., 2011a,b; Schmale et al., 2011). These fields, referred to as focus fields, feature an increased amplitude and a low frequency and are superimposed to the drive fields and the selection field. The use of focus fields offers different possibilities for the FOV enlargement. A first option is a focus-field-based shift of the sampling region to one or multiple fixed spatial positions (Knopp et al., 2012; Kaethner et al., 2014). Such a static focus field approach either shifts the FOV to a specific region of an object or divides the FOV into a predefined number of smaller partial FOVs or patches (see Section 5.4) that are sampled one after another (Ahlborg et al., 2016; Knopp et al., 2015). An interesting aspect of this kind of approach is that the position of each patch can be chosen either in an arbitrary way or according to a defined grid structure. This flexibility allows for adapting the patch positions to complex structures such as an arterial tree. Another possibility is given by a simultaneous variation of the focus field and the drive fields (Nothnagel and Sánchez-Gonzáles, 2015; Rahmer et al., 2013b, 2014b). Through this, an arbitrary dynamic shift of the data acquisition path within the entire FOV becomes possible. An illustration of the focus field approaches is given by Fig. 3.

The sole use of such additional focus fields already offers various possibilities to enlarge the FOV. However, it remains a challenging task, when objects with an axial length longer

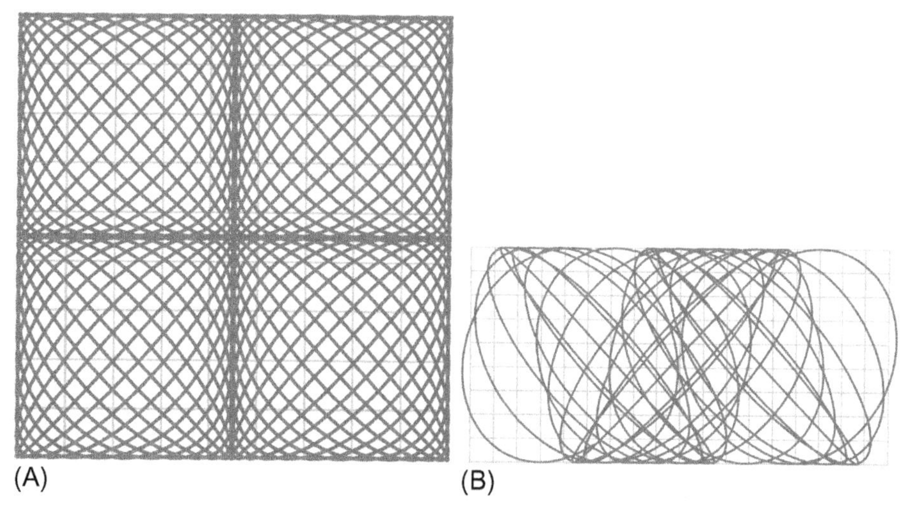

FIG. 3 Focus fields used in addition to the drive field and the selection field allow for a further enlargement of the FOV within the technical and medical limitations. A focus field can shift the imaging area either to static predefined positions (A) or dynamically through the entire FOV (B).

than the scanner dimensions are planned to be scanned. In addition to a focus-field-based enlargement of the FOV an enlargement in the axial direction can be achieved by a mechanical movement of the imaging object in this direction. This movement of the object causes an elongation of the data acquisition path (Kaethner et al., 2015). The additional movement in the axial direction results in an elongated data acquisition path similar to a spiral or a helix structure (Kalender et al., 1990). Since the course of the 2D acquisition path in MPI can vary, the progression and FOV coverage of an elongated trajectory strongly depends on this choice (Kaethner et al., 2015; Knopp et al., 2009a). In contrast to CT, this means that the resulting path can cover the FOV in an inhomogeneous way. Thus, it becomes an important step in the design of the imaging sequence to prevent signal loss or the occurrence of artifacts (Kaethner et al., 2015, 2017). In addition to a mechanical movement, the trajectory elongation in MPI can also be realized by applying a linear focus field in axial direction. An example of such an elongated trajectory based on a 2D Lissajous trajectory is shown in Fig. 4.

2.4 Current Limitations

Currently realized and commercially available MPI systems are designed for research purposes and preclinical studies on small animals such as mice or rats (Bruker and Magnetic Insight). The main limitations despite technical challenges for an upscaling to larger animal or even human-sized systems are physiological effects caused by the frequency and the amplitude of the oscillating magnetic fields that are used for the signal generation (see Section 2.1) and the volume sampling of the FOV (see Section 2.3). Due to these effects, the achievable FOV size within the limitations is relatively small. The main two effects are a stimulation of the peripheral nerves and a tissue heating caused by an energy deposition due to the used electromagnetic fields (Reilly, 1991).

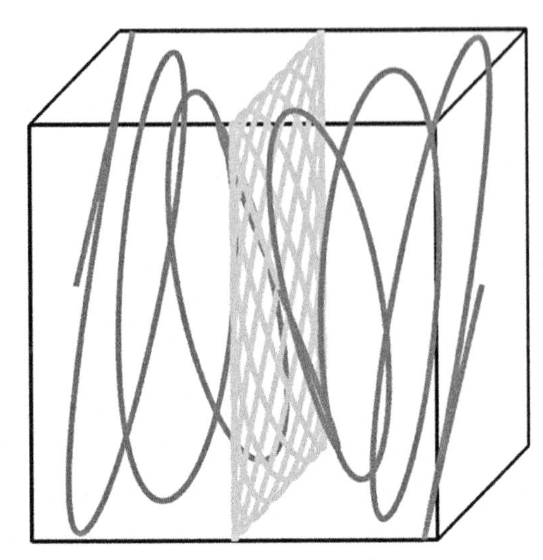

FIG. 4 When a 2D Lissajous trajectory *(light gray)* is superimposed with an axially oriented focus field or mechanical movement, the acquisition path is elongated *(dark gray)*.

A comprehensive formulation of the physiological safety limits is still part of ongoing research (Doessel and Bohnert, 2013; Saritas et al., 2013; Schmale et al., 2015). The magnetic fields currently used in small animal MPI systems feature a frequency range between 10 and 100 kHz and the peak amplitudes vary between 10 and 100 mT (Panagiotopoulos et al., 2015). In order to ensure an operation within the safety limits, the choice of the field frequency must be considered when choosing the amplitude and vice versa. Current systems for small animal imaging allow for a maximal frequency of 25 kHz and a peak amplitude of up to 10 mT (Saritas et al., 2013; Doessel and Bohnert, 2014). However, for a potential human-sized system, these parameters need to be adapted (Schmale et al., 2015; Doessel and Bohnert, 2014; Saritas et al., 2012; Rahmer et al., 2014a).

Since both, the frequency and the amplitude, need to be below the safety limits, an encoding of an FOV is limited in size and thus requires more time. To still allow for a relatively fast encoding of a sufficiently large FOV, the use of focus fields (see Section 2.3) is a promising approach.

3 MAGNETIC NANOPARTICLES

Nanoparticles are usually defined as objects ranging in size from 1 to 100 nm. They can be based on many different metallic materials, such as zinc, titanium, gold, or silver, and are established in different industrial branches and medical applications. One of the most common applications is the use of nanoparticles in colors. Gold nanoparticles, as colloidal dispersions, have been utilized by artists because of their vibrant colors produced by the interaction with visible light. Modern applications include the use as solar cells and fuel cells for energy generation and storage, in diverse materials of everyday use such as cosmetics, sunscreen, or clothes and as contrast agents or drug carrier in medical applications.

MPI is a tracer-based medical imaging modality, which uses MNPs as tracer material. The investigation of the physical particle characteristics is of paramount importance in MPI, because the main aspect influencing the imaging is the combination of imaging sequence and tracer material. As a result, a strong research focus lies on the investigation of physical properties of the tracer material, that is, the MNPs. This includes the analysis of the final stage of tracer material and intermediated stages during synthesis.

In Section 3.1 some basic information on the synthesis of MNPs will be introduced. Section 3.2 will give an overview of the magnetic properties of MNPs and their relevance for MPI. In Section 3.3, different analytic tools will be presented, which can be used to evaluate the outcome of the synthesis. Especially when it comes to medical applications, the characteristics of MNPs, which among others affect the biocompatibility and the reproducibility of the particles, need to be considered. In Section 3.4, different analysis techniques of MNPs in MPI are presented. This includes the mathematical analysis of particle theory and the measurement of certain characteristics via magnetic particle spectroscopy (MPS).

3.1 Synthesis

Nanosized iron oxide particles are used as tracer material in MPI applications. The particles are so small that only a single magnetic domain forms throughout the entire object. This allows the particles to behave in a very specific way, called superparamagnetic, that is, the nonlinear magnetization curve looks paramagnetic with a high saturation magnetization (see Section 3.2.1). Therefore, the particles are often called superparamagnetic iron oxide nanoparticles (SPIONs). In general, the MNPs consist of a superparamagnetic core and a biocompatible coating. In the last decade, Resovist (Bayer Schering Pharma AG, Berlin) has been established as a gold-standard for MPI tracers, because it is an approved contrast agent for MRI liver applications. Resovist is made of SPIONs covered with carboxy dextran in a water-based isotonic suspension. It has been used off-label in MPI, although only a few percent of the nanoparticles in the suspension actually contribute to the MPI signal. Therefore, efforts have been put into the development of tailored nanoparticles for MPI.

A frequently used method to synthesize the superparamagnetic *core* of the tracer material is the coprecipitation of ferrous and ferric ions in order to produce *magnetite* (Fe_3O_4). The method is based on the nucleation of an initial product of Fe^{2+} and Fe^{3+}. In the second step, the iron oxide, which is black in color, is formed:

$$\text{1. Step} — Fe^{2+} + 2Fe^{3+} + 8OH^- \rightarrow Fe(OH)_2 + 2\,Fe(OH)_3 \tag{1}$$

$$\text{2. Step} — Fe(OH)_2 + 2\,Fe(OH)_3 \rightarrow Fe_3O_4 + 4H_2O \tag{2}$$

The reaction has to be protected against oxidation, for instance, with a protective gas like nitrogen.

The coprecipitation can be characterized as a bottom-up process (Lüdtke-Buzug, 2012). It starts on the atomic level with the two different iron salts. The thermodynamics of the coprecipitation require a ratio of 1:2 for Fe^{2+} and Fe^{3+} at a pH between 8 and 14. If the ratio between Fe^{2+} and Fe^{3+} is changed, the mainly formed product is not magnetite but maghemite (Fe_2O_3), because the Fe^{2+} can be oxidized easily. However, the magnetization of maghemite is not sufficient for being a suitable MPI tracer material.

The synthesis can be processed as follows. Two solutions have to be prepared: The first contains the base, for instance, ammonia hydroxide or sodium hydroxide, and the second solution is a mixture of the iron chlorides. The basic solution has to be heated to 70°C while stirring mechanically in the protection atmosphere, to prevent the Fe^{2+} salt of oxidation. In the next step, the iron salt mixture has to be added to the basic solution. The reaction starts immediately with the nucleation. To allow the growing process, the mixture must be stirred at 70°C for 45 min. After cooling down to room temperature, the particles have to be decanted, using a permanent magnet to remove the unreacted components and washed with deionized water. The particles have to be dispersed in tetramethylammonium hydroxide (TMAH) to achieve a stable colloidal dispersion. However, a disadvantage of this method is the low biocompatibility of the product. The particles are solved in the toxic TMAH, and further, the iron oxide cores have no coating that prevents them from agglomeration.

Concerning the stability of the colloidal dispersion, the usage of polymers as surface-complexing agents is preferable. *Coatings* are used for several purposes. For example, the coating allows water solubility, prevents the particles from agglomeration, and increases biocompatibility. It is described in (Khandar et al., 2015) that MNPs with a smaller hydrodynamic diameter down to approximately 15 nm usually result in a longer circulation in the blood, whereas larger coatings in the area around 200–500 nm will increase the particle uptake in the spleen. Within the synthesis chain, the coating material must be added to the iron salt solution. Often, dextran and dextran derivatives such as carboxy dextran or silica are used.

A variation of this method is to add a coating material for the particle cores to the reaction mixture at the beginning of the synthesis. In addition, the iron has to be prepared for the nucleation process and the crystal growth, and the base must be added slowly. These steps must be supported by suitable cooling. The nucleation process of magnetite is easier when the pH value of the solution is lower than 11, while the growth is easier when the pH value is higher than 11. Therefore, the base, for instance ammonia, is added slowly. The second step, the formation of the magnetite then proceeds, as described above, under heating of the reactants. The advantage of this method is the *biocompatibility* of the outcome, because the solvent is water, and no transfer or exchange of the solvent is necessary in comparison with the thermal decomposition method that is described below (Massart, 1981; Gupta and Gupta, 2005; Laurent et al., 2008; Nune et al., 2009; Wu et al., 2008; Ali et al., 2016; Hasan, 2015; Kim et al., 2001).

The thermal decomposition, a high-temperature method, is a method to achieve monodisperse particles, since particles of the same size give the same signal and therefore allow for an optimal imaging result. MNPs of different size would not all contribute to the MPI signal, and therefore, it would be necessary to administer larger amounts of MNPs. One advantage of thermal decomposition is the opportunity of particle-core-size control. For an optimal imaging result in MPI, the size distribution should be monodisperse. The core size was theoretically derived to be optimal around a size of 30 nm (Gleich and Weizenecker, 2005). In further investigations, this size was experimentally refined to a core size of about 20–25 nm, whereas the currently best performing MNPs feature a core size of approximately 25 nm (Ferguson et al., 2015).

The method includes iron complexes that are decomposed in the presence of high-temperature boiling solvents and of surfactants. In a first step, an iron-oleate complex must be formed, usually iron pentacarbonyl in the presence of oleic acid is used. The Fe^{3+} chloride has to be dispersed in water and the sodium oleate in hexane. With the addition of ethanol, the solution has to be heated to 65°C under stirring. After 4h of reaction time, the two phases have to be separated. After drying and removing the hexane, the iron oleate can be used in the next step for the thermal decomposition at a very high boiling temperature. Therefore, a nonpolar solvent, like diethylether, has to be heated up to approximately 300°C for 1.5h. The solvent has to be removed, and the SPIONs have to be centrifugalized. The product must be washed several times, and finally, the particles have to be dispersed in hexane with an addition of oleic acid (Krishnan, 2016; Rossetti et al., 1983; Murray et al., 1993, 2001). The particles have to be stored at low temperature (4°C) for further use, because the particles are not stable at room temperature.

3.2 Magnetic Properties

The combination of the magnetic iron core with its magnetic characteristics and the magnetically neutral coating makes SPIONs a suitable MPI tracer material. The characteristic nonlinear magnetization behavior can be described by the Langevin theory of paramagnetism (see Section 3.2.2). In order to understand, how a theory describing a paramagnetic material can also be applied to a ferromagnetic material, which is then referred to as superparamagnetic, it is important to understand the magnetic properties of ferromagnetic materials.

3.2.1 *Ferromagnetism, Superparamagnetism, and Hysteresis Effects*

Iron, nickel, and cobalt are well-known examples for ferromagnetic materials. In ferromagnetism, the dipole moments form strong interactions between each another, building domains also called Weiss domains. Within a domain, the dipole moments are aligned. Due to this, the magnetization in a domain is maximal and thus in saturation. The boundaries of the Weiss domains are called Bloch walls. If a ferromagnetic material is exposed to an external magnetic field, the Bloch walls of the ferromagnetic material start to move in such a way that those Weiss domains, in which the magnetization is aligned in the direction of the magnetic field, are increased in size. This increases the overall magnetization of the material. Ultimately, if the external magnetic field is increased further, there can only be one Weiss domain left, leaving the material in magnetic saturation. If the external magnetic field is turned off in such a state of saturation, the Bloch walls do not return to their initial position, leaving a residual magnetization of the material, known as remanence. The characteristic S-shape of the hysteresis loop of ferromagnetic materials can be seen in Fig. 5 (Erbe, 2014; Mayergoyz, 1986; Biederer, 2012).

The size of the Weiss domains depends on the corresponding material. If the size of a particle is smaller than a critical diameter, it is energetically more efficient for spherical particles to consist of only one Weiss domain (Kronmüller and Parkin, 2007). If an MPI tracer particle with a size of not more than 30nm is assumed (Gleich and Weizenecker, 2005), this one-domain structure leaves the particle with one large particle magnetic moment. Due to their coating, which also ensures biocompatibility, the particles do not agglomerate and thus can be assumed as noninteracting. With this, the behavior of the magnetic moment of those one-domain particles can be compared with the behavior of the noninteracting atomic magnetic moments of paramagnetic materials. Because of this, those particles are referred to as superparamagnetic (Bean and Livingston, 1959).

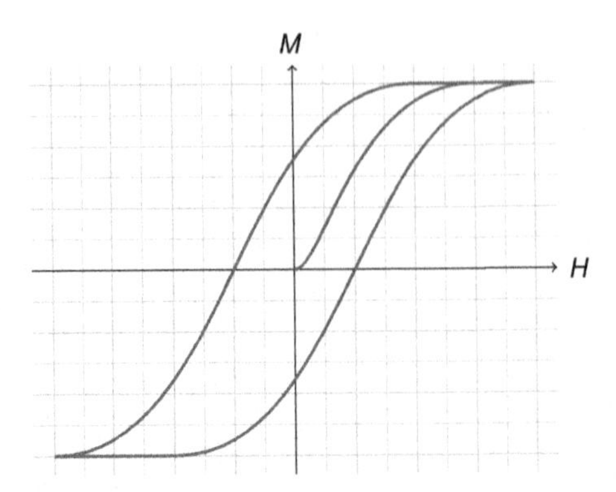

FIG. 5 Hysteresis curve of a ferromagnetic material. Only during the first magnetization, the magnetization curve begins at the origin. After switching off the magnetic field, a remanence is left. Depending on the orientation of the magnetic field, this remanence can be either positive or negative.

3.2.2 *Langevin Theory of Paramagnetism*

In a simplified model, the nonlinear magnetization behavior of SPIONs can be described by Langevin theory of paramagnetism as long as anisotropy, a particle magnetic directional dependence, particle interactions, and hysteresis effects are neglected.

When particles are exposed to an external magnetic field, the magnetic moments align in the direction of the magnetic field and change their magnetization direction. This magnetization process can be described as a function of the magnetic field. For ideal particles, the magnetization shows a very characteristic behavior. With increasing field strength, the magnetization rises to a saturation magnetization and decreases back to zero, when the field strength decreases. The saturation effect of the particles magnetization can mathematically be described by the Langevin function:

$$\mathcal{L}(\xi) := \begin{cases} \left(\coth(\xi) - \frac{1}{\xi}\right) & \xi \neq 0 \\ 0 & \xi = 0 \end{cases} \tag{3}$$

The dependency of the particles magnetization M is then described by

$$M(H) = cm\mathcal{L}(\beta H) \quad \text{with } \beta := \frac{\mu_0 m}{k_B T^P} \tag{4}$$

and k_B being the Boltzmann constant, T^P being the particle temperature, c being the particle concentration, μ_0 being the permeability of free space, $m := \|\boldsymbol{m}\|_2$ being the modulus of the magnetic moment of a single particle, and H being the magnetic field strength. The average of all the magnetic moments of the particles in a tracer is referred to as the mean magnetic moment, $\bar{m}$, which is defined as

$$\bar{m} := \frac{1}{N^P} \sum_{j=0}^{N^P-1} m_j \tag{5}$$

with N^P being the number of particles. The steepness of the magnetization curve determines the achievable spatial resolution in MPI and is mostly dependent on the size of the iron core

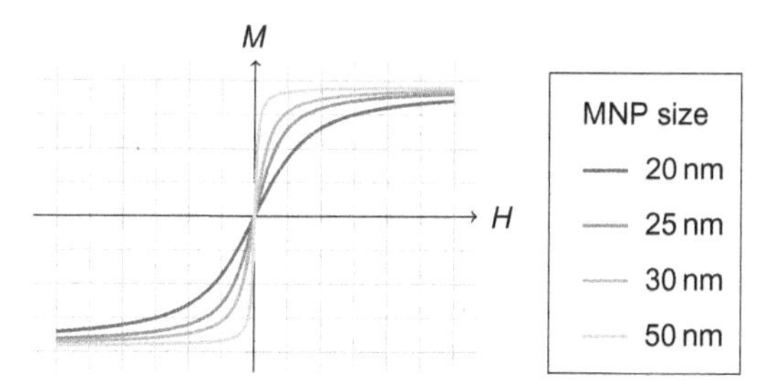

FIG. 6 Magnetization behavior of a distribution of SPIONs according to the Langevin theory for various particle diameters.

(Knopp and Buzug, 2012; Chikazumi and Charap, 1964; Caizer, 2003; Weizenecker et al., 2007; Graeser et al., 2015b). Fig. 6 displays the magnetization behavior of a distribution of SPIONs according to the Langevin theory for various particle diameters.

3.2.3 Relaxation Processes

Two relaxation processes, the Brownian relaxation and the Néel relaxation, cause the decrease of the magnetization of MNPs while turning down the external magnetic field. Both processes show characteristic relaxation times and cause a stochastic distribution of the magnetic moments of the MNPs (Odenbach, 2009).

Aligned magnetic moments of the particles start to relax as soon as the external magnetic field is turned off. If the relaxation is due to thermal movement and the entire particle including its magnetic moment is rotating around its physical axis, the process is called Brownian relaxation. The relaxation time needed for this Brownian rotation is denoted as the Brownian relaxation time constant τ_B and can be described as

$$\tau_B = \frac{3\eta V_{hyd}}{k_B T^P} \quad (6)$$

with the hydrodynamic volume V_{hyd} of the particle and the viscosity η (Lange, 2001; Brown, 1963; Payet et al., 1998).

If only the magnetic moment of the particle rotates while the orientation of the particle itself remains fixed, the rotation of the particle is due to the Néel relaxation. One characteristic for the Néel relaxation is that it depends on an energy barrier that needs to be overcome by the magnetic moment in order to rotate inside the crystal lattice. Without an external magnetic field, the particles are aligned parallel or antiparallel to the anisotropy axis of the particle. When the thermal energy $k_B T^P$ is sufficiently high, the energy barrier for the orientation of the magnetic moments can be overcome. The relaxation time needed for this rotation is denoted as the Néel relaxation time constant τ_N and can be described as

$$\tau_N = \tau_0 \exp \frac{K V_c}{k_B T^P} \quad (7)$$

with anisotropy constant K, the volume of the core V_c, and the constant factor τ_0 in the order of $(10^{-11} - 10^{-9})\ \mathrm{s}^{-1}$ (Odenbach, 2009; Lange, 2001; Néel, 1955; Rogge, 2015).

3.3 Analytical Tools

For the analysis of MNPs, several different measurement techniques are available. To evaluate the outcome of the synthesis, information on the structure, the chemical composition, and the particle size distribution are of particular interest. There are also several application-oriented analysis tools available, and with the focus on MPI, measurements gathering information on the magnetic properties of the MNPs are interesting as well. A small selection of methods used in the characterization will be presented in the following. This list of methods raises no claim to completeness. Most of the methods can also be used to describe further parameters than those named here.

3.3.1 Regarding Structure, Chemical Composition, and Size Determination

One of the very important parameters in the analysis of MNPs is the determination of the iron concentration of a sample. There are several techniques available.

The most common method to determine the iron concentration in an MNP sample is the photometric determination of the iron content of the sample using *Prussian blue* or *phenanthroline* staining. Both methods require the destruction of the particle coating as a first step, since both chemicals react with the iron cores of the sample. Thus, the MNP samples are compounded with hydrochloric acid to dissolve the coating. In a next step, the solution is treated with ascorbic acid to reduce all of the Fe^{2+} to Fe^{3+} compounds. After this, the prepared sample solution is stained forming complexes of Prussian blue or phenanthroline. If iron is present in the solution, the Prussian blue complex generates a bright blue staining of the solution, while the reaction of the phenanthroline leads to a bright red staining. Both solutions can then be analyzed with a photometer. For the determination of the iron concentration, the light transmission of the sample is correlated to calibration curves of a standard sample with known concentration (Falbe and Regitz, 1995; Hollemann et al., 2007; Bencze, 1948).

Another possibility to determine the iron concentration is *inductively coupled plasma mass spectrometry*. Due to its high sensitivity, it can give information on the purity of the sample. During the measurement process, the sample is atomized and ionized in hot inductively coupled plasma. The concentration is then determined by a comparison of the measured ion intensity with a calibration curve created by dilutions of a standard solution with known concentration (Ammann, 2007).

With its high lateral resolution, *transmission electron microscopy* (TEM) is one of the key techniques to investigate the nanoparticle core size and the core size distribution. In a TEM microscope, a high-energy electron beam illuminates and thus interacts with a thin sample, usually smaller than 100nm, producing a range of signals, which can be used for imaging, diffraction analysis and chemical analysis (Williams and Carter, 1996). TEM is mostly used to investigate the projected physical size and shape of the MNPs, but it also allows for an analysis of the crystal structure, the chemical composition, the atomic bonding, and the coordination on a subnanometer scale. Due to a direct 2D imaging of the nanoparticles, one of the advantages of TEM is that the interpretation of the size and shape data has no need for further modeling (Pyrz and Buttrey, 2008; Levitin et al., 2014; de Montferrand et al., 2013; Baaziz et al., 2014).

Two techniques using the elastic scattering of X-rays and neutrons by a material are conventional *X-ray diffraction* and *neutron diffraction*. With the help of X-ray, the crystallographic structure of MNPs can be determined; neutron scattering also gives information on the nuclear and the magnetic structures of MNPs. The information about the scattered material is gathered from the intensity and position of the diffraction peaks in the diffraction pattern, showing the angle-dependent diffraction intensity. This information can then be used to determine the crystal structure of the MNP sample (Snyder et al., 1999; Bacon, 1981).

Another important step in the analysis is the determination of the hydrodynamic diameter using *dynamic light scattering*. The principle of dynamic light scattering is the exposition of suspended particles to a laser beam and the detection of the time-dependent fluctuations of the scattered light. Due to the Brownian motion of particles, this detected light intensity changes over time. The timescale of these fluctuations can be used to determine the particles diffusion coefficient using an intensity autocorrelation function. Considering measurement parameters as temperature and viscosity of the dispersing agent, the Stokes-Einstein relation (Kurzweil et al., 2009) can then be used to calculate the hydrodynamic diameter. It is advantageous that with dynamic light scattering the determination of the particles size is possible using the complete particle in the dispersion agent including shell and hydration layer. However, assumptions on particles and their size distribution make the results only reliable for monodisperse particle solutions. Thus, for nonmonodisperse samples, the proportions of large particles tend to get overestimated (Strutt, 1871; Einstein, 1905). A possibility to increase the sensitivity of measurements is the use of photon cross correlation spectroscopy. Due to the use of the cross correlation function, it is possible to differentiate between single-scattered light and multiple-scattered light, which makes it possible to measure solutions with higher concentrations (Witt et al., 2004).

A technique for the fractionation or separation of MNPs according to the hydrodynamic diameter is *asymmetrical field flow-field fractionation*. For the fractionation of the MNPs, a carrier liquid is pumped through a channel with an ultrafiltration membrane. Thus, a longitudinal flow of the carrier liquid along the channel and a perpendicular cross flow through the ultrafiltration membrane are created. When the MNP sample is injected into the carrier liquid, the cross flow pushes the particles toward the ultrafiltration membrane. At the same time, the diffusion of the particles points into the opposite direction, thus causing a steady-state concentration distribution. Due to their higher Brownian motion, the farther the particles are away from the ultrafiltration membrane, the smaller they are. Due to the parabolic flow profile of the longitudinal flow, small components of the sample will be transported to the channel outlet faster than larger components, thus establishing the separation of the different fractionations of the MNP solution due to their hydrodynamic diameter (Giddings, 1973, 1993; Williams and Caldwell, 2011).

For a quantitative and a qualitative analysis of the iron-containing compounds, for example, maghemite or magnetite, *Mössbauer spectroscopy* can be used. The ratio of maghemite and magnetite is of special interest in the context of MPI, since both iron ores can occur during synthesis, but only magnetite is used in MPI (see Section 3.1). A mechanical movement modulates a gamma-ray beam, so that it passes through the MNP sample with varying energy. Since the absorption of gamma rays by the MNP sample depends on the gamma-ray energy, a highly sensitive transmission spectrum as a function of the gamma-ray energy can be recorded. Such a spectrum can then be displayed as histograms of accumulated gamma-ray count events as a function of the source velocity and gives precise information about the atoms contained in the MNP sample (Cranshaw et al., 1985).

3.3.2 *Regarding Applications*

The separation time of MNPs can be determined using *magnetic separation* of the MNPs depending on their hydrodynamic diameter. During those measurements, the MNP suspension is placed in a slit in a cuvette, which has a permanent magnet placed at the end of the slit. Due to the field gradient, the particles get magnetized and move toward the magnet. The changing particle concentration can be monitored by the detection of the changing light intensity of nonpolarized light, which is transmitted through the cuvette and later matched to calibrated reference concentrations. The separation time is then defined as the time from the beginning of the measurement and either the time when the particle concentration has decreased to 50% of the initial particle concentration or the time when the particle concentration converges to zero (Schaller et al., 2008).

The ability of an MNP sample to influence the nuclear magnetic resonance (NMR) T_1 and T_2 relaxation times of protons can be measured using *NMR relaxivity*. When protons are exposed to a static magnetic field, a magnetic moment is induced by the self-rotation of their spins. This magnetic moment is aligned either parallel or antiparallel to the applied magnetic field. When a short radiofrequency pulse is applied, more spins flip into the energetically unfavored antiparallel direction. After the pulse, two different relaxation processes cause the spins to return to their original state and leading to the loss of the transversal net magnetization, namely, the longitudinal or spin-lattice relaxation with the time T_1 and the transversal or spin-spin relaxation with the time T_2 (Roch et al., 1999). The T_1 and T_2 relaxation times are of special interest, when the suitability of MNPs for a multimodal approach (see Section 6.3) between MPI and MRI is investigated.

3.3.3 *Regarding the Magnetic Behavior*

For the investigation of the alternating current (AC) susceptibility of MNPs, *AC magnometry* is used, also referred to as AC susceptibility measurements, which can be either temperature- or frequency-dependent. For temperature-dependent AC susceptibility measurements, information about the equilibrium and nonequilibrium magnetic properties of the particles is gathered. For frequency-dependent AC susceptibility measurements, the measured dynamic magnetic response from the sample can be pictured as a complex dynamic susceptibility, and thus, the complex susceptibility can be recorded as a function of the excitation frequency. Furthermore, information on the hydrodynamic size of the particles and the single-domain size (see Section 3.2.1) can be extracted from the AC susceptibility spectra (Martien, 2010; Youssif et al., 2000).

Several magnetic properties of MNPs can be investigated using *magnetorelaxometry*. During a magnetorelaxometry measurement, the response of the magnetic moments of the particles is recorded as a function of time after a weak external field is switched off. The analysis of this relaxation behavior reveals information on several structural and magnetic properties of MNPs. Considering this, magnetorelaxometry can be used to gain information on the mean magnetic moment, the characteristic relaxation times, the particle core size and size distribution, the anisotropy constant, the hydrodynamic particle size and size distribution, and the dipolar interactions. It can also be used to determine the MNP concentration or give information about the binding state in various matrices (Ludwig et al., 2005, 2007).

Another instrument to determine the particle size distribution and the magnetic moments of the MNPs is a magnetic particle spectrometer (MPS), of which a detailed description is given in Section 3.4.

3.4 Particle Analysis in MPI

To study the specific behavior of particles for MPI applications (Bauer et al., 2016b; Panagiotopoulos et al., 2015), several analysis tools are possible (Ludwig et al., 2013). On the one hand, a theoretical analysis can be used to understand the principle particle behavior while changing specific properties precisely. On the other hand, a measurement-based analysis is necessary to evaluate theoretical findings and gain detailed knowledge about the actual particle behavior.

3.4.1 *Theoretical Analysis*

There are several known characteristics of SPIONs that may have an influence on the MPI signal (see Section 3.3). However, the dynamic behavior of SPIONs is not yet fully understood and needs further investigation. The knowledge is important for MPI to improve imaging sequences, image reconstruction, and the synthesis process of the particles itself with the purpose of increasing the particle performance. Since the particle synthesis often cannot produce arbitrary particles with required properties, there is a need for theoretical models.

In order to gain a full understanding of the influence of an MPI imaging sequence on the used tracer, it is necessary to model the characteristics of individual particles and the combination and interaction of the particle ensemble. Typically, stochastic differential equations are used for detailed theoretical investigations. Of course, different precision levels are possible to model the particles dependent on the characteristic that should be investigated.

A very simple way to model the particle magnetization is the Langevin function (see Section 3.2.2). It allows for an approximation of the magnetization in dependence of the imaging sequence and may be used for simulation studies of calculated model-based system matrices (see Section 5.2.1). To include hysteresis effects (see Section 3.2.3), the Debye relaxation model may be used that includes a possible time lag between drive field and particle magnetization (see Section 4.1.1) (Goodwill et al., 2011; Wawrzik et al., 2015; Bente et al., 2015; Croft et al., 2012).

For a detailed understanding of the particle behavior in a dynamic magnetization process, more complex models need to be considered. To model the Néel relaxation, the Landau-Lifschitz-Gilbert equation (Weizenecker et al., 2010) or the Stoner-Wohlfarth model can be used (Tannous and Gieraltowski, 2008). Typically, the Fokker-Planck equation is solved to introduce Brownian relaxation (Yoshida and Enpuku, 2009; Yasumuri et al., 1963; Rogge et al., 2013). Additionally, the modeling of the shape anisotropy (Weizenecker et al., 2012) and the crystal anisotropy (Graeser et al., 2015b) are important factors to consider.

Next to the modeling of the parameters of each individual particle, the modeling of the entirety of particles is necessary. For example, the size distribution of a tracer material can be determined using multicore models (Yoshida et al., 2013; Eberbeck et al., 2011). Further, the interaction between particles can have an important influence on the particle signal (Them, 2017).

In addition to the mentioned research topic, it is important to understand the behavior of particles when complex trajectories are applied. The dynamic behavior of the particles strongly depends on the time- and spatial-dependent trajectory (Graeser et al., 2015a).

It should be noted that the more complex the model, the higher the computational effort to simulate the particles behavior. Hence, the Fokker-Planck equation is hardly used to model complex dynamics. This is one reason why the model for system matrices currently is based on the rather simple Langevin function.

3.4.2 *Measurement-Based Analysis With Magnetic Particle Spectrometry*

In order to provide information on the properties of synthesized MNPs, several measurement tools are necessary. As described in Section 3.3, different analysis methods exist that allow the investigation of the general magnetic properties of the MNPs. These tools provide important information for the use of the MNPs in MPI. However, since the dynamic behavior of MNPs is not fully understood, a dedicated test for the MPI suitability is inevitable.

For a detailed analysis of MNPs for MPI, the development of MPS was initiated. An MPS operates with the same signal encoding principle as an MPI scanning device (see Section 2.1), but without a spatial encoding, that is, the application of a selection field. The first publication on spectrometry in MPI describes the general layout of the device and shows first measurements of the magnetic moment of Resovist for a one-dimensional excitation (Biederer et al., 2009). Since then, several MPS were constructed for different purposes (Weaver et al., 2008; Schilling et al., 2013; Viereck et al., 2017; Kuhlmann et al., 2015; Draack et al., 2017; Behrends et al., 2016; Graeser et al., 2017; Chen et al., 2016, 2017). Furthermore, the measurement chamber is quite small, and therefore, the receive coils are close to the measurement chamber. The sensitivity of an MPS is consequentially very high and allows the measurement of more harmonics than an MPI scanner. Additionally, the SNR of MPS data is very high because less noise is present.

Usually, an MPS is constructed to emulate the behavior of particles in an MPI imaging device. However, the investigation of different physical properties requires a suitable measurement setup. As a result, different unique MPS have been developed, for example, an MPS to measure the Brownian motion of MNPs (Weaver et al., 2008, 2013; Rauwerdink and Weaver, 2010b; von Gladiss et al., 2015b), MPS implementing different drive field frequencies (Graeser et al., 2013; Schilling et al., 2013; Viereck et al., 2017; Kuhlmann et al., 2015), and MPS to analyze the heating properties of MNPs (Draack et al., 2017; Behrends et al., 2016).

To analyze MNPs for their behavior in different environments, effects such as temperature and viscosity can be investigated. The temperature of the particles is particularly interesting when it comes to the prediction of measurements in a flow-like situation that causes a cooling. In Weaver et al. (2009), a method was proposed that allows for an estimation of the temperature based on a measured calibration curve. Here, the curve is determined by the ratio of the fifth and the third harmonic, which varies, when the applied field and the temperature are changed. This technique was refined in Rauwerdink and Weaver (2010a), where the influence of a selection field in addition to the drive field was included. In Draack et al. (2017)), it was experimentally shown with MPS measurements that the harmonic spectra of MNPs depend on the temperature and that this influence on the spectra also depends on the environment of the particles, that is, if the particles are in a fluidal solution or in a matrix-like structure. With these first MPS experiments, a basis was built to carry out experiments toward combined diagnostic and therapeutic approaches based on hyperthermia in the field of MPI (see Section 6.5). As mentioned above, the viscous effect is another external factor that influences the MNPs, that is, the slope of the harmonics (see Section 2.1) (Rauwerdink and Weaver, 2010a).

For many MPI scanners based on an FFP encoding, complex trajectories generated by different drive fields (see Section 3.3) are implemented (Knopp et al., 2009a; Graeser et al., 2017). The particle behavior caused by these trajectories is different to the behavior caused by a single drive field frequency. In Graeser et al. (2015a), it was shown that the particle parameters can even be tailored to specific trajectories. If the trajectories feature a fast rotation of the magnetic field vector, it is recommended to use isotropic particles with a small hydrodynamic diameter. If on the other hand, only a slow rotation of the magnetic field vector is caused by the chosen trajectory, the hydrodynamic diameter can be chosen much larger without causing a significant loss in signal strength. Hence, spectrometers with different drive field frequencies have been introduced (Graeser et al., 2017; Chen et al., 2016, 2017).

The possibility of generating complex drive field sequences as, for example, described in Knopp et al. (2009a) offers to investigate the dependency of the particle magnetization on a chosen trajectory. For many MPI scanners based on an FFP encoding, complex trajectories generated by different drive fields (see Section 2.3) are implemented. The particle behavior caused by these trajectories is different to the behavior caused by a single drive field frequency. Hence, spectrometers with different drive field frequencies have been introduced (Graeser et al., 2017; Chen et al., 2016, 2017). In Graeser et al. (2015a), it was shown that the particle parameters can even be tailored to specific trajectories. If the trajectories feature a fast rotation of the magnetic field vector, it is recommended to use isotropic particles with a small hydrodynamic diameter. If on the other hand, only a slow rotation of the magnetic field vector is caused by the chosen trajectory, the hydrodynamic diameter can be chosen much larger without causing a significant loss in signal strength.

If these MPS are extended by the possibility to introduce a DC offset (Biederer et al., 2010; Graeser et al., 2017), the measurement of hybrid system matrices is possible (see Section 5.2.1). In Fig. 7, the model and the implementation of a multidimensional spectrometer (Graeser et al., 2017) are shown.

4 MPI SCANNER

Today, a variety of MPI scanner prototypes are developed by various research groups. The main challenges are the achievement of a high spatial and/or temporal resolution, an efficient field generation, patient safety, and upscaling to larger scanners. Typically, the design of the developed prototypes and preclinical scanners is either focused on the improvement of sensitivity or acquisition time.

In Section 5.1, the main technical components needed to build an MPI scanner will be introduced. Afterward, different approaches to realize an MPI scanner will be presented. In Section 4.2, the first and still the most widely used MPI scanner configuration will be

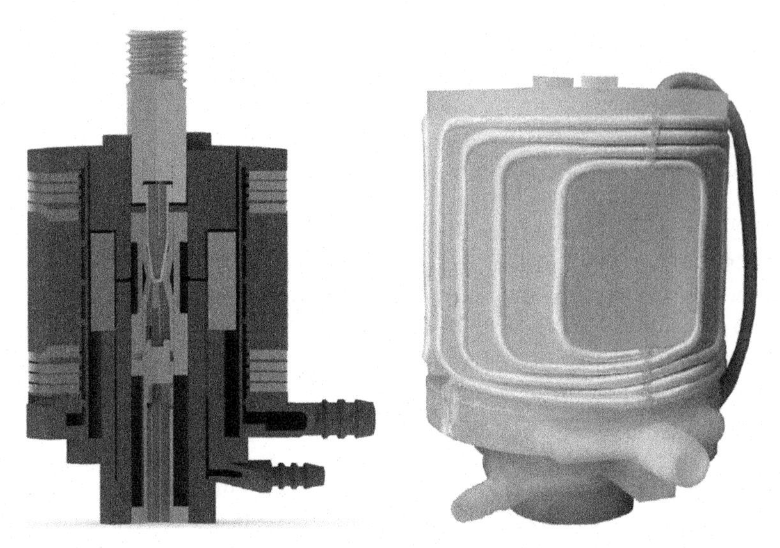

FIG. 7 *Left*: The model of a multidimensional spectrometer. *Right*: The implementation of the field generator. *From Graeser, M., von Gladiss, A., Weber, M., Buzug, T.M., 2017. Two dimensional magnetic particle spectrometry. Phys. Med. Biol. 62(9), 3378–3391. doi:10.1088/1361-6560/aa5bcd.*

introduced. Section 4.3 will present a solution for better patient accessibility. The sensitivity of MPI can be improved by using an FFL instead of an FFP to spatially encode the signal (see Section 4.4). A promising approach to enlarge the FOV of the scanner in one direction is the so-called traveling wave MPI, which will be explained in Section 4.5. In the end, Section 4.6 will give examples for commercially developed scanners.

4.1 Basic Technical Components of an MPI Scanner

In the past years of the development of MPI, a variety of different scanner topologies have been investigated. There are many possible ways to build an MPI scanner, but there are some basic technical components necessary for all these systems. As described in Section 2, a drive field for particle excitation (see Section 4.1.1) and a selection field for spatial encoding (see Section 4.1.2) are needed.

4.1.1 The Drive Field and Receive Path

The drive field and receive path includes all components for particle excitation and receiving the induced particle signal. Since higher frequencies are used for the excitation of particles and even higher harmonics occur in the receive chain, one has to take into account the skin effect. Induced eddy currents in solid wires lead to the phenomenon that the current mainly flows at the conductor surface and therefore the resistance of the wire increases. Using Litz wire, which is made out of several thousand copper strands each several tens of micrometers in diameter and each strand isolated, is therefore preferable for winding the drive field coils and connecting the components (Knopp and Buzug, 2012). The following main components of the drive field path of an MPI scanner are described, which can be seen in Fig. 8 (Schmale et al., 2009; Buzug et al., 2012; Knopp and Buzug, 2012).

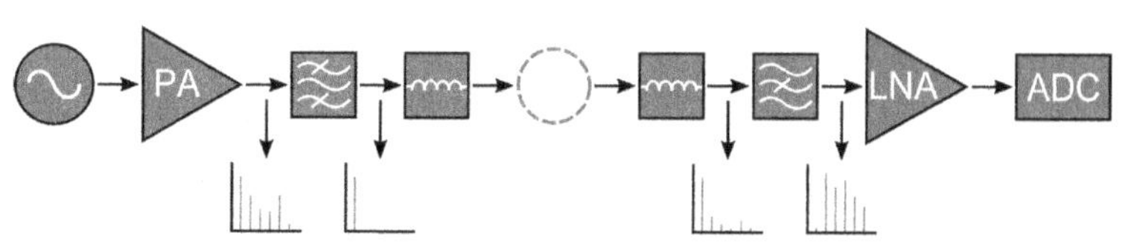

FIG. 8 Schematic flow diagram of the basic technical components of an MPI drive field and receive path. The spectra occurring at the indicated positions in the signal path are displayed.

At first, a *signal generator* is needed, which generates a sinusoidal signal with an excitation frequency usually around 25 kHz. Investigations showed that frequencies in between 10 and 100 kHz should be used for safety reasons, but this is also depend on the amplitude (Doessel and Bohnert, 2013). For large scanner setups, high field amplitudes are necessary, and then, high frequencies are necessary to prevent nerve stimulation (see Section 2.4). The signal is generated digitally on a PC containing a digital-to-analog converter or with a signal generator. It is essential to generate a signal of high purity that can be assured by, for example, an additional analog filter that smoothes the generated signal or a high sending sampling rate and small quantization steps within the digital-to-analog converter.

A *power amplifier (PA)* amplifies the generated signal. It is essential to amplify it as linearly as possible, so that the signal is of high spectral purity. This means it contains the excitation frequency and negligible energy in higher harmonics, since higher harmonics would interfere with the receive signal. However, also amplifiers of high quality generate harmonic distortions (Schmale et al., 2009; Buzug et al., 2012).

Therefore, a *band-pass filter* is required to filter all frequencies except the excitation frequency in order to minimize the disturbances on the receive signal. The band-pass filter compensates for the harmonic distortion caused by the digital-to-analog converter and the part of nonlinear amplification of the power amplifier. It consists of at least a capacitor and an inductor in series and a resistor in parallel. The quality of the band-pass filter can be increased by higher-order filters (Knopp and Buzug, 2012), most often a third-order Butterworth or a Chebyshev II filter are used (Buzug et al., 2012).

The field generator consists of the *transmit coil (TxC)* and the *receive coil (RxC)*. The current flowing in the transmit coil generates the drive field, which excites the nanoparticles. In the receive coil, the particle signal is induced, because of the change of the particle magnetization. For generating the drive field and receiving the induced particle signal, it is also possible to use the same coils instead of depicted transmit and receive units. The most common way to generate a homogeneous drive field is using Helmholtz coil pairs. In order to reach a bore geometry of the scanner setup and fit the drive field coils into the focus field coils, saddle-shaped coils are preferable (Knopp and Buzug, 2012; Erbe et al., 2013). Not only the particle signal but also the excitation signal couples directly from the transmit coil into the receive coil.

Therefore, behind the receive coil a *band-stop filter* is needed, which suppresses the excitation frequency. The induced signal, corresponding to the particle response, contains higher harmonics of the excitation frequency. Since the excitation signal, which couples into the transmit coil, is much larger than the induced particle signal, the excitation frequency needs to be removed from the signal. The disadvantage of using a filter is that also the first harmonic of the particle signal gets suppressed. Other approaches to get rid of the excitation frequency in the receive signal are active cancellation of this frequency component or the use of gradiometer coils (Graeser et al., 2013; Zheng et al., 2013).

A *low-noise amplifier (LNA)* amplifies the detected particle signal since it has very small amplitudes in the range of few pico- to nanovolts. It needs to be amplified over a broad frequency range starting at a frequency of two times the excitation frequency up to several MHz. The amplitude of the incoming signal should not exceed the linear amplification range, which makes the aforementioned band-stop filter essential.

The received signal is digitized by an *analog-to-digital converter (ADC)* in order to be displayed, stored, and used for image reconstruction.

4.1.2 The Selection Field Path

The selection field to spatially encode the particle signal can be generated by either permanent magnets or electromagnetic coils. An opposing pair of permanent magnets generates an FFP, if the same poles are facing each other. In general, much higher magnetic field gradients can be achieved with permanent magnets than with electromagnets. Neodymium magnets (NdFeB) are most often used for this purpose, since they provide strong magnetic fields (Knopp and Buzug, 2012; Goodwill et al., 2009). Permanent magnets, arranged in a Halbach array are able to generate a uniform magnetic field (Bagheri et al., 2015). Also, an FFL configuration can be realized by using permanent magnets (Konkle et al., 2013b; Weber et al., 2015c). However, they neither are tunable nor can be turned off, and either the magnets or the sample has to be moved mechanically in order to move the FFP or FFL. This is why selection field configurations by use of electromagnetic coils are most often preferred, but it is also possible to combine permanent magnets and electromagnetic coils (Weizenecker et al., 2009). The magnetic flux density of electromagnetic coils can be amplified by use of ferromagnetic material inside the coils. The most common way to generate the gradient field featuring an FFP with electromagnetic coils is by use of a Maxwell coil configuration, where the current in each coil flows in opposite directions. For generating an FFL at least two Maxwell coil pairs are needed (Knopp et al., 2010d).

In the case of using electromagnets, the schematic signal flow diagram of the selection field can be seen in Fig. 9. A direct current is generated by the DC source. A band-stop filter, filtering the excitation frequency, which couples from the transmit coil of the drive field path into the selection field coils, protects the DC source against the AC (Buzug et al., 2012).

When using additional focus fields to enlarge the FOV (see Section 2.3), either dedicated coils carrying a low-frequency current or most often the same coils as for the selection field can be used. Then, the low-frequency signal of the focus field is superimposed to the DC current of the selection field.

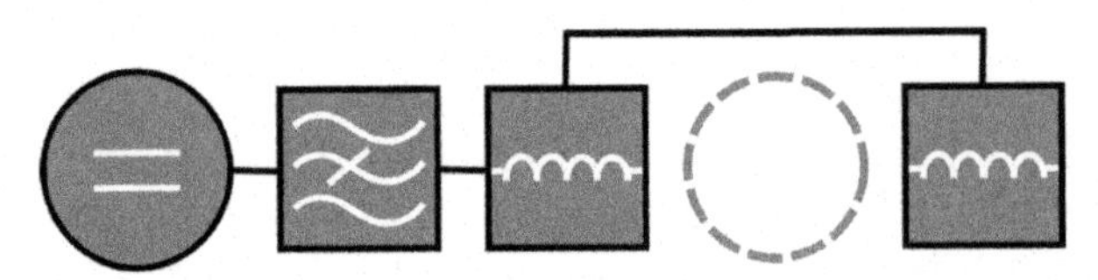

FIG. 9 Schematic flow diagram of the basic technical components of an MPI selection field path.

4.2 FFP Scanner With Closed Bore Geometry

The first MPI scanner was developed at the Philips Research Laboratories, Hamburg, and introduced by Gleich and Weizenecker (2005). The scanner consists of two permanent magnets to generate an FFP selection field with a magnetic field gradient of 3.4 $T\ m^{-1}\ \mu_0^{-1}$ (the unit $T\ \mu_0^{-1}$ indicates the magnetic field strength) and three opposing pairs of drive field coils for each direction in space with an amplitude of 10 mT. The drive field coils generate a homogeneous field inside the scanning volume and excite the particles. These homogenous fields are also used to move the FFP. Additionally, the object can be mechanically moved by using a robot to enlarge the FOV. Two receive coils are used to record the induced signal. First experiments using a phantom that consists of holes with a diameter of 0.5 mm filled with Resovist were performed at 52×52 data points, within a scanning volume of $9.4\times9.4\ mm^2$. A schematic drawing of an FFP scanner with closed-bore geometry and the first in vivo images taken with this scanner can be seen in Fig. 10.

Up till now, several MPI scanner devices with bore geometry were developed either aiming at three-dimensional real-time imaging (Weizenecker et al., 2009), extending the FOV (Schmale et al., 2011), or improving the sensitivity (Goodwill et al., 2012c). The scanners differ

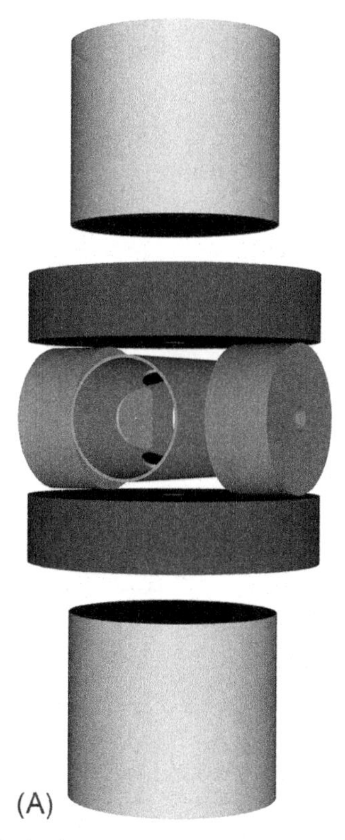

FIG. 10 (A) Schematics of the principal scanner setup developed by Philips (according to Gleich and Weizenecker, 2005).

(Continued)

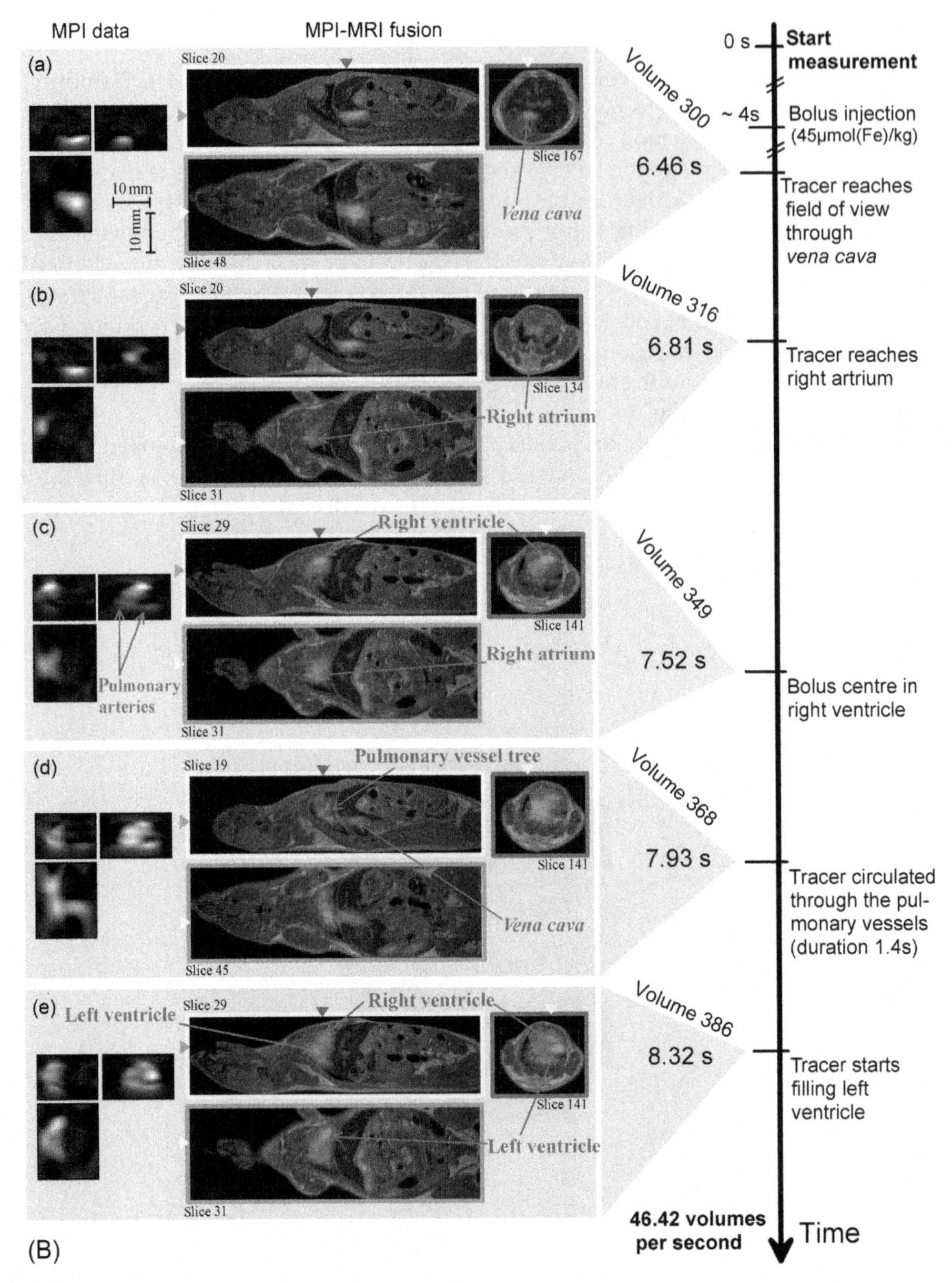

FIG. 10, CONT'D (B) In vivo measurements of a beating mouse heart. Resovist was injected into the tail vein of the mouse. The scanner achieves an FOV of $20.4 \times 12 \times 16.8\,\text{mm}^3$ and a temporal resolution of 21.5 ms. The MPI images are fused with static MRI images *With permission taken from Weizenecker, J., Gleich, B., Rahmer, J., Dahnke, H., Borgert, J., 2009. Three-dimensional real-time in vivo magnetic particle imaging. Phys. Med. Biol. 54(5), L1–L10. https://doi.org/10.1088/0031-9155 /54/5/L01.*

in their applied excitation: either a one-dimensional excitation is used in combination with a high gradient field, causing a slow image acquisition, because the FOV is small and several patches need to be recorded, but with this a very good image resolution can be obtained (Goodwill et al., 2012c; Vogel et al., 2014a), or a two- or even three-dimensional excitation is applied but with smaller gradient fields resulting in very fast image acquisition, because more dimensional excitation enables a higher sampling rate and less patches are needed or even only one FOV is sufficient to cover the whole region of interest (Gleich et al., 2008).

4.3 FFP Scanner With a Single Sided Geometry

Compared with the closed-bore geometry, described in Section 4.2 with which one can only image objects as large as the bore, with a single-sided geometry, the sample size is not limited. A basic 1D single-sided scanner consists of two concentrically arranged transmit coils and a receive coil. A direct current that flows in opposite directions through the transmit coils generates an FFP in the FOV above the coil setup. The FFP can be moved up and down by applying an additional oscillating drive field to acquire 1D MPI data. Due to the arrangement of the coils, the penetration depth of the FFP is limited. This means that the field of view can only cover shallow parts of an object and that the resolution decreases with increasing distance to the coils. The first single-sided scanner was presented by Sattel et al. (2009) and was able to acquire image data up to 15 mm in depth. Later, first 2D images were presented (Gräfe et al., 2016); an additional pair of D-shaped transmit coils underneath the former described coil configuration carrying an AC current moves the FFP parallel to the scanner surface. A dedicated pair of receive coils with the same D-shaped geometry, but smaller, were added to the scanner. Here, an FOV of $15 \times 30\,\text{mm}^2$ and a penetration depth of about 19 mm could be achieved. For 3D, a second pair of D-shaped transmit and receive coils, rotated by 90 degrees, is needed (Fig. 11).

FIG. 11 The principle coil geometry of a single-sided MPI scanner. It consists of two concentrically aligned transmit coils and a dedicated receive coil. For 3D imaging, additional D-shaped transmit and receive coils need to be added.

4.4 FFL Scanner

As described in Section 2.2, the spatial encoding of the received particle signal can be performed by use of an FFL instead of an FFP. It yields a better SNR, and therefore, a better sensitivity by a factor of 10 can be reached (Weizenecker et al., 2008), since more particles contribute to the signal. However, a major drawback is the much larger power consumption of the in (Weizenecker et al., 2008) proposed FFL scanner compared with an FFP scanner of equal size and gradient strength. Simulation studies showed that it consumes 1000 times the power (Knopp et al., 2009b), but further investigations on the scanner design showed promising improvements, and the power consumption could be significantly reduced (Knopp et al., 2009b, 2010b), which makes the use of an FFL feasible.

The first experimental setup was presented by Knopp et al. (2010d). A static FFL is generated by two orthogonal Maxwell coil pairs. The currents of each of the opposing coils flow in opposite directions. Thus, each of the coil pairs generates an FFP. The superposition of both magnetic fields leads to an FFL along the bore axis of the scanner. This first FFL scanner obtained an FOV of $28 \times 28\,mm^2$. Instead of Maxwell coils, permanent magnets can also be used, with which higher magnetic field gradients can be achieved. By use of such a static FFL, the object or the whole scanner setup needs to be moved in order to scan the object (Knopp et al., 2010d; Goodwill et al., 2012b; Weber et al., 2015c).

Usually, the full FOV is scanned electronically by a translation and rotation of the FFL, which can be realized much faster than moving the sample or the scanner. An additional homogeneous field generated by a Helmholtz coil pair perpendicular to the static FFL direction is needed to shift the FFL (Knopp et al., 2010d; Konkle et al., 2013b). It is also possible to generate, rotate, and shift the FFL fully electronically (Knopp et al., 2009b; Erbe et al., 2011; Bente et al., 2015). Additional permanent magnets can be used to amplify the gradient strength (Bente et al., 2015). With this concept, the power consumption of an FFL scanner becomes comparable with those of equal size and gradient strength using an FFP. Using an FFL with a single-sided scanner, geometry was investigated by Tonyushkin (2017).

4.5 Traveling Wave Scanner

An alternative approach to the use of drive fields for shifting the FFP through the FOV is the use of a magnetic gradient field configuration, known as a dynamic linear gradient array that hands off the FFP from one part of the configuration to another (Vogel et al., 2014a). The configuration is designed to enable both, an efficient FFP generation and a fast movement of the FFP through the FOV. This movement of the FFP, similar to a drive field approach, is based on applying an oscillating current that in contrast to the usual drive field approach also features a phase shift. This phase shift increases from one element of the gradient array to another. The phase difference between adjacent elements is adjusted in a way that the current distribution describes a wave in the axial direction of the scanner. Thus, the approach is referred to as traveling wave MPI. By this, two FFPs can be generated simultaneously and shifted linearly in axial direction. In order to achieve a multidimensional coverage of the FOV and not just one scanned line, the imaging sequence can be adjusted so that these lines can either be moved through the FOV (Vogel et al., 2014a) or superimposed with an oscillating current and a rotation field to cover a full 3D FOV (Vogel et al., 2015). The acquisition path of the latter approach shows resemblance to the path of an elongated 2D trajectory generated by drive fields and an additional axial movement. Due to its flexibility and scanning speed, the technique enables real-time data acquisition (Vogel et al., 2016). Further, this technique of an FOV enlargement does not violate the medical constrains.

The technical realization consists of several electromagnetic coils forming a dynamic linear gradient. A sinusoidal current is applied with a phase shift between the coils such that one wavelength equals the length of the coil array. This way two FFPs are created, one having a positive and the other a negative gradient slope. The magnetic field wave is traveling linearly along the long axis, such that the FFPs are moved along this way and generate the MPI signal. With two additional perpendicular saddle-shaped coils, a depicted offset field can shift the scanning line in the 3D volume. Since one FFP has a positive slope and the other a negative, they are shifted in opposite directions and can therefore scan different lines of the volume at once (Vogel et al., 2014a).

4.6 Industrial Scanner Development

Since MPI was invented by the researchers at the Philips Research Laboratories, Hamburg, the first scanner prototypes were built there as well. First, a preclinical demonstrator (Gleich and Weizenecker, 2005; Gleich et al., 2008), later a clinical demonstrator, was under construction. The clinical demonstrator has a bore diameter of 34cm in vertical and 45cm in horizontal direction to fit a human and an FOV of 20cm in diameter (Panagiotopoulos et al., 2015; Rahmer et al., 2017a). For scaling up MPI, two problems are being faced. First, the power consumption of a whole-body scanner will be much larger as for preclinical scanners, and second, peripheral nerve stimulation and tissue heating become a severe issue, because larger field strengths are needed for sufficient image resolution (Borgert et al., 2013).

A preclinical MPI scanner, allowing three-dimensional real-time imaging performing an acquisition rate of 46 frames per second, is provided by Bruker BioSpin GmbH, Ettlingen, Germany. An open bore of 12cm in diameter enables imaging of small animals (Bruker Biospin GmbH, n.d.).

Another preclinical MPI scanner, which applies one excitation frequency, is offered by Magnetic Insight Inc., Alameda, the United States (Magnetic Insight, Inc., 2017).

5 IMAGE RECONSTRUCTION

Image reconstruction in MPI can be performed in several ways. This is also a consequence of the different approaches of scanning devices developed so far. The principle based on the physical phenomenon utilized in MPI is the same for all scanners, but the technical realization differs (see Section 4) due to different scanning schemes as described in Section 2.3. The detailed knowledge of the trajectory and the corresponding particle behavior may be used to improve the image reconstruction for specific scanning devices or applications.

In Section 5.1, the image reconstruction problem in MPI will be introduced with respect to the physical principle of MPI and technical limitations of the actual data acquisition. Sections 5.2 and 5.3 will describe two main approaches of performing the image reconstruction including a discussion of the respective advantages and drawbacks. Finally, in Section 5.4, image reconstruction challenges in MPI will be briefly discussed.

5.1 The Reconstruction Problem

In MPI, the presence of magnetic nanoparticles within the FOV is detected by applying magnetic fields and measuring the resulting voltage induced in dedicated receive coils (see Section 2). It can be concluded from the resulting signal (u in time space or $\hat{u}$ in Fourier space) if particles are present, but a spatial correlation has to be established to reconstruct the actual position of the particles. From the physical theory, it is known that the signal induced by a particle distribution can be derived via the reciprocity principle:

$$u(t) = -\mu_0 \int_\Omega \frac{\partial}{\partial t} \boldsymbol{M}(\boldsymbol{r},t) \sigma^{Rx}(\boldsymbol{r}) \mathrm{d}^3 r \tag{8}$$

with $\sigma^{Rx}(\boldsymbol{r})$ the receive coil sensitivity, μ_0 the permeability of free space, and Ω the FOV and $\boldsymbol{M}(\boldsymbol{r},t)$ the time-dependent and spatially dependent magnetization of the tracer. The corresponding Fourier-transformed signal of Eq. (8) reads

$$\hat{u}_k = -\frac{\mu_0}{T} \int_\Omega \int_0^T \frac{\partial}{\partial t} \boldsymbol{M}(\boldsymbol{r},t) \sigma^{Rx}(\boldsymbol{r}) e^{\frac{-2\pi i k t}{T}} \mathrm{d}t \mathrm{d}^3 r. \tag{9}$$

Since the magnetization $\boldsymbol{M}(\boldsymbol{r},t)$ is proportional to the particle concentration, that is, $\boldsymbol{M}(\boldsymbol{r},t) = \overline{m} c(\boldsymbol{r})$, a mapping of the concentration $c(\boldsymbol{r})$ onto the induced voltage signal is given. Due to linearity, the signal of two individual particle distributions is equivalent to the signal of the combined particle distributions, that is,

$$u_{p_1}(t) + u_{p_2}(t) = u_{p_{1,2}}(t). \tag{10}$$

Fig. 12 illustrates this relationship.

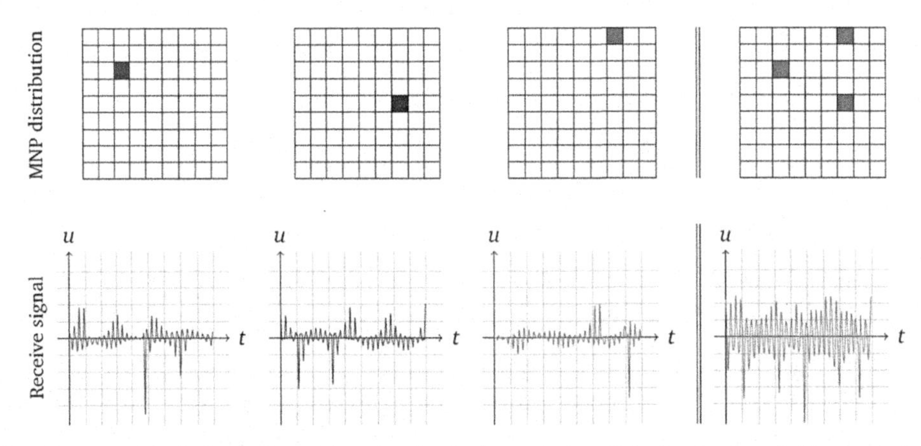

FIG. 12 The sum of the individual receiving signals of independent particle distributions (left) is equivalent to the receiving signal of the combined particle distribution.

A severe issue, which needs to be faced, is that data loss may occur due to noise. Higher harmonics of the base frequency f^B are produced by the nonlinearity of the magnetic behavior of the particles (see Section 3). Dependent on the noise level of a scanning device, these harmonics cannot be resolved and are therefore lost for image reconstruction. As higher harmonics contain the information of small structures, this loss of information results in a loss of spatial resolution. Typically, empty measurements are performed in MPI to cope for background signal (Graeser et al., 2012; Them et al., 2015).

The image reconstruction in MPI can be differentiated in measurement-based and model-based approaches (Grüttner et al., 2013b). Here, the methods will be differentiated mathematically. Based on the presented theory in Eqs. (8) and (9), it is possible to solve the reconstruction problem in two ways that are described in Sections 5.2 and 5.3. However, the suitability of the reconstruction methods is dependent on several factors, because first, the complexity of the received signal is strongly dependent on the applied trajectory (see Section 2.3); second, the signal may be altered due to technical implementations such as filters or amplifiers (see Section 4); and third, the particle behavior varies between different tracer materials (see Section 3).

5.2 Reconstruction via Solving a Linear System of Equations

One possibility to perform the image reconstruction in MPI is to use the knowledge of the relationship of Eq. (10); any particle distribution signal can be expressed as the linear combination of the independent particle distribution signals. Thus, if the signal of every individual position in the FOV is known, any other combination can be calculated from the known signals.

This relationship can be used to build a linear system of equations (LSE). Therefore, a series of measurements with the MPI system are performed that calibrate and store the signal of individual positions in a system matrix $S \in R^{K \times R}$ with K the number of stored frequencies and R the number of spatial positions. The system matrix includes the information of the complex interaction of trajectory and tracer material (Graeser et al., 2015b; Rahmer et al., 2009; Rahmer et al., 2012). This allows for a calibration of every possible combination of tracer material and trajectory, especially trajectories that are generated via different fast drive fields that contain mixing frequencies (see Section 2.3). However, it also means that for every new combination of tracer and trajectory, a new system matrix has to be acquired. Section 5.2.1 gives an overview of different possibilities to acquire a system matrix.

With the system matrix, the signal of any combination of particle positions can be calculated via

$$Sc = u. \tag{11}$$

To solve the LSE, different mathematical concepts can be used (see Section 5.2.2).

5.2.1 System Matrix Acquisition

The system matrix in MPI calibrates the imaging system with respect to the used imaging sequence and used tracer. In order to do this, the sequence-dependent signal of a defined tracer sample is acquired for each voxel position in the FOV. Fig. 13 illustrates this concept.

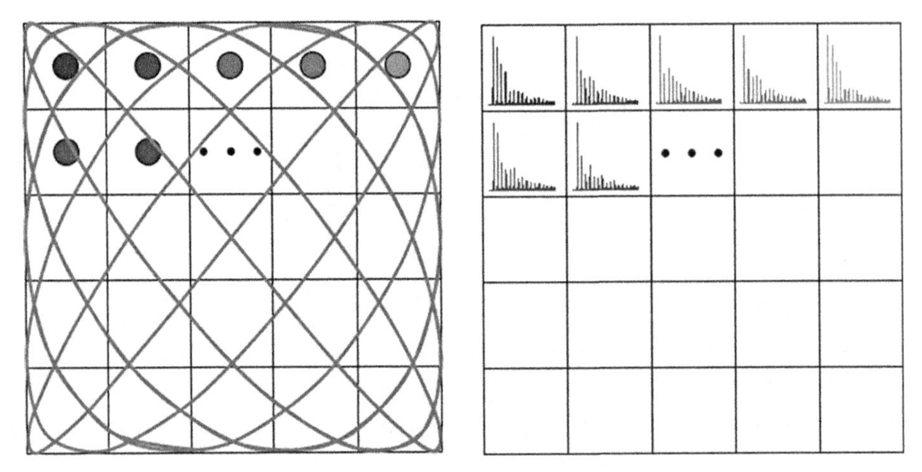

FIG. 13 Principle concept of the system matrix acquisition: a tracer sample with known volume and tracer concentration is placed sequentially at every position in the FOV. For every sample position, the corresponding receive signal is acquired and stored in frequency space as one entry in the system matrix.

The most intuitive way to acquire a system matrix is to place a defined sample at every voxel position (e.g., by using a robot to shift the sample accurately) within the FOV and measure the response when applying the sequence (Weizenecker et al., 2009). This method is often named measurement-based or robot-based system matrix acquisition and includes all possible influences on the signal generation. However, due to the movement of the sample, this kind of calibration is very time-consuming. As calculated in Grüttner et al. (2013b), a calibration for an FOV of 64^3 voxels requires approximately 3 days.

Alternatively, it is possible to calculate the system matrix according to a physical model. This has been successfully demonstrated by modeling the magnetic fields applied in MPI with Eq. (8) and using the Langevin theory of paramagnetism (see Section 3.2.2) to model the particle behavior (Knopp et al., 2010a). As stated in Grüttner et al. (2013b), the generation of such a model-based system matrix for an FOV of 64^3 voxels can be sped up to only a few minutes. This is possible, because the model is quite simple and thus computationally efficient. If sophisticated models are used (see Section 3.4.1), the calculation time may even exceed the measurement time of the robot-based approach (Graeser et al., 2015b).

As the measurement-based approach has its strength in the accuracy by including the system behavior and the model-based approach in the calibration time, a combination of both approaches is a desirable. The idea of this hybrid approach is to omit the movement of the tracer sample, which is the reason for most of the calibration time. This is achieved by varying a magnetic offset field to emulate the spatial correspondence between trajectory and particle sample (Grüttner et al., 2013b). This method can be performed with the actual imaging system (Halkola et al., 2012, 2013) or with an MPS (Graeser et al., 2012; Grüttner et al., 2011; von Gladiss et al., 2017).

5.2.2 *Solving the Linear System of Equations*

The LSE can be solved for c via different algorithms that can be combined with application-dependent prior knowledge to improve the image reconstruction results.

Using a system matrix, the reconstruction problem can be formulated as in Eq. (11). Because measured data always contain noise, the aim is to find an approximated solution with a minimal error. As shown in (Knopp et al., 2010c), the resulting inverse problem is ill-posed, and consequentially, a regularization is needed. The current state-of-the-art regularization in MPI is the Tikhonov regularization (Knopp et al., 2010c). To improve the image reconstruction, it is possible to use different weighting schemes (Knopp et al., 2010c; Szwargulski et al., 2015b). The resulting formula of the least-squares approach to solve the LSE reads

$$\| \boldsymbol{S}\boldsymbol{c} - \hat{\boldsymbol{u}} \|_{W}^{2} + \lambda \| \boldsymbol{c}_{2}^{2} \rightarrow \underset{c}{\operatorname{argmin}} \tag{12}$$

with $\boldsymbol{W}$ the weighting matrix and λ a regularization parameter (Knopp et al., 2010c). Numerous algorithms are possible to derive the mathematical solution, such as a singular value decomposition, conjugate gradient, or the Kaczmarz method (Knopp et al., 2010c).

The mathematical methods may result in solutions that are physical impossible. For example, a negative particle concentration is physically impossible. Hence, it is necessary to introduce a nonnegativity constraint for the reconstruction of c (Weizenecker et al., 2007). Furthermore, the resulting image reconstruction is complex valued. Since the imaginary part contains noise, it can be neglected (Weizenecker et al., 2007).

The reconstruction concept may be further improved by using application-specific prior knowledge. For example, the MPI tracer very often is injected into the blood. Hence, it can be assumed that the resulting image includes blood vessel structures, that is, connected structures with sharp edges. Using this knowledge, the regularization can be adapted using a nonnegative fused lasso method (Storath et al., 2017) or a method based on the minimum mean-square error (Siebert et al., 2016).

5.3 Reconstruction via a Direct Approach

MPI has one of its strengths in the fast signal encoding allowing the examination of fast processes, such as blood flow quantification (Rahmer et al., 2013a). Consequentially, there is desire for a real-time image reconstruction to visualize the images instantaneously to use them as a direct feedback, for example, in vascular interventions (Haegele et al., 2012).

One possibility for a fast image reconstruction is the direct mapping of the measurement signal to the corresponding spatial position. This method is referred to as x-space reconstruction (Goodwill and Conolly, 2010, 2011; Goodwill et al., 2012a), because a mapping of the time signal (t) is performed to the spatial domain (x). This is possible, because the trajectory of the FFP is known by design (see Section 2.3). Assuming that the coil sensitivity σ^{Rx} and the drive fields are homogenous (which is true for many MPI systems), Eq. (8) can be reformulated to

$$u(t) = -\mu_0 \sigma^{Rx} \dot{H}^{A}(t) \tilde{c}(x) \tag{13}$$

and solved for the particle concentration via

$$\tilde{c}(x) = \frac{u(t)}{-\mu_0 \sigma^{Rx} \dot{H}^{A}(t)}. \tag{14}$$

A detailed description of the mathematical derivations can be found in Grüttner et al. (2013b).

The simplicity of the model is intriguing but comes with costs; currently, sufficient image reconstruction results are published for sequences with one fast drive field only. This includes FFP scanners (Goodwill et al., 2012c), FFL scanners (Weizenecker et al., 2008; Goodwill et al., 2012b; Bente et al., 2015), and traveling wave MPI scanners (Vogel et al., 2014a) (see Section 4.5 for technical realization). The extension to complex trajectories with multiple drive fields, for example, the Lissajous trajectory, is still a topic of ongoing research (Cordes et al., 2016; Alacaoglu et al., 2016).

To ensure data consistency with the model of Eq. (14), some pre- and postprocessing steps are necessary: the correction for relaxation effects (Bente et al., 2015; Croft et al., 2012) (see Section 3.2.3), the exclusion of singularities (Knopp et al., 2016), and the correction for low frequencies that are affected by the receive filter (see Section 4.1.1). Additionally, a deconvolution can be performed to compensate for the point-spread function (Goodwill and Conolly, 2011).

It has been shown that assuming the same imaging sequence, the system matrix approach (see Section 5.2) and the x-space approach are mathematically (Grüttner et al., 2013b) and experimentally (Knopp et al., 2016) equivalent.

5.4 Particular Challenges

With the development of larger and more sophisticated scanning devices (see Section 4), new image reconstruction problems arise. In the following, main problems are briefly described. The higher the energy of the harmonics, the lower the signal intensity of the harmonics. This implies that noise disturbs higher harmonics more than lower harmonics, which results in the loss of spatial resolution. Denoising can be achieved by a frequency selection based on the SNR of the system (Knopp and Hofmann, 2016) a subtraction of constant background signal (Them et al., 2015) or the enhancement of the quality of the system matrix via frequency domain filters (Weber et al., 2015a).

To benefit from the fast imaging capability of MPI, it is necessary to decrease the time effort to reconstruct via a dedicated system matrix. As frequency components of a system matrix are linked to Chebyshev polynomials (Rahmer et al., 2009, 2012), they have a sparse representation. Therefore, methods of compressed sensing (CS) can be applied. In (Lampe et al., 2012), the reconstruction time was decreased using several sparse representations of the signal, and in Knopp and Weber (2013), CS was used to reduce the calibration effort. The combination of sparse reconstruction and CS was investigated in von Gladiss et al. (2015a). Further, CS algorithms have been applied to MPI signals in Maass et al. (2016a, b). A further possibility to decrease the calibration effort is given by exploiting symmetries in the MPI system matrix (Weber and Knopp, 2015; Grüttner et al., 2013a). In Knopp and Hofmann (2016), it was shown that with a subtle data reduction, a system-matrix-based online reconstruction is possible.

The signal encoding in MPI is typically performed by either several 1D scans or a Lissajous trajectory (see Section 2.3). Since the influence of the trajectory on the resulting signal is very high, the investigation of suitable trajectories is important. Several theoretical (Knopp et al., 2009a; Szwargulski et al., 2015a) and experimental (Werner et al., 2017; Graeser et al., 2015a) comparisons have been published. Furthermore, alternative concepts with focus on an enlarged FOV are investigated (Rahmer et al., 2011b, 2013b; Kaethner et al., 2015).

An interesting technique to enhance the resolution of reconstructed images is the use of superresolution algorithms (Omer et al., 2015, 2014; Timmermeyer et al., 2013).

The coverage in MPI is limited by technical and medical aspects (see Section 2.4). To enlarge the FOV, the usage of FOV-patches was introduced (Rahmer et al., 2013b, 2011b), that is, small FOVs that are combined to one image. There are different possibilities to reconstruct patches in combination with system matrices, either separately (Ahlborg et al., 2016) or in a joint system of equation (Ahlborg et al., 2016; Knopp et al., 2015). The most crucial point is to introduce a certain overscan for an artifact reduction (Weber et al., 2015b). The patch reconstruction is used for the x-space reconstruction as well (Lu et al., 2013).

6 APPLICATIONS OF MPI

Typically, SPIONs are administered in a blood vessel and allow for a visualization of blood flow as shown with the first in vivo mouse images (see Fig. 10) in 2009 (Weizenecker et al., 2009). Since most MPI scanners have no depth limitation (see Section 4) and the tissue of the patient does not attenuate or shield the receive signal, it is possible to image the SPIONs everywhere in the object of interest.

MNPs can be bound to a large variety of different molecules and materials. As a result, the field of possible applications of MPI is very large. Additionally, the high sensitivity, the possibility of quantitative imaging, and the fast signal encoding in MPI offer applications that are not possible with current medical imaging technologies.

In the following, an overview of recently published ideas in the field of medical imaging with MPI will be given. This section is organized into diagnostics and interventions with MPI (Section 6.1), color MPI (Section 6.2), multimodality with MPI (Section 6.3), cellular imaging with MPI (Section 6.4), thermotherapy with MPI (Section 6.5), and magnetic manipulation and drug targeting with MPI (Section 6.6).

6.1 Diagnostics and Interventions With MPI

The MNPs can be easily administered in the arterial system. There is a wide range of (pre) medical applications for vascular interventions. It is possible to replace state-of-the-art procedures as the digital subtraction angiography, which suffers from the use of ionizing radiation, in the catheter lab (Sedlacik et al., 2016). One key advantage of MPI is the simultaneous high spatial and temporal resolution. That is why, MPI is especially valuable for cardiovascular applications. Images of beating mice hearts could be acquired and reconstructed with different MPI systems (Weizenecker et al., 2009; Vogel et al., 2016).

When MNPs are administered to the blood, there is just a small time window of several minutes to image the particles during their circulation in blood before they accumulate in the liver and spleen (Khandhar et al., 2013). However, for many applications, it is necessary to have larger circulation times. A possibility to achieve this is to incorporate the MNPs directly in red blood cells (Antonelli et al., 2013). As a result, the signal can be measured within the blood hours after the injection (Rahmer et al., 2013a).

MPI can visualize MNPs at any location in the FOV. Thus, if the particles can be administered or transported to the location of interest, MPI can be used to examine the medical condition in situ. It could already been shown that MPI allows imaging of tumors (Yu et al., 2017) or specific organs (Dieckhoff et al., 2017). As described in Section 3.1, MNPs can be modified with respect to their coating or by adding ligands. These ligands can be very different and are used for an active targeting of the MNPs; for example, they can be bound to cancerous tissue.

Another possible application for MPI is breast cancer staging. Here, the MNPs are injected into close vicinity of the cancerous tissue inside the breast. They are transported via the lymphatic system and accumulate in the sentinel lymph node, which is the first lymph node behind the tumor. The sentinel lymph node has to be removed in order to examine whether metastases are present outside the actual tumor region. Using a single-sided coil arrangement (see Section 5.3), the sentinel lymph node can be precisely localized and then removed with minimal invasive surgery (Finas et al., 2012; Sattel et al., 2009; Douek et al., 2014). This technique allows to remove as much tissue as necessary and as less tissue as possible. MPI would be an improvement in breast cancer treatment as current state-of-the-art methods use either radioactive material or blue dye, which has a very low resolution and therefore leads to the removal of healthy tissue and large surgical wounds.

Further, MPI can support physicians during an intervention by showing the interventional instruments in real time (Haegele et al., 2012), when they are equipped with a special MNP coating (see Fig. 14). Thus, with MPI, an angioplasty is possible without the use of X-ray and an iodized contrast agent (Salamon et al., 2016).

6.2 Color MPI

The basic concept in MPI allows for the detection if particles are present at a certain location. A very attractive possibility is to color this information by discriminating different particles properties (Rahmer et al., 2015). This is possible, because the magnetization of the

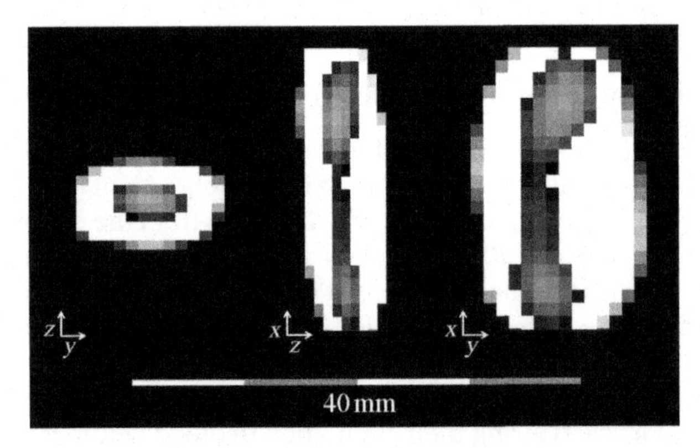

FIG. 14 Color MPI can be used to differentiate between a varnished catheter *(colored)* and a filled lumen *(gray scale)*. This is possible, because the physical properties of MNPs in varnish are different from the properties of MNPs in a fluid. *From Haegele, J., Vaalma, S., Panagiotopoulos, N., Barkhausen, J., Vogt, F.M., Borgert, J., Rahmer, J., 2016. Multi-color magnetic particle imaging for cardiovascular interventions. Phys. Med. Biol. 61(16), N415–N426. https://doi.org/10.1088/0031-9155/61/16/N415.*

particles has a strong influence on the MPI signal (see Section 2.3). With color MPI, it is possible to discriminate between characteristics such as tracer material (Rahmer et al., 2015; Coene et al., 2017), temperature (Rauwerdink et al., 2009), or viscosity of the particle matrix (Rauwerdink and Weaver, 2010a). For the image reconstruction, it is possible to extend the LSE by combining the system matrices of the different tracers (see Section 5.2).

6.3 Multimodality With MPI

MPI suffers from the lack of direct anatomical information of the subjects. Hence, it is desirable to combine MPI with other imaging technologies to gain the missing anatomical information. An obvious choice would be the combination with MRI, because both MPI and MRI use magnetic fields. It is possible to perform the imaging in different devices and combine the information in a postprocessing step (Kaul et al., 2015; Salamon et al., 2016). Since the object needs to be transferred between two different scanners, the spatial confidence is a severe issue, and in vivo real-time imaging is not feasible. A device combining both imaging modalities in a single system was introduced by Vogel et al. (2014b), in which a traveling wave MPI scanner was inserted to a low-field MRI system, showing promising two-dimensional phantom measurement results. By Franke et al. (2016), a fully integrated MPI-MRI system for static 3D imaging was presented. MPI and MRI need very different magnetic field topologies, and the magnetic field strength needed for MRI is so large that the SPIONs get into saturation. Therefore, simultaneous imaging is hard to implement, and a sequential data acquisition was realized. Since both modalities share the same FOV, a high spatial registration accuracy can be achieved. To correlate the information of both modalities, fiducial markers can be used (Werner et al., 2016).

Another possible combination of MPI is with ultrasound (US). Additional to the morphology, US can be used for flow imaging and even thermal therapies. A design study of an MPI compatible US transducer has been published already (Kranemann et al., 2017).

Additionally, it might be useful to develop tracer materials that are visible not only with MPI but also in combination with other imaging modalities as well. In Arami et al. (2015), such tracer materials, being suitable for MPI and several other imaging techniques, have been investigated.

6.4 Cellular Imaging With MPI

It is also possible to perform a labeling of cells with MNPs ex vivo and monitor their behavior in vivo after reinjection (Ittrich et al., 2013). It has been successfully demonstrated that MPI can be used for stem cell tracking (Bulte et al., 2011, 2015; Them et al., 2016; Zheng et al., 2015, 2016).

6.5 Thermo-Therapy With MPI

Fast oscillating magnetic fields cause two reactions of the MNPs: first, a change of the magnetization direction and, second, a movement of the particles, referred to as Néel and Brownian relaxation (see Section 3.2.3). The energy needed to change the magnetization direction is converted into heat. And the friction in a viscous medium caused by a mechanical movement of the particles also leads to heating of the particles. Using specific frequencies and amplitudes, it is possible to use this effect for a controlled heating of the particles. Such a heating might for instance be used to destroy cancerous tissue. Further, knowledge gained by the heating process, can be used to diagnose physiological processes such as blood coagulation (Murase et al., 2014). However, since imaging and heating of particles are very different properties, it is necessary to tailor the particles to the application at hand (Bauer et al., 2016a). As shown in Stehning et al. (2016)), it is possible to monitor the heating of nanoparticles with color MPI (see Section 6.2).

6.6 Magnetic Manipulation and Drug Targeting With MPI

The magnetic fields that are used in MPI for the imaging of MNPs can also be used to manipulate magnetic material. In a first study, it could be shown that the fields applied in MPI can be used to move and hold a magnetic object while imaging its location (Nothnagel et al., 2016). With this technique, it is possible to manipulate catheters to support physicians in an intervention (Rahmer et al., 2017b). In Rahmer et al. (2017a), it was shown that even a spatially selective magnetic manipulation is possible for screws and screwlike objects, by using the focus fields of an MPI scanner. It is also possible that MPI might be used to steer the particles itself.

Not only macroscopic devices but also MNPs can be manipulated by applying magnetic forces. Magnetic drug targeting is a very promising approach to deliver therapeutic agents more efficiently. Drugs such as chemotherapeutics are bound to MNPs. The MNPs can be directed by external magnetic fields toward a targeted volume, for example, a tumor or inflammation. This allows lower dosages of drugs, and healthy tissue is less affected. First human trials, in which a strong permanent magnet holds the chemotherapeutic agent bound to MNPs in a shallow tumor, were successfully performed (Lübbe et al., 1996). Despite using permanent magnets (Pankhurst et al., 2003; Gitter and Odenbach, 2010; Nacev et al., 2011; Sarwar et al., 2012), the use of electromagnets was further investigated aiming at directing the MNPs more precisely and make the steering controllable (Alexiou et al., 2000; Probst et al., 2010). Since the applied force is dependent on the particle size, not only MNPs but also microparticles are under investigation for magnetic drug targeting (Häfeli et al., 1995; Cherry et al., 2009). However, the manipulation of MNPs toward deeper tissue regions remains challenging. For in vivo measurements, a tomographic imaging technique is needed to monitor and control the movement of the particles. Here, MPI has great potential to offer this missing information. As a first step, in Kuboyabu et al. (2016), an accumulation of particles at a certain position could be imaged with MPI. First setups are being investigated in which an MPI scanner is integrated into a coil setup for steering the MNPs, where steering and imaging is alternately operated (Mahmood et al., 2015). The combination of magnetic drug targeting and MPI is an ongoing and very promising field of research.

Acknowledgments

Funding by the German Research Foundation (DFG) under Grant Number BU 1436/7-1 and Grant Number BU 1436/9-1, by the European Union Seventh Framework Programme (FP7/2007–2013) under Grant Agreement No. 604448, and by the Federal Ministry of Education and Research under Grant Number 13GW0071D is gratefully acknowledged.

Further, the authors thank Anselm von Gladiss for fruitful discussion, helpful remarks, and correction of the manuscript and Thomas Friedrich for further corrections.

References

Ahlborg, M., Kaethner, C., Knopp, T., Szwargulski, P., Buzug, T.M., 2016. Using data redundancy gained by patch overlaps to reduce truncation artifacts in magnetic particle imaging. Phys. Med. Biol. 61 (12), 4583–4598. https://doi.org/10.1088/0031-9155/61/12/4583.

Alacaoglu, A., Ozaslan, A.A., Demirel, O.B., Saritas, E.U., 2016. In: Nonlinear scanning in x-space MPI. International Workshop on Magnetic Particle Imaging, Book of Abstracts. vol. 74.

Alexiou, C., Arnold, W., Klein, R.J., Parak, F.G., Hulin, P., Bergemann, C., Erhardt, W., Wagenpfeil, S., Lübbe, A.S., 2000. Locoregional cancer treatment with magnetic drug targeting. Cancer Res. 60 (23), 6641–6648.

Ali, A., Zafar, H., Zia, M., Ul Haq, I., Phull, A.R., Ali, J.S., Hussain, A., 2016. Synthesis, characterization, applications, and challenges of iron oxide nanoparticles. Nanotechnol. Sci. Appl. 9, 49–67. https://doi.org/10.2147/NSA.S99986.

Ammann, A.A., 2007. Inductively coupled plasma mass spectrometry (ICP MS): a versatile tool. J. Mass Spectrom. 42 (4), 419–427. https://doi.org/10.1002/jms.1206.

Antonelli, A., Sfara, C., Rahmer, J., Gleich, B., Borgert, J., Magnani, M., 2013. Red blood cells as carriers in magnetic particle imaging. Biomed. Tech./Biomed. Eng. 58 (6), 517–525. https://doi.org/10.1515/bmt-2012-0065.

Arami, H., Khandhar, A., Tomitaka, A., Yu, E., Goodwill, P., Conolly, S., Krishnan, K.M., 2015. In vivo multimodal magnetic particle imaging (MPI) with tailored magneto/optical contrast agents. Biomaterials 52 (1), 251–261. https://doi.org/10.1016/j.biomaterials.2015.02.040.

Baaziz, W., Pichon, B.P., Fleutot, S., Liu, Y., Lefevre, C., Greneche, J.-M., Toumi, M., Mhiri, T., Begin-Colin, S., 2014. Magnetic iron oxide nanoparticles: reproducible tuning of the size and nanosized-dependent composition, defects, and spin canting. J. Phys. Chem. C 118 (7), 3795–3810. https://doi.org/10.1021/jp411481p.

Bacon, G.E., 1981. Neutron Diffraction Scattering. Taylor and Francis, London.

Bagheri, H., Kierans, C.A., Nelson, K.J., Andrade, B.A., Wong, C.L., Frederick, A.L., Hayden, M.E., 2015. In: A novel scanner architecture for MPI. International Workshop on Magnetic Particle Imaging, Book of Abstracts. vol. 107.

Bauer, L.M., Situ, S.F., Griswold, M.A., Samia, A.C., 2016a. High-performance iron oxide nanoparticles for magnetic particle imaging—guided hyperthermia (hMPI). Nano 8 (24), 12162–12169. https://doi.org/10.1039/c6nr01877g.

Bauer, L.M., Situ, S.F., Griswold, M.A., Samia, A.C.S., 2016b. Magnetic particle imaging tracers: state-of-the-art and future directions. J. Phys. Chem. Lett. 6 (13), 2509–2517. https://doi.org/10.1021/acs.jpclett.5b00610.

Bean, C.P., Livingston, J.D., 1959. Superparamagnetism. J. Appl. Phys. 30 (4), 120–129. https://doi.org/10.1063/1.2185850.

Behrends, A., Buzug, T.M., Neumann, A., 2016. In: Magnetic particle spectrometer for the analysis of magnetic particle heating applications. International Workshop on Magnetic Particle Imaging, Book of Abstracts. vol. 47.

Bencze, B., 1948. Die photometrische Bestimmung des Eisens mit o-Phenanthrolin. Anal. Chem. 128 (2), 179–185. https://doi.org/10.1007/BF00556480.

Bente, K., Weber, M., Graeser, M., Sattel, T.F., Erbe, M., Buzug, T.M., 2015. Electronic field free line rotation and relaxation deconvolution in magnetic particle imaging. IEEE Trans. Med. Imaging 34 (2), 644–651. https://doi.org/10.1109/TMI.2014.2364891.

Biederer, S., 2012. Magnet-Partikel-Spektrometer—Entwicklung eines Spektrometers zur Analyse superparamagnetischer Eisenoxid-Nanopartikel für Magnetic-Particle-Imaging. Vieweg+Teubner Verlag.

Biederer, S., Knopp, T., Sattel, T.F., Lüdtke-Buzug, K., Gleich, B., Weizenecker, J., Borgert, J., Buzug, T.M., 2009. Magnetization response spectroscopy of superparamagnetic nanoparticles for magnetic particle imaging. J. Phys. D: Appl. Phys. 42 (20), 205007. https://doi.org/10.1088/0022-3727/42/20/205007.

Biederer, S., Sattel, T.F., Kren, S., Erbe, M., Knopp, T., Lüdtke-Buzug, K., Buzug, T.M., 2010. In: A spectrometer using oscillating and static fields to measure the suitability of super-paramagnetic nanoparticles for magnetic particle imaging. World Molecular Imaging Congress. World Molecular Imaging Congress. vol. 96.

Borgert, J., Schmidt, J.D., Schmale, I., Bontus, C., Gleich, B., David, B., Weizenecker, J., Jockram, J., Lauruschkat, C., Mende, O., Heinrich, M., Halkola, A., Bergmann, J., Woywode, O., Rahmer, J., 2013. Perspectives on clinical magnetic particle imaging. Biomed. Tech./Biomed. Eng. 58 (6), 551–556. https://doi.org/10.1515/bmt-2012-0064.

Brown Jr., W.F., 1963. Thermal fluctuations of a single-domain particle. Phys. Rev. 130 (5), 1677. https://doi.org/10.1103/PhysRev.130.1677.

Bruker Biospin GmbH, n.d. https://www.bruker.com/nc/news-records/single-view/article/bruker-announces-first-customer-installation-of-its-preclinical-magnetic-particle-imaging-mpi-scan.html (Accessed 2 May 2017).

Bulte, J.W.M., Walczak, P., Gleich, B., Weizenecker, J., Markov, D.E., Aerts, H.C.J., Boeve, H., Borgert, J., Kuhn, M., 2011. In: MPI cell tracking: what can we learn from MRI? SPIE Medical Imaging: Biomedical Applications in Molecular, Structural, and Functional Imaging. vol. 79605z. https://doi.org/10.1117/12.879844.

Bulte, J.W.M., Walczak, P., Janowski, M., Krishnan, K.M., Arami, H., Halkola, A., Gleich, B., Rahmer, J., 2015. Quantitative "hot-spot" imaging of transplanted stem cells using superparamagnetic tracers and magnetic particle imaging. Tomography 1 (2), 91–97. https://doi.org/10.18383/j.tom.2015.00172.

Buzug, T.M., 2008. Computed Tomography: From Photon Statistics to Modern Cone-Beam CT. Springer, Berlin Heidelberg. https://doi.org/10.1007/978-3-540-39 408-2.

Buzug, T.M., Bringout, G., Erbe, M., Gräfe, K., Graeser, M., Grüttner, M., Halkola, A., Sattel, T.F., Tenner, W., Wojtczyk, H., Haegele, J., Vogt, F.M., Barkhausen, J., Lüdtke-Buzug, K., 2012. Magnetic particle imaging: introduction to imaging and hardware realization. Z. Med. Phys. 22 (4), 323–334. https://doi.org/10.1016/j.zemedi.2012.07.004.

Caizer, C., 2003. T_2 law for magnetite-based ferrofluids. J. Phys. Condens. Matter 15 (6), 765–776. https://doi.org/10.1088/0953-8984/15/6/303.

Chen, X., Behrends, A., Graeser, M., Neumann, A., Buzug, T.M., 2016. In: Optimizing the coil setup for a three-dimensional magnetic particle spectrometer. International Workshop on Magnetic Particle Imaging, Book of Abstracts. vol. 59.

Chen, X., Graeser, M., Behrends, A., von Gladiss, A., Buzug, T.M., 2017. In: First measured result of the 3D magnetic particle spectrometer. International Workshop on Magnetic Particle Imaging, Book of Abstracts. vol. 123.

Cherry, E.M., Maxim, P.G., Eaton, J.K., 2009. Particle size, magnetic field, and blood velocity effects on particle retention in magnetic drug targeting. Med. Phys. 37 (1), 175–182. https://doi.org/10.1118/1.3271344.

Chikazumi, S., Charap, S.H., 1964. Physics of Magnetism. Wiley, New York.

Coene, A., Leliaert, J., Liebl, M., Löwa, N., Steinhoff, U., Crevecoeur, G., Dupré, L., Wiekhorst, F., 2017. Multi-color magnetic nanoparticle imaging using magnetorelaxometry. Phys. Med. Biol. 62 (8). https://doi.org/10.1088/1361-6560/aa5e90.

Cordes, A., Kaethner, C., Ahlborg, M., Buzug, T.M., 2016. In: x-Space deconvolution for multidimensional lissajous-based data-acquisition schemes. International Workshop on Magnetic Particle Imaging, Book of Abstracts. vol. 74.

Cranshaw, T.E., Dale, B.W., Longworth, G.O., Johnson, C.E., 1985. Mössbauer Spectroscopy and its Applications. Cambridge University Press, Cambridge.

Croft, L.R., Goodwill, P.W., Conolly, S.M., 2012. Relaxation in x-space magnetic particle imaging. IEEE Trans. Med. Imaging 31 (12), 2335–2342. https://doi.org/10.1007/978-3-642-24133-8_240.

de Montferrand, C., Hu, L., Milosevic, I., Russier, V., Bonnin, D., Motte, L., Brioude, A., Lalatonne, Y., 2013. Iron oxide nanoparticles with sizes, shapes and compositions resulting in different magnetization signatures as potential labels for multiparametric detection. Acta Biomater. 9 (4), 6150–6157. https://doi.org/10.1016/j.actbio.2012.11.025.

Dieckhoff, J., Kaul, M.G., Mummert, T., Jung, C., Salamon, J., Adam, G., Knopp, T., Ludwig, F., Balceris, C., Ittrich, H., 2017. In vivo liver visualizations with magnetic particle imaging based on the calibration measurement approach. Phys. Med. Biol. 62 (9), 3470–3482. https://doi.org/10.1088/1361-6560/aa562d.

Doessel, O., Bohnert, J., 2013. Safety considerations for magnetic fields of 10mT to 100mT amplitude in the frequency range of 10kHz to 100kHz for magnetic particle imaging. Biomed. Tech./Biomed. Eng. 58 (6), 611–621. https://doi.org/10.1515/bmt-2013-0065.

Doessel, O., Bohnert, J., 2014. In: Considerations on safety limits for magnetic fields used in magnetic particle imaging. International Workshop on Magnetic Particle Imaging, Book of Abstracts. vol. 14.

Douek, M., Klaase, J., Monypenny, I., Kothari, A., Zechmeister, K., Brown, D., Wyld, L., Drew, P., Garmo, H., Agbaje, O., Pankhurst, Q., Anninga, B., Grootendorst, M., ten Haken, B., Hall-Craggs, M.a., Purushotham, A., Pinder, S., 2014. Sentinel node biopsy using a magnetic tracer versus standard technique: the SentiMAG multicentre trial. Ann. Surg. Oncol. 21 (4), 1237–1245. https://doi.org/10.1245/s10434-013-3379-6.

Draack, S., Viereck, T., Kuhlmann, C., Schilling, M., Ludwig, F., 2017. Temperature-dependent MPS measurements. Int. J. Magnet. Particle Imaging 3 (1), 1703018. https://doi.org/10.18416/ijmpi.2017.1703018.

Eberbeck, D., Wiekhorst, F., Wagner, S., Trahms, L., 2011. How the size distribution of magnetic nanoparticles determines their magnetic particle imaging performance. Appl. Phys. Lett. 98 (18), 182502. https://doi.org/10.1063/1.3586776.

Einstein, A., 1905. Über die von der molekularkinetischen Theorie der Wärme geforderte Bewegung von in ruhenden Flüssigkeiten suspendierten Teilchen. Ann. Phys. 322 (8), 549–560. https://doi.org/10.1002/andp.19053220806.

Erbe, M., 2014. Field Free Line Magnetic Particle Imaging. Springer Vieweg. https://doi.org/10.1007/978-3-658-05337-6.

Erbe, M., Knopp, T., Sattel, T.F., Biederer, S., Buzug, T.M., 2011. Experimental generation of an arbitrary rotated field-free line for use in magnetic particle imaging. Med. Phys. 38 (9). https://doi.org/10.1118/1.3626481.

Erbe, M., Sattel, T.F., Buzug, T.M., 2013. Improved field free line magnetic particle imaging using saddle coils. Biomed. Tech./Biomed. Eng. 58 (6), 577–582. https://doi.org/10.1515/bmt-2013-0030.

Falbe, J., Regitz, M., 1995. Römpp-Lexikon Chemie. Thieme, Stuttgart.

Ferguson, R.M., Khandar, A.P., Kemp, S.J., Arami, H., Saritas, E.U., Croft, L.R., Konkle, J., Goodwill, P.W., Halkola, A., Rahmer, J., Borgert, J., Conolly, S.M., Krishnan, K.M., 2015. Magnetic particle imaging with tailored iron oxide nanoparticle tracers. IEEE Trans. Med. Imaging 34 (5), 1077–1084. https://doi.org/10.1109/TMI.2014.2375065.

Finas, D., Baumann, K., Sydow, L., Heinrich, K., Gräfe, K., Buzug, T.M., Lüdtke-Buzug, K., 2012. Detection and distribution of superparamagnetic nanoparticles in lymphatic tissue in a breast cancer model for magnetic particle imaging. Dtsch. Ges. Biomed. Tech. Jahrestagung Band 57 (Suppl. 1), 81–83. https://doi.org/10.1515/bmt-2012-4158.

Franke, J., Heinen, U., Lehr, H., Weber, A., Jaspard, F., Ruhm, W., Heidenreich, M., Schulz, V., 2016. System characterization of a highly integrated preclinical hybrid MPI-MRI scanner. IEEE Trans. Med. Imaging 35 (9), 1993–2004. https://doi.org/10.1109/TMI.2016.2542041.

Giddings, J.C., 1973. The conceptual basis of field-flow fractionation. J. Chem. Educ. 50 (10), 667. https://doi.org/10.1021/ed050p667.

Giddings, J.C., 1993. Field-flow fractionation: analysis of macromolecular, colloidal, and particulate materials. Science 260 (5113), 1456–1465.

Gitter, K., Odenbach, S., 2010. Experimental investigations on a branched tube model in magnetic drug targeting. J. Magn. Magn. Mater. 323, 1413–1416. https://doi.org/10.1016/j.jmmm.2010.11.061.

Gleich, B., Weizenecker, J., 2005. Tomographic imaging using the nonlinear response of magnetic particles. Nature 435 (7046), 1214–1217. https://doi.org/10.1038/nature03808.

Gleich, B., Weizenecker, J., Borgert, J., 2008. Experimental results on fast 2D-encoded magnetic particle imaging. Phys. Med. Biol. 53 (6), N81–N84. https://doi.org/10.1088/0031-9155/53/6/N01.

Goodwill, P.W., Conolly, S.M., 2010. The x-space formulation of the magnetic particle imaging process: one-dimensional signal, resolution, bandwidth, SNR, SAR, and magnetostimulation. IEEE Trans. Med. Imaging 29 (11), 1851–1859. https://doi.org/10.1109/TMI.2010.2052284.

Goodwill, P.W., Conolly, S.M., 2011. Multi-dimensional x-space magnetic particle imaging. IEEE Trans. Med. Imaging 30 (9), 1581–1590. https://doi.org/10.1109/TMI.2011.2125982.

Goodwill, P.W., Konkle, J.J., Zheng, B., Saritas, E.U., Conolly, S.M., 2012b. Projection x-space magnetic particle imaging. IEEE Trans. Med. Imaging 31 (5), 1076–1085. https://doi.org/10.1109/TMI.2012.2185247.

Goodwill, P.W., Lu, K., Zheng, B., Conolly, S.M., 2012c. An x-space magnetic particle imaging scanner. Rev. Sci. Instrum. 83 (3), 033708. https://doi.org/10.1063/1.3694534.

Goodwill, P.W., Saritas, E.U., Croft, L.R., Kim, T.N., Krishnan, K.M., Schaffer, D.V., Conolly, S.M., 2012a. x-Space MPI: magnetic nanoparticles for safe medical imaging. Adv. Mater. 24, 3870–3877. https://doi.org/10.1002/adma.201200221.

Goodwill, P.W., Scott, G.C., Stang, P.P., Conolly, S.M., 2009. Narrowband magnetic particle imaging. IEEE Trans. Med. Imaging 28 (8), 1231–1237. https://doi.org/10.1109/TMI.2009.2013849.

Goodwill, P.W., Tamrazian, A., Croft, L.R., Lu, C.D., Johnson, E.M., Pidaparthi, R., Ferguson, R.M., Khandhar, A.P., Krishnan, K.M., Conolly, S.M., 2011. Ferro-hydrodynamic relaxometry for magnetic particle imaging. Appl. Phys. Lett. 98 (26), 262502. https://doi.org/10.1063/1.3604009.

Graeser, M., Bente, K., Buzug, T.M., 2015b. Dynamic single-domain particle model for magnetite particles with combined crystalline and shape anisotropy. J. Phys. D: Appl. Phys. 48 (27), 275001. https://doi.org/10.1088/0022-3727/48/27/275001.

Graeser, M., Bente, K., Neumann, A., Buzug, T.M., 2015a. Trajectory dependent particle response for anisotropic mono domain particles in magnetic particle imaging. J. Phys. D: Appl. Phys. 49 (4), 045007. https://doi.org/10.1088/0022-3727/49/4/045007.

Graeser, M., Biederer, S., Grüttner, M., Wojtczyk, H., Sattel, T.F., Tenner, W., Bringout, G., Buzug, T.M., 2012. Determination of system functions in magnetic particle imaging. Springer Proc. Phys. 140, 59–64. https://doi.org/10.1007/978-3-642-24133-8_10.

Graeser, M., Knopp, T., Grüttner, M., Sattel, T.F., Buzug, T.M., 2013. Analog receive signal processing for magnetic particle imaging. Med. Phys. 40 (4), 042303. https://doi.org/10.1118/1.4794482.

Graeser, M., von Gladiss, A., Weber, M., Buzug, T.M., 2017. Two dimensional magnetic particle spectrometry. Phys. Med. Biol. 62 (9), 3378–3391. https://doi.org/10.1088/1361-6560/aa5bcd.

Gräfe, K., von Gladiss, A., Bringout, G., Ahlborg, M., Buzug, T.M., 2016. 2D images recorded with a single-sided magnetic particle imaging scanner. IEEE Trans. Med. Imaging 35 (4), 1056–1065. https://doi.org/10.1109/TMI.2015.2507187.

Grüttner, M., Gräser, M., Biederer, S., Sattel, T.F., Wojtczyk, H., Tenner, W., Knopp, T., Gleich, B., Buzug, J.B.u.T.M., 2011. In: 1D-image reconstruction for magnetic particle imaging using a hybrid system function. IEEE Nuclear Science Symposium and Medical Imaging Conference, pp. 2545–2548. https://doi.org/10.1109/NSSMIC.2011.6152687.

Grüttner, M., Knopp, T., Franke, J., Heidenreich, M., Rahmer, J., Halkola, A., Kaethner, C., Borgert, J., Buzug, T.M., 2013b. On the formulation of the image reconstruction problem in magnetic particle imaging. Biomed. Tech./Biomed. Eng. 58 (6), 583–591. https://doi.org/10.1515/bmt-2012-0063.

Grüttner, M., Sattel, T.F., Griese, F., Buzug, T.M., 2013a. In: System matrices for field of view patches in magnetic particle imaging. SPIE Medical Imaging: Biomedical Applications in Molecular, Structural, and Functional Imaging. vol. 86721A. https://doi.org/10.1117/12.2002424.

Gupta, A.K., Gupta, M., 2005. Synthesis and surface engineering of iron oxide nanoparticles for biomedical applications. Biomaterials 26 (18), 3995–4021. https://doi.org/10.1016/j.biomaterials.2004.10.012.

Haegele, J., Rahmer, J., Gleich, B., Borgert, J., Wojtczyk, H., Panagiotopoulos, N., Buzug, T.M., Barkhausen, J., Vogt, F.M., 2012. Magnetic particle imaging: visualization of instruments for cardiovascular intervention. Radiology 265 (3), 933–938. https://doi.org/10.1148/radiol.12120424.

Haegele, J., Vaalma, S., Panagiotopoulos, N., Barkhausen, J., Vogt, F.M., Borgert, J., Rahmer, J., 2016. Multi-color magnetic particle imaging for cardiovascular interventions. Phys. Med. Biol. 61 (16), N415–N426. https://doi.org/10.1088/0031-9155/61/16/N415.

Häfeli, U.O., Sweeney, S.M., Beresford, B.A., Humm, J.L., Macklis, R.M., 1995. Effective targeting of magnetic radioactive ^{90}Y-microspheres to tumor cells by an externally applied magnetic field. Preliminary in vitro and in vivo results. Nucl. Med. Biol. 22 (2), 147–155.

Halkola, A., Buzug, T.M., Rahmer, J., Gleich, B., Bontus, C., 2012. System calibration unit for magnetic particle imaging: focus field based system function. Springer Proc. Phys. 140, 27–31. https://doi.org/10.1007/978-3-642-24133-8_5.

Halkola, A., Rahmer, J., Gleich, B., Borgert, J., Buzug, T.M., 2013. In: System calibration unit for magnetic particle imaging: system matrix. International Workshop on Magnetic Particle Imaging. IEEE Xplore Digital Library. https://doi.org/10.1109/IWMPI.2013.6528343.

Hasan, S., 2015. A review on nanoparticles: their synthesis and types. Res. J. Recent Sci. 2277-25024 (ISC-2014), 9–11.

Hollemann, A.F., Wiberg, N., Wiberg, E., 2007. Holleman-Wiberg: Lehrbuch der Anorganischen Chemie. De Gruyter.

Ittrich, H., Peldschus, K., Raabe, N., Kaul, M., Adam, G., 2013. Superparamagnetic iron oxide nanoparticles in biomedicine: applications and developments in diagnostics and therapy. Fortschr. Geb. Röntgenstrahlen Bildgeb. Verfahr. 185 (12), 1149–1166. https://doi.org/10.1055/s-0033-1335438.

Kaethner, C., Ahlborg, M., Bringout, G., Weber, M., Buzug, T.M., 2015. Axially elongated field-free point data acquisition in magnetic particle imaging. IEEE Trans. Med. Imaging 34 (2), 381–387. https://doi.org/10.1109/TMI.2014.2357077.

Kaethner, C., Ahlborg, M., Knopp, T., Sattel, T.F., Buzug, T.M., 2014. Efficient gradient field generation providing a multi-dimensional arbitrary shifted field-free point for magnetic particle imaging. J. Appl. Phys. 115 (4), 044910. https://doi.org/10.1063/1.4863177.

Kaethner, C., Cordes, A., Hänsch, A., Buzug, T.M., 2017. Artifact analysis for axially elongated lissajous trajectories in magnetic particle imaging. Int. J. Magnet. Particle Imaging 3 (1), 1703001. https://doi.org/10.18416/ijmpi.2017.1703001.

Kalender, W.A., 2011. Computed Tomography: Fundamentals, System Technology, Image Quality, Applications. Publicis Publishing, Erlangen.

Kalender, W.A., Seissler, W., Klotz, E., Vock, P., 1990. Spiral volumetric CT with single-breath-hold technique, continuous transport, and continuous scanner rotation. Radiology 176 (1), 181–183. https://doi.org/10.1148/radiology.176.1.2353088.

Kaul, M.G., Weber, O., Heinen, U., Reitmeier, A., Mummert, T., Jung, C., Raabe, N., Knopp, T., Ittrich, H., Adam, G., 2015. Combined preclinical magnetic particle imaging and magnetic resonance imaging: initial results in mice. Fortschr. Geb. Röntgenstrahlen Bildgeb. Verfahr. 187 (5), 347–352. https://doi.org/10.1055/s-0034-1399344.

Khandar, A.P., Ferguson, R.M., Arami, H., Kemp, S.J., Krishnan, K.M., 2015. Tuning surface coatings of optimized magnetite nanoparticle tracers for in vivo magnetic particle imaging. IEEE Trans. Magn. 51 (2), 5300304. https://doi.org/10.1109/TMAG.2014.2321096.

Khandhar, A.P., Ferguson, R.M., Arami, H., Krishnan, K.M., 2013. Monodisperse magnetite nanoparticle tracers for in vivo magnetic particle imaging. Biomaterials 34 (15), 3837–3845. https://doi.org/10.1016/j.biomaterials.2013.01.087.

Kim, D.K., Zhang, Y., Voit, W., Rao, K.V., Muhammed, M., 2001. Synthesis and characterization of surfactant-coated superparamagnetic monodispersed iron oxide nanoparticles. J. Magn. Magn. Mater. 225, 30–36.

Knopp, T., Biederer, S., Sattel, T., Weizenecker, J., Gleich, B., Borgert, J., Buzug, T.M., 2009a. Trajectory analysis for magnetic particle imaging. Phys. Med. Biol. 54 (2), 385–391. https://doi.org/10.1088/0031-9155/54/2/014.

Knopp, T., Biederer, S., Sattel, T.F., Rahmer, J., Weizenecker, J., Gleich, B., Borgert, J., Buzug, T.M., 2010a. 2D model-based reconstruction for magnetic particle imaging. Med. Phys. 37 (2), 485–491. https://doi.org/10.1118/1.3271258.

Knopp, T., Buzug, T.M., 2012. Magnetic Particle Imaging: An Introduction to Imaging Principles and Scanner Instrumentation. Springer, Berlin. https://doi.org/10.1007/978-3-642-04199-0.

Knopp, T., Erbe, M., Biederer, S., Sattel, T.F., Buzug, T.M., 2010b. Efficient generation of a magnetic field-free line. Med. Phys. 37 (7), 3538–3540. https://doi.org/10.1118/1.3447726.

Knopp, T., Erbe, M., Sattel, T.F., Biederer, S., Buzug, T.M., 2010d. Generation of a static magnetic field-free line using two Maxwell coil pairs. Appl. Phys. Lett. 97, 092505. https://doi.org/10.1063/1.3486118.

Knopp, T., Erbe, M., Sattel, T.F., Biederer, S., Buzug, T.M., 2011. A Fourier slice theorem for magnetic particle imaging using a field-free line. Inverse Prob. 27 (9), 095004. https://doi.org/10.1088/0266-5611/27/9/095004.

Knopp, T., Hofmann, M., 2016. Online reconstruction of 3D magnetic particle imaging data. Phys. Med. Biol. 61 (11), N257–N267. https://doi.org/10.1088/0031-9155/61/11/N257.

Knopp, T., Kaethner, C., Ahlborg, M., Buzug, T.M., 2016. In: x-Space and Chebyshev reconstruction in magnetic particle imaging: a first experimental comparison. International Workshop on Magnetic Particle Imaging, Book of Abstracts. vol. 75.

Knopp, T., Rahmer, J., Sattel, T.F., Biederer, S., Weizenecker, J., Gleich, B., Borgert, J., Buzug, T.M., 2010c. Weighted iterative reconstruction for magnetic particle imaging. Phys. Med. Biol. 55 (6), 1577–1589. https://doi.org/10.1088/0031-9155/55/6/003.

Knopp, T., Sattel, T.F., Biederer, S., Buzug, T.M., 2009b. Field-free line formation in a magnetic field. J. Phys. A Math. Theor. 43. https://doi.org/10.1088/1751-8113/43/1/012002.

Knopp, T., Sattel, T.F., Buzug, T.M., 2012. Efficient magnetic gradient field generation with arbitrary axial displacement for magnetic particle imaging. IEEE Magn. Lett. 3, 6500104. https://doi.org/10.1109/L MAG.2011.2181341.

Knopp, T., Them, K., Kaul, M., Gdaniec, N., 2015. Joint reconstruction of non-overlapping magnetic particle imaging focus-field data. Phys. Med. Biol. 60 (8), L15–L21. https://doi.org/10.1088/0031-9155 /60/8/L15.

Knopp, T., Weber, A., 2013. Sparse reconstruction of the magnetic particle imaging system matrix. IEEE Trans. Med. Imaging 32 (8), 1473–1480. https://doi.org/10.1109/TMI.2013.2258029.

Konkle, J.J., Goodwill, P.W., Carrasco-Zevallos, O., Conolly, S.M., 2013a. Projection reconstruction magnetic particle imaging. IEEE Trans. Med. Imaging 32 (2), 338–347. https://doi.org/10.1109/TMI.2012.2 227121.

Konkle, J.J., Goodwill, P.W., Saritas, E.U., Zheng, B., Lu, K., Conolly, S.M., 2013b. Twenty-fold acceleration of 3D projection reconstruction MPI. Biomed. Tech. 58 (6), 565–576. https://doi.org/10.1515/bmt-2012-0062.

Kranemann, T.C., Ersepke, T., Schmitz, G., 2017. Towards the integration of an MPI compatible ultrasound transducer. Int. J. Magnet. Particle Imaging 3 (1), 1703016. https://doi.org/10.18416/ijmpi.2017.1703016.

Krishnan, K.M., 2016. Fundamentals and Applications of Magnetic Material. Oxford Press, ISBN: 978-0-19-957044-7.

Kronmüller, H., Parkin, S., 2007. General micromagnetic theory. In: Handbook of Magnetism and Advanced Magnetic Materials. John Wiley & Sons, New York.

Kuboyabu, T., Ohki, A., Banura, N., Murase, K., 2016. Usefulness of magnetic particle imaging for monitoring the effect of magnetic targeting. Open J. Med. Imaging 6, 33–41. https://doi.org/10.4236/ojmi.2016.62004.

Kuhlmann, C., Khandhar, A.P., Ferguson, R.M., Kemp, S., Wawrzik, T., Schilling, M., Krishnan, K.M., Ludwig, F., 2015. Drive-field frequency dependent MPI performance of single-core magnetite nanoparticle tracers. IEEE Trans. Magn. 51 (2), 6500504. https://doi.org/10.1109/TMAG.2014.2329772.

Kurzweil, P., Frenzel, B., Gebhard, F., 2009. Physik Formelsammlung : mit Erläuterungen und Beispielen aus der Praxis für Ingenieure und Naturwissenschaftler. Springer-Verlag.

Lampe, J., Bassoy, C., Rahmer, J., Weizenecker, J., Voss, H., Gleich, B., Borgert, J., 2012. Fast reconstruction in magnetic particle imaging. Phys. Med. Biol. 57 (4), 1113–1134. https://doi.org/10.1088/0031-9155/57/4/1113.

Lange, J., 2001. Spezifisch bindende magnetische Nanopartikel als Signalgeber in magnetischen Relaxationsmessungen. Universitätsbibliothek Greifswald.

Laurent, S., Forge, D., Port, M., Roch, A., Robic, C., Vander Elst, L., Muller, R.N., 2008. Magnetic iron oxide nanoparticles: synthesis, stabilization, vectorization, physicochemical characterizations, and biological applications. Chem. Rev. 108, 2064–2110. https://doi.org/10.1021/cr068445e.

Levitin, E.Y., Kokodiy, N.G., Timanjuk, V.A., Vedernikova, I.O., Chan, M.T., 2014. Measurements of the size and refractive index of Fe_3O_4 nanoparticles. Inorg. Mater. 50 (8), 817–820.

Lu, K., Goodwill, P.W., Saritas, E.U., Zheng, B., Conolly, S.M., 2013. Linearity and shift invariance for quantitative magnetic particle imaging. IEEE Trans. Med. Imaging 32 (9), 1565–1575. https://doi.org/10.1109/TMI.2013.2257177.

Lübbe, A.S., Bergemann, C., Riess, H., Schriever, F., Reichardt, P., Possinger, K., Matthias, M., Dörken, B., Herrmann, F., Gürtler, R., Hohenberger, P., Haas, N., Sohr, R., Sander, B., Lemke, A.J., Ohlendorf, D., Huhnt, W., Huhn, D., 1996. Clinical experiences with magnetic drug targeting: a phase I study with 4′-epidoxorubicin in 14 patients with advanced solid tumors. Cancer Res. 56, 4686–4693.

Lüdtke-Buzug, K., 2012. Magnetische Nanopartikel—Von der Synthese zur klinischen Anwendung. Chem. Unserer Zeit 46 (1), 32–39. https://doi.org/10.1002/ciuz.201200558.

Ludwig, F., Eberbeck, D., Löwa, N., Steinhoff, U., Wawrzik, T., Schilling, M., Trahms, L., 2013. Characterization of magnetic nanoparticle systems with respect to their magnetic particle imaging performance. Biomed. Tech./Biomed. Eng. 58 (6), 535–545. https://doi.org/10.1515/bmt-2013-0013.

Ludwig, F., Heim, E., Schilling, M., 2007. Characterization of superparamagnetic nanoparticles by analyzing the magnetization and relaxation dynamics using fluxgate magnetometers. J. Appl. Phys. 101 (11), 113909. https://doi.org/10.1063/1.2738416.

Ludwig, F., Mäuselein, S., Heim, E., Schilling, M., 2005. Magnetorelaxometry of magnetic nanoparticles in magnetically unshielded environment utilizing a differential fluxgate arrangement. Rev. Sci. Instrum. 76 (10), 106102. https://doi.org/10.1063/1.2069776.

Maass, M., Bente, K., Ahlborg, M., Medimagh, H., Phan, H., Buzug, T.M., Mertins, A., 2016a. In: Compression of FFP system matrix with a special sampling rate on the Lissajous trajectory. International Workshop on Magnetic Particle Imaging, Book of Abstracts. vol. 56.

Maass, M., Bente, K., Ahlborg, M., Medimagh, H., Phan, H., Buzug, T.M., Mertins, A., 2016b. Optimized compression of MPI system matrices using a symmetry-preserving secondary orthogonal transform. Int. J. Magnet. Particle Imaging 2 (1), 1607002. https://doi.org/10.18416/ijmpi.2016.1607002.

Magnetic Insight, Inc., 2017. http://www.magneticinsight.com/momentum-imager.

Mahmood, A., Dadkhah, M., Ok Kim, M., Yoon, J., 2015. A novel design of an MPI-Based guidance system for simultaneous actuation and monitoring of magnetic nanoparticles. IEEE Trans. Magn. 51 (2). https://doi.org/10.1109/TMAG.2014.2358252.

Martien, D., 2010. AC susceptibility, Technical Report. LOT-Oriel Group, Europe.

Massart, R., 1981. Preparation of aqueous magnetic liquids in alkaline and acidic media. IEEE Trans. Magn. 17 (2), 1247–1248. https://doi.org/10.1109/TMAG.1981.1061188.

Mayergoyz, I.D., 1986. Mathematical models of hysteresis. Phys. Rev. Lett. 56 (15), 1518–1521. https://doi.org/10.1103/PhysRevLett.56.1518.

Murase, K., Song, R., Hiratsuka, S., 2014. Magnetic particle imaging of blood coagulation. Appl. Phys. Lett. 104 (25), 252409. https://doi.org/10.1063/1.4885146.

Murray, C.B., Norris, D.J., Bawendi, M.G., 1993. Synthesis and characterization of nearly monodisperse CdE (E=sulfur, selenium, tellurium) semiconductor nanocrystallites. J. Am. Chem. Soc. 115 (19), 8706–8715. https://doi.org/10.1021/ja00072a025.

Murray, C.B., Sun, S., Gaschler, W., Doyle, H., Betley, T.A., Kagan, C.R., 2001. Colloidal synthesis of nanocrystals and nanocrystal superlattices. IBM J. Res. Dev. 45 (1), 47–56. https://doi.org/10.1147/rd.451.0047.

Nacev, A., Beni, C., Bruno, O., Shapiro, B., 2011. The behaviors of ferromagnetic nano-particles in and around blood vessels under applied magnetic fields. J. Magn. Magn. Mater. 323, 651–668. https://doi.org/10.1016/j.jmmm.2010.09.008.

Néel, L., 1955. Some theoretical aspects of rock-magnetism. Adv. Phys. 4 (14), 191–243. https://doi.org/10.1080/00018735500101204.

Nothnagel, N., Rahmer, J., Gleich, B., Halkola, A., Buzug, T.M., Borgert, J., 2016. Steering of magnetic devices with a magnetic particle imaging system. IEEE Trans. Biomed. Eng. 63 (11), 2286–2293. https://doi.org/10.1109/TBME.2016.2524070.

Nothnagel, N.D., Sánchez-Gonzáles, J., 2015. Measurement of system functions with extended field-of-view. IEEE Trans. Magn. 51 (2), 5100504. https://doi.org/10.1109/TMAG.2014.2326253.

Nune, S.K., Gunda, P., Thallapally, P.K., Lin, Y.Y., Forrest, M.L., Berkland, C.J., 2009. Nanoparticles for biomedical imaging. Expert Opin. Drug Deliv. 6 (11), 1175–1194. https://doi.org/10.1517/17425240903229031.

Odenbach, S., 2009. Colloidal Magnetic Fluids: Basics, Development and Application of Ferrofluids. Springer. https://doi.org/10.1007/978-3-540-85387-9.

Omer, O.A., Medimagh, H., Buzug, T.M., 2015. In: High resolution magnetic particle imaging with low density trajectory. International Workshop on Magnetic Particle Imaging. IEEE Xplore Digital Library. https://doi.org/10.1109/IWMPI.2015.7106995.

Omer, O.A., Wojtczyk, H., Buzug, T.M., 2014. Simultaneous reconstruction and resolution enhancement for magnetic particle imaging. IEEE Trans. Magn. 51 (2), 6500804. https://doi.org/10.1109/TMAG.2014.2330553.

Panagiotopoulos, N., Duschka, R.L., Ahlborg, M., Bringout, G., Debbeler, C., Graeser, M., Kaethner, C., Lüdtke-Buzug, K., Medimagh, H., Stelzner, J., Buzug, T.M., Barkhausen, J., Vogt, F.M., Haegele, J., 2015. Magnetic particle imaging: current developments and future directions. Int. J. Nanomedicine 10 (1), 3097–3114. https://doi.org/10.2147/ijn.s70488.

Pankhurst, Q.A., Conolly, J., Jones, S.K., Dobson, J., 2003. Application of magnetic nanoparticles in biomedicine. J. Phys. D: Appl. Phys. 36, R167–R181.

Payet, B., Vincent, D., Delaunay, L., Noyel, G., 1998. Influence of particle size distribution on the initial susceptibility of magnetic of magnetic fluids in the Brown relaxation range. J. Magn. Magn. Mater. 186 (1–2), 168–174. https://doi.org/10.1016/S0304-8853(98)00082-1.

Probst, R., Lin, J., Komaee, A., Nacev, A., Cummins, Z., Shapiro, B., 2010. Planar steering of a single ferrofluid drop by optimal minimum power dynamic feedback control for four electromagnets at a distance. J. Magn. Magn. Mater. 323, 885–896. https://doi.org/10.1016/j.jmmm.2010.08.024.

Pyrz, W.D., Buttrey, D.J., 2008. Particle size determination using TEM: a discussion of image acquisition and analysis for the novice microscopist. Langmuir 24 (20), 11350–11360. https://doi.org/10.1021/la801367j.

Rahmer, J., Antonelli, A., Sfara, C., Tiemann, B., Gleich, B., Magnani, M., Weizenecker, J., Borgert, J., 2013a. Nanoparticle encapsulation in red blood cells enables blood-pool magnetic particle imaging hours after injection. Phys. Med. Biol. 58 (12), 3965–3977. https://doi.org/10.1088/0031-9155/58/12/3965.

Rahmer, J., Borgert, J., Gleich, B., Schmale, I., Bontus, C., Gressmann, J., Vollertsen, C., 2014a. In: Strategies for fast MPI within the limits determined by nerve stimulation. International Workshop on Magnetic Particle Imaging, Book of Abstracts. vol. 16.

Rahmer, J., Gleich, B., Bontus, C., Schmale, I., Schmidt, J., Kanzenbach, J., Woywode, O., Weizencker, J., Borgert, J., 2011a. In: Rapid 3D in vivo magnetic particle imaging with a large field of view. Proceedings of the International Society for Magnetic Resonance in Medicine. vol. 19, pp. 3285.

Rahmer, J., Gleich, B., Bontus, C., Schmidt, J., Schmale, I., Borgert, J., Woywode, O., Buzug, T.M., 2011b. In: Results on rapid 3D magnetic particle imaging with a large field of view. Proceedings of the International Society for Magnetic Resonance in Medicine. vol. 19, pp. 629.

Rahmer, J., Gleich, B., Bontus, C., Schmidt, J., Schmale, I., Borgert, J., Woywode, O., Halkola, A., Buzug, T.M., 2014b. In: Automated derivation of sub-volume system functions for 3D MPI with fast continuous focus field variation. International Workshop on Magnetic Particle Imaging, Book of Abstracts. vol. 97.

Rahmer, J., Gleich, B., Weizenecker, J., Halkola, A., Bontus, C., Schmidt, J., Schmale, I., Woywode, O., Buzug, T.M., Borgert, J., 2013b. In: Fast continuous motion of the field of view in magnetic particle imaging. International Workshop on Magnetic Particle Imaging. IEEE Xplore Digital Library. https://doi.org/10.1109/IWMPI.2013.6528353.

Rahmer, J., Halkola, A., Gleich, B., Schmale, I., Borgert, J., 2015. First experimental evidence of the feasibility of multi-color magnetic particle imaging. Phys. Med. Biol. 60 (5). https://doi.org/10.1088/0031-9155/60/4/1775.

Rahmer, J., Stehning, C., Gleich, B., 2017a. Spatially selective remote magnetic actuation of identical helical micromachines. Sci. Robot. 2 (3). https://doi.org/10.1126/scirobotics.aal2845.

Rahmer, J., Weizenecker, J., Gleich, B., Borgert, J., 2009. Signal encoding in magnetic particle imaging. BMC Med. Imaging 9 (4). https://doi.org/10.1186/1471-2342-9-4.

Rahmer, J., Weizenecker, J., Gleich, B., Borgert, J., 2012. Analysis of a 3-D system function measured for magnetic particle imaging. IEEE Trans. Med. Imaging 31 (6), 1289–1299. https://doi.org/10.1109/TMI.2 012.2188639.

Rahmer, J., Wirtz, D., Bontus, C., Borgert, J., Gleich, B., 2017b. Interactive magnetic catheter steering with 3D real-time feedback using multi-color magnetic particle imaging. IEEE Trans. Med. Imaging. https://doi.org/10.1109/TMI.2017.2679099.

Rauwerdink, A.M., Hansen, E.W., Weaver, J.B., 2009. Nanoparticle temperature estimation in combined ac and dc magnetic fields. Phys. Med. Biol. 54 (19), L51–L55. https://doi.org/10.1088/0031-9155/54/19/L01.

Rauwerdink, A.M., Weaver, J.B., 2010a. Viscous effects on nanoparticle magnetization harmonics. J. Magn. Magn. Mater. 322 (6), 609–613. https://doi.org/10.1016/j.jmmm.2009.10.024.

Rauwerdink, A.M., Weaver, J.B., 2010b. Measurement of molecular binding using the Brownian motion of magnetic nanoparticle probes. Appl. Phys. Lett. 96, 033702. https://doi.org/10.1063/1.3291063.

Reilly, J.P., 1991. Magnetic field excitation of peripheral nerves and the heart: a comparison of thresholds. Med. Biol. Eng. Comput. 29 (6), 571–579. https://doi.org/10.1007/BF02446087.

Roch, A., Muller, R.N., Gillis, P., 1999. Theory of proton relaxation induced by superparamagnetic particles. J. Chem. Phys. 110 (11), 5403. https://doi.org/10.1063/1.478435.

Rogge, H., 2015. In-Silico Analysis of Superparamagnetic Nanoparticles: Physics of the Contrast Agent in Magnetic Particle Imaging. Infinite Science Publishing, Lübeck.

Rogge, H., Erbe, M., Buzug, T.M., Lüdtke-Buzug, K., 2013. Simulation of the magnetization dynamics of diluted ferrofluids in medical applications. Biomed. Tech./Biomed. Eng. 58 (6), 601–609. https://doi.org/10.1515/bmt-2013-0034.

Rossetti, R., Nakahara, S., Brus, L.E., 1983. Quantum size effects in the redox potentials, resonance Raman spectra, and electronic spectra of CdS crystallites in aqueous solution. J. Chem. Phys. 79, 1086. https://doi.org/10.1063/1.445834.

Salamon, J., Hofmann, M., Jung, C., Kaul, M.G., Werner, F., Them, K., Reimer, R., Nielsen, P., vom Scheidt, A., Adam, G., Knopp, T., Ittrich, H., 2016. Magnetic particle/magnetic resonance imaging: in-vitro MPI-guided real time catheter tracking and 4D angioplasty using a road map and blood pool tracer approach. PLoS One 11 (6), e0156899. https://doi.org/10.1371/journal.pone.0156899.

Saritas, E.U., Goodwill, P.W., Zhang, G.Z., Conolly, S.M., 2013. Magnetostimulation limits in magnetic particle imaging. IEEE Trans. Med. Imaging 32 (9), 1600–1610. https://doi.org/10.1109/TMI.2013.22 60764.

Saritas, E.U., Goodwill, P.W., Zhang, G.Z., Yu, W., Conolly, S.M., 2012. Safety limits for human-size magnetic particle imaging systems. Springer Proc. Phys. 140, 325–330.

Sarwar, A., Nemirovski, A., Shapiro, B., 2012. Optimal Halbach permanent magnet designs for maximal pulling and pushing nanoparticles. J. Magn. Magn. Mater. 324, 742–754. https://doi.org/10.1016/j.jmmm.2011.09.008.

Sattel, T.F., Knopp, T., Biederer, S., Gleich, B., Weizenecker, J., Borgert, J., Buzug, T.M., 2009. Single-sided device for magnetic particle imaging. J. Phys. D: Appl. Phys. 42 (2), 022001. https://doi.org/10.1088/0022-3727/42/2/022001.

Schaller, V., Kräling, U., Rusu, C., Petersson, K., Wipenmyr, J., Krozer, A., Wahnström, G., Sanz-Velasco, A., Enoksson, P., Johansson, C., 2008. Motion of nanometer sized magnetic particles in a magnetic field gradient. J. Appl. Phys. 104 (9), 093918. https://doi.org/10.1063/1.3009686.

Schilling, M., Ludwig, F., Kuhlmann, C., Wawrzik, T., 2013. Magnetic particle imaging scanner with 10-kHz drive-field frequency. Biomed. Tech./Biomed. Eng. 58 (6), 557–563. https://doi.org/10.1515/bmt-2013-0014.

Schmale, I., Gleich, B., Kanzenbach, J., Rahmer, J., Schmidt, J., Weizenecker, J., Borgert, J., 2009. An introduction to the hardware of magnetic particle imaging. IFMBE Proc. 25 (2), 450–453. https://doi.org/10.1007/978-3-642-03879-2_127.

Schmale, I., Gleich, B., Rahmer, J., Bontus, C., Schmidt, J., Borgert, J., 2015. MPI safety in the view of MRI safety standards. IEEE Trans. Magn. 51 (2), 6502604. https://doi.org/10.1109/TMAG.2014.2322940.

Schmale, I., Rahmer, J., Gleich, B., Kanzenbach, J., Schmidt, J.D., Bontus, C., Woywode, O., Borgert, J., 2011. First phantom and in vivo MPI images with an extended field of view. In: SPIE Medical Imaging: Biomedical Applications in Molecular, Structural, and Functional Imaging. vol. 7965. https://doi.org/10.1117/12.877339.

Sedlacik, J., Frölich, A., Spallek, J., Forkert, N.D., Faizy, T.D., Werner, F., Knopp, T., Krause, D., Fiehler, J., Buhk, J.-H., 2016. Magnetic particle imaging for high temporal resolution assessment of aneurysm hemodynamics. PLoS One 11 (8), e0160097. https://doi.org/10.1371/journal.pone.0160097.

Siebert, H., Maass, M., Ahlborg, M., Buzug, T.M., Mertins, A., 2016. In: MMSE MPI reconstruction using background identification. International Workshop on Magnetic Particle Imaging, Book of Abstracts. vol. 58.

Snyder, R.L., Fiala, J., Bunge, H.J., 1999. Defect and Microstructure Analysis by Diffraction. Oxford University Press Inc., Oxford.

Stehning, C., Gleich, B., Rahmer, J., 2016. Simultaneous magnetic particle imaging (MPI) and temperature mapping using multi-color MPI. Int. J. Magnet. Particle Imaging 2 (2), 1612001. https://doi.org/10.18416/ijmpi.2016.1612001.

Storath, M., Brandt, C., Hofmann, M., Knopp, T., Salamon, J., Weber, A., Weinmann, A., 2017. Edge preserving and noise reducing reconstruction for magnetic particle imaging. IEEE Trans. Med. Imaging 36 (1), 74–85. https://doi.org/10.1109/TMI.2016.2593954.

Strutt, J., 1871. On the light from the sky, its polarization and colour. J. Philos. Mag. Ser. 41 (271). https://doi.org/10.1080/14786447108640452.

Szwargulski, P., Ahlborg, M., Kaethner, C., Buzug, T.M., 2015a. Trajectory analysis using patches for magnetic particle imaging. IEEE Trans. Magn. 51 (2), 6501104. https://doi.org/10.1109/TMAG.2014.2350152.

Szwargulski, P., Rahmer, J., Ahlborg, M., Kaethner, C., Buzug, T.M., 2015b. Experimental evaluation of different weighting schemes in magnetic particle imaging reconstruction. Curr. Direct. Biomed. Eng. 1 (1), 206–209. https://doi.org/10.1515/cdbme-2015-0052.

Tannous, C., Gieraltowski, J., 2008. The Stoner–Wohlfarth model of ferromagnetism. Eur. J. Phys. 29 (3), 475–487. https://doi.org/10.1088/0143-0807/29/3/008.

Them, K., 2017. On magnetic dipole-dipole interactions of nanoparticles in magnetic particle imaging. Phys. Med. Biol. 62 (14), 5623–5639. https://doi.org/10.1088/1361-6560/aa70ca.

Them, K., Kaul, M.G., Jung, C., Hofmann, M., Mummert, T., Werner, F., Knopp, T., 2015. Sensitivity enhancement in magnetic particle imaging by background subtraction. Trans. Med. Imaging 35 (3), 893–900. https://doi.org/10.1109/TMI.2015.2501462.

Them, K., Salamon, J., Szwargulski, P., Sequeira, S., Kaul, M.G., Lange, C., Ittrich, H., Knopp, T., 2016. Increasing the sensitivity for stem cell monitoring in system-function based magnetic particle imaging. Phys. Med. Biol. 61 (9), 3279–3290. https://doi.org/10.1088/0031-9155/61/9/3279.

Timmermeyer, A., Wojtczyk, H., Tenner, W., Bringout, G., Grüttner, M., Graeser, M., Sattel, T.F., Halkola, A., Buzug, T.M., 2013. In: Super-resolution approach in magnetic particle imaging—evaluation of effectiveness at various noise levels. Proceedings der 44. Jahrestagung der Deutschen Gesellschaft für Medizinische Physik.

Tonyushkin, A., 2017. Single-sided hybrid selection coils for field-free line magnetic particle imaging. Int. J. Magnet. Particle Imaging 3 (1). https://doi.org/10.18416/ijmpi.2017.1703009.

Viereck, T., Kuhlmann, C., Draack, S., Schilling, M., Ludwig, F., 2017. Dual-frequency magnetic particle imaging of the Brownian particle contribution. J. Magn. Magn. Mater. 427, 156–161. https://doi.org/10.1016/j.jmmm.2016.11.003.

Vogel, P., Lother, S., Rückert, M.A., Kullmann, W.H., Jakob, P.M., Fidler, F., Behr, V.C., 2014b. MRI meets MPI: a bimodal MPI-MRI tomograph. IEEE Trans. Med. Imaging 33 (10), 1954–1959. https://doi.org/10.1109/TMI.2014.2327515.

Vogel, P., Rückert, M.A., Klauer, P., Kullmann, W.H., Jakob, P.M., Behr, V.C., 2014a. Traveling wave magnetic particle imaging. IEEE Trans. Med. Imaging 33 (2), 400–407. https://doi.org/10.1109/TMI.2013.2285472.

Vogel, P., Rückert, M.A., Klauer, P., Kullmann, W.H., Jakob, P.M., Behr, V.C., 2015. Rotating slice scanning mode for traveling wave magnetic particle imaging. IEEE Trans. Magn. 51 (2), 6501503. https://doi.org/10.1109/TMAG.2014.2335255.

Vogel, P., Rückert, M.A., Klauer, P., Kullmann, W.H., Jakob, P.M., Behr, V.C., 2016. First in vivo traveling wave magnetic particle imaging of a beating mouse heart. Phys. Med. Biol. 61 (18), 6620–6634. https://doi.org/10.1088/0031-9155/61/18/6620.

von Gladiss, A., Ahlborg, M., Knopp, T., Buzug, T.M., 2015a. Compressed sensing and sparse reconstruction in magnetic particle imaging. IEEE Trans. Magn. 51 (2), 6501304. https://doi.org/10.1109/TMAG.2014.2326432.

von Gladiss, A., Graeser, M., Lüdtke-Buzug, K., Buzug, T.M., 2015b. Contribution of brownian rotation and particle assembly polarisation to the particle response in magnetic particle spectrometry. Curr. Direct. Biomed. Eng. 1 (1), 298–301. https://doi.org/10.1515/cdbme-2015-0074.

von Gladiss, A., Graeser, M., Szwargulski, P., Knopp, T., Buzug, T.M., 2017. Hybrid system calibration for multidimensional magnetic particle imaging. Phys. Med. Biol. 62 (9), 3392–3406. https://doi.org/10.1088/1361-6560/aa5340.

Wawrzik, T., Yoshida, T., Schilling, M., Ludwig, F., 2015. Debye-based frequency-domain magnetization model for magnetic nanoparticles in magnetic particle spectroscopy. IEEE Trans. Magn. 51 (2), 5300404. https://doi.org/10.1109/TMAG.2014.2332371.

Weaver, J.B., Rauwerdink, A.M., Hansen, E.W., 2009. Magnetic nanoparticle temperature estimation. Med. Phys. 36 (5), 1822–1829. https://doi.org/10.1118/1.3106342.

Weaver, J.B., Rauwerdink, A.M., Sullivan, C.R., Baker, I., 2008. Frequency distribution of the nanoparticle magnetization in the presence of a static as well as a harmonic magnetic field. Med. Phys. 35 (5), 1988–1994. https://doi.org/10.1118/1.2903449.

Weaver, J.B., Rauwerdink, K.M., Rauwerdink, A.M., Perreard, I.M., 2013. Magnetic spectroscopy of nanoparticle Brownian motion measurement of microenvironment matrix rigidity. Biomed. Tech./Biomed. Eng. 58 (6), 547–550. https://doi.org/10.1515/bmt-2013-0012.

Weber, A., Knopp, T., 2015. Symmetries of the 2D magnetic particle imaging system matrix. Phys. Med. Biol. 60 (10), 4033–4044. https://doi.org/10.1088/0031-9155/60/10/4033.

Weber, A., Weizenecker, J., Heinen, U., Heidenreich, M., Buzug, T.M., 2015a. Reconstruction enhancement by denoising the magnetic particle imaging system matrix using frequency domain filter. IEEE Trans. Magn. 51 (2), 7200105. https://doi.org/10.1109/TMAG.2014.2332612.

Weber, A., Werner, F., Weizenecker, J., Buzug, T.M., Knopp, T., 2015b. Artifact free reconstruction with the system matrix approach by overscanning the field-free-point trajectory in magnetic particle imaging. Phys. Med. Biol. 61 (2), 475–487. https://doi.org/10.1088/0031-9155/61/2/475.

Weber, M., Bente, K., von Gladiss, A., Buzug, T.M., 2015c. In: MPI with a mechanically rotated FFL. International Workshop on Magnetic Particle Imaging. IEEE Xplore Digital Library. https://doi.org/10.1109/IWMPI.2015.7107026.

Weizenecker, J., Borgert, J., Gleich, B., 2007. A simulation study on the resolution and sensitivity of magnetic particle imaging. Phys. Med. Biol. 52 (21), 6363–6374. https://doi.org/10.1088/0031-9155/52/21/001.

Weizenecker, J., Gleich, B., Borgert, J., 2008. Magnetic particle imaging using a field free line. J. Phys. D: Appl. Phys. 41 (10), 105009. https://doi.org/10.1088/0022-3727/41/10/105009.

Weizenecker, J., Gleich, B., Rahmer, J., Borgert, J., 2010. Particle dynamics of mono-domain particles in magnetic particle imaging. Magnet. Nanopart. 3–15. https://doi.org/10.1142/9789814324687_0001.

Weizenecker, J., Gleich, B., Rahmer, J., Borgert, J., 2012. Micro-magnetic simulation study on the magnetic particle imaging performance of anisotropic mono-domain particles. Phys. Med. Biol. 57 (22), 7317–7327. https://doi.org/10.1088/0031-9155/57/22/7317.

Weizenecker, J., Gleich, B., Rahmer, J., Dahnke, H., Borgert, J., 2009. Three-dimensional real-time in vivo magnetic particle imaging. Phys. Med. Biol. 54 (5), L1–L10. https://doi.org/10.1088/0031-9155/54/5/L01.

Werner, F., Gdaniec, N., Knopp, T., 2017. First experimental comparison between the Cartesian and the Lissajous trajectory for magnetic particle imaging. Phys. Med. Biol. 62 (9), 3407–3421. https://doi.org/10.1088/1361-6560/aa6177.

Werner, F., Jung, C., Hofmann, M., Werner, R., Salamon, J., Säring, D., Kaul, M.G., Them, K., Weber, O.M., Mummert, T., Adam, G., Ittrich, H., Knopp, T., 2016. Geometry planning and image registration in magnetic particle imaging using bimodal fiducial markers. Med. Phys. 43 (6), 2884. https://doi.org/10.1118/1.4948998.

Williams, D.B., Carter, C.B., 1996. Transmission Electron Microscopy. Springer US. https://doi.org/10.1007/978-0-387-76501-3.

Williams, S.K.R., Caldwell, K.D., 2011. Field-Flow Fractionation in Biopolymer Analysis. Springer Science & Business Media.

Witt, W., Geers, H., Aberle, L., 2004. In: Measurement of particle size and stability of nanoparticles in opaque suspensions and emulsions with photon cross correlation spectroscopy. PARTEC, International Congress for Particle Technology, Nürnberg.

Wu, W., Quanguo, H., Changzhong, J., 2008. Magnetic iron oxide nanoparticles: synthesis and surface functionalization strategies. Nanoscale Res. Lett. 3 (11), 397–415. https://doi.org/10.1007/s11671-008-9174-9.

Yasumuri, I., Reinen, D., Selwood, P.W., 1963. Anisotropic behaviour in superparamagnetic systems. J. Appl. Phys. 34 (12), 3544–3549. https://doi.org/10.1063/1.1729255.

Yoshida, T., Enpuku, K., 2009. Simulation and quantitative clarification of AC susceptibility of magnetic fluid in nonlinear brownian relaxation region. Jpn. J. Appl. Phys. 48 (12), 127. https://doi.org/10.1143/JJAP.48.127002.

Yoshida, T., Othman, N.B., Enpuku, K., 2013. Characterization of magnetically fractionated magnetic nanoparticles for magnetic particle imaging. J. Appl. Phys. 114 (17), 173908. https://doi.org/10.1063/1.4829484.

Youssif, M.I., Bahgat, A.A., Ali, I.A., 2000. AC magnetic susceptibility technique for the characterization of high temperature superconductors. Egypt. J. Solids 23 (2), 231–250.

Yu, E.Y., Bishop, M., Zheng, B., Ferguson, R.M., Khandhar, A.P., Kemp, S.J., Krishnan, K.M., Goodwill, P.W., Conolly, S.M., 2017. Magnetic particle imaging: a novel in vivo imaging platform for cancer detection. Nano Lett. 17 (3), 1648–1654. https://doi.org/10.1021/acs.nanolett.6b04865.

Zheng, B., Vazin, T., Goodwill, P.W., Conway, A., Verma, A., Saritas, E.U., Schaffer, D., Conolly, S.M., 2015. Magnetic particle imaging tracks the long-term fate of in vivo neural cell implants with high image contrast. Sci. Rep. 5 (14055). https://doi.org/10.1038/srep14055.

Zheng, B., von See, M.P., Yu, E., Gunel, B., Lu, K., Vazin, T., Schaffer, D.V., Goodwill, P.W., Conolly, S.M., 2016. Quantitative magnetic particle imaging monitors the transplantation, biodistribution, and clearance of stem cells in vivo. Theranostics 6 (3), 291–301. https://doi.org/10.7150/thno.13728.

Zheng, B., Yang, W., Massey, T., Goodwill, P., Conolly, S., 2013. In: High-power active interference suppression in magnetic particle imaging. International Workshop on Magnetic Particle Imaging. IEEE Xplore Digital Library. https://doi.org/10.1109/IWMPI.2013.6528381.

CHAPTER

10

Nanoparticles and Nanosized Structures in Diagnostics and Therapy

Lisa J. Jacob, Hans-Peter Deigner*,†*

*Institute of Precision Medicine, Furtwangen University, Villingen-Schwenningen, Germany

†Fraunhofer EXIM/IZI, Rostock-Leipzig, Germany

Nanoparticles can be used for various medical applications. Dependent on material, size, and shape, they can exhibit an enormous number of properties, and the number of their applications can be further multiplied by using the high surface to volume ratio for functionalization. It is not surprising that physicians, health-care providers, regulatory authorities, and pharmaceutical industry recognized this huge potential and implemented nanoparticles in all fields of diagnostic and therapy, from basic laboratory applications like enzyme-linked immunosorbent assay (ELISA) and fluorescence microscopy over in vivo applications, for example, contrast agents for computed tomography (CT) and magnetic resonance imaging (MRI), to the design of precision cancer nanodrugs. This chapter gives an overview on various type-specific functionalities that can be used for therapy and medicine and later presents selected examples for applications in different fields of medicine.

1 DIFFERENT FUNCTIONALITIES OF NANOPARTICLES

Nanoparticles are defined as a particular material for which at least one dimension is within the range of 1–100 nm (Valcárcel Cases and López-Lorente, 2014). Properties of nanoparticles are different from that of bulk material (Valcárcel Cases and López-Lorente, 2014). Reason therefore is, besides the large surface to volume ratio, the quantum-size effect (Valcárcel Cases and López-Lorente, 2014; Chen et al., 2016). Nanoparticles are derived either by bottom-up or by top-down methods from all kinds of materials (Vijayakumar, 2013; Prasad, 2012).

1.1 Optical Functionalities

Optical properties of nanoparticles depend either on light scattering or on optical transitions between energy states. To our notion, optical functionalities of nanoparticles are those

Precision Medicine
https://doi.org/10.1016/B978-0-12-805364-5.00010-X

exploited most in the field of diagnostic and medicine. This section deals with the basic optical principles that are harnessed in nanoparticle applications and presents types of nanoparticles closely related.

When electromagnetic radiation (light) hits matter in, it is either absorbed or scattered. The interplay of absorption and scattering of specific wavelengths makes our world appear in color. A tree appears green, because the leaves reflect all colors of the sunlight, except green. During absorption, the electromagnetic energy is transformed into internal energy and leads to excitation of the material. Pathways for relaxation are vibrational relaxation, internal conversion, and intersystem crossing. This energy transformation and the spectra of scattered light can provide information or can be used for therapy (Prasad, 2012; Thomas, 2006).

1.1.1 *Surface Plasmon Resonance*

Surface plasmon resonance (SPR) appears when a light beam hits a metal surface. Then, electromagnetic waves are produced and travel along the surface and into the metal. That can be imagined as a drop that falls into an even water surface. In nanoparticles, the propagation along the surface and the propagation into the metal are limited, due to the small size of the particles (e.g., colloidal gold). Thus, incident light is reflected with one important exception. When the frequency of incident electromagnetic wave and the natural frequency of valence electrons are the same, the light can be fully absorbed and excites the particle (Valcárcel Cases and López-Lorente, 2014; Chen et al., 2016; Vijayakumar, 2013). The energy of the electromagnetic radiation is transferred into vibration of valence electrons. Thus, SPR is often described as collective oscillation of valence electrons. SPR is the reason why metal nanoparticles have strong absorption peaks that are dependent on composition, shape, and size of the nanoparticles.

Gold nanoparticles are one example for plasmonic nanocarriers (Chen et al., 2016). They either consist completely of gold or have a dielectric core that is covered with gold (nanoshells) (Chen et al., 2016). Gold colloids are obtained by reduction of Au salts in the presence of mild reducing agents. Aggregation is usually prevented by electrostatic repulsion of adsorbed surface groups. Aggregation in solvents leads to change of color from red to blue. Thus, binding of gold nanoparticle-labeled entities to their respective target that leads to aggregation of the nanoparticles can easily be detected (aggregation assays). This favors the use of gold nanoparticles as detection molecules in immunoassays, especially lateral-flow immunoassays (LFIAs). Their absorption properties make gold nanoparticles perfect for optical signal readout. In therapy, gold nanoparticles can be used for photothermal cancer therapy, radiofrequency therapy, angiogenesis therapy, and rheumatoid arthritis therapy and as vectors for drugs, genes, and imaging agents (Boisselier and Astruc, 2009).

1.1.2 *Raman Scattering*

Raman scattering is inelastic scattering, which means that the kinetic energy of an incident photon is increased (Stokes Raman scattering) or reduced (anti-Stokes Raman scattering) during interaction (Fig. 1). By measuring the energy difference between incident electromagnetic radiation and scattered electromagnetic radiation, information about the vibrational energy and frequencies can be obtained. These energies are molecule-specific and can be used for imaging. Some nanoparticles have resonance-enhanced Raman signatures that can be used for contrast generation.

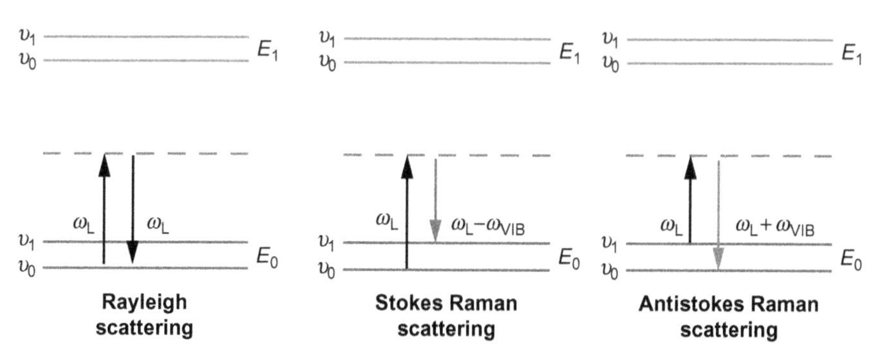

FIG. 1 Scheme of the Rayleigh and linear spontaneous Raman scattering.

For example, carbon nanotubes are rolled graphene sheets with strong optical absorption in the near-infrared (NIR) range and resonance-enhanced Raman signatures for Raman detection/imaging (Boisselier and Astruc, 2009). Single-walled carbon nanotubes (SWCN) have a diameter of 1–2 nm and a length ranging from 50 nm to 1 cm (Boisselier and Astruc, 2009) and are considered as one-dimensional. Multiwalled carbon tubes exhibit weaker optical properties, but due to their larger size (diameter 10–100 nm), they are suitable as vectors for biomolecules and drugs (Chen et al., 2016; Yong et al., 2013).

1.1.3 Fluorescence

Fluorescent properties of nanoparticles stem from the quantum confinement effect. If an electron from the valence band is excited to the conductive band, usually, an electron-hole pair (exciton) is produced that is divided by the so-called Bohr radius. In the case of nanoparticles, the size of the particle is smaller than the Bohr radius, and energy levels (valence band and conduction band) become more discrete. In consequence, the bandgap has a defined distance, and as the energy of the bandgap corresponds to the energy released when the electron-hole pair reunites, the emitted wavelength (color of fluorescence) can be adjusted. Thus, particles with various colors, sharp emission peak, and low photobleaching can be designed.

One example for such particles is quantum dots. Quantum dots are fluorescent nanocrystals made from semiconductor materials. When used as fluorescent probes, quantum dots emitting at different wavelengths (colors) can be excited by a single UV light. This is especially advantageous for multiplex applications. Quantum dots have a core shell structure. The shell firstly prevents toxic effects from the semiconductor materials used to produce quantum dots (CdSe, CdS, CdTe, ZnS, InP, InAs, GaP, and GaAs), secondly provides solubility in water, and thirdly can be used for functionalization. The stability of this shell is one of the major concerns that hinder quantum dots to enter the field of biological applications as it prevents toxic effects of heavy metals. Toxicity studies in primates showed no signs of toxicity after injection of phospholipid-micelle-encapsulated CdSe-CdS-ZnS quantum dots for 90 days (Yong et al., 2013). But findings of cadmium and selenium especially in the liver, spleen, and kidneys make long-term studies crucial before clinical studies in humans can be taken into account (Yong et al., 2013). Recently, there are efforts to produce cadmium- and selenide-free quantum dots, for example, $CuInS_2$ with comparable quantum yields that could overcome the toxicity problems (Prasad, 2012). So, the main application of quantum dots is up to date in bioanalytical fields.

They are used, preferably in multiplex applications, as bioprobes for western blots, ELISA, or flow cytometry (Kaganman, 2006; Ornberg et al., 2005). Due to their high quantum yield in a range of 600–900nm, quantum dots can as well be used in NIR fluorescence in vivo imaging (Resch-Genger et al., 2008).

Carbon dots (under 10nm in all dimensions) share favorable optical properties of quantum dots (adjustable emission wavelength, low photobleaching, and bright luminescence) (Chen et al., 2016), but in contrast to quantum dots, they exhibit low toxicity and can be rapidly cleared from the body (Chen et al., 2016). Thus, they are preferable candidates for in vivo applications. Carbon nanotubes can even be used as NIR fluorescent molecules with an emission range from 800 to 2000 nm (Jaque et al., 2014).

Dye-doped silica nanoparticles are another kind of fluorescent nanoparticles, where the fluorescence is based on a high number of fluorophores encapsulated, rather than on the quantum confinement effect. The encapsulation in silica nanoparticles hinders the interaction of the fluorophores with aqueous environment, where they usually suffer from low fluorescence quantum yield, degradation, and insufficient photostability (Hahn et al., 2011).

The last and relatively new class of fluorescent nanoparticles is upconverting nanoparticles that absorb NIR light and emit light at a higher energy. Thus, the nanoparticles doped with rare-earth ions are especially suitable for near-infrared fluorescence (NIRF) imaging. One example is yttrium oxide (Y_2O_3) nanoparticles doped with erbium and yttrium that have good photostability in the NIR combined with low toxicity (Hahn et al., 2011).

1.1.4 Förster Resonance Energy Transfer

Förster resonance energy transfer (FRET) is a nonradiative energy transfer between two molecules that can cause a change in their emission (Valcárcel Cases and López-Lorente, 2014; Prasad, 2012). That means energy is transferred between two molecules, not by photons, but by dipole-dipole interactions (Valcárcel Cases and López-Lorente, 2014; Prasad, 2012; Agasti et al., 2010). Therefore, the following conditions have to be fulfilled:

(i) There has to be an overlap between the emission spectrum of the donor and the excitation spectra of the acceptor.

(ii) The two molecules must be in close proximity (1–10nm), as the quantum efficiency of energy transfer changes with the sixth power of the distance between the donor and acceptor.

$$E = \frac{1}{\left(1 + \left(\frac{R}{R_0}\right)^6\right)}$$

E is the quantum efficiency of the energy transfer
R is the distance between donor and acceptor
R_0 is the distance at which 50% of the energy is transferred

(iii) The donor and the acceptor have to be in a suitable orientation to each other (Valcárcel Cases and López-Lorente, 2014; Prasad, 2012).

In the case of fluorescent molecules, the change in energy can then lead to a change in emission. The emission of the donor molecule is reduced, and the acceptor molecule can either perform enhanced emission or adsorb the complete energy, which results in quenching of the donor fluorescence (Valcárcel Cases and López-Lorente, 2014; Prasad, 2012).

The change in emission, for example, the distance between two fluorescence-tagged molecules, one with a donor and one with an acceptor, can be monitored in vitro. This is also referred to as quantitative FRET microscopy and, for example, is used to monitor cell-cell interactions (Prasad, 2012).

1.2 Magnetic Functionalities

This section provides a brief overview on the different kinds of magnetism and outlines the related nanoparticles.

1.2.1 *Paramagnetism*

In paramagnetic materials, every single atom behaves as noninteracting, individual, and randomly orientated molecular magnet that has a magnetic dipole moment. In the presence of an external magnetic field, the molecular magnets align to it depending on its individual strengths. Total alignment is referred to as magnetic saturation and can only occur at low temperatures and in strong magnetic fields. When the external magnetic field is removed, the molecular magnets will randomly orientate (Rumenapp et al., 2012).

1.2.2 *Ferromagnetism*

In ferromagnetic materials, a strong exchange interaction energy leads to a regular orientation of the magnetic moments in so-called Weiss domains. In the case of an external magnetic field, the Weiss domains align to the field and thus the total magnetic moment and the magnetization M. After removal of the external field, the magnetization can be partly or fully retained. If the size of the particle is smaller than a critical size, only one Weiss domain per particle is present, and internal dipole moments point to the same direction. Then, the magnetization can only be changed by relaxation of the bulk due to Brownian relaxation or Néel relaxation. In contrast to paramagnetic particles, a relatively weak external field is sufficient for magnetic saturation (Rumenapp et al., 2012).

1.2.3 *Superparamagnetism*

Superparamagnetism occurs, when the ferromagnetic single domain is smaller than a second critical size. In this case, thermal fluctuations outcompete dipole-dipole interactions and make the magnetization flip randomly. Consequently, superparamagnetic particles have, like paramagnetic particles, no net magnetization. But in contrast to paramagnetic particles, the external magnetic field does not need to be very strong to achieve magnetic saturation (Rumenapp et al., 2012).

Magnetic nanoparticles can be divided into three types: oxides, metallic metals, and alloys. Metals are ferromagnetic (dependent on shape and size), whereas oxides have paramagnetic or superparamagnetic properties. Alloys usually combine the magnetic properties of the metals they are composed of (Sobczak-Kupiec et al., 2016). Most popular are iron oxide nanoparticles that have unique magnetic properties and are nontoxic and well tolerated in living organisms. Magnetic nanoparticles can be used as contrast agents in MRI as their spin relaxation time is different from that of water protons (Rumenapp et al., 2012).

Some immunoassays that use magnetic nanoparticles as detection molecules have also been developed (Sobczak-Kupiec et al., 2016).

1.3 Thermal Functionalities

There are two ways on how heat can be induced by nanoparticles. The first is to use magnetic nanoparticles and induce heat by applying oscillating magnetic fields. Responsible for the produced heat are different mechanisms, including hysteresis loss, Néel relaxation and Brown relaxation. Thermal functionalities of magnetic nanoparticles have already been used to treat brain tumors in patients (Jaque et al., 2014).

The second way to induce heat by nanoparticles is photothermal. As described before, if nanoparticles are illuminated, a part of the energy is absorbed, and the other part is scattered. The absorbed energy can then be released in different pathways. If vibrational relaxation takes place, heat is produced. To achieve effective heating, nanoparticles with large absorption efficiencies and low quantum yields are required, so that the light-to-heat conversion efficiency is large. If photothermal heating of nanoparticles is intended to be used in biological tissue, it is important that the absorption of the particle lies within the first (from 700 to 980 nm) or the second (1000 to 1400 nm) biological window to avoid unintended tissue heating during excitation of the nanoparticles (Jaque et al., 2014).

1.4 Targeting Functionalities

Targeting of distinct tissues, cells, organelles, or organs can improve therapeutics and diagnostics in four different ways. Firstly, therapeutics can be directly delivered to specific biological sites. Thus, any toxic side effect is limited to the targeted biological site, and damage of healthy tissue/cells is minimized. Secondly, the local concentration of therapeutic or diagnostic agents is increased in the targeted biological site. This leads to a more effective treatment in case therapeutics and to improved sensitivity and specificity of diagnostic methods. Thirdly, it enables delivery of drugs to specific cell organelles. This is a basic requirement for some kinds of therapy; for example, in the case of gene therapy, the drug must be delivered to the nucleus. Lastly, targeting can be used to transfer therapeutics across biological borders (Prasad, 2012).

Kinds of targeting are passive targeting, field-directed targeting, intrinsic targeting, and active targeting (Prasad, 2012).

Passive targeting is mainly used for cancer diagnostic and therapy and relies on the enhanced permeability and retention (EPR) effect.

1.4.1 Enhanced Permeability and Retention

Due to the high metabolic activity of tumor cells, tumors are forced to generate new blood vessels for sufficient supply with nutrients (angiogenesis). In addition, vascular endothelium often has "leaks" and is more permeable than normal tissue. This effect together with reduced or absent clearing mechanism leads to the accumulation of nanoparticles (20–200 nm in diameter) in tumor tissue and is referred to as EPR effect (Valcárcel Cases and López-Lorente, 2014; Prasad, 2012; Jaque et al., 2014).

When the EPR effect shall be used for targeting, nanoparticles must be designed in a manner that they escape the reticuloendothelial system (RES) to enhance the circulation time. The RES marks foreign substances with opsonins, followed by clearance from the blood stream. Especially hydrophobic and charged molecules are cleared very fast. To bypass the RES, nanoparticles can be coated with polyethylene glycol. This was shown to be useful to prolong the circulation time of different kinds of particles (Jaque et al., 2014) and resulted in more nanoparticle accumulation in tumors.

Field-directed targeting applies an external (e.g., magnetic) field to direct nanoparticles containing magnetic materials to the cell, tissue, organ, or organelles of interest. Intrinsic targeting means that a molecular probe or a drug accumulates in a selected biological site. And finally, if a nanoparticle is marked with a molecular address, this is called active targeting. These target molecules can be small molecules, antibodies, aptamers, or integrin-receptor targeting peptides (Fig. 2).

1.5 Pharmakokinetic/Pharmacodynamic Functionalities

Nanoparticles can carry nonpolar or polar agents, such as drugs, diagnostic, or imaging agents. These agents can be incorporated into or attached to the surface of the nanoparticles. Dependent on the strategy chosen, temporally controlled release can be enabled. This can improve half-life and biodistribution of these agents.

Suitable nanoparticles for encapsulation of drugs and imaging agents are, for example, lipid nanoparticles. They consist of molecules with a polar head group and a nonpolar long alkyl tail. Micelles or liposomes self-assemble in aqueous environment. To improve the structural stability, polymeric phospholipid structures have been developed. Another approach to improve structural stability is to produce micelles from lipids that are solid at higher temperatures. As lipids are ubiquitous in the body, lipid nanoparticles are highly biocompatible and biodegradable. By conjugation of polyethylene glycol to the surface, the half-life of lipid nanoparticles can be prolonged further.

Solid polymeric nanoparticles can be used to physically encapsulate or conjugate agents to the surface (Prasad, 2012). Special attention should be paid to biodegradable polymeric nanoparticles. Degradation can be induced, for example, by intracellular enzymes, pH, or external thermal activation, and thus, they are perfectly suitable for timed drug or gene delivery (Chen et al., 2016; Prasad, 2012). Materials for such particles are, for example, polyamino

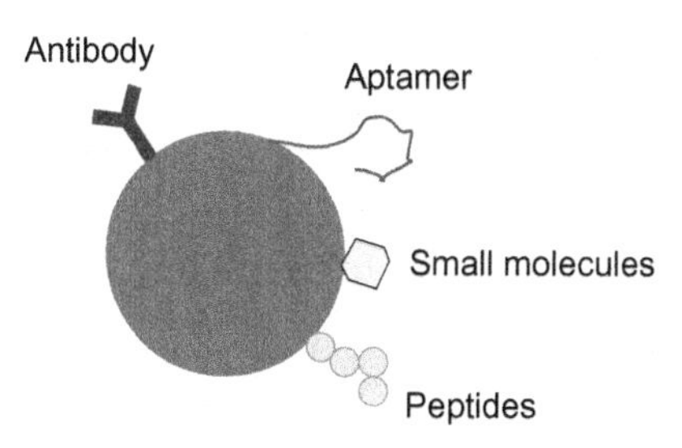

FIG. 2 Different target-binding molecules.

esters, polyanhydrides, polylactides, and chitosan. Core-shell structures with a hydrophilic core for incorporation of hydrophobic drugs can be produced by amphiphilic copolymers that self-assemble upon contact with aqueous environments (Jaque et al., 2014).

1.6 Multiple Functionalities

The functionalities presented above are remarkable each alone, but the major advantage to use nanoparticles is that they can be designed to combine multiple favorable functionalities. For example, magnetic nanoparticles can be used as contrast agent for MRI and at the same time for magnetothermal therapy of tumors. Another example is to attach drugs to nanoparticles and to monitor their delivery by using their optical properties. This wealth of possibilities inspired the creation of a new sector in medicine: the therapy theranostics.

2 NANOPARTICLE-BASED DIAGNOSTIC THERAPY AND THERANOSTICS

Sensitive methods for detection of biomarkers, such as proteins and nucleic acids, are important for diagnostics, therapy decision, and therapy control (also as part of personalized medicine). When biomarkers are identified, sensors with high sensitivity, selectivity, and stability are needed. Such sensing systems are composed of two functional components (Agasti et al., 2010). A recognition element for binding the target analyte and a transduction element that generates the signal output (Agasti et al., 2010). Nanoparticles have potential to improve both. They can be functionalized and used as bioprobes to detect the analyte, and their various functionalities open new possibilities for signal transduction. An important new direction of nanomedicine called "theranostics" combines diagnostic and therapeutic functionalities of nanoparticles.

2.1 Fluorescence-Linked Immunosorbent Assay

ELISA is an immunoassay used for medical diagnostics and quality control measures in different fields of industry and as an analytic tool in biomedical research. Its purpose is the detection and quantification of antigens in a given sample. In a basic setup, an antibody labeled with an enzyme binds to the antigen of interest. The exact procedure is not explained here. The detection principle is illustrated in Fig. 3A. The linked enzyme catalyzes a color change through cleavage of a chromogenic substrate. The visible color can then be detected by colorimetric readout and provide quantitative and/or qualitative information about the antigen or antibody present in the sample (Gan and Patel, 2013).

In fluorescence-linked immunosorbent assay (FLISA) fluoresce-labeled antibodies or analytes are used for detection. In the so-called FLISA, there is no need for addition of a substrate (see Fig. 3B), and the fluorescence can be directly detected in a microplate reader. However, organic fluorescent dyes have several disadvantages: they have to be excited at different wavelengths, have an asymmetrical unsharpened emission peak (Resch-Genger et al., 2008), and suffer from photobleaching (Beloglazova et al., 2014). Therefore, quantum dots should be used as fluorescent labels.

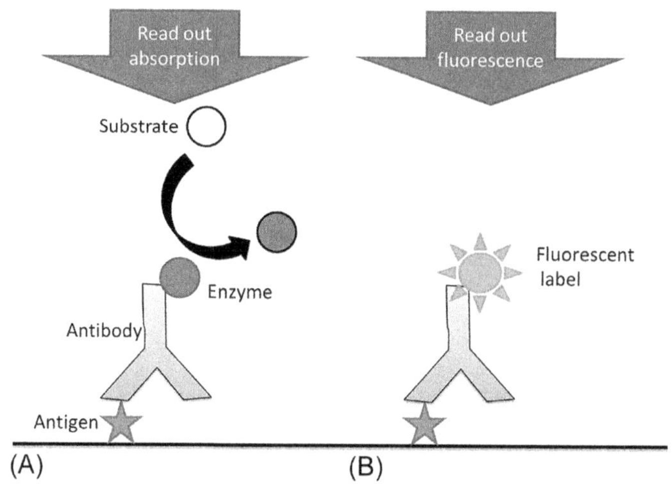

FIG. 3 Scheme of detection principle used in (A) direct ELISA and (B) FLISA.

Quantum-dot-based FLISA has several advantages compared with ELISA:

(i) Due to higher signal-to-noise ratio and high brightness of quantum dots, the limit of detection can be decreased in FLISA compared with ELISA.
(ii) The readout is not time-dependent. As this system is not based on the enzymatic activity to produce substrate, there will be no substrate saturation after a certain time. As quantum dots do not suffer from photobleaching, the readout must not be performed in a district time span. That means as well that results from different FLISAs are better comparable.
(iii) **The** FLISA is more robust than the ELISA. As the FLISA does not depend on enzyme activity, it is less vulnerable to temperature or pH.

In research, several groups used quantum dots for the design of FLISAs. Lv et al. (2017) showed that a quantum-dot-based FLISA is superior compared with an ELISA for detection of c-reactive protein, and Zhang et al. (2017) developed a quantum-dot-based FLISA that detects morphine and Yao et al. (2017) to detect ochratoxin A. However, to date, FLISA is not commercially available.

An additional feature of FLISA is that simultaneous detection of more than one target can be realized if two or more different antibodies are labeled with different colors. Using the favorable fluorescent properties of quantum dots as fluorescent labels, especially the narrow symmetrical fluorescent peaks and the possibility to excite them simultaneously at the same wavelength (Beloglazova et al., 2014; Resch-Genger et al., 2008; Oh et al., 2015), multiplexing is only limited by the color separation suitability of the imaging system.

Considering these advantages, it is surprising that there are only a low number of publications about FLISA-based on quantum dots for multiplexed detection yet. Goldman et al. (2004) used a quantum-dot-based multiplex sandwich fluoroimmunoassay for the simultaneous detection of four different toxins (cholera toxin, ricin, Shiga-like toxin 1, and staphylococcal enterotoxin B). Beloglazova et al. (2014) labeled four different mycotoxins with quantum dots and developed a multiplex fluorescent immunoassay related to a FLISA format. Le et al. (2016) used a dual-quantum-dot FLISA to detect carbadox and olaquindox metabolites simultaneously.

2.2 Western Blot

Another application of quantum-dot bioprobes is the western blot. Again, these bioprobes are especially suitable for multiplexed detection (as already described in the last section). This was also realized by the industry. Thermo Fisher Scientific offers quantum-dot-based fluorescent immunodetection kits for duplex detection (Fig. 4).

2.3 Lateral Flow Immunoassays

LFIAs are used for a simple, rapid, low-cost detection of analytes in different fields. In diagnostics, these assays are used to detect specific components in sample matrices like blood, urine, and other biological fluids. Analytic targets include hormones, viruses, bacteria, cancer and cardiac biomarkers, and small molecules, like therapeutic drugs and drugs of abuse (Lee et al., 2013; McMeekin, 2003).

In general, lateral-flow devices share the same composition and performance characteristics: they comprise a porous membrane (e.g., nitrocellulose) onto which a capture molecule is immobilized (solid support). Usually, the capture molecule is an antibody (e.g., sandwich assay) or analyte molecule (e.g., competitive assay) (McMeekin, 2003). For test performance, a sample is applied to the sample pad and then migrates (due to capillary diffusion) through the conjugate pad, where the liquid sample dissolves dried bioprobes and buffer components (McMeekin, 2003). After the sample encounters the labeled probes, a specific interaction occurs, and the resulting complexes move through the membrane toward the test line where they are immobilized through the capture molecules (Lee et al., 2013). Excess/unbound conjugates migrate further and are immobilized at the control line. This process takes only minutes and produces a band in the test/control window of the device (Lee et al., 2013). The control line indicates that there was enough sample volume and the test is valid (Lee et al., 2013). The rest of the sample volume is adsorbed in the adsorbent pad (Fig. 5).

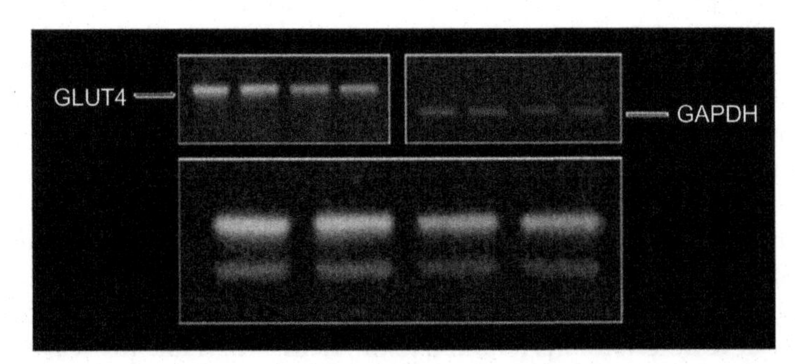

FIG. 4 Duplex detection of GAPDH and GLUT4 with WesternDot (Thermo Fisher Scientific, 2017).

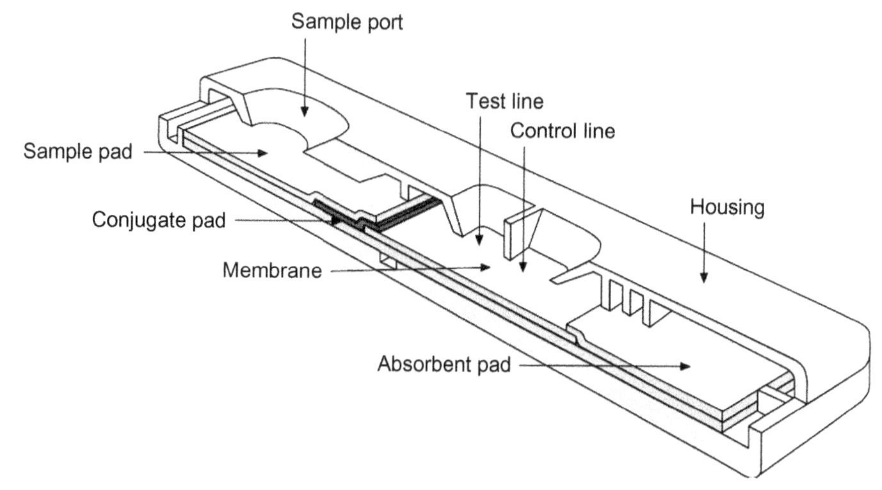

FIG. 5 Scheme of a lateral-flow test strip (EMD Millipore Corporation, 2013).

Different types of nanoparticles can be used as bioprobes (Huang et al., 2016):

1. Colored nanoparticles like colloidal gold, carbon, and selenium particles
2. Luminescent nanoparticles like quantum dots, upconverting phosphor nanoparticles, and dye-loaded shell-type nanoparticles
3. Magnetic nanoparticles

Lateral-flow assays can be designed as competitive and noncompetitive. For a competitive lateral-flow assay, an antigen or corresponding hapten is immobilized at the test line. A positive assay shows one test line (Salieb-Beugelaar and Hunziker, 2014; Fig. 6).

For a sandwich lateral-flow assay, two antibodies against two different epitopes of the same analyte are needed. One of them is immobilized on the test line, and the other is marked with a particle. In this design, a positive test shows two lines (Fig. 7). To enable LFIAs for detection of more than one analyte, there are in general two possibilities.

The simplest solution to detect more than one analyte is to add one additional test line for each additional target that shall be detected. This approach is already commercially applied (e.g., SD Bioline Influenza Ag A/B/A(H1N1) Pandemic rapid test kit). According to Li et al., however, expanding the multiplex capacity of LFIA by adding more test lines is extremely limited since the capillary flow time is proportional to the exponential of the distance. That means that the time to reach the furthest test line increases with the number of test lines.

The second possibility is multiplexing at a single test line. Therefore, two (or more) different capture molecules are immobilized at one test line, and two (or more) different antibodies marked with different reporter particles are used for the generation of two (or more) different signals. Fig. 8A (left) schematically shows how the test strips look like if the test is negative for two targets. In this case, the main part of the bioprobes binds to the analyte immobilized at the test line, and the excess bioprobes bind to the control line. Two lines with a mixed signal from red and green emitting bioprobes (orange in color) appear. Fig. 8B (right) shows a scheme of a test strip after running a sample positive for target 1 and target 2. In this case, the main part

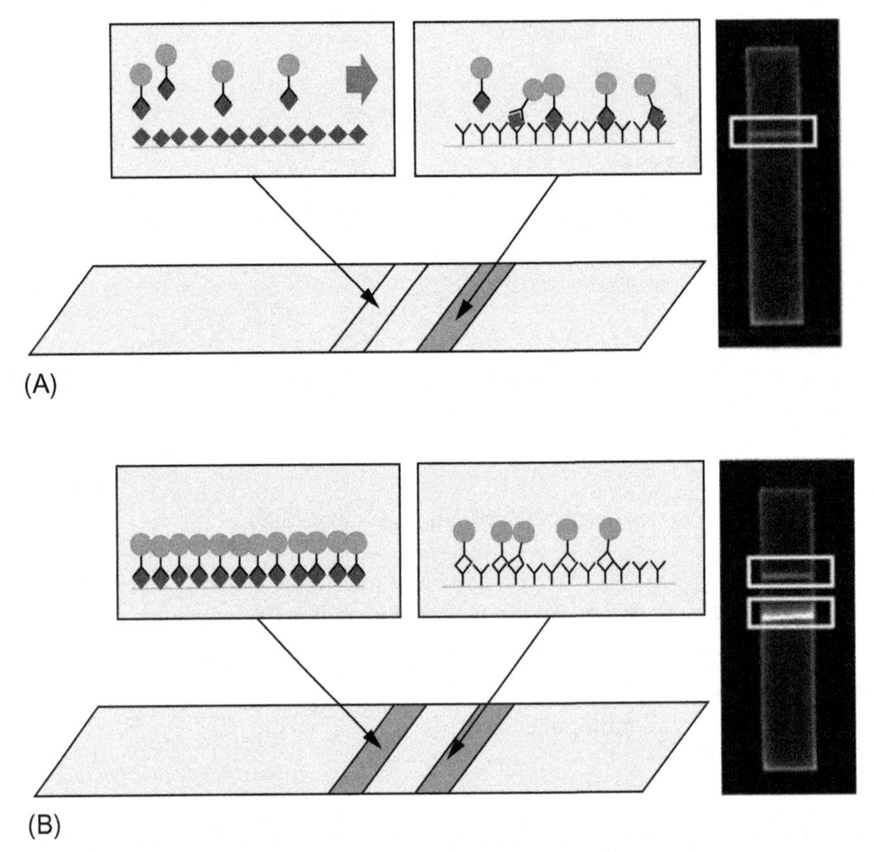

FIG. 6 Scheme of a positive (A) and a negative (B) competitive lateral-flow test.

of the bioprobes binds to the analyte in the sample, and no or fewer bioprobes bind to the antigen at the test line. Consequently, there will be no or little signal detectable from the test line, and only one line (control line) with signal from red and green emitting bioprobes will appear. In case the test is only positive for target 1, only red emitting bioprobes for target 2 bind to the test line, and only a red fluorescence signal can be detected. In the case that only target 2 is present in the sample, the signal from the test line is green.

A combination of multiple test lines and signal separation from each test line multiplies the number of analytes that can be detected from one LFIA. If, for example, four test lines are used and triplex readout from each test line is possible, 12 analytes could be detected in only one test. That would be an important step in the direction of affordable personalized medicine.

2.3.1 *LFIAs With Colored Nanoparticles*

Gold-nanoparticle-based LFIAs are widely used for screening applications. These assays are stable and easy to use and can be read out by the bare eye, instantly after the test is performed. However, these LFIAs have only relatively low sensitivity. An LFIA reader or

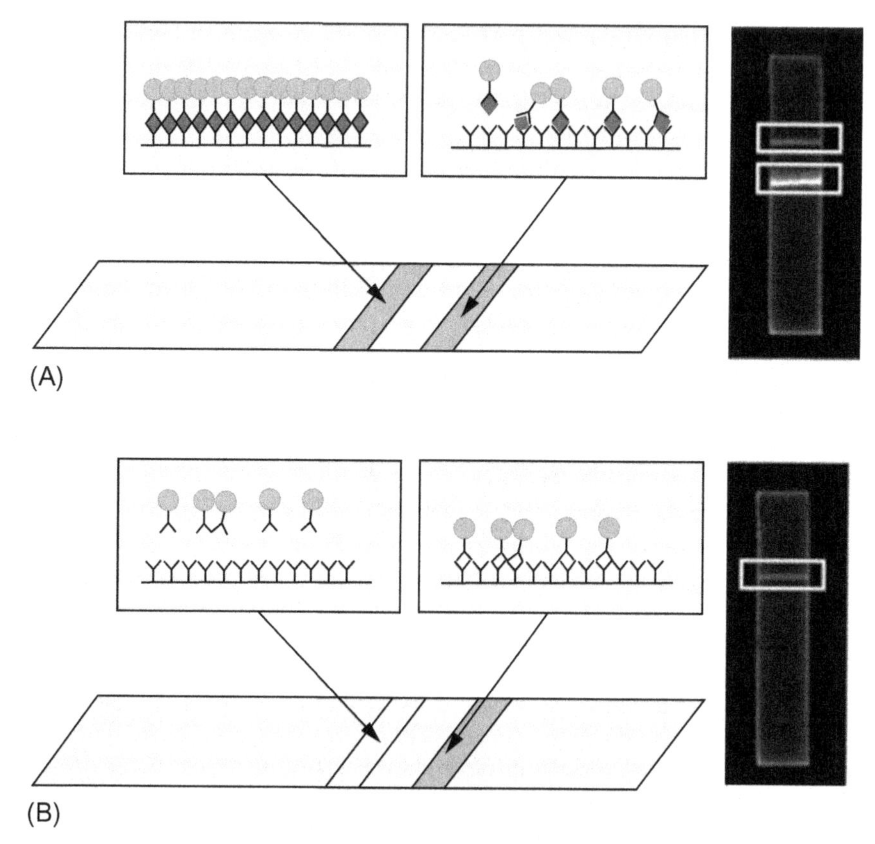

FIG. 7 Scheme of a positive (A) and a negative (B) sandwich lateral-flow test.

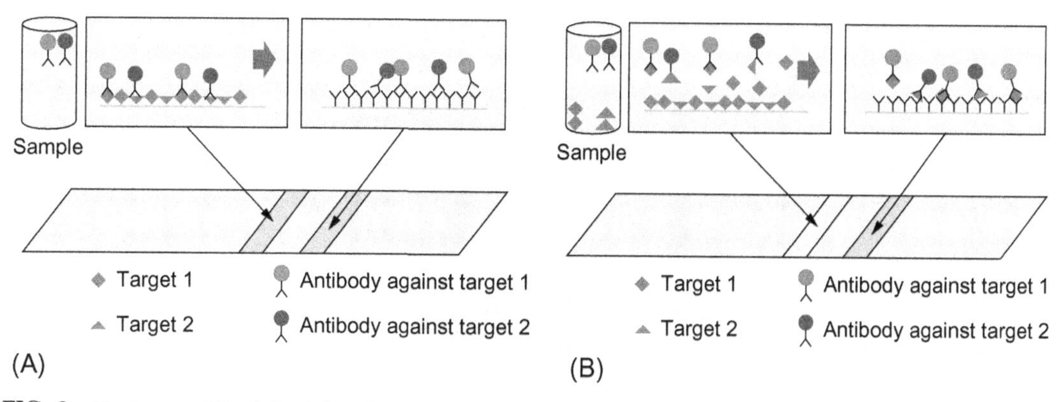

FIG. 8 Dual competitive lateral-flow immunoassay negative (A) and positive (B) for two targets.

enzyme-based signal enhancement can partly overcome this issue and enable quantification (Huang et al., 2016).

Carbon nanoparticles are especially useful for LFIAs. Their black color produces a good signal-to-noise ratio. Thus, assays that detect targets in picomolar range can be evaluated by the bare eye (Huang et al., 2016). Noguera et al. (2011) developed a carbon-based LFIA that detects five different virulence factors of Shiga-toxin-producing *Escherichia coli* from the PCR product with comparable sensitivity and specificity with q-PCR. Linares et al. (2012) compared colored nanoparticles most recently used and found that black carbon particles give the best limit of detection before silver-coated gold nanoparticles, gold nanoparticles, and polystyrene beads.

2.3.2 LFIAs With Fluorescent Nanoparticles

The use of fluorescent particles for LFIAs usually requires a reader system for signal detection. However, they are very promising candidates for LFIA development, because if quantification is intended, a reader system has to be used anyway. By using fluorescent bioprobes, the limit of detection can be decreased. In addition, fluorescent bioprobes are superior for multiplexed LFIAs.

Quantum dots are especially suitable for LFIAs. As they can be excited, independent of their emission wavelength with UV light, reader systems can be very simple. For example, Lee et al. (2013) designed a low-cost device for LFIA readout that uses basically a LED for illumination and a smartphone for imaging. Petryayeva and Algar (2016) showed that not only imaging but also excitation of quantum dots that can be used as bioprobes is possible by smartphone without external light source. Therefore, they used a simple and low-cost 3D-printed accessory to create a dark environment for fluorescence measurements and to support the smartphone in alignment with reflectors and filters for excitation and imaging of emission (Petryayeva and Algar, 2016). Petryayeva and Algar (2016) showed as well that quantum dots are superior over conventional dyes for this purpose. Taranova et al. (2015) designed the "traffic light assay" that detects three different kinds of antibiotics in milk. This LFIA showed better sensitivity compared with ELISA that used the same bioprobes, and quantification of each antibiotic was possible after imaging with a CCD camera and digital processing (Taranova et al., 2015; Fig. 9).

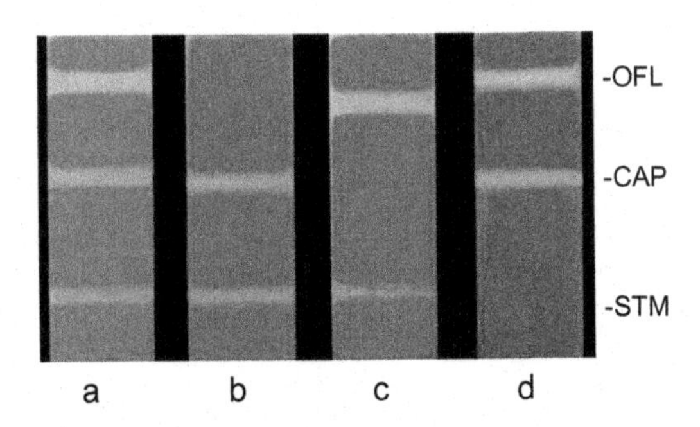

FIG. 9 "Traffic light" immunochromatographic test based on multicolor quantum dots for the simultaneous detection of several antibiotics in milk (Taranova et al., 2015).

Qi et al. (2016) quantified two sepsis markers (c-reactive protein, CRP and procalcitonin, PCT) simultaneously using quantum dots with different emission wavelength for bioprobe synthesis. Different lines are used for the detection of the two analytes. This leads to the total amount of three lines (test line 1 for analyte 1 and test line 2 for analyte 2 and control line). Qi et al. decided to use a competitive design for CRP detection and a sandwich design for PCT detection. For PCT, they achieved a detection range of 0.3 and 200 ng/mL and a minimum detection limit of 0.1 ng/mL and for CRP a detection range of 50 ng/mL and 250 µg/mL and a limit of detection of 1 ng/mL (Qi et al., 2016).

As LFIAs are in vitro diagnostics, the quantum dots have no contact with the patient, and their potential toxicity is no issue.

Dye-doped polystyrene nanoparticles enhance the fluorescence properties of free dye molecules and can be used for LFIAs as well. Xie et al. (2014) used fluorescein isothiocyanate-doped polystyrene NPs that performed better in the detection of *E. coli* O157:H7 when compared with colloidal gold.

2.3.3 *LFIAs With Magnetic Nanoparticles*

When fluorescence or absorbance is measured for the evaluation of the LFIAs, only the nanoparticles at the membrane surface are recorded, and ca. 90% stay undetected. The use of magnetic nanoparticles enables the (time-independent) readout of the cross-sectional area of test and control line and bypasses optical artifacts that can occur in more conventional LFIAs. Magnetic LFIAs use mainly two strategies for quantification. One is based on measuring the magnetization saturation of the cross section of the test line in an oscillating magnetic field and the other measures the magnetic flux generated in different zones of the LFIA. Both methods need complex readout systems but show higher sensitivities compared with classical LFIAs (Huang et al., 2016).

Wang et al. developed a superparamagnetic iron oxide-particle-based LFIA that detects the spores of *Bacillus anthracis*. For readout, a portable magnetic assay reader was designed. The sensitivity is that high that 200 spores can be detected per milligram milk powder and 130 spores per milligram soil. The detection of *B. anthracis* spores to prevent bioterrorism is an excellent example of how portable ultrasensitive, rapid, and robust assays can be applied (Wang et al., 2013).

2.4 Immunohistochemistry

Immunohistochemistry (IHC) is a well-established method for identification of cell types and specific cell-surface markers. Classical methods use antibodies that are marked with organic dyes as bioprobes. The disadvantages of using organic dyes, namely, photobleaching, susceptibility for environmental conditions, and distinct excitation wavelength for every dye, have been discussed before. Quantum dots show brighter fluorescence, are photostable, can be excited independent of the emission wavelength with UV light below 400 nm, and are thus particularly suitable for multiplex IHC.

Xing et al. published an excellent paper, where they showed up the benefits of using quantum dots (QDs) for IHC and provided protocols for bioprobe synthesis and tissue staining. Exemplarily, they stained prostate cancer tissue with red and green emitting bioprobes for p53 and EGR-1 and produced images where the distribution of the biomarkers is mapped (Xing et al., 2007).

Besides the improvement of classical IHC, nanoparticles serve as intracellular sensors. Casanova et al. (2009) used single europium-doped nanoparticles to measure the temporal pattern of reactive oxygen species inside cells. The emission of silica-coated 20–40 nm europium nanoparticles depends on their reduction state. A decrease in luminescence is related to the photoreduction of Eu^{3+} to Eu^{2}. Reoxidation can be induced by various oxidants produced in cells, especially hydrogen peroxide (H_2O_2) (Casanova et al., 2009). H_2O_2 is known to play an important role in signaling mechanisms like proliferation, differentiation, apoptosis, and migration, and thus, there is big interest in a time-resolved intracellular measuring method. Using the europium-doped nanoparticles and an inverted microscope, it could be shown that different signaling processes involve distinct timing in H_2O_2 production (Alexandrou, 2017).

A completely different approach is microrheology that can measure the viscoelasticity of cells. To study if cells are either viscoelastic liquids or soft solids can be informative to understand the formation of metastasis (Berret, 2017). Berret (2016) used micron-sized wires formed from iron oxide nanoparticles that can be manipulated remotely by an external magnetic field. Rotational magnetic spectroscopy measures their motion as response to the magnetic fields and can thus make conclusions about viscoelastic properties of the cells (Berret, 2017). Using the method of rotational magnetic spectroscopy, Berret (2017) was able to show that living cell interior is best described as a viscoelastic liquid rather than an elastic soft gel. That means that circulating cancer cells are more flexible than first expected (Berret, 2017).

2.5 Liquid Biopsy

Tissue biopsy is a potent way for diagnostics of various diseases, especially different types of cancer, but the invasive technique has the disadvantages that it is quite costly, impractical for repeated testing, and unavailable to some cancer types. That's why, liquid biopsy has evolved. It is noninvasive, inexpensive, and accessible to large populations and allows for repeated testing for real-time monitoring of disease stage and treatment response. Liquid biopsy aims to detect, for example, cancer biomarkers like circulating tumor cells, vesicles, nucleic acids, and protein in liquid samples like blood and urine. The first liquid biopsy test has been approved by the US Food and Drug Administration in June 2016. This test detects mutations in the epidermal growth factor gene (Huang et al., 2017).

Myung et al. (2016) give a detailed overview on how gold nanoparticles, magnetic nanoparticles, quantum dots, graphene and graphene oxide nanoparticles, and other nanoparticles can be used for liquid biopsy.

2.6 Nanoparticles as Bioprobes In Vivo

The implementation of bioimaging into diagnostic and therapy control has drastically improved the quality of health care. NIRF imaging, X-ray and CT, positron emission tomography (PET), MRI, and ultrasound are used to diagnose and monitor various diseases. Nanoparticles can serve as potent contrast agents to enhance the imaging performance and as bioprobes to identify distinct structures.

2.6.1 NIRF Imaging

Nanoparticles with absorption in the biological windows are used for in vivo imaging of tumor cells. Nanoparticles that are accumulated in the tumor either due to EPR effector or due to antibody-based targeting can be imaged. A promising application of NIRF imaging would be, for example, in breast-conserving surgery. Pleijhuis et al. (2011) showed the potential of NIRF imaging for pre- and intraoperative tumor localization, margin status assessment, and detection. They used tissue-simulating breast phantoms with implemented tumor-like fluorescence-tagged agarose inclusions and a custom-made NIRF camera system to mimic a breast-conserving surgery and found that the system was suitable to detect a "tumor" in 21 mm depth (Pleijhuis et al., 2011; Fig. 10).

In this case, indocyanine green was used as fluorescence dye, but quantum dots, dye-doped silica nanoparticles, and upconverting nanoparticles have shown good performance in NIRF (Hahn et al., 2011).

2.6.2 X-ray and CT

X-ray and CT images are based on the different attenuation of X-ray in tissues. An optimal contrast agent for CT and X-ray has a high X-ray opacity (measured by mass attenuation coefficient) and produces only a small fraction of ionization (measured by mass energy absorption coefficient). The resulting contrast of X-ray probes is dependent on the energy of incident X-ray. Thus, contrast agents that perform well in projectional radiographic imaging are not necessarily suitable for CT imaging. The research of new nanoparticles used as contrast agents in CT and X-ray imaging is mainly focused on iodinated and gold nanoparticles. Iodinated compounds are clinically applied for a long time, and research in this area is focused on the enhancement of localized iodine concentrations by incorporation of iodinated organic compounds into nanoparticles. Used and tested in vivo are, for example, emulsions, liposomes, lipoprotein, nanomilled insoluble compounds, and polymeric nanoparticles. In the recent years, different kinds of gold nanoparticles have attracted attention as contrast agents in CT and X-ray imaging. It was shown that gold nanoparticles perform equal or superior compared with iodinated compounds in several applications. A strong pro for gold nanoparticles is the advanced knowledge about their fate, transport, and toxicology (Hahn et al., 2011).

2.6.3 Positron Emission Tomography

PET measures the emission from radioisotopes in the form of positrons and thus needs no external excitation. As it does not provide information about the anatomy, it is usually coupled with another imaging technology like CT or MRI. PET can detect local concentrations of radionuclide tracer with high sensitivity and thus uncover single abnormal cells labeled with only a few trace isotopes. Thereby, the penetration depth is unlimited. PET is especially used in cancer imaging, because molecular changes can be detected before macromolecular changes occur and disease progression after treatment can be assessed. ^{18}F, ^{11}C, ^{15}O, ^{13}N, ^{64}Cu, ^{124}I, ^{68}Ga, ^{82}Rb, and ^{86}Y are isotopes that can be chelated onto or incorporated within nanoparticles (Hahn et al., 2011).

As PET is usually coupled with another imaging technique, contrast agents should be detectable with both techniques, for example, PET and CT or PET and MRI. A review from

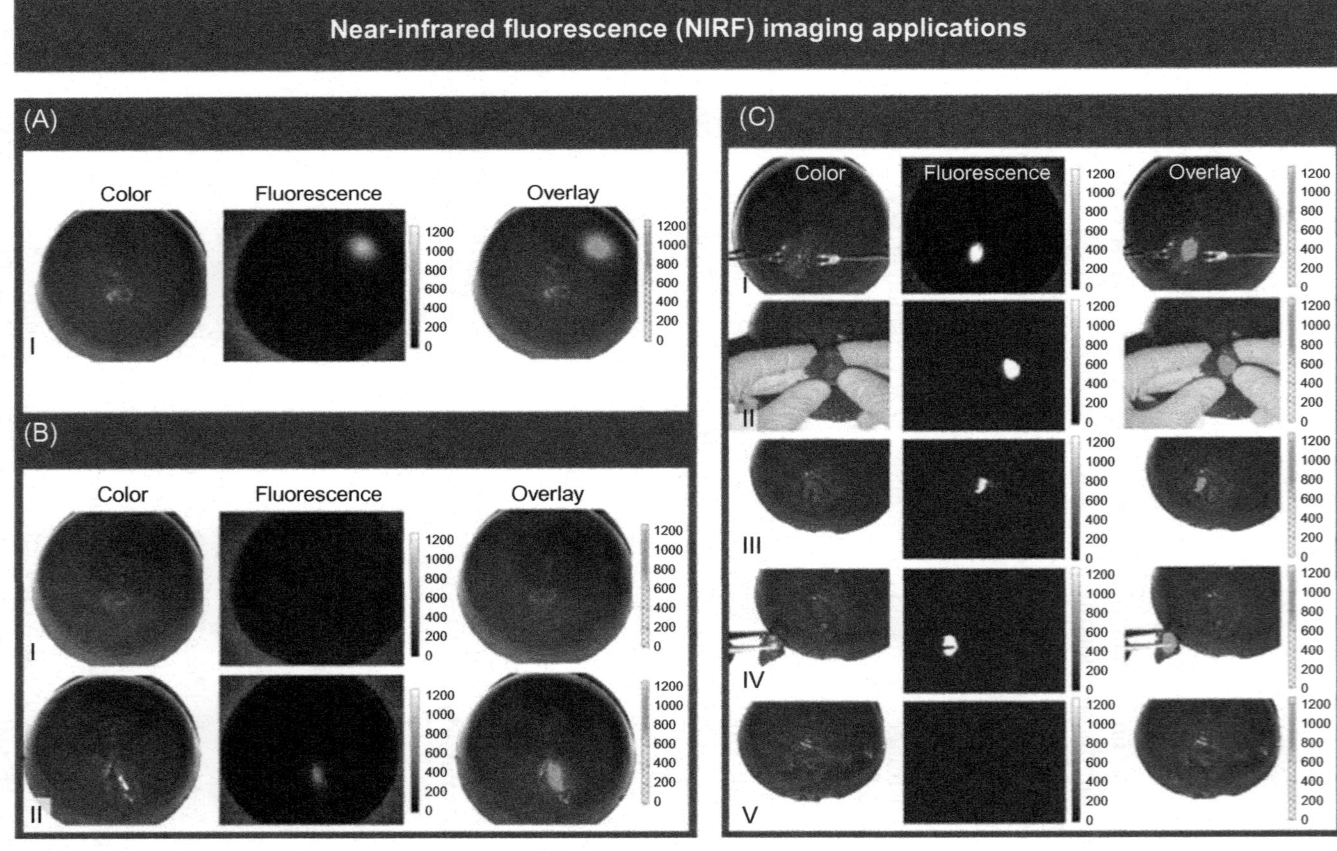

FIG. 10 Overview of NIRF application in breast-conserving surgery: (A) preoperative tumor localization, (B) intraoperative tumor localization, and (C) NIRF-guided surgery.

Lahooti et al. about dual nanosized contrast agents in PET/MRI came to the conclusion that PEGylated nanoparticles radiolabeled with ^{64}Cu or ^{68}Ga and hydrodynamic size of 40 nm showed the best targeting results, when functionalized with peptides or engineered antibodies (Lahooti et al., 2016).

2.6.4 Magnetic Resonance Imaging

MRI is a nonradiative technique that measures relaxation rates of water protons and provides physiological and pathological information, especially about soft tissues with high water content. Two types of MRI contrast agents exist: one that increases the T1 signal in T1-weighted images, which results in a positive/brighter contrast, and one that reduces the T2 signal in T2-weighted image, which results in a negative/dark contrast.

The most popular material studied for T2 contrast agents are superparamagnetic iron oxide nanoparticles. These kinds of particles have been used clinically for diagnosis of liver diseases and for lymph-node imaging, angiography, and blood-pool imaging. Besides clinical applications, iron nanoparticles can be used to study biological processes. A drawback of iron oxide nanoparticles is the lack of protocols for reliable monodisperse production, dispersion in aqueous media, and functionalization for biological targeting.

Popular nanoparticles used as T1 contrast agents for MRT are gadolinium-based. Gd(III) has been incorporated in silica and perfluorocarbon nanoparticles, carbon nanotubes, and nanodiamonds. Thereby, incorporation of Gd(III) into nanoparticles resulted in a 10-fold increase in relaxivity (Hahn et al., 2011).

2.6.5 Ultrasound

Ultrasound imaging relies on the emission and receipt of sound waves with frequency higher than 20 kH. Contrast in ultrasound is generated by reflection and refraction of sound waves that differ by the ability of sound to propagate through media. Since sound travels poorly through gas, several designs of microbubbles have been developed that are used clinically to enhance the echogenicity of the vasculature and organ-specific regions. Gases like air, perfluorocarbons, or nitrogen are encapsulated by surfactants, proteins, and/or polymer shells. Potential for improvements lies in the weak stability of these microbubbles. In fact, most contrast agents used in ultrasound have a size between 150–1000 nm and can hardly be classified as real nanoparticles (Hahn et al., 2011).

2.7 Nanopharmacotherapy

The implementation of nanoparticles as a delivery system for drugs will change the pharmaceutical industry. When drugs are incorporated or attached to nanoparticles, their pharmacokinetics can be custom-designed. Their efficacy can be enhanced, their tolerability can be improved, the bioavailability of water-insoluble drugs can be enabled, large payloads can be delivered, and therapeutics can be protected from physiological barriers or RES.

The highest percentage of nanodrugs approved by the FDA includes lipid- or polymer-based drugs that rely on the EPR effect to deliver drugs into tumor tissues. This is not surprising. The metabolic pathways of lipids are well known and considered as safe, and the same holds for many biodegradable polymers. The risk of adverse side effects is low when there are no targeting molecules attached to the surface. So, the effort for clinical validation is comparably small. Difficulties among targeting with targeting ligands include the specificity of targeting molecules to a certain cell type, the binding affinity, and the purity of the ligand. In addition, a suitable method for conjugation to the nanoparticle that does not alter the size and/or charge of the nanoparticle disadvantageously has to be found. Despite these difficulties, the potential of targeted drug delivery, based on targeting molecules on the surface, should not be underestimated. Drugs like cetuximab, rituximab, trastuzumab, and bevacizumab show that the concept works, and the future aim should be to specifically target other than tumor tissues to avoid adverse side effects and/or reduce the required dose given.

In fact, the whole field of nanopharmacotherapy is huge but goes beyond the scope of this chapter that, however, includes aspects of nanotheranostics. An excellent overview can be found in "Introduction to nanomedicine and nanobioengineering" from Prasad (2012).

2.8 Photothermal Therapy

Temperature is one of the most important parameters that determine viability and dynamics of biological systems. To heat a part of the body above normal temperature for a defined period is referred to as thermal treatment. In the case of humans, this means heating of the body above 37°C. Today, thermal treatment is mainly used to support the effect of drugs in target locations. The underlying effect is despite large interest in the recent years, it is still not clearly understood (Jaque et al., 2014).

The aim of nanoparticle-based photothermal therapy is to provide a method for controlled and localized heating. As mentioned before to achieve effective heating, nanoparticles with large absorption efficiencies and low quantum yields are required, so that the light-to-heat conversion efficiency is large. Metallic nanoparticles, quantum dots, carbon nanotubes, graphene-based nanoparticles, rare earth-doped nanoparticles, and organic nanoparticles have so far fulfilled these criteria and been used for photothermal therapy. According to Jaque et al., gold nanoparticles and carbon nanotubes are the most understood, controlled, and developed nanoparticles for phototherapy at this point of time. An important point that is not addressed yet is that the heat generated should also be monitored in the tissue. As a solution, new nanoparticles for photothermal therapy can be designed, or two different kinds of nanoparticles can be used simultaneously, one for heating and the other for temperature monitoring. Further improvements in the field of photothermal therapy are expected in the fields of lasers that can excite nanoparticles through biological windows.

2.9 Nanoparticles and Theranostics

Theranostics is a portmanteau that consists of therapy and diagnostics. By combining of diagnostic and therapy, several synergetic effects occur. Diseases can be detected and treated in their earliest stage, when the disease is most likely to be curable or treatable. Further, therapy can be controlled directly after treatment, by repeated imaging. Thus, it is also possible to personalize the treatment. As theranostics is multifunctional in nature, multifunctional nanoparticles deliver the perfect platform. Most promising are therapeutic agents that are coupled with an enhanced imaging functionality, like fluorescence, MRI, CT, etc. (Muthu et al., 2014).

For example, Lee et al. (2014) developed doxorubicin-loaded poly(lactic-*co*-glycolic acid)-gold half-shell nanoparticles covered with PEG, antideath receptor-4 monoclonal antibody, and protein G for targeting (DR4-DOX-PLGA-Au H-S nanoparticles; Figs. 11 and 12).

Fig. 12 shows the construction of these particles. Lee et al. showed that these particles were able to enter human colon cancer (DLD-1) and doxorubicin-resistant DLD-1 (DLD-1/DOX) cell lines. They further showed that the doxorubicin-resistant cells were vulnerable to doxorubicin when delivery by nanoparticle was combined with photothermal treatment of the cells (Lee et al., 2014). This is an excellent example of a theranostic system as it combines drug delivery and photothermal functionality (Fig. 13).

Robinson et al. used a theranostic approach that combined Raman imaging and photothermal therapy for tumor treatment in mice. They injected PEG-coated SWCN into mice to image the tumor and treat it photothermally. They achieved complete tumor elimination without any toxic side effects in the following six months (Robinson et al., 2010).

Theranostics did not enter the clinical field yet. A possible reason therefore can be that clinical approval might be difficult and costly. Nanotheranostics is still a young field of research; however, huge progress can be expected in the next years.

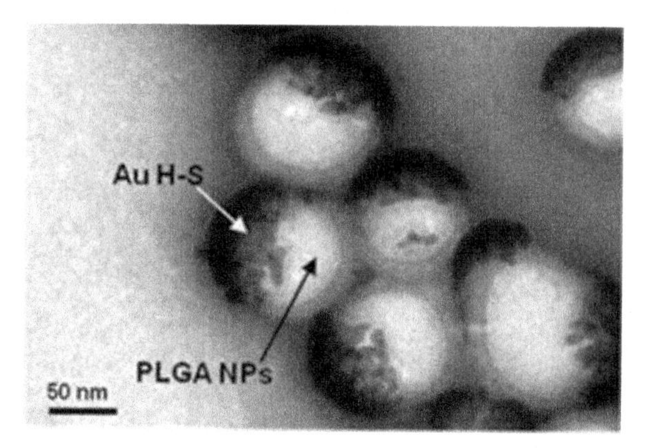

FIG. 11 TEM of DR4-DOX-PLGA-Au H-S nanoparticles (Lee et al., 2014).

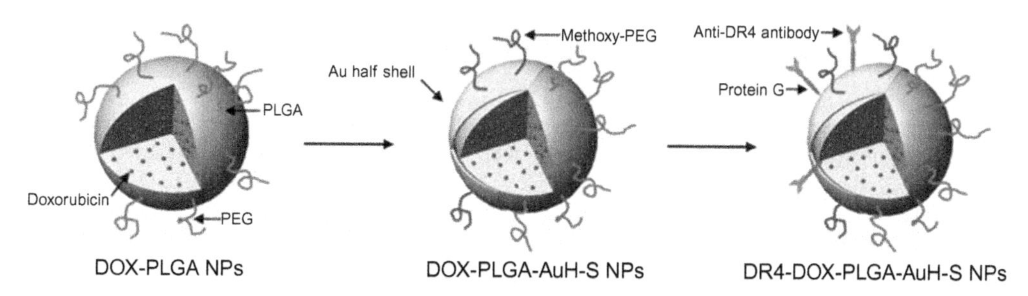

FIG. 12 Schematic diagrams showing the construction of DR4-DOX-PLGA-Au H-S nanoparticles (Lee et al., 2014).

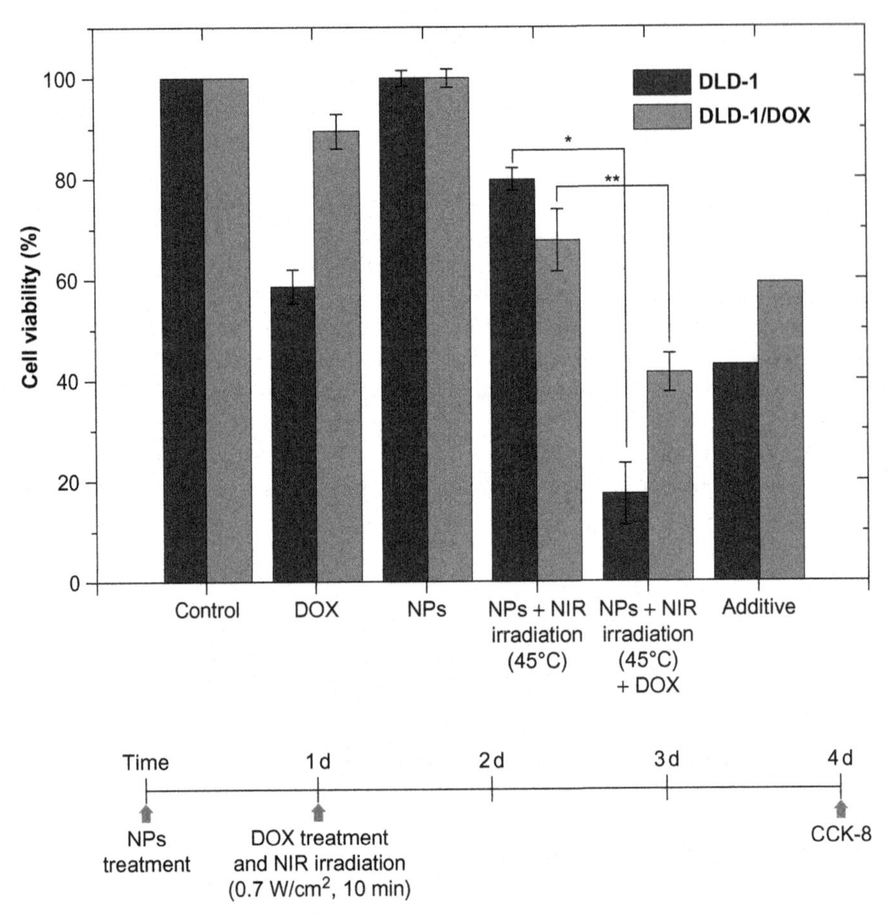

FIG. 13 Viability of DLS-1 and DLD1/DOX cells with chemophotothermal therapy. DLD-1 and DLD/DOX cells were treated with doxorubicin (DOX) only, nanoparticles without (NPs) or with NIR (NPs + NIR irradiation), or nanoparticles with NIR and doxorubicin treatment (NPs + NIR irradiation + DOX).

References

Agasti, S.S., Rana, S., Park, M.-H., Kim, C.K., You, C.-C., Rotello, V.M., 2010. Nanoparticles for detection and diagnosis. Adv. Drug Deliv. Rev. 62 (3), 316–328.

Alexandrou, A., 2017. Reactive oxygen species detection with nanosensor particles: in vitro and in vivo applications. In: Nanomedicine: Nanoparticles in Nanomedicine. Nanotexnology, Thessaloniki.

Beloglazova, N.V., Speranskaya, E.S., Wu, A., Wang, Z., Sanders, M., Goftman, V.V., et al., 2014. Novel multiplex fluorescent immunoassays based on quantum dot nanolabels for mycotoxins determination. Biosens. Bioelectron. 62, 59–65.

Berret, J.F., 2016. Microrheology of viscoelastic solutions studied by magnetic rotational spectroscopy. Int. J. Nanotechnol. 13 (8/9), 597.

Berret, J.-F., 2017. Magnetic colloids for micro-rheology and biophysics applications. In: Nanomedicine Session—Clinical Nanomedicine & Nanotoxicity. Nanotexnology, Thessaloniki.

Boisselier, E., Astruc, D., 2009. Gold nanoparticles in nanomedicine: preparations, imaging, diagnostics, therapies and toxicity. Chem. Soc. Rev. 38 (6), 1759–1782.

Casanova, D., Bouzigues, C., Nguyên, T.-L., Ramodiharilafy, R.O., Bouzhir-Sima, L., Gacoin, T., et al., 2009. Single europium-doped nanoparticles measure temporal pattern of reactive oxygen species production inside cells. Nat. Nanotechnol. 4 (9), 581–585.

Chen, G., Roy, I., Yang, C., Prasad, P.N., 2016. Nanochemistry and nanomedicine for nanoparticle-based diagnostics and therapy. Chem. Rev. 116 (5), 2826–2885.

EMD Millipore Corporation, 2013. Rapid Lateral Flow Test Strips Considerations for Product Development. EMD Millipore Corporation, Burlington, MA.

Gan, S.D., Patel, K.R., 2013. Enzyme immunoassay and enzyme-linked immunosorbent assay. J. Invest. Dermatol. 133 (9), e12.

Goldman, E.R., Clapp, A.R., Anderson, G.P., Uyeda, H.T., Mauro, J.M., Medintz, I.L., et al., 2004. Multiplexed toxin analysis using four colors of quantum dot fluororeagents. Anal. Chem. 76 (3), 684–688.

Hahn, M.A., Singh, A.K., Sharma, P., Brown, S.C., Moudgil, B.M., 2011. Nanoparticles as contrast agents for in-vivo bioimaging: current status and future perspectives. Anal. Bioanal. Chem. 399 (1), 3–27.

Huang, X., Aguilar, Z.P., Xu, H., Lai, W., Xiong, Y., 2016. Membrane-based lateral flow immunochromatographic strip with nanoparticles as reporters for detection: a review. Biosens. Bioelectron. 75, 166–180.

Huang, X., O'Connor, R., Kwizera, E.A., 2017. Gold Nanoparticle Based Platforms for Circulating Cancer Marker Detection. Nanotheranostics 1 (1), 80–102.

Jaque, D., Martínez Maestro, L., del Rosal, B., Haro-Gonzalez, P., Benayas, A., Plaza, J.L., et al., 2014. Nanoparticles for photothermal therapies. Nanoscale 6 (16), 9494–9530.

Kaganman, I., 2006. Quantum dots go with the flow. Nat. Methods 3 (9), 662–663.

Lahooti, A., Sarkar, S., Laurent, S., Shanehsazzadeh, S., 2016. Dual nano-sized contrast agents in PET/MRI: a systematic review. Contrast Media Mol. Imaging 11 (6), 428–447.

Le, T., Zhu, L., Yu, H., 2016. Dual-label quantum dot-based immunoassay for simultaneous determination of carbadox and olaquindox metabolites in animal tissues. Food Chem. 199, 70–74.

Lee, L.G., Nordman, E.S., Johnson, M.D., Oldham, M.F., 2013. A low-cost, high-performance system for fluorescence lateral flow assays. Biosensors (Basel) 3 (4), 360–373.

Lee, S.-M., Kim, H.J., Kim, S.Y., Kwon, M.-K., Kim, S., Cho, A., et al., 2014. Drug-loaded gold plasmonic nanoparticles for treatment of multidrug resistance in cancer. Biomaterials 35 (7), 2272–2282.

Linares, E.M., Kubota, L.T., Michaelis, J., Thalhammer, S., 2012. Enhancement of the detection limit for lateral flow immunoassays: evaluation and comparison of bioconjugates. J. Immunol. Methods 375 (1–2), 264–270.

Lv, Y., Wu, R., Feng, K., Li, J., Mao, Q., Yuan, H., et al., 2017. Highly sensitive and accurate detection of C-reactive protein by CdSe/ZnS quantum dot-based fluorescence-linked immunosorbent assay. J. Nanobiotechnol. 15 (1), 35.

McMeekin TA, editor. Detecting Pathogens in Food. first ed.. publ. Cambridge: Woodhead Publishing; 2003. (Woodhead Publishing in Food Science and Technology.

Muthu, M.S., Leong, D.T., Mei, L., Feng, S.-S., 2014. Nanotheranostics—application and further development of nanomedicine strategies for advanced theranostics. Theranostics 4 (6), 660–677.

Myung, J.H., Tam, K.A., Park, S.-J., Cha, A., Hong, S., 2016. Recent advances in nanotechnology-based detection and separation of circulating tumor cells. Wiley Interdiscip. Rev. Nanomed. Nanobiotechnol. 8 (2), 223–239.

Noguera, P., Posthuma-Trumpie, G.A., van Tuil, M., van der Wal, F.J., de Boer, A., Moers, A.P.H.A., et al., 2011. Carbon nanoparticles in lateral flow methods to detect genes encoding virulence factors of Shiga toxin-producing *Escherichia coli*. Anal. Bioanal. Chem. 399 (2), 831–838.

Oh, S.-D., Duong, H.D., Rhee, J.I., 2015. Simple and sensitive progesterone detection in human serum using a CdSe/ZnS quantum dot-based direct binding assay. Anal. Biochem. 483, 54–61.

Ornberg, R.L., Harper, T.F., Liu, H., 2005. Western blot analysis with quantum dot fluorescence technology: a sensitive and quantitative method for multiplexed proteomics. Nat. Methods 2 (1), 79–81.

Petryayeva, E., Algar, W.R., 2016. A job for quantum dots: use of a smartphone and 3D-printed accessory for all-in-one excitation and imaging of photoluminescence. Anal. Bioanal. Chem. 408 (11), 2913–2925.

Pleijhuis, R.G., Langhout, G.C., Helfrich, W., Themelis, G., Sarantopoulos, A., Crane, L.M.A., et al., 2011. Near-infrared fluorescence (NIRF) imaging in breast-conserving surgery: assessing intraoperative techniques in tissue-simulating breast phantoms. Eur. J. Surg. Oncol. 37 (1), 32–39.

Prasad, P.N., 2012. Introduction to Nanomedicine and Nanobioengineering. Wiley Series in Biomedical Engineering and Multi-Disciplinary Integrated Systems, Wiley, Hoboken, NJ.

Qi, X., Huang, Y., Lin, Z., Xu, L., Yu, H., 2016. Dual-quantum-dots-labeled lateral flow strip rapidly quantifies procalcitonin and C-reactive protein. Nanoscale Res. Lett. 11 (1), 167.

Resch-Genger, U., Grabolle, M., Cavaliere-Jaricot, S., Nitschke, R., Nann, T., 2008. Quantum dots versus organic dyes as fluorescent labels. Nat. Methods 5 (9), 763–775.

Robinson, J.T., Welsher, K., Tabakman, S.M., Sherlock, S.P., Wang, H., Luong, R., et al., 2010. High performance in vivo near-IR (1 μm) imaging and photothermal cancer therapy with carbon nanotubes. Nano Res. 3 (11), 779–793.

Rumenapp, C., Gleich, B., Haase, A., 2012. Magnetic nanoparticles in magnetic resonance imaging and diagnostics. Pharm. Res. 29 (5), 1165–1179.

Salieb-Beugelaar, G.B., Hunziker, P.R., 2014. Towards nano-diagnostics for rapid diagnosis of infectious diseases—current technological state. Eur. J. Nanomedicine 6 (1), 12–14.

Sobczak-Kupiec, A., Venkatesan, J., Alhathal AlAnezi, A., Walczyk, D., Farooqi, A., Malina, D., et al., 2016. Magnetic nanomaterials and sensors for biological detection. Nanomedicine 12 (8), 2459–2473.

Taranova, N.A., Berlina, A.N., Zherdev, A.V., Dzantiev, B.B., 2015. 'Traffic light' immunochromatographic test based on multicolor quantum dots for the simultaneous detection of several antibiotics in milk. Biosens. Bioelectron. 63, 255–261.

Thermo Fisher Scientific, 2017. Simultaneous detection of GLUT4 and GAPDH. Thermo Fisher Scientific, Waltham, MA. July 28, Available from https://www.thermofisher.com/content/dam/LifeTech/migration/images/cell-analysis/data.par.63014.image.500.214.1.dat-western-glut4-gapdh-gif.gif.

Thomas, M.E., 2006. Optical Propagation in Linear Media: Atmospheric Gases and Particles, Solid State Components, and Water. Johns Hopkins University/Applied Physics Laboratory Series in Science and Engineering. Oxford University Press, Oxford, NY. Available from, http://search.ebscohost.com/login.aspx?direct=true&scope=site&db=nlebk&db=nlabk&AN=167544.

Valcárcel Cases, M., López-Lorente, Á.I., 2014. Comprehensive analytical chemistry. In: Gold Nanoparticles in Analytical Chemistry. vol. 66. Elsevier, Amsterdam.

Vijayakumar, S., 2013. Gold Nanoparticles for Medical Applications. Lap Lambert Academic Publishing, Saarbrücken.

Wang, D.-B., Tian, B., Zhang, Z.-P., Deng, J.-Y., Cui, Z.-Q., Yang, R.-F., et al., 2013. Rapid detection of Bacillus anthracis spores using a super-paramagnetic lateral-flow immunological detection system. Biosens. Bioelectron. 42, 661–667.

Xie, Q.-Y., Wu, Y.-H., Xiong, Q.-R., Xu, H.-Y., Xiong, Y.-H., Liu, K., et al., 2014. Advantages of fluorescent microspheres compared with colloidal gold as a label in immunochromatographic lateral flow assays. Biosens. Bioelectron. 54, 262–265.

Xing, Y., Chaudry, Q., Shen, C., Kong, K.Y., Zhau, H.E., Chung, L.W., et al., 2007. Bioconjugated quantum dots for multiplexed and quantitative immunohistochemistry. Nat. Protoc. 2 (5), 1152–1165.

Yao, J., Xing, G., Han, J., Sun, Y., Wang, F., Deng, R., et al., 2017. Novel fluoroimmunoassays for detecting ochratoxin A using CdTe quantum dots. J. Biophotonics 10 (5), 657–663.

Yong, K.-T., Law, W.-C., Hu, R., Ye, L., Liu, L., Swihart, M.T., et al., 2013. Nanotoxicity assessment of quantum dots: from cellular to primate studies. Chem. Soc. Rev. 42 (3), 1236–1250.

Zhang, C., Han, Y., Lin, L., Deng, N., Chen, B., Liu, Y., 2017. Development of quantum dots-labeled antibody fluorescence immunoassays for the detection of morphine. J. Agric. Food Chem. 65 (6), 1290–1295.

CHAPTER

11

Interaction of Nanoparticles With Biomolecules, Protein, Enzymes, and Its Applications

Navneet Phogat[*,†], *Matthias Kohl*[*], *Imran Uddin*[‡], *Afroz Jahan*[§]

[*]Institute of Precision Medicine, Furtwangen University, Villingen-Schwenningen, Germany
[†]University of Tübingen, Tübingen, Germany [‡]Aligarh Muslim University, Aligarh, UP, India
[§]Faculty of Pharmacy, Integral University, Lucknow, India

1 INTRODUCTION

1.1 Nanotechnology

Nanotechnology is a technology, science, and engineering conducted at the nanoscale that ranges from 1 to 100 nm. Nanoscience and nanotechnology are the analysis and use of tremendously minute items that can be used throughout all the other science arenas, such as biology, materials science, physics, chemistry, and engineering. Nanoparticles (NPs) are those particles that ranges with a diameter >100 nm; are gradually used in miscellaneous purposes, involving drug carrier systems; and pass tissue barriers such as the blood-brain barrier.

NPs whose one of the sizes is in the range of 1–100 nm act as a link between molecular or atomic structures and bulk materials (Kaushik et al., 2010). NPs gain amazing and significant possessions owing to their large surface area with limitless dangling bonds, small dimensions, and higher reactivity over their bulk companions (Bogunia-Kubik and Sugisaka, 2002; Daniel and Astruc, 2004; Zharov et al., 2005). Researchers have been well conscious of the ability of biological entities to decrease metal precursors, but the mechanisms are still unaware. The advancement of effectual green synthesis applying natural capping, reducing, and stabilizing agents without the usage of venomous, lavish chemicals and high-energy incorporation has attracted scientists near to biological techniques (Mukherjee et al., 2001; Mohanpuria et al., 2008; Korbekandi et al., 2009; Luangpipat et al., 2011; Dhillon et al., 2012; Arumugam et al., 2015).

https://doi.org/10.1016/B978-0-12-805364-5.00011-1

Rapid industrialization, urbanization, and population explosion are resulting in deterioration of earth atmosphere, and a huge amount of hazardous and unwanted substances are being released. It is now high time to learn about the secrets that are present in the nature and its natural products that lead to advancements in the synthesis processes of NPs. Furthermore, NPs are widely applied to human contact areas, and there is a growing need to develop processes for synthesis that do not use harsh toxic chemicals. Therefore, green/biological synthesis of NPs is a possible alternative to chemical and physical methods.

The first question related with the production of green nanomaterials is "why are biologically synthesized NPs so interesting and gaining importance nowadays?" The unique properties of the NPs synthesized by biological methods are preferred over nanomaterials produced from physicochemical methods. NPs may be synthesized following physicochemical methods (Singh et al., 2015). However, these methods are capital intensive with many problems including the use of toxic solvents, the generation of hazardous by-products, and the imperfection of the surface structure (Li et al., 2011). Chemical methods are generally composed of more than one chemical species or molecules that could increase the particle reactivity and toxicity and might harm human health and the environment due to the composition ambiguity and the lack of predictability (Li et al., 2011).

The particles produced by green synthesis differ from those using physicochemical approaches. Green synthesis, a bottom-up approach, is like chemical reduction where an expensive chemical reducing agent is replaced by extract of a natural product such as leaves of trees/crops or fruits for the synthesis of metal or metal oxide NPs. Biological entities possess a huge potential to produce NPs. Biogenic reduction of metal precursors to corresponding NPs is eco-friendly (Jayaseelana et al., 2012), sustainable (Gopinath et al., 2014), free of chemical contamination (Chandran et al., 2006; Huang et al., 2007), and less expensive (Mittal et al., 2013) and can be used for mass production (Iravani, 2011). Moreover, the biological production of NPs allows the recycle of expensive metal salts like gold and silver contained in waste streams. These metals have limited resources and have fluctuating prices (Wang et al., 2009). We may get green NPs with the desired properties. The biological molecules, mostly proteins, enzymes, sugars, and even whole cells, that stabilize NPs easily allow NPs to interact with other biomolecules and thus increase the antimicrobial activity by improving the interactions with microorganisms (Botes and Cloete, 2010). The biological formation of NPs permits easy separation of the NPs from the reaction media or upconcentration by centrifugation (Sintubin et al., 2011). Biogenic silver NPs when compared with chemically produced NPs showed 20 times higher antimicrobial activity (Sintubin et al., 2011). The choice of plant extracts to produce NPs is based on the added value of the biological material itself. The algal cells of *Spirulina platensis* were chosen because in addition to possessing reducing agent it also exhibits pharmaceutical and nutraceutical properties (Govindaraju et al., 2008).

Unicellular bacteria and extracts of multicellular eukaryotes in the reaction processes reduce metal precursors into NPs of desire shapes and sizes (Kaushik et al., 2010). In addition to this, biological entities possess capping and stabilizing agents required in as growth terminator and for inhibiting aggregation/agglomeration process (Kharissova et al., 2013). The nature of biological entities and its concentrations in combination with organic reducing agents influence the size and shape of NPs (Aromal et al., 2012). Moreover, size and shape of NPs strongly depend on the growth medium parameters such as pH, temperature, salt concentration, and exposure time (Gericke and Pinches, 2006; Dwivedi and Gopal, 2010).

Bioreduction of metal precursors takes place either in vitro or in vivo for the synthesis of nanomaterials. However, enzymes; proteins; sugars; and phytochemicals, like flavonoids, phenolic, and terpenoids cofactors, mainly act as reducing and stabilizing agents (Kaushik et al., 2010; Kharissova et al., 2013).

The in vivo production of NPs has been reported using bacteria, yeast, fungi, algae, and plants (Torresdy et al., 2003; Narayanan and Sakthivel, 2010; Duran et al., 2011; Lloyd et al., 2011; Kharissova et al., 2013). Mostly, biological extracts are used for in vitro synthesis, which involves the purification of bioreducing agents and mixing it into an aqueous solution of the relevant metal precursor in controlled manner. The reaction occurs spontaneously at room temperature (Rajakumara et al., 2012), but sometimes, additional heating and stirring are needed (Sankar et al., 2014). Among the biological entities mentioned above, plants or their extracts seem to be the best agents because they are easily available and suitable for mass production of NPs and their waste products are eco-friendly (Li et al., 2011) unlike some microbial extracts.

Despite a great deal of research in nanotechnology using physicochemical approaches, syntheses of silver (Ag) and gold (Au) NPs are widely exploited using green synthesis. However, a relatively modest number of studies have attempted to elucidate the biosynthesis and potential applications of other metallic and semiconductor NPs. This review presents an overview on biosynthesis of various metallic and semiconductor NPs, namely, Cu, Fe, Pd, Ru, PbS, CdS, CuO, CeO_2, Fe_3O_4, TiO_2, and ZnO NPs with an emphasis on their applications in biotechnology.

2 TYPES OF NPs

2.1 Liposomes

Liposome is composed of concentric bilayered vesicles in which a membranous lipid bilayer totally encircles an aqueous volume chiefly composed of natural or synthetic phospholipids. Liposomes are characterized in terms of size, surface charge, and number of bilayers. Liposome reveals numerous advantages in terms of amphiphilic character, biocompatibility, and ease of surface modification rendering it a suitable candidate delivery system for biotech drugs. Liposomes have been widely used in the field of biology, biochemistry, and medicine. These alter the pharmacokinetic profile of loaded drug particularly in case of proteins and peptides and can be smoothly altered by surface attachment of polyethylene glycol (PEG) units creating it as stealth liposomes and therefore increasing its circulation half-life (Barnard, 2006; Pattni et al., 2015).

2.2 Nanocrystals and Nanosuspension

Nanocrystals are masses of more than thousands of molecules that associate in a crystalline form, combined with pure drug with only a skinny coating covered with surfactant or mixture of surfactants. Nanocrystals can resolve the typical problems of poorly soluble drugs like reduced bioavailability, inappropriate absorption pattern, and problems of preparing the parenteral dosage. This has numerous benefits; dissimilar extent of loading may be low carrier-based NPs. Nanocrystals for steric and electrostatic surface stabilization involve a less number of surfactants. Moreover, if dissolution is sufficiently slow, administration of high drug levels with depot release can be attained (Yeap et al., 2017).

As pure drug is used and no carrier is needed, eliminating potential toxicity issues associated with the carrier molecule. Nanocrystal technology can be utilized for many dosage forms. Nanoparticles proposed the potential for directing the mucosa of the gastrointestinal tract following oral administration and targeting the cells of the mononuclear phagocytic system (MPS) to treat toxicities of the MPS such as leishmaniasis and fungal mycobacterial infections, thus serving as a favorable delivery system for similar drugs, tacrolimus, amphotericin B, etc. The size of nanocrystals permits for nontoxic and efficient passage through capillaries. Potential of nanocrystals can be contingent by the FDA endorsement of Rapamune comprising sirolimus that is an immunosuppressant drug to prevent graft rejection in children after liver transplantation. Emend, which contains aprepitant MK 869, is used in the healing of emesis associated with the cancer chemotherapy (Rabinow, 2004; Junghanns and Müller, 2008).

2.3 Solid Lipid Nanoparticles

At the beginning of the 1990s, solid lipid nanoparticles (SLN) were established as a substitute carrier system to liposomes, emulsions, and polymeric nanoparticles (PNPs) as a colloidal carrier system for controlled drug delivery. The reason behind their development is the combination of benefits from different carrier systems like liposomes and PNPs. SLN have been established and examined for parenteral, pulmonal, and dermal application routes (Mukherjee et al., 2009).

SLN contain a solid lipid matrix, where the drug is usually combined, with an average diameter as low as 1 μm. To evade accumulation and to stabilize the dispersion, dissimilar surfactants are used that have an accepted generally recognized as safe position. High-pressure homogenization also produces NPs, as described for nanosuspensions. SLN have been studied as new transfection agents using cationic lipids for the matrix lipid composition. SLN as cationic for gene transfer can be organized using the similar cationic lipids as for liposomal transfection agents. In comparison with SLN or nanosuspensions, PNPs contain a biodegradable polymer. Biocompatibility is vital for potential application as drug and gene delivery, tissue engineering, and new immunization policies.

Most biodegradable polymers involve synthetic polyesters like polycyanoacrylate or poly(D,L-lactide) and connected polymers like poly(lactic acid) PLA or poly(lactide-*co*- glycolide) to give a few examples. Newest advances also include natural polymers like chitosan, gelatin, and sodium alginate to overwhelm toxicological problems with the synthetic polymers. PNPs suggest a noteworthy enhancement over traditional, oral, and intravenous methods of administration in terms of efficacy and efficiency.

In drug delivery, PNPs have various benefits; they usually increase the stability of several volatile pharmaceutical agents. They can be easily and cheaply fabricated in great amounts by a multitude of methods. Nanospheres are studied as a matrix system in which the matrix is homogeneously dispersed. It should be mentioned that besides these spherical vesicular systems, nanocapsules are known, where a polymeric membrane surrounds the drug in a matrix core. The choice of polymer and the ability to modify drug release from PNPs have prepared them as ideal contenders for delivery of vaccines, contraceptives, and cancer therapy and delivery of targeted antibiotics. Additionally, PNPs can be fused into various issues related to drug delivery, such as tissue engineering, and into drug delivery for genus other than humans (Banik et al., 2016).

2.4 Dendrimers

Dendrimers as unique classes of polymers are extremely branched macromolecules whose size and shape can be accurately controlled. Dendrimers are fabricated from monomers using either convergent or divergent step-growth polymerization. Two representations of polyamidoamine are based on dendrimer. The well-defined structure, monodispersity of size, surface functionalization capability, and stability are properties of dendrimers that make them appealing drug carrier candidates. Drug molecules can be incorporated into dendrimers via either encapsulation or complexation. Dendrimers are being investigated for both drug and gene delivery, as carriers for penicillin, and for use in anticancer therapy. Dendrimers that are used in drug delivery incorporate more than one polymers: polyamidoamine (PAMAM), polyethylenimine (PEI), chitin, melamine, poly(L-glutamic acid) (PG), poly(propyleneimine), and poly(ethylene glycol) (PEG) (Dykes, 2001).

2.5 Silicon-Based Structures

Etching, photolithography, and deposition techniques fabricate silicon-based structures frequently used in the manufacture of microelectromechanical systems and semiconductors. Silicon-based materials generally used for drug delivery are porous silicon and silica or silicon dioxide. Architectures involve solidified nanopores, platinum-containing nanopores, nanoneedles, and porous NPs. The density and diameter of the nanopores can be accurately controlled to achieve a constant drug delivery rate through the pores. Porous hollow silica nanoparticles (PHSNPs) are fabricated in a suspension containing sacrificial nanoscale templates, for example, calcium carbonate. Silica precursors, such as sodium silicate, are added into the suspension, which is then dried and calcinated creating a central of the template material coated with a porous silica shell. The template material is dissolved in a wet carve bath, left behind the porous silica shell. Formation of drug carriers includes the mingling of the PHSNPs with the drug molecule. Consequently, drying the mixture merges the drug particles to the surface of the silica NPs. The porous hollow NPs exhibit an ample more desirable steady release. Numerous examples of treatments that have been examined for use with silicon-based delivery systems comprise porous silicon inserted with platinum as an antitumor agent, silicon nanopores for antibody delivery, calcified porous silicon designed as an artificial growing issue, and porous silica NPs containing antibiotics, enzymes, and DNA (Meng et al., 2016).

2.6 Carbon Nanostructures

Carbon structures are of two types, single-walled nanotubes (SWNTs) and multiwalled nanotubes (MWNTs), and C60 fullerenes are common configurations. The size, geometry, and surface characteristics of these structures make them appealing for drug carrier usage. SWNTs and C60 fullerenes have diameters of 1 nm, around half the diameter of the average DNA helix. MWNTs have diameters ranging from several nanometers to tens of nanometers dependent on the number of walls in the structure. Fullerenes and carbon nanotubes (CNTs) are characteristically fabricated using laser ablation, electric arc discharge, chemical vapor deposition, or combustion processes (Avouris et al., 2008; Eatemadi et al., 2014).

Surface-functionalized CNTs can be affected inside mammalian cells and when connected to peptides could be used as vaccine delivery structures. With the use of molecular dynamics (MD) simulations, the flow of water molecules through CNTs has been modeled and implies their potential use as minor molecule transporters. Other simulations have convoluted the transport of DNA through CNTs, indicating likely use as a gene delivery tool. For example, temperature-stabilized hydrogels for drug delivery uses include CNTs. Fullerenes have also shown drug-targeting ability. Furthermore, experiments with fullerenes have also displayed that they demonstrate antioxidant and antimicrobial behavior (Sun et al., 2002).

2.7 Metal Nanostructures

Hollow metal nanoshells are examined for drug delivery applications. Characteristic fabrication methods include templating of the thin metal shell about a core material, for example, a silica NP. Typical metals comprise silver, gold, platinum, and palladium. When linked to or embedded inside polymeric drug carriers, metal NPs might be used as thermal release initiates when exposed with infrared light or excited by an alternating magnetic field. Biomolecular conjugation methods of metals comprise bifunctional linkages, lipophilic interaction, salinization, electrostatic interaction, magnetism, and nanobead interactions (Cedervall et al., 2007).

3 SURFACE CHARGE OF NP

NP surface charge is important reason in protein interaction. It has been reported that the protein adsorption increases by increasing the surface charge of NPs. Positively charged NPs appeal to adsorb proteins with isoelectric points (p*I*) <5.5, for example, albumin, while the negative surface charge enhances the adsorption of proteins with $pI > 5.5$, for example, IgG (R. F. Service, 2006). Gessner et al. (Borm and Kreyling, 2004) while using negatively charged PNPs observed that while increasing surface charge density, increase in plasma protein adsorption is observed. Further analyses from the similar group with polystyrene NPs reveal that positively charged particles predominantly adsorb proteins with $pI < 5.5$, such as albumin, whereas negatively charged particles such as IgG adsorb proteins with $pI > 5.5$.

Bradley et al. (Jahanshahi and Babaei, 2008) described binding of complement (C1q) to anionic liposomes. It has been reported that significant plasma protein binds to vesicles containing cationic lipids (Rawat et al., 2006). This arises from electrostatic interactions among the cationic lipids and the negatively charged plasma proteins. Surface charge denatures the adsorbed proteins. In a current analysis on the gold NPs with positive, negative, and neutral ligands, it was discovered that proteins denature in the existence of charged ligands, either positive or negative, although the neutral ligands retain the normal structure of proteins (Health and Safety Executive, 2005).

4 NP MATERIAL

The analysis of the plasma proteins bound to single-walled carbon nanotubes (SWCNT) and nanosized silica specified different patterns of adsorption. Serum albumin was found to be the most abundant protein coated on SWCNT but not on silica NP. TiO_2, SiO_2, and ZnO NP of similar surface charge bind to different plasma proteins (Saptarshi et al., 2013). After intravenous administration, blood is the first physiological atmosphere, which a nanomaterial "perceives." Blood plasma comprises several 1000 diverse proteins with 12 orders-of- magnitude differences in the concentration of these proteins (Cohignac et al., 2014). In addition to the proteins, lipids are also present in blood plasma. Thus, upon vaccination of NPs inside the blood, a competition occurs between dissimilar biological molecules to adsorb on the surface of the NPs. In the initial stage, most abundant proteins are adsorbed on the surface, though, over the time, NPs would be replaced by higher-affinity proteins (Vroman's effect) (Hirsh et al., 2013; Vroman et al., 1980).

The organization and composition of the protein corona depends on the physicochemical properties of the nanomaterial (size, configuration, shape, surface functional groups, and surface charges), the type of the physiological environment (blood, interstitial fluid, cell cytoplasm, etc.), and the duration of exposure. Protein corona alters the size and interfacial composition of a nanomaterial, giving it a new biological uniqueness, that is, what is seen by cells. The biological uniqueness defines the physiological response comprising agglomeration, cellular uptake, circulation lifetime, signaling, kinetics, transport, accumulation, and toxicity.

5 NP AND PROTEIN CORONA

Protein corona is complicated and "universal" plasma protein corona for all nanomaterials and that the relative densities of the adsorbed proteins do not associate with their relative abundances in plasma. Thus, the structure of the protein corona depends on many parameters and is unique to each nanomaterial.

5.1 Structure and Composition of Corona

Recently, some minor traces of lipids have also been reported, and most adsorbed biomolecules on the surface of NPs in blood plasma are proteins. NP is governed by protein-NP binding affinities and protein interactions by the adsorption of proteins on the surface of NP. Proteins that adsorb with high affinity form the "hard" corona, containing of compactly bound proteins that do not readily desorb, and proteins that adsorb with low affinity form the "soft" corona, consisting of loosely bound proteins. On the bases on their exchange times, soft and hard corona can be defined. Hard corona normally shows much larger exchange times in the order of several hours (Monopoli et al., 2011).

A theory is that the hard corona proteins interrelate directly with the nanomaterial surface while the soft corona proteins interact with the hard corona via weak protein-protein interactions (Walkey et al., 2012). Commonly observed that even at low plasma concentrations, there is a complete surface coverage of corona layer (Monopoli et al., 2011). Still, the adsorbed corona does not completely mask the surface of NP or its functional groups. In a report on dextran-coated superparamagnetic iron oxide nanoparticles (SPIONs), the formation of the protein corona and the incubation of SPIONs in plasma did not significantly alter the circulation life span (Simberg et al., 2009).

The thickness of protein corona can be a factor of many parameters such as protein concentration, particle size, and surface properties of particle. The coronas on these NPs are extremely thick to be combined of only a single layer of adsorbed protein and are composed of multiple layers because of the plasma protein hydrodynamic diameter of round 3–15 nm. Simberg et al. proposed a model for the protein corona; it consists of "primary binders" that identify the nanomaterial surface directly and "secondary binders" that associate with the primary binders via protein-protein interactions. This multilayered structure is significant for the physiological response as the secondary binders may change the activity of the primary binders or "mask" them, preventing their interaction with the surrounding environment (Simberg et al., 2009).

In a recent analysis, Walkey and Chan precise a subset of 125 plasma proteins, called adsorbome, that were identified in protein corona of at least one nanomaterial. This list will maybe increase due to further studies in the future. Since about 20 years ago, results assembled over many studies showed that a "standard" plasma protein corona is composed of nearly 2–6 proteins adsorbed with high and low abundance (Walkey and Chan, 2012).

5.2 Probing the Interactions of Proteins and NPs

Cedervall et al. have acquired a significant step in this matter of PNAS on the development of approaches for searching the association of proteins to NPs: such association is almost forever an initial step of NPs to enter in a biological fluid. When we consider the interactions of NPs with a living system, protein-coated particles, adsorption of the proteins onto the particle surface can lead to exposure of unique "cryptic" peptide epitopes, transformed conformation, and disturbed function that occurs due to structural effects (local high concentration) or avidity effects that result from the adjacent spatial repetition of the same protein (Cedervall et al., 2007). Cedervall et al. focused on the specific binding rates and affinities of different plasma-related proteins to NPs in which a growth in this field is always appreciated. Proteins compete for the NP surface, leading to an adsorbed protein layer or "corona" that largely defines the biological uniqueness of the particle. Considering these effects, we can conclude that NPs have a very large surface-to-volume ratio, so that large surface areas are available for protein binding for even small particles.

The importance of protein-NP interactions in nanomedicine and nanotoxicity has begun to appear recently with the development of the idea of the NP-protein "corona." This dynamic layer of proteins (and other biomolecules) adsorbs to NP surfaces immediately upon contact with organic systems. Therefore, in highly selective protein adsorption, particles can reach subcellular sites, which results in significant novel potential effects for NPs on protein interactions and cellular behavior (Cedervall et al., 2007; Gessner et al., 2000; Karmali and Simberg, 2011).

Within the therapeutic device community, it is now well putative that the adsorption of biomolecules transforms material surfaces such as proteins in a biological environment (Gessner et al., 2000). Recently studied protein interactions with planar surfaces drew attention to the fact that distortion of the protein might appear upon adsorption (Lundqvist et al., 2011).

Conversely, the significance of the adsorbed protein layer in mediating interactions with organic organisms has been time-consuming to develop in the case of NP-protein interactions.

Recent studies proposed that nanomaterial surfaces, which have larger surface area than flat ones, are more agreeable to determine adsorbed protein identity and residence times (Jansch et al., 2012). Recent reports proposed the concept of the "nanoparticle-protein corona" as the developing collection of proteins that associate with NPs in biological fluids, that is, the "biologically relevant entity" interacts with cells (Mahmoudi et al., 2011).

6 NPs HAVE A VERY LARGE SURFACE-TO-VOLUME RATIO

Cedervall et al. made a significant observation that large variation in dissociation rates for proteins on NPs suggests that diverse sets of proteins could be identified as part of the particle-protein corona depending on the experimental procedures and times. Likewise, the outcome of identification experiments will be influenced by the concentrations of particles and biological fluid. The total protein concentration in bodily fluids, particularly in intracellular environments, can be equal to 35% (0.35 g/mL). Therefore, signifying thousand different proteins spans an expansive range of concentrations. So, in a typical biological environment, there will be competition among the proteins for the existing NP surface area.

Human serum albumin (HSA) and fibrinogen may dominate on the particle surface for short interval of time but will consequently be transferred by minor abundance proteins with higher affinity and slower kinetics. The uniqueness and importance of the work by Cedervall et al. is that it inventively adapts familiar approaches to examine the issue of NP-protein interactions; the resulting coating of the particles by a protein layer is called corona. This issue will be increasing the significance to the fields of nanomedicine, nanotoxicology that identifies the potential for toxic interactions between living tissues, and submicrometer- or nanometer-scale objects in a way that might differ qualitatively from more recognizable, larger-scale particles with which living organisms have evolved (Cedervall et al., 2007).

It is now well-known within the medical device community that the adsorption of biomolecules alters material surfaces such as proteins in a biological atmosphere (Monopoli et al., 2011). An early study revealed that distortion of the protein might happen upon adsorption when protein interactions with planar surfaces occur (Lundqvist et al., 2011). Contrariwise, the significance of the adsorbed protein layer in mediating interactions with living systems is arduous to appear in these NP-protein interactions, while studies of protein adsorption to NPs are beginning to appear (Rocker et al., 2009; Dobrovolskaia et al., 2009; Slack and Horbett, 1995). In the nanotoxicology literature, considering the significance of the detailed structure of the outer surface of the adsorbed protein layer, any changes in protein structure (protein-solution interface) have not yet been broadly appreciated, even though this is the primary surface in contact with cells (Monopoli et al., 2011; Cedervall et al., 2007).

Obviously, a change occurs from a flat surface to particles. The particles become smaller (finally approaching the size of the proteins themselves); the composition and organization of the associated protein will change from the simple in case of flat surfaces. We might expect this to lead to quite different biological consequences. There is the potential that very small NPs with highly curved surfaces can suppress protein adsorption to the point where it no longer appears. An effect like being selective to larger proteins proposes a route to differential control of protein adsorption (Goppert and Muller, 2005). In addition, flat surfaces can only affect biological process via integrins (cell surface receptors), whereas NPs can enter cells and access a vast range of extra biological processes. The recent study evaluates that nanomaterial surfaces having larger surface area than flat ones are more amenable to study to conclude residence times of adsorbed proteins and its identity (Walkey and Chan, 2012; Jansch et al., 2012).

7 EFFECTS ON PROTEIN CONFORMATION OF BINDING TO NPs

The chains of amino acids are called proteins. Protein shapes, structure, and function can be determined by the exact arrangement of the amino acids. The principle units of protein secondary structure are α-helices and β-sheets, and the three dimensional arrangement of these is the tertiary structure (structures are shown in Fig. 1). The native conformation of a protein is tightly controlled by the shape complementarity of the hydrophobic residues that allow close packing of the cores (De et al., 2007; Lynch and Dawson, 2008). Proteins are yet marginally stable because the valuable interactions that rule the native structure are compensated by a large entropy loss associated with accepted from a large ensemble of states to a more restricted set of conformations, in addition to the repulsive electrostatic interactions present in the native stats (Lynch and Dawson, 2008; Maiorano et al., 2010). Thus, interaction with a surface can simply interrupt the native conformation and, therefore, the protein function. This has consequences for the biological influence of NPs.

For many years, the effect of the surface chemistry of biomaterials on the protein adsorption procedure has been a topic of great interest and significance (Lynch and Dawson, 2008; Petersdorf, 1974). Protein adsorption to numerous materials has been widely reviewed. It has been found that features such as electrostatic interactions, hydrophobic interactions, and specific chemical interactions between the protein and the adsorbent play vital functions. Selective adsorption of proteins on numerous synthetic adsorbents has been analyzed under different conditions, for example, solution pH and protein concentration, and for numerous

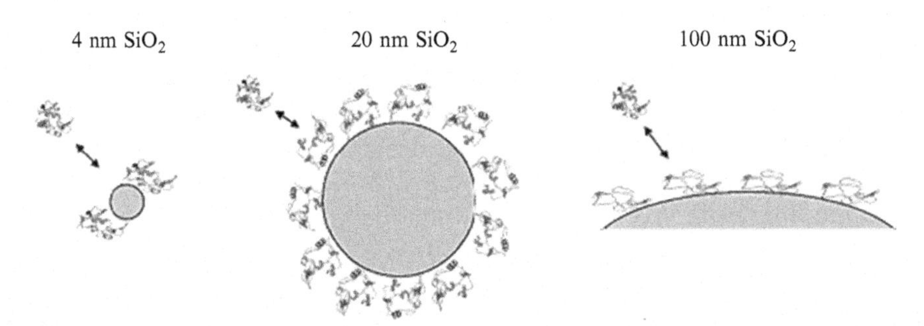

FIG. 1 Schematic of lysozyme adsorption on silica particles with different sizes. Stronger protein particle interactions exist in the case of larger nanoparticles, resulting in more protein unfolding and less enzymatic activity (Vertegel et al., 2004).

proteins, the mechanism of selective adsorption has been recognized to electrostatic interactions (Lynch and Dawson, 2008; Rocker et al., 2009).

8 INCREASED PROTEIN STABILITY/ACTIVITY UPON BINDING TO NPs (ENZYMES)

In common, the loss of secondary structure and consequential changes in the activity of proteins upon binding to NPs can be realized as a disadvantage or a possible source of NP injuriousness; it has a potential positive consequence too. Promising applications of NPs comprise increasing protein stability regarding enzyme degradation, and the activity of enzymes is enhancing via immobilization at surfaces (Lynch and Dawson, 2008; Bogunia-Kubik and Sugisaka, 2002).

Enzymes, for example, *Pseudomonas cepacia* Lipase (PCL) and *Candida rugosa* Lipase (CRL), had been adsorbed to nanostructured polystyrene (PS) and polymethyl methacrylate (PMMA) by simple addition of the lipase mixture to the PNPs below protein-friendly situations (pH 7.6). Adsorption leads to enhanced functioning in terms of activity and selectivity with respect to that exposed by lipases adsorbed on the similar nonnanostructured carriers, in addition to improved enantioselectivity and pH and thermal stability.

For example, the highly curved surface of C60 fullerenes has exposed to enhance enzyme stability in intensely denaturing environments to a higher extent than flat supports. The half-life of a model enzyme soybean peroxidase, adsorbed on fullerenes at 95°C, is ~ 2.5-fold higher in comparison with the enzyme that is adsorbed onto graphite flakes and ~ 13-fold higher than that of the natural enzyme. Similar explanations were found with extra nanoscale supports involving Au NPs and silica. The capability to improve protein stability by interfacing them with nanomaterials might influence several arenas ranging from the plan of diagnostics, sensors, and nanocomposites to drug delivery (Lynch and Dawson, 2008; Daniel and Astruc, 2004; Zharov et al., 2005).

9 PLANT MEDIATED SYNTHESIS OF METAL NPs

Plant-mediated synthesis of metal NP attains prominence because of its austerity, rapid rate of synthesis of NPs of miscellaneous morphologies, and removal of intricate continuance of cell cultures and eco-friendliness (Shiv Shankar et al., 2003). The mechanism for synthesis of NPs in standard residues is the same for both microorganisms and plants. Metal salts including metal ion are firstly reduced to atoms by the help of a reducing agent. The attained atoms then nucleate in minor clusters that flourish into particles. Shankar had analyzed the presence of proteins and secondary metabolites in the water-soluble fractions of geranium leaves and assumed that the terpenoids contribute to the decrease of silver ions and oxidized to carbonyl groups. Fourier-transform infrared spectroscopy (FTIR) analysis of the study proposed that ester C=O group of chlorophyll represents as a reducing agent, whereas protein involved in the outward capping of gold NPs is synthesized using geranium leaf extract (Phogat et al., 2016; Shankar et al., 2004).

10 INFLUENCE OF THE SIZE OF NPs ON THE ACTIVITY OF ADSORBED ENZYMES

The NP size is the vital parameter that affects the interaction between protein and NPs. Silica NPs with different diameters (4, 20, and 100 nm) were used to examine the size impact on the structure and enzymatic activity of adsorbed lysozyme, whose dimensions are equivalent in size to the 4 nm NPs (Treuel and Nienhaus, 2012; Vertegel et al., 2004). Both adsorption patterns and protein structure functions are reliant on the size of the NPs. The development of molecular complexes was examined for adsorption on 4 nm silica (Shang et al., 2007).

The circular dichroism (CD) results verified that the loss in α-helix content is intensely dependent on the size of the NPs. The great loss of α-helicity was examined for the lysozyme adsorbed on larger NPs. The lysozyme activity adsorbed on silica NPs is minor as compared with free protein. The fraction of movement loss associates well with the reduction in α-helix content. Consequently, the outcomes indicate that the size of the NPs, perhaps donations of surface curvature, influences adsorbed protein structure and function (Fig. 1).

The similar phenomenon was spotted when ribonuclease A (Shang et al., 2007) and human carbonic anhydrase (Lundqvist et al., 2011) remained absorbed onto silica NP surfaces with diverse sizes. Urea denaturation analyses showed that the thermodynamic stability of ribonuclease A was decreased upon adsorption onto the NPs, with greater decrease on larger NPs. Therefore, the larger NPs tend to cause unfolding of adsorbed proteins (Shang et al., 2007).

11 NP-APTAMER BIOCONJUGATES FOR CANCER TARGETING

The recent advances in the nanotechnology regarding NP-drug conjugates or NP-biomolecule conjugates, where a biomolecule can be aptamer or protein, are emerging as potential tools for treatment of diseases like cancer (particularly the primary-stage tumor). These advancements are of topical interest in health sector. But still, these advancements need improvement, where the major concern is the uptake of these therapeutic NPs. These therapeutic NPs should be absorbed (uptake) by cell or tissue in a disease-specific manner. Regarding the implementation of these therapeutic NPs, particularly NP-aptamer bioconjugate, there are four major points to be considered: (i) the NPs should be approved by the Food and Drug Administration (FDA) for a clinical use. (ii) Aptamers have an upper hand in comparison with other biomolecules, because they are nonimmunogenic and have remarkable stability in wide pH range (4–9). These aptamers can be used for surface functionalization of the encapsulated NP (NP-drug) for the recognition of the domain of the antigen. The antigen should be a well-characterized one, expressing on the cells to be targeted. (iii) The NPs should have resistibility to macrophages and nontargeted cells, because this will enhance their duration at the localized site of delivery. (iv) Encapsulation of Np with appropriate drug, for example, docetaxel (Dtxl), when targeted to prostate cancer (PCa) in a slow-release manner and then Dtxl has enhanced cytotoxicity and antitumor activity.

Farokhzad et al. synthesized the Dtxl-encapsulated NP-aptamer bioconjugates (Dtxl-NP-Apt) for targeting the Pca. First, they encapsulated the NP with Dtxl. Then, these encapsulated Dtxl-NP-Apt were formulated with biocompatible and biodegradable poly(D,L-lactic-*co*-glycolic

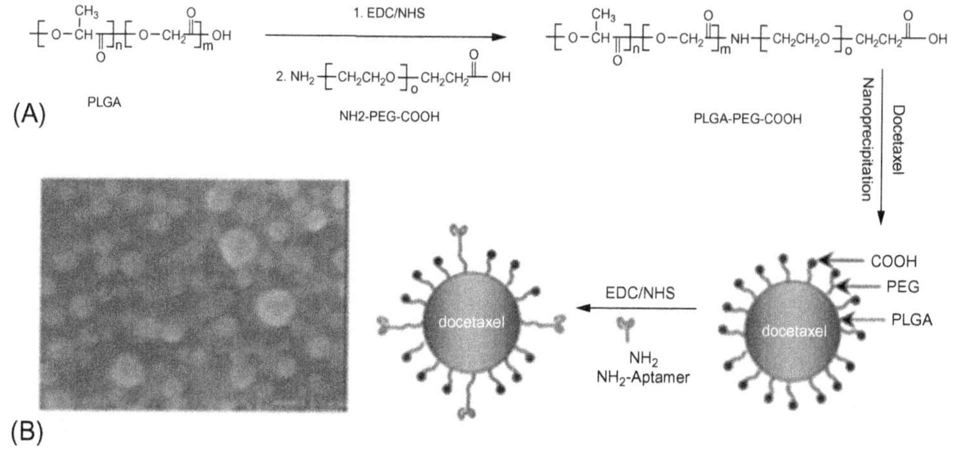

FIG. 2 The development of Dtxl-encapsulated pegylated PLGA NP-Apt bioconjugates. (A) Schematic representation of the synthesis of PLGA-PEG-COOH copolymer and strategy of encapsulation of Dtxl. We developed Dtxl-encapsulated, pegylated NPs by the nanoprecipitation method. These particles have a negative surface charge attributable to the carboxylic acid on the terminal end of the PEG. The NPs were conjugated to amine-functionalized A10 PSMA Apt by carbodiimide coupling chemistry. (B) Representative scanning electron microscopy image of resulting Dtxl-encapsulated NPs is shown. EDC, 1-ethyl-3-(3-dimethylaminopropyl)-carbodiimide; NHS, *N*-hydroxysuccinimide (Farokhzad et al., 2006).

acid)-*block*-poly(ethylene glycol) (PLGA-*b*-PEG) copolymer to make them biocompatible, and later on, these were surface functionalized with A 10 2′-fluoropyrimidine RNA aptamers that recognize the extracellular domain of the prostate-specific membrane antigen (PSMA). PSMA is a well-characterized antigen on the surface of LNCaP prostate epithelial cells. The schematic representation of synthesis of Dtxl-NP-Apt is shown in Fig. 2 (Farokhzad et al., 2006).

The results show that the in vitro toxicity of the aptamer-conjugated NPs was significantly enhanced in comparison with the nontargeted NPs (without aptamer). The significant *P*-value ($P<.0004$) was reported in this case (Farokhzad et al., 2006).

12 ANALYTICAL METHODS TO STUDY PROTEIN-NP INTERACTIONS

Many methods are widely used to study the NP-protein interactions, which includes spectroscopy methods, X-ray, centrifugation, differential centrifugation sedimentation (DCS), isothermal titration calorimetry (ITC), capillary electrophoresis, and mass spectrometry (MS).

12.1 Spectroscopy Methods

12.1.1 UV/Vis

The absorption spectra of NP or proteins change on binding of protein to the NP. This change in absorption spectra can be used to evaluate the binding specificities. The shift and broadening of the absorption spectra of NP-protein complex depend on the three main factors, including NP size, aggregation state, and dielectric constant of medium. The advantages of UV/Vis spectroscopy are that it is a fast, flexible, and straightforward method. The drawback of UV/Vis spectroscopy is that it is nonconclusive; also, the quantitative results are difficult to achieve. That's why it is always necessary to implement the other complementary spectroscopy and structural investigation methods along with UV/Vis spectroscopy (Shang et al., 2007; Mahmoudi et al., 2011; Kharazian et al., 2016).

12.1.2 Fluorescence

In fluorescence spectroscopy, the electrons are excited from ground state (low energy level) to excited state (higher energy level) and then relaxed back to the ground state. This phenomenon can be radiative or nonradiative recombination. In radiative, the photons are emitted during relaxation of the excited electrons to the ground state, which is called as fluorescence. In nonradiative, the phonons are created. For example, amino acids, tryptophan, tyrosine, and phenylalanine fall under the category of radiative recombination. Now, the point is that in NP-protein interaction, either one, both, or none can be radiative. This happens due to the labeling, which can change the structure, conformation, or binding affinity of the protein in NP-protein interaction. This is also one of the most challenging aspects, not only for NP-protein interaction but also for the interaction of the NP with other biomolecules. Binding of the NP with protein or other biomolecule can be studied by steady-state or time-resolved fluorescence spectroscopy. For example, in a study, Carlheinz Röcker et al. studied the human albumin adsorbed onto small (10–20nm) polymer-coated FePt and CdSe/ZnS NPs, through fluorescence correlation spectroscopy and time-resolved fluorescence quenching experiments, where they found that protein resides on the particle for 100s (Rahman et al., 2013; You et al., 2007; Röcker et al., 2009; Kharazian et al., 2016).

12.1.3 FTIR and Raman Spectroscopy

Raman and FTIR mainly reveals two types of information: (i) surface property of the NP-protein (NP-biomolecule) complex and (ii) detection of the protein binding of the protein (biomolecule) on the surface of the NP. FTIR is implemented more in case of the secondary structure of proteins. In comparison with FTIR, Raman spectroscopy has two main advantages: (i) It can measure the Np-biomolecule (protein) complex in aqueous solution too; (ii)it involves the concept of the localized vibrations of the electron-rich groups and double or triple bonds, due to which it produces the strong Raman bands than single-bond or electron-poor groups. That's why Raman spectroscopy has greater spectral simplicity than IR (Rahman et al., 2013; Kharazian et al., 2016).

12.1.4 Circular Dichroism

CD is implemented to study the conformational changes of the protein (biomolecule), which happens during the NP-protein (NP-biomolecule) interaction. Particularly, regarding proteins, CD is more useful regarding secondary structure of proteins, because the α-helix and β-pleated sheets have their own CD in the UV region. CD has its own advantage and limitations, which are as follows: (i) It can detect up to 50 α-helices, but no information about the amino acid residue involved in binding; (ii) it can detect secondary structure in the UV region; (iii) it can detect the tertiary structure in the near-UV region; and (iv) it cannot detect the complex protein mixtures, but can provide the useful information about the structural changes (Rahman et al., 2013).

12.1.5 *Centrifugation*

Centrifugation process doesn't provide that much information about the structural and conformational changes. It is basically a method that is mainly implemented for washing to remove the excess proteins. This method has its own limitations: (i) Volume during washing can affect the outcome; (ii) protein should be removed during washing, but if they are present in abundance, they can be identified as false; and (iii) also, there can be false identification of the large proteins that settle to the bottom. Although because of its own limitation, the method cannot be ignored, because this method is quite easy and straightforward to separate the large and small molecules, with less volume (little material). Also, the method is quite successful in case of large molecules, for example, major proteins like immunoglobulins and fibrinogens (Aggarwal et al., 2009; Kharazian et al., 2016).

12.1.6 *Differential Centrifugation Sedimentation*

DCS is used to measure the NP size distribution, where the NP size ranges from 2 nm to 80 μm. Major advantages of DCS are as follows: (i) ultrahigh resolution capability, (ii) measure peaks of different sizes, and (iii) also measure of small additional peaks, which can be compared with SEM. The study of Zeljkakrptic et al. on gold NP functionalized with organic ligands (peptides and PEG thiol) shows that DCS can be used to measure even the small thickness variation of as small as 0.1 nm on particle. The study of another group, Angela Jedlovszky et al., reveals that DCS is not capable to provide the quantitative structural data of protein-SPION complexes, but it can provide the valuable qualitative information (Jedlovszky-Hajdu et al., 2012). The study of Waczyk et al. interprets that DCS can measure the size distribution in semiqualitative way in a protein complex mixture (Walczyk et al., 2010; Kharazian et al., 2016).

12.1.7 *Isothermal Titration Calorimetry*

ITC is used to study the affinity of protein binding. The study of Lindman et al. shows that ITC is a straightforward technique to study the association reactions. The study of Mrinmoy et al. shows that ITC can be used to provide the free energy and enthalpy of association. This study was carried out on functionalized gold NPs to study the gold NP-histone and gold NP-cytochrome interactions (Lindman et al., 2007; De et al., 2007).

12.1.8 *Mass Spectrometry*

These days, MS is an analytic tool to identify the protein coronas, by applying a gel-based methodology, where the proteolytic enzymes are used to digest the protein into smaller peptides. Then, these smaller peptides are ionized and separated based on mass-to-charge ratio. Stefan Tenzer et al. studied corona (blood plasma-silica NPs), where silica NPs were of different sizes (20, 30, and 100 nm) using MS and 1D and 2D gel electrophoresis. Their study reveals that protein adsorption on NPs does not correlate with protein size or charge (Sparkman, 2001; Tenzer et al., 2011).

12.1.9 *NMR*

Hellstrand et al. (2009) showed by size-exclusion chromatography and nuclear magnetic resonance (NMR) that copolymer NPs bind lipophilic molecules like cholesterol, triglycerides, or phospholipids from human plasma. The lipid- and protein-binding patterns correspond closely with the composition of high-density lipoprotein (HDL). Apolipoproteins have been identified as binding to many other NPs, suggesting that lipid and lipoprotein binding is a general feature of NPs under physiological conditions.

Stayton et al. (2003) have studied by solid-state NMR technique in situ secondary structure determination of statherin peptides on hydroxyapatite (HAP) surfaces. The molecular insight provided by these studies has also led to the design of biomimetic fusion peptides that utilize nature's crystal-recognition mechanism to display accessible and dynamic bioactive sequences from the HAP surface.

13 SIMULATION METHODS TO STUDY NP-PROTEIN INTERACTIONS

Simulation methods exhibit some drawbacks: (i) They cannot distinguish individual proteins, because they measure the average of adsorbed proteins; (ii) simulation methods are ex situ methods, so the measurements do not express the in vivo measurements. This might be possible that results can be near about, but not exactly the same to in vivo. Advantages of simulation methods are (i) determining the time-dependent behavior of atoms and molecules and (ii) determining the conformation and configurational changes. Some techniques like X-ray crystallography are helpful for the simulation methods to determine the protein structures. Simulation methods can be divided based on methodology in two parts: (i) molecular simulation and (ii) algorithm. Based on molecular simulation, they have three categories: (i) ab initio quantum mechanical (QM) methods, (ii) all-atom empirical force-field methods, and (iii) coarse grained (united atoms). Based on the algorithm, they have three categories: (i) molecular mechanics (MM), (ii) Monte Carlo (MC), and (iii) molecular dynamics calculations (MDC) (Mahmoudi et al., 2011).

QM methods are based on the Schrödinger wave equation. These methods involve only a few tens of atoms, while for protein corona problems, they need tens of thousands of atoms; that's why they are not suitable for protein corona problems. The other drawback is that they are quite expensive.

These methods involve tens of thousands of atoms, so, they are applicable for protein corona problems. The benefits of the all-atom empirical force-field methods are the following: (i) It can determine the behaviour of peptide-surface interactions and protein-surface interactions, (ii) it can include the effect of solvation and ions in solvent, and (iii) it is less expensive, saving the computational cost (Latour, 2008; Kharazian et al., 2016).

13.1 Au(111)-His (and His-Derived Peptide) Interactions

The simulation study of Zhen Xu et al., on Au(111)-His and Au(111)-His-derived peptides, shows that amino acids adsorb on Au(111) after 3ns and Au(111) easily adsorb the Gly-His peptide of four amino acids. The nitrogen and carboxyl groups of His and His-derived peptides are involved in the Au(111)-peptide interactions, due to which they get adsorbed on the surface of Au(111). Fig. 3 represents the initial dynamics of the Au(111)-His (and His-derived peptide) interaction simulation system (Kharazian et al., 2016).

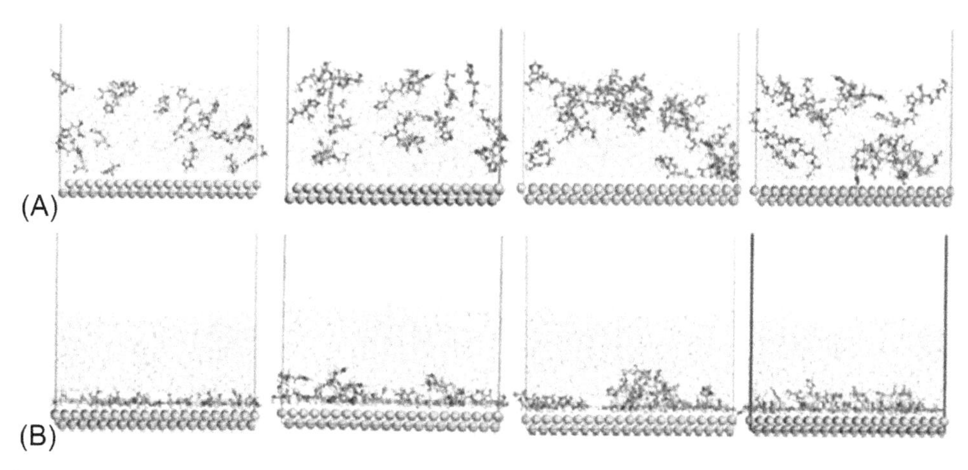

FIG. 3 Snapshots of the configurations of four systems at (A) $t=0$ ns and (B) $t=3$ ns (Kharazian et al., 2016).

13.2 Hydroxylated Fullerene-Ubiquitin

Radic et al. studied the effect of hydroxylated fullerenes on ubiquitin, through computational methods of MDs simulations and docking studies to identify the active sites of fullerene and fullerenol NPs on ubiquitin, which are shown in Fig. 4A and B (Kharazian et al., 2016).

MD and DMD simulations, which were implemented by placing C_{60} NPs around the protein (Fig. 4C and E), reveal two points: (i) Native structure of protein was maintained; (ii) clusters of NPs were formed near two similar binding sites. They compared the further simulation results to the experimental results, which were generated by the techniques fluorescence quenching, ITC, and CD spectroscopy. The following results were obtained: (i) NPs bound to the ubiquitin via hydrogen bonding; (ii) fullerenols can introduce misfolding. Fig. 4 has RMSD, d_{CM}, and N_C (residues) results, which are as follows: (i) the RMSD results of protein indicate that protein fluctuations appear at 2–3 and 4Å, where fluctuations were larger at 4Å. (ii) The d_{CM} (the distance between center of mass of the protein and NP) shows that NP stayed on the surface with $d_{CM} > 15$Å. RMSD, N_C, and d_{CM} are represented in Fig. 5 (Kharazian et al., 2016).

13.3 CNT-Protein Interactions

Large-scale MDs simulations of interactions of CNT with several proteins show that the CNTs plug into the hydrophobic core of protein and disrupt the original function of the proteins (Fig. 6). The CNTs interact with the hydrophobic residues of the protein (Zuo et al., 2010; Kharazian et al., 2016).

13.4 NP-Albumin Interactions

Ding et al. simulated the effect of four classes of NPs (hydrophilic, hydrophobic, positively charged, and negatively charged) on HSA protein, to compute the potential of mean force

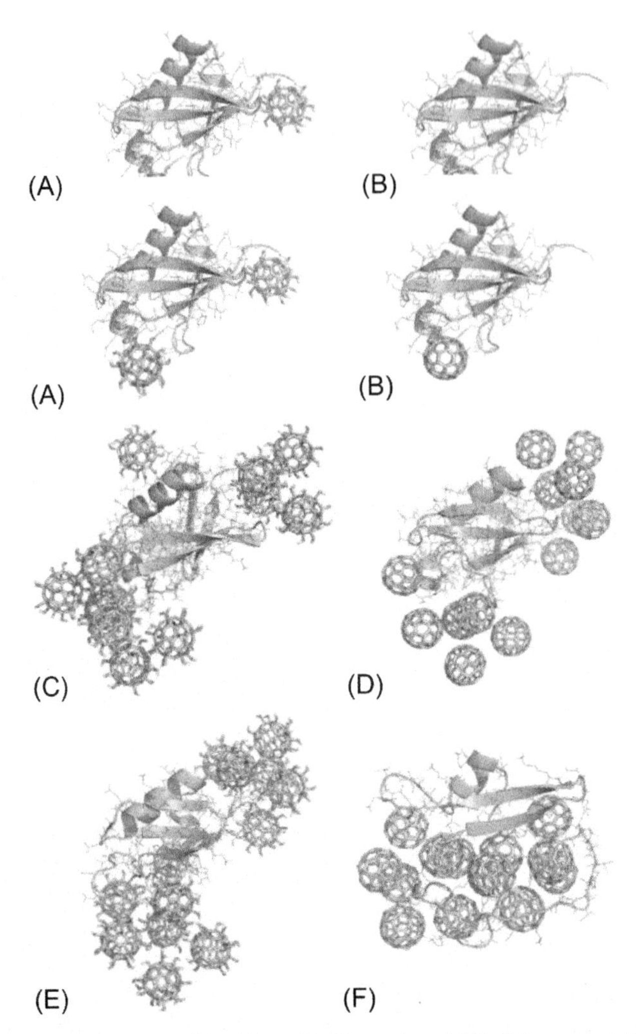

FIG. 4 The binding structures between ubiquitin and fullerene-based NPs. The computational modeling includes molecular docking (A and B), MD simulations with explicit solvent (C and D), and DMD simulations with implicit solvent (E and F). The panels (A, C, and E) correspond to the results for fullerenol C60(OH)20 binding, and panels (B, D, and F) illustrate the binding with fullerene C60 (Kharazian et al., 2016).

(PMF) of NP-protein interaction as a function of distance. Two significant effects were found: (i) PMF of hydrophilic NPs increases with decrease in distance; (ii) there was no direct interaction between HSA and NPs (Fig. 7; Ding and Ma, 2014; Kharazian et al., 2016).

(iii) Due to the repulsive electric interactions, HSA cannot be adsorbed on the anionic and hydrophilic NPs. Here, PMF is quite high, where the PMF of anionic NPs is little higher than the hydrophilic particles. (iv) HSA can be adsorbed on the cationic and hydrophobic NPs due to low PMF, where PMF falls below zero. (v) The energy depth reveals about the hard and soft corona. Hard corona has higher energy depth than the soft corona. The energy depth of

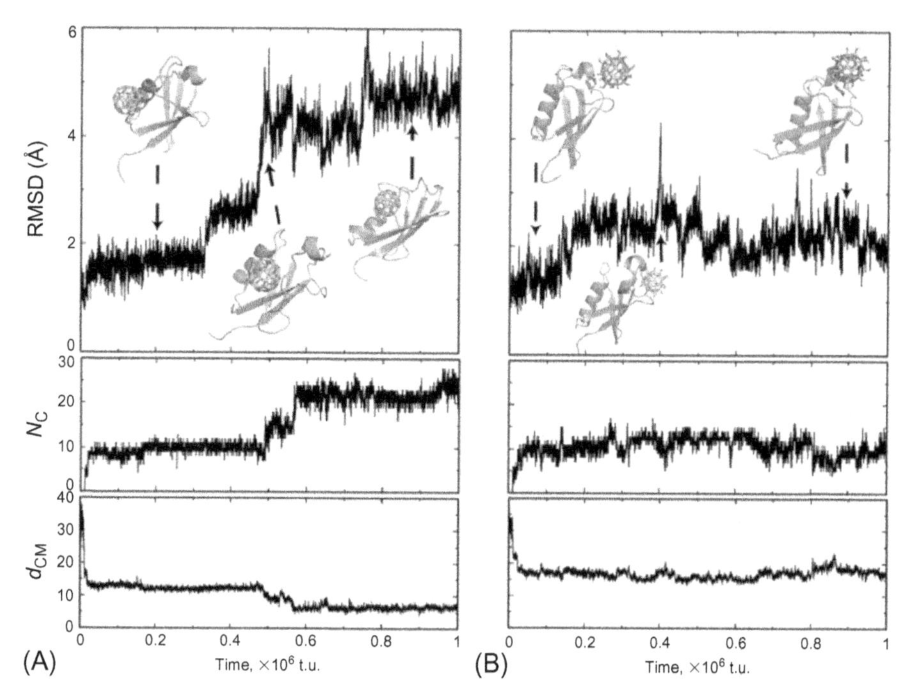

FIG. 5 Differential conformational dynamics of ubiquitin were observed upon binding fullerene C_{60} (A) and fullerenol C60(OH)20 (B). The RMSD of ubiquitin, the number of residues in contact with the nanoparticle (NC), and the intermolecular distance between the corresponding centers of mass (d_{CM}) were monitored as the function of simulation time (Kharazian et al., 2016).

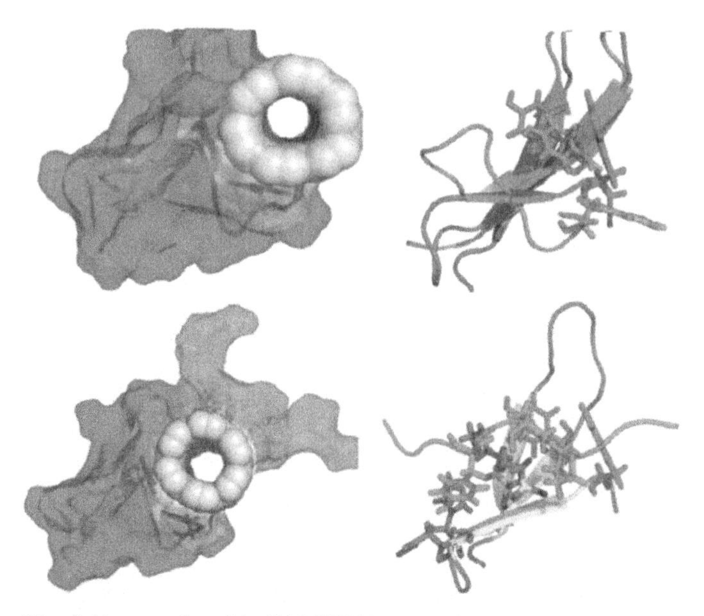

FIG. 6 Structures of (top left) a complex of the YAP65WW protein domain and the single-walled carbon nanotube (SWCNT) showing the insertion of the SWCNT into the protein domain. WW domain (top right) in the complex superposed with its native state. Protein-SWCNT complex (bottom left) with the adsorbed proline-rich motif (PRM) and (bottom right) domain binding with the PRM (PDB code 1JMQ). The (6,6) armchair SWCNT has a diameter of 8.08 (Zuo et al., 2010; Kharazian et al., 2016).

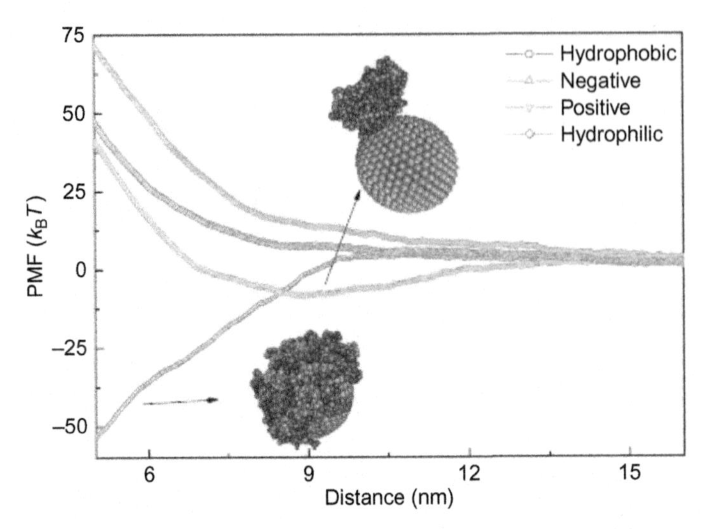

FIG. 7 Potential of mean force (PMF) of adsorption as a function of the distance (Z) from center of mass of nanoparticle to that of HSA protein where the NP size is 10 nm. The graphs show the effect of nanoparticle surface property on protein adsorption (pH is fixed as 7.4, and charge density is 0.2 e/nm^2) (Ding and Ma, 2014; Kharazian et al., 2016).

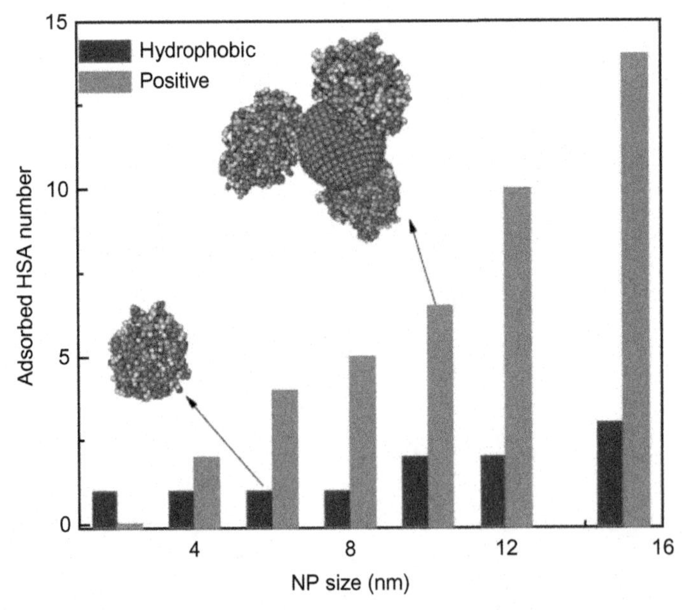

FIG. 8 The number of adsorbed HSA as functions of the (hydrophobic and positively charged) nanoparticle sizes where pH is 7.4 (Ding and Ma, 2014; Kharazian et al., 2016).

hydrophobic NP (5 K_BT), which may be the soft corona, is lower than the energy depth of cationic NP (8 K_BT), which may be the hard corona. (vi) One HSA can be adsorbed on the hydrophilic surface, because after adsorption, the surface of NP changed from hydrophobic to partial hydrophilic. On the other hand, multiple HSA are adsorbed on the cationic NPs (Fig. 8; Ding and Ma, 2014; Kharazian et al., 2016).

References

Aggarwal, P., Hall, J.B., McLeland, C.B., Dobrovolskaia, M.A., McNeil, S.E., 2009. Nanoparticle interaction with plasma proteins as it relates to particle biodistribution, biocompatibility and therapeutic efficacy. Adv. Drug Deliv. Rev. 61, 428.

Aromal, S.A., Vidhu, V.K., Philip, D., 2012. Green synthesis of well-dispersed gold nanoparticles using *Macrotyloma uniflorum*. Spectrochim. Acta Part A 85, 99–104.

Arumugam, A., Karthikeyan, C., Hameed, A.S.H., Gopinath, K., Gowri, S., Karthika, V., 2015. Synthesis of cerium oxide nanoparticles using *Gloriosa superba* L. leaf extract and their structural, optical and antibacterial properties. Mater. Sci. Eng. 49, 408–415.

Avouris, P., Freitag, M., Perebeinos, V., 2008. Carbon-nanotube photonics and optoelectronics. Nat. Photonics 2, 341–350.

Banik, B.L., Fattahi, P., Brown, J.L., 2016. Polymeric nanoparticles: the future of nanomedicine. WIREs Nanomed. Nanobiotechnol. 8, 271–299. https://doi.org/10.1002/wnan.1364.

Barnard, A.S., 2006. Nanohazards: knowledge is our first defence. Nat. Mater. 5, 245–248.

Bogunia-Kubik, K., Sugisaka, M., 2002. From molecular biology to nanotechnology and nanomedicine. Biosystems 65, 123–138.

Borm, P.J.A., Kreyling, W., 2004. Toxicological hazards of inhaled nanoparticles-potential implications for drug delivery. J. Nanosci. Nanotechnol. 4 (6), 1–11.

Botes, M., Cloete, T.E., 2010. The potential of nanofibers and nanobiocides in water purification. Crit. Rev. Microbiol. 36, 68–81.

Cedervall, T., Lynch, I., Lindman, S., Berggård, T., Thulin, E., Nilsson, H., Dawson, K.A., Linse, S., 2007. Understanding the nanoparticle-protein corona using methods to quantify exchange rates and affinities of proteins for nanoparticles. Proc. Natl. Acad. Sci. USA 104 (7), 2050–2055.

Chandran, S.P., Chaudhary, M., Pasricha, R., Ahmad, A., Sastry, M., 2006. Synthesis of gold nanotriangles and silver nanoparticles using *Aloe vera* plant extract. Biotechnol. Prog. 22, 577–583.

Cohignac, V., Landry, M.J., Boczkowski, J., Lanone, S., 2014. Autophagy as a possible underlying mechanism of nanomaterial toxicity. Nanomaterials 4, 548–582. https://doi.org/10.3390/nano4030548.

Daniel, M.C., Astruc, D., 2004. Gold nanoparticles: assembly, supramolecular chemistry, quantum-size-related properties, and applications toward biology, catalysis, and nanotechnology. Chem. Rev. 104 (1), 293–346.

De, M., You, C.-C., Srivastava, S., Rotello, V.M., 2007. Biomimetic interactions of proteins with functionalized nanoparticles: a thermodynamic study. J. Am. Chem. Soc. 129, 10747–10753.

Dhillon, G.S., Brar, S.K., Kaur, S., Verma, M., 2012. Green approach for nanoparticle biosynthesis by fungi: current trends and applications. Crit. Rev. Biotechnol. 32, 49–73.

Ding, H.-M., Ma, Y.-Q., 2014. Computer simulation of the role of protein corona in cellular delivery of nanoparticles. Biomaterials 35 (2014), 8703.

Dobrovolskaia, M.A., Patri, A.K., Zheng, J., Clogston, J.D., Ayub, N., Aggarwal, P., Neun, B.W., Hall, J.B., McNeil, S.E., 2009. Interaction of colloidal gold nanoparticles with human blood: effects on particle size and analysis of plasma protein binding profiles. Nanomedicine 5, 106–117.

Duran, N., Marcato, P.D., Durán, M., Yadav, A., Gade, A., Rai, M., 2011. Mechanistic aspects in the biogenic synthesis of extracellular metal nanoparticles by peptides, bacteria, fungi, and plants. Appl. Microbiol. Biotechnol. 90, 1609–1624.

Dwivedi, A.D., Gopal, K., 2010. Biosynthesis of silver and gold nanoparticles using *Chenopodium album* leaf extract. Colloids Surf. A 369, 27–33.

Dykes, G.M., 2001. Dendrimers: a review of their appeal and applications. J. Chem. Technol. Biotechnol. 76 (9), 903–918.

Eatemadi, A., Daraee, H., Karimkhanloo, H., Kouhi, M., Zarghami, N., Akbarzadeh, A., Abasi, M., Hanifehpour, Y., Joo, S.W., 2014. Carbon nanotubes: properties, synthesis, purification, and medical applications. Nanoscale Res. Lett. 9 (1), 393.

Farokhzad, O.C., Cheng, J., Teply, B.A., Sherifi, I., Jon, S., Kantoff, P.W., Richie, J.P., Langer, R., 2006. Targeted nanoparticle-aptamer bioconjugates for cancer chemotherapy in vivo. PNAS 103 (16), 6315–6320.

Gericke, M., Pinches, A., 2006. Biological synthesis of metal nanoparticles. Hydrometallurgy 83, 132–140.

Gessner, A., Waicz, R., Lieske, A., Paulke, B.R., Mader, K., Müller, R.H., 2000. Nanoparticles with decreasing surface hydrophobicities: influence on plasma protein adsorption. Int. J. Pharm. 196, 245–249.

Gopinath, K., Shanmugam, V.K., Gowri, S., Senthilkumar, V., Kumaresan, S., Arumugam, A., 2014. Antibacterial activity of ruthenium nanoparticles synthesized using *Gloriosa superba* L. leaf extract. J. Nanostruct. Chem. 4, 83.

Goppert, T.M., Muller, R.H., 2005. Adsorption kinetics of plasma proteins on solid lipid nanoparticles for drug targeting. Int. J. Pharm. 302, 172–186.

Govindaraju, K., Basha, S.K., Kumar, V.G., Singaravelu, G., 2008. Silver, gold and bimetallic nanoparticles production using single-cell protein (*Spirulina platensis*). J. Mater. Sci. 43, 5115–5122.

Health and Safety Executive, 2005. Health Effects of Particles Produced for Nanotechnologies. Health and Safety Executive, London.

Hellstrand, E., Lynch, I., Andersson, A., Drakenberg, T., Dahlback, B., Dawson, K.A., Linse, S., Cedervall, T., 2009. Complete high-density lipoproteins in nanoparticle corona. FEBS J. 276, 3372–3381.

Hirsh, S.L., McKenzie, D.R., Nosworthy, N.J., Denman, J.A., Sezerman, O.U., Bilek, M.M.M., 2013. The Vroman effect: competitive protein exchange with dynamic multilayer protein aggregates. Colloids Surf. B: Biointerfaces 103, 395–404.

Huang, J., Li, Q., Sun, D., Lu, Y., Su, Y., Yang, X., et al., 2007. Biosynthesis of silver and gold nanoparticles by novel sundried *Cinnamomum camphora* leaf. Nanotechnology 18, 105104–105115.

Iravani, S., 2011. Green synthesis of metal nanoparticles using plants. Green Chem. 13, 2638–2650.

Jahanshahi, M., Babaei, Z., 2008. Protein nanoparticle: a unique system as drug delivery vehicles. Afr. J. Biotechnol. 7 (25), 4926–4934.

Jansch, M., Stumpf, P., Graf, C., Ruhl, E., Muller, R.H., 2012. Adsorption kinetics of plasma proteins on ultrasmall superparamagnetic iron oxide (USPIO) nanoparticles. Int. J. Pharm. 428, 125–133.

Jayaseelana, C., Rahumana, A.A., Kirthi, A.V., Marimuthua, S., Santoshkumara, T., Bagavana, A., et al., 2012. Novel microbial route to synthesize ZnO nanoparticles using aqueous extract of flowers of Casia alata and particles characterization. Int. J. Nanomat. Biostruct. 4, 66–71.

Jedlovszky-Hajdu, A., Bombelli, F.B., Monopoli, M.P., Tomba, E., Dawson, K.A., 2012. Surface coatings shape the protein corona of SPIONs with relevance to their application in vivo. Langmuir 28 (42), 14983.

Junghanns, J.-U.A.H., Müller, R.H., 2008. Nanocrystal technology, drug delivery and clinical applications. Int. J. Nanomed. 3 (3), 295–310.

Karmali, P.P., Simberg, D., 2011. Interactions of nanoparticles with plasma proteins: implication on clearance and toxicity of drug delivery systems. Expert Opin. Drug. Deliv. 8, 343–357.

Kaushik, N., Thakkar, M.S., Snehit, S., Mhatre, M.S., Rasesh, Y., Parikh, M.S., 2010. Biological synthesis of metallic nanoparticles. Nanomed. Nanotechnol. Biol. Med. 6, 257–262.

Kharazian, B., Hadipour, N.L., Ejtehadi, M.R., 2016. Understanding the nanoparticle–protein corona complexes using computational and experimental methods. Int. J. Biochem. Cell Biol. 75, 162–174.

Kharissova, O.V., Dias, H.V.R., Kharisov, B.I., Perez, B.O., Victor, M., Perez, J., 2013. The greener synthesis of nanoparticles. Trends Biotechnol. 31, 240–248.

Korbekandi, H., Iravani, S., Abbasi, S., 2009. Production of nanoparticles using organisms. Crit. Rev. Biotechnol. 29, 279–306.

Latour, R.A., 2008. Molecular simulation of protein-surface interactions: benefits, problems, solutions, and future directions. Biointerphases 3, FC2.

Li, X., Xu, H., Chen, Z.S., Chen, G., 2011. Biosynthesis of nanoparticles by microorganisms and their applications. J. Nanomater. 2011, 1–16.

Lindman, S., Lynch, I., Thulin, E., Nilsson, H., Dawson, K.A., Linse, S., 2007. Systematic investigation of the thermodynamics of HSA adsorption to *N-iso*-propylacrylamide/*N-tert*-butylacrylamide copolymer nanoparticles. Effects of particle size and hydrophobicity. Nano Lett. 7, 914–920.

Lloyd, J.R., Byrne, J.M., Coker, V.S., 2011. Biotechnological synthesis of functional nanomaterials. Curr. Opin. Biotechnol. 22, 509–515.

Luangpipat, T., Beattie, I.R., Chisti, Y., Haverkamp, R.G., 2011. Gola nanoparticles produced in a microalga. J. Nanopart. Res. 13, 6439–6445.
Lundqvist, M., Stigler, J., Cedervall, T., Berggard, T., Flanagan, M.B., Lynch, I., Elia, G., Dawson, K., 2011. The evolution of the protein corona around nanoparticles: a test study. ACS Nano 5, 7503–7509.
Lynch, I., Dawson, K.A., 2008. Protein–nanoparticle interactions. Nanotoday 3 (1–2), 40.
Mahmoudi, M., Lynch, I., Ejtehadi, M.R., Monopoli, M.P., Bombelli, F.B., Laurent, S., 2011. Protein nanoparticle interactions: opportunities and challenges. Chem. Rev. 111, 5610–5637.
Maiorano, G., Sabella, S., Sorce, B., Brunetti, V., Malvindi, M.A., Cingolani, R., Pompa, P.P., 2010. Effects of cell culture media on the dynamic formation of protein-nanoparticle complexes and influence on the cellular response. ACS Nano 4, 7481–7491.
Meng, L., He, X., Gao, J., Li, J., Wei, Y., Yan, J., 2016. A novel nanofabrication technique of silicon-based nanostructures. Nanoscale Res. Lett. 11, 504.
Mittal, A.K., Chisti, Y., Banerjee, U.C., 2013. Synthesis of metallic nanoparticles using plant extracts. Biotechnol. Adv. 31, 346–356.
Mohanpuria, P., Rana, N.K., Yadav, S.K., 2008. Biosynthesis of nanoparticles: technology concepts and future applications. J. Nanopart. Res. 10, 507–517.
Monopoli, M.P., Walczyk, D., Campbell, A., Elia, G., Lynch, I., Bombelli, F.B., Dawson, K.A., 2011. Physical-chemical aspects of protein corona: relevance to in vitro and in vivo biological impacts of nanoparticles. J. Am. Chem. Soc. 133, 2525–2534.
Mukherjee, P., Ahmad, A., Mandal, D., Senanpati, S., Sainkar, S.R., Khan, M.I., Parischcha, R., Ajay Kumar, P.V., Alam, M., Kumar, R., Sastry, M., 2001. Fungus-mediated synthesis of silver nanoparticles and their immobilization in the mycelial matrix: a novel biological approach to nanoparticle synthesis. Nano Lett. 1 (10).
Mukherjee, S., Ray, S., Thakur, R.S., 2009. Solid lipid nanoparticles: a modern formulation approach in drug delivery system. Indian J. Pharm. Sci. 71 (4), 349–358.
Narayanan, K.B., Sakthivel, N., 2010. Biological synthesis of metal nanoparticles by microbes. Adv. Colloid Interface Sci. 156, 1–13.
Pattni, B.S., Chupin, V.V., Torchilin, V.P., 2015. New developments in liposomal drug delivery. Chem. Rev. 115 (19), 10938–10966. https://doi.org/10.1021/acs.chemrev.5b00046.
Petersdorf, R.G., 1974. Chills and fever. In: Wilson, J.D., Braunwald, E., Isselbacher, K.J., et al. (Eds.), Harrison's Principles of Internal Medicine, 12th ed. McGraw-Hill, New York.
Phogat, N., Khan, S.A., Shankar, S., Ansary, A.A., Uddin, I., 2016. Fate of inorganic nanoparticles in agriculture. Adv. Mater. Lett. 7 (1), 3–13.
R. F. Service, 2006. Science policy: priorities needed for nano-risk research and development. Science 314, 45.
Rabinow, B.E., 2004. Nanosuspensions in drug delivery. Nat. Rev. Drug Discov. 3, 785–796. https://doi.org/10.1038/nrd1494.
Rahman, M., Laurent, S., Tawil, N., Yahia, L., Mahmoudi, M., 2013. Protein-Nanoparticle Interactions, Vol. 15 of Springer Series in Biophysics. Springer, Berlin, HeidelbergISBN: 978-3-642-37554-5.
Rajakumara, G., Rahumana, A.A., Roopan, S.M., Khannac, V.G., Elangoa, G., Kamaraja, C., et al., 2012. Fungus-mediated biosynthesis and characterization of TiO_2 nanoparticles and their activity against pathogenic bacteria. Spectrochim. Acta Part A 9, 123–129.
Rawat, M., Singh, D., Saraf, S., Saraf, S., 2006. Nanocarriers: promising vehicle for bioactive drugs. Biol. Pharm. Bull. 29 (9), 1790–1798.
Rocker, C., Potzl, M., Zhang, F., Parak, W.J., Nienhaus, G.U., 2009. A quantitative fluorescence study of protein monolayer formation on colloidal nanoparticles. Nat. Nanotechnol. 4, 577–580.
Sankar, R., Rizwana, K., Shivashangari, K.S., Ravikumar, V., 2014. Ultra-rapid photocatalytic activity of *Azadirachta indica* engineered colloidal titanium dioxide nanoparticles. Appl. Nanosci. 5, 731–736.
Saptarshi, S.R., Duschl, A., Lopata, A.L., 2013. Interaction of nanoparticles with proteins: relation to bio-reactivity of the nanoparticle. J. Nanobiotechnol. 11, 11–26. https://doi.org/10.1186/1477-3155-11-26.
Shang, W., Nuffer, H.H., Dordick, J.S., Siegel, R.W., 2007. Unfolding of ribonuclease A on silica nanoparticle surfaces. Nano Lett. 7, 1991–1995.
Shankar, S.S., Rai, A., Ankamwar, B., Singh, A., Ahmad, A., Sastry, M., 2004. Biological synthesis of triangular gold nanoprisms. Nat. Mater. 3, 482.
Shiv Shankar, S., Ahmad, A., Sastry, M., 2003. Geranium leaf assisted biosynthesis of silver nanoparticles. Biotechnol. Prog. 19, 1627–1631.

Simberg, D., Park, J., Karmali, P.P., Zhang, W., Merkulov, S., McCrae, K., Bhatia, S.N., Sailor, M., Ruoslahti, E., 2009. Differential proteomics analysis of the surface heterogeneity of dextran iron oxide nanoparticles and the implications for their in-vivo clearance. Biomaterials 30, 3926–3933.

Singh, A., Singh, N.B., Hussain, I., Singh, H., Singh, S.C., 2015. Plant–nanoparticles interaction: an approach to improve agricultural practices and plant productivity. Int. J. Pharm. Sci. Inven. 4, 25–40.

Sintubin, L., Gusseme, D.B., Meeren, V.P., Pycke, B.F.G., Verstraete, W., Boon, N., 2011. The antibacterial activity of biogenic silver and its mode of action. Appl. Microbiol. Biotechnol. 91, 153–162.

Slack, S.M., Horbett, T.A., 1995. The Vroman effect. ACS Symp. Ser. 602, 112–128.

Sparkman, O., 2001. Mass spectrometry desk reference. Chem. Educ. Today 78 (2001), 46556.

Stayton, P.S., Drobny, G.P., Shaw, W.J., Long, J.R., Gilbert, M., 2003. Molecular recognition at the protein-hydroxyapatite interface. Crit. Rev. Oral Biol. Med. 14, 370–376.

Sun, Y.-P., Fu, K., Lin, Y., Huang, W., 2002. Functionalized carbon nanotubes: properties and applications. Acc. Chem. Res. 35 (12), 1096–1104.

Tenzer, S., Docter, D., Rosfa, S., Wlodarski, A., Rekik, A., Knauer, S.K., Bantz, C., Nawroth, T., Bier, C., Sirirattanapan, J., et al., 2011. Nanoparticle size is a critical physicochemical determinant of the human blood plasma corona: a comprehensive quantitative proteomic analysis. ACS Nano 5, 7155–7167.

Torresdy, J.L.G., Gomez, E., Videa, J.R.P., Parsons, J.G., Troiani, H., Yacaman, J.M., 2003. Alfalfa sprouts: a natural source for the synthesis of silver nanoparticles. Langmuir 19, 1357–1361.

Treuel, L., Nienhaus, G.U., 2012. Toward a molecular understanding of nanoparticle–protein interactions. Biophys. Rev. 4 (2), 137–147. https://doi.org/10.1007/s12551-012-0072-0.

Vertegel, A.A., Siegel, R.W., Dordick, J.S., 2004. Silica nanoparticle size influences the structure and enzymatic activity of adsorbed lysozyme. Langmuir 20, 6800–6807.

Vroman, L., Adams, A.L., Fischer, G.C., Munoz, P.C., 1980. Interaction of high molecular-weight kininogen, factor-Xii, and fibrinogen in plasma at interfaces. Blood 55, 156–159.

Walczyk, D., Bombelli, F.B., Monopoli, M.P., Lynch, I., Dawson, K.A., 2010. What the cell "sees" in bionanoscience. J. Am. Chem. Soc. 132 (16), 5761–5768.

Walkey, C.D., Chan, W.C., 2012. Understanding and controlling the interaction of nanomaterials with proteins in a physiological environment. Chem. Soc. Rev. 41, 2780–2799.

Wang, X.W., Zhang, L., Ma, C.L., Song, R.Y., Hou, H.B., Li, D.L., 2009. Enrichment and separation of silver from waste solutions by metal ion imprinted membrane. Hydrometallurgy 100, 82–86.

Yeap, E.W.Q., Ng, D.Z.L., Prhashanna, A., Somasundar, A., Acevedo, A.J., Xu, Q., Salahioglu, F., Garland, M.V., Khan, S.A., 2017. Bottom-up structural design of crystalline drug-excipient composite microparticles via microfluidic droplet-based processing. Cryst. Growth Des. 17 (6), 3030–3039. https://doi.org/10.1021/acs.cgd.6b01701.

You, C.-C., Miranda, O.R., Gider, B., Ghosh, P.S., Kim, I.-B., Erdogan, B., Krovi, S.A., Bunz, U.H.F., Rotello, V.M., 2007. Detection and identification of proteins using nanoparticle–fluorescent polymer 'chemical nose' sensors. Nat. Nanotechnol 2, 318.

Zharov, V.P., Letfullin, R.R., Galitorskava, E.N., 2005. Microbubbles-overlapping mode for laser killing of cancer cells with absorbing nanoparticle clusters. J. Phys. D Appl. Phys. 38 (15).

Zuo, G., Huang, Q., Wei, G., Zhou, R., Fang, H., 2010. Plugging into proteins: poisoning protein function by a hydrophobic nanoparticle. ACS Nano 4, 7508–7514.

Further Reading

Fleischer, C.C., Payne, C.K., 2014. Nanoparticle–cell interactions: molecular structure of the protein corona and cellular outcomes. Acc. Chem. Res. 47 (8), 2651–2659. https://doi.org/10.1021/ar500190q.

CHAPTER

12

Modeling and Simulating Carcinogenesis

Jenny Groten, Anusha Venkatraman, Roland Mertelsmann
Albert-Ludwigs-University Freiburg, Freiburg im Breisgau, Germany

A doctrine of nature can only contain so much science proper as there is in it of applied mathematics.
Immanuel Kant

1 INTRODUCTION

1.1 Background, Rationale and Approach

The objectives of current cancer research, from cancer vaccines to data sharing and the promotion of innovative and investigational approaches, reflect the urge to rethink oncologic research landscape (Lowy and Collins, 2016). With big data and information sharing taking the center stage, the focus now lies on data collection, analysis, evaluation, and implementation. One example is the massive decoding of the human cancer genome through next-generation *DNA* sequencing (Hayes and Kim, 2015). On the other hand, a major shift where patient care is edging away from evidence-based medicine (*EBM*) to a personalized medicine (*PM*) (Sugarman, 2012) is taking place simultaneously. This movement is a major turning point in the history of cancer research, which has so long focused on detecting regularities and defining classifications. Here, the role of bioinformatics becomes crucial to enable big data and *PM* to go hand in hand, allowing novel and innovative perspectives. The knowledge and data regarding the carcinogenesis process and treatment have increased dramatically and have reached unmanageable complexity. Common lab experiments and clinical trial tools can no longer provide adequate opportunities to investigate carcinogenesis and cancer treatment in either whole or in part of the endless number of subunits and possible interactions, all constantly changing over time ("panta rhei"). The result is the reductionist approach leading to delusive selective insights into complex biologic processes, bringing to mind the parable

☆ Parts of this paper have been previously published (Groten et al., 2016b).

Precision Medicine
https://doi.org/10.1016/B978-0-12-805364-5.00012-3

of the blind monks examining an elephant, where each examines a part and fails to recognize the entire creature (Wikipedia, 2016) This approach inadequately takes into account cellular heterogeneity and "noise," two fundamental characteristics of cellular and system behavior (Walker and Southgate, 2009).

Mertelsmann and Georg provided a distinct approach with the help of a virtual game called "Mitosis Game." To address a wider audience, they modeled an interactive simulation, based on evolutionary algorithms, to provide a tool to describe and actively model evolutionary processes that can eventually even lead to malignancy. For simplicity, as seen on Fig. 1, they reduced relevant cell mechanisms to ten fundamental intrinsic parameters called "hallmarks of evolution" and six external environmental parameters.

The controller can adjust different tools, altering all conditions, to grow a cell population and observe cell progression.

1.2 Analytical and Simulation Models

Bioinformatics, implementing mathematical models and information theory, addresses the complexity of data, correlations, and calculations. These in silico experiments and analyses are in-depth and time-saving. *Analytic models* can be distinguished from *simulation models.* Analytic models, such as regression analyses for statistical purposes, have experienced broad implementation during the last decade (Pelossof et al., 2015). In contrast, simulation models have hardly received attention, despite a broad and fruitful application in disciplines such as engineering, economics, and some aspects of biology, where it has become the state-of-the-art solution for understanding and optimizing processes and complex systems (Suthaharan, 2016). In both these model types, *machine learning* can be applied by means of self-teaching systems equipped with basic parameters, fundamental conditions, and feedback mechanisms to evaluate target parameters. It shows the significant advantage of optimizing processes

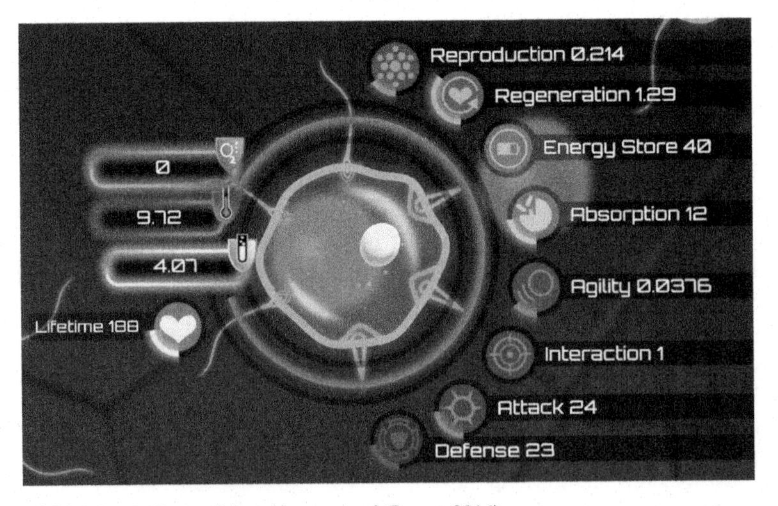

FIG. 1 Surface of the Mitosis Game (Mertelsmann and Georg, 2016).

without entire knowledge of actual mechanisms, parameters, and measurements, therefore increasing the probability of unexpected outcomes (Riedmiller et al., 2009). Simulation models provide various benefits over analytic models, with the additional dimension of *time* probably being the most important. This allows observation and calculation of a development over time and enables evaluation of data at any selected point in time. Moreover, a simulation represents a tool to visualize processes, thereby increasing comprehension of mechanisms and operations. An additional feature is the possibility to alter an ongoing process by adjusting any parameter at any discretionary time directly achieving tangible results. Simulations also allow a "middle-out" approach, instead of common "top-down" or "bottom-up" models, focusing on the cell as the "basic unit of life" (Walker and Southgate, 2009).

1.3 Modeling Evolution

A typical example of the exploitation of simulations including machine-learning tools is in investigating evolutionary processes, which are determined by probability and chance by nature and bear great potential to evolve unpredictable outcomes (Groten et al., 2016a).

In sight of the widely accepted hypothesis of carcinogenesis as an evolutionary development (Almendro et al., 2013; Beerenwinkel et al., 2015; Cairns, 1975; Klein, 2013; Greaves and Maley, 2012; Hanahan and Weinberg, 2000; Merlo et al., 2006; Nowell, 1976; Vogelstein et al., 2013; Willyard, 2016), it seems reasonable to establish the use of machine learning, particularly simulation models, in cancer research. Recent emphasis on the pivotal role of chance in the development of malignant diseases (Tomasetti and Vogelstein, 2015; Luzzatto and Pandolfi, 2015) fosters the perception of simulation models as the logical next step.

While current cancer research focuses on data generation, "bottom-up" analysis ("big pixel"), the next major step promises to be a view from a metalevel ("big picture," "middle-out") by exploring data analysis through simulation, hopefully leading to better understanding and novel concepts of cancer prevention, diagnosis, control, and therapy.

With evolution being a complex multiparametric process with a large timescale, mathematical models are widely utilized as a solution to the analysis and simulation of evolutionary developments (Mackey et al., 2015).

Evolution bears various phenomena that are connected via various correlations. To reveal the impact of each factor and the related interaction with other factors, one would have to test each evolutionary parameter separately in a wet-lab experiment, which is impossible to perform as the possibilities are too immense. In silico approaches in evolutionary biology represent methods consisting of simulated organisms and populations, which are observed and tested in *synthetic experiments*. In these experiments, the evolutionary conditions of the organisms and the evolutionary process can be characterized precisely and set one at a time to test the individual effects on the genome and the organization of both the organism and the population (Batut et al., 2013). This way, simulated organisms can be observed during competition and reproduction, while phenomena like "linkage, epistasis, demographic and environmental variability, and behavior" (Mackey et al., 2015) are considered. The possibility of applying random, hand-written, or former-evolved genomes and setting, for example, the mutation rate, the fitness measure, or the spatial arrangement of a population gives total control. The implication of synthetic experiments provides various advantages, such as limitless repetition to gain statistical power. Furthermore, one can observe as many generations as necessary, and the events, in both genotype and phenotype, can be recorded simultaneously. In synthetic experiments, depictions vary from mathematical functions to graphs, sequences of nucleotides, two-dimensional (*2D*) and three-dimensional (*3D*) simulations, and computer games.

1.4 Modeling Cancer

The aforementioned mathematical models and information-theoretical simulations have been applied to oncology in several approaches. Advances include the detection of metamarkers in breast cancer (Blokh et al., 2007), investigations of *DNA*-methylation processes (De Carvalho et al., 2012), revelation of the role of micro*RNA* and proteins in prostate cancer (Alshalalfa et al., 2013), the examination of gene-gene and gene-environment correlations in bladder cancer (Fan et al., 2011), and deviation of transcriptional profiles in cancer cells (Blokh and Stambler, 2017). Furthermore, oligoparametric simulation models have been developed to investigate targeted therapy of cancer, with possibilities of cell death and Boolean states of mutations to symbolize resistance (Komarova and Wodarz, 2014).

A spatial *3D* model was developed by Waclaw et al. (2015) to elucidate how cell dispersal and turnover could contribute to rapid cell mixing inside a tumor (Waclaw et al., 2015). They first modeled metastasis as an expansion of cancer cells, which have left their primary site, assumed to have acquired all necessary driver gene mutations in advance. Then they let cells replicate stochastically according to the number of surrounding empty spaces (Waclaw et al., 2015), whereas a surrounded cell does not replicate at all.

1.5 Simulating and Visualizing Cancer

To address the need for in silico simulation models, we have developed a visually attractive and interactive and at the same time plausible and qualitatively valid simulation model of carcinogenesis and cancer treatment, developed from experimental data. Thereby, the overall objective is to offer a novel tool for basic, clinical, and therapeutic research and also a teaching tool to make in silico research applicable to a wide audience.

The present paper's focus lies on the collection and review of relevant data, the formation process of developed simulations, and the first qualitative validation process. In this context, the term *validation* is used to document the close resemblance of the qualitative prediction of the in silico simulation and *real-life* biological and clinical data extracted from the relevant literature.

To accomplish this aim, we have addressed the following research objectives:

1. Identify the essential "hallmarks of cancer" through literature review. The term "hallmarks of cancer" was adopted from Hanahan and Weinberg and defines fundamental phenotypic characteristics of cancer cells (Hanahan and Weinberg, 2000, 2011).
2. Review the genetic hallmarks of cancer with a focus on gene expression.
3. Review the principles of evolution, entropy, and chance in cancer, to develop a new concept of phenotypic cancer hallmarks.

4. Evaluate and synthesize the elaborated hallmark capabilities.
5. Develop mathematical models and algorithms to describe the hallmark characteristics of carcinogenesis and to develop a computational simulation of carcinogenesis.
6. Test, validate, and adapt the algorithms and correlations via repetitive simulation phases.
7. Simulate hematopoietic tissue homeostasis, the establishment of hematopoiesis after stem cell transplantation and leukemogenesis.
8. Reevaluate the "hallmark concept" in light of the simulation results and identify key hallmarks of carcinogenesis.

2 MATERIALS AND METHODS

Mathematical models were developed using the review of pertinent cancer research literature. Programming languages applied here were Python, Java, and JavaScript. For the exact codes, see Groten et al. (2016b).

2.1 Balancing—Visual Validation

The first validation step, which was performed during the development of the simulation, was a tool called "balancing," where "balancing" is a game design strategy describing iterative observation, testing, and comparison with known evidence and subsequent adaptation until the resulting processes and developments seem plausible and visually resemble those depicted in pertinent literature (Schell, 2015).

Here, "balancing" defines a strategy normally applied in game design and describes a process of iterative observation, testing, and comparison with known evidence and subsequent adaptation of the settings of the simulation, until the resulting processes and developments visually resemble the processes and mechanisms depicted in pertinent literature and seem mainly plausible. This is preferably implemented by several people (here, all authors contributed) from different perspectives.

2.2 Regression Analysis—Statistical Validation

The second validation step was a statistical analysis of possible correlations between different parameters by computational regression analysis. For this, we applied the function for ridge regression (Suthaharan, 2016; Pedregosa et al., 2011). Therefore, evolutionary sections of the simulation were not considered. We also parameterized the investigated pathways and defined cell characteristics.

Post preloop, settings with all possible combinations of one, two, and three pathways set to the maximum value were simulated for 6000 ticks, which equaled 120 days in real time. To achieve better statistical power, they were repeated six times each. Every simulation round started with six stem cells with pathways to be altered set to maximum and all others set to average. During the simulation, cell parameters were measured at predefined points in time. The results were analyzed employing the equation of ridge regression (Pedregosa et al., 2011).

3 RESULTS

3.1 Modeling and Simulation

3.1.1 *Simulating Carcinogenesis and Cancer Therapy*

APPROACHES—PRIMARY MODELS

The first versions of our simulation of carcinogenesis are primarily based on the "Mitosis Game" by Mertelsmann and Georg (2016). The new simulations are meant to serve as an educational tool and an approach to allow scientists to explore the applicability of computational models in oncology, in basic research, and in clinical trials. Therefore, the focus of the new models lies in the improvement of comprehensibility by reduction of complexity and increasing transparency of the simulated processes.

SIMULATING CANCER TREATMENT

The first approach consists of the simulation of tumor growth, treatment response, and resistance. On the surface of the simulation, there is a petri dish representing the primary tumor, containing continuously growing cells. Cell characteristics and behavior can be influenced by 10 different targeted therapies. Their application can be controlled by adjusting the "+" and "−" buttons placed around the petri dish, with "+" increasing allocated therapeutic dosage and "−" decreasing it. Each targeted therapy is accompanied by a description of the attacked cellular pathway, the "hallmark," frequently altered in a malignant cell (Hanahan and Weinberg, 2000, 2011; Fig. 2).

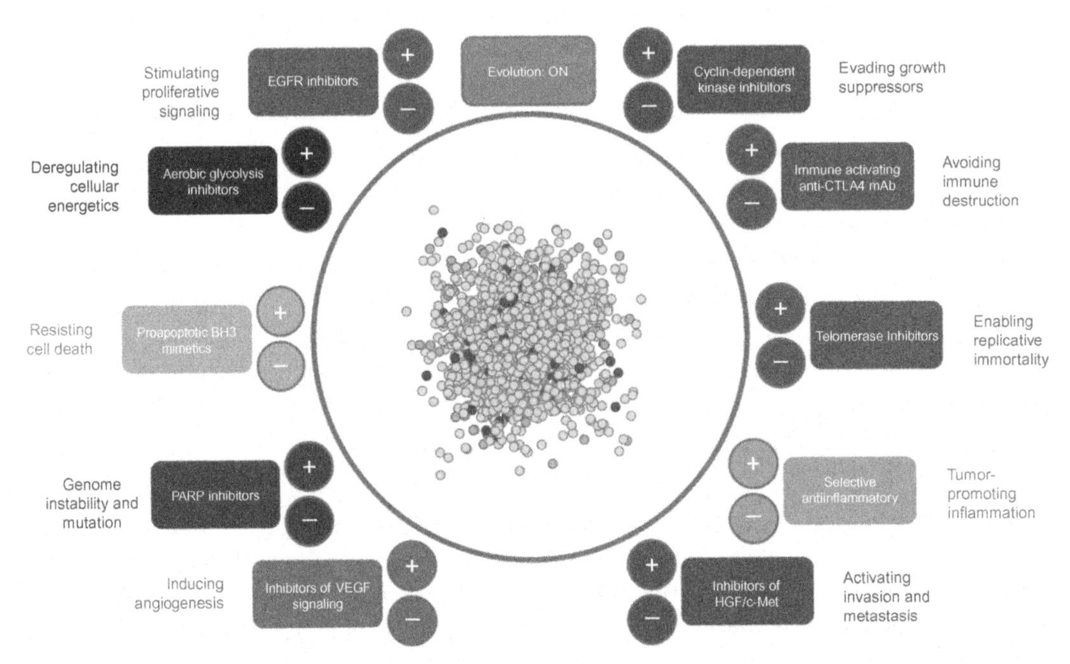

FIG. 2 Surface of the simulation of cancer therapy (Georg et al., 2016). For color figure, please refer to online version of this chapter.

SIMULATING CARCINOGENESIS

Since the main aim of carcinogenesis and cancer treatment simulation was visualization and interactivity, the first approach depicted above was still far too complex and opaque for visual validation. Hence, the treatment option was eliminated, and instead, cancer growth was simulated taking into account the impact of different "hallmarks" (Hanahan and Weinberg, 2000, 2011).

Due to higher transparency and closer resemblance to wet-lab experiments, the previously developed pseudo-*3D* surface of the simulation (not shown) was altered to a *2D* visualization, which allows the visibility of every single cell and its behavior (Fig. 3).

The full simulation model is provided at: http://mertelsmann.psiori.com/.

Instead of the "Drug" buttons applied in the first simulation, the "Hallmark" buttons influence these features, resulting to cell behavior and tumor growth.

LOGICAL BACKGROUND AND PROGRAMMING The whole simulation is written in JavaScript and executable in every common web browser. EaselJS is applied as an additional library for visualization purposes.

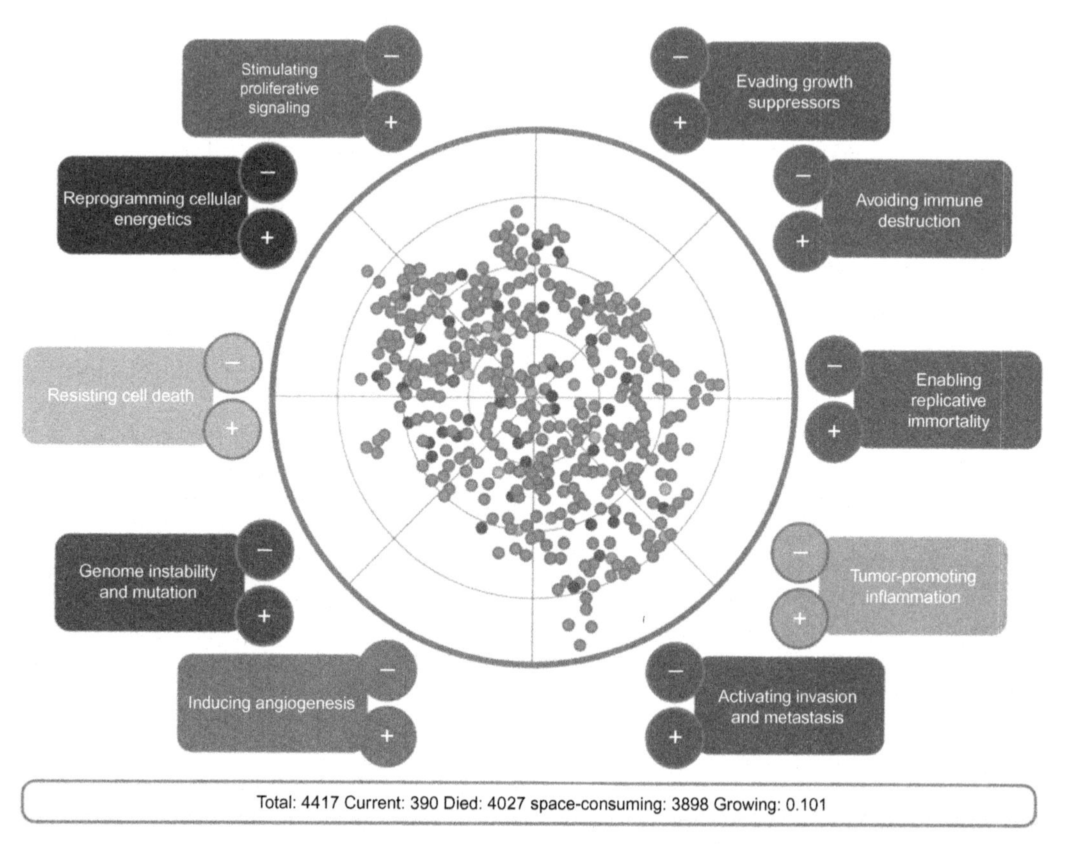

FIG. 3 Surface of the simulation of carcinogenesis II (Georg et al., 2016). For color figure, please refer to online version of this chapter.

At the beginning, there is an arbitrary number of start cells, adjustable in code (currently set at 6) that subsequently divides asymmetrically into one new daughter cell and one renewed parent cell. Each evolving cell is equipped with six different characteristics: spectral color (reflecting the relative activation of each pathway), position, velocity, direction vector, Hayflick limit, and a Boolean state of immune attack. For each newly built cell, these components are tested and calculated in chronological order depending on the intensity of cellular pathway alteration.

Color Each pathway is represented by a spectral color value calculated as the weighted sum of the color values (*RGB*, vector 3 with values from 0 to 255) of all pathways, depending on their percentage of mutation (a state between 1 and 5). This value builds the cell's primary color and is inspired by the work of Weber et al. (2011).

Further color coding is used for the states "imminent apoptosis" (light blue) and "immune attack" (dark gray/red contour).

Position The simulation interface is defined via a *2D* coordinate system. The first cells are approximately equidistant from visible coordinate system's center point. The cell position of each evolving cell is calculated using the former position of the parent cell. The new daughter cell will be placed at a spot around the parent cell, at a random angle at distance of a cell diameter.

Velocity Each cell's speed is defined as pixel per tick ranging from 0.1 to 1 and depending on the value of "activating invasion and metastasis." The starting speed of a cell is 0.5 pixel/tick. After one tick, the speed of a cell undergoes exponential decrease till it runs toward zero.

Direction Vector The direction vector is a two-coordinate vector, which determines the direction of cell movement. It is allocated randomly to every evolving cell and remains unchanged for the whole lifetime of a cell.

Hayflick-Limit Hayflick limit defines the number of possible cell divisions and depends on the length of chromosomal telomeres, which decreases in standard cells with every cell division. In the simulation, the default Hayflick limit of a normal stem cell is 72 as an approximation of the realistic number between 50 and 70 (Shay and Wright, 2000). Each cell evolving from cell division is assigned with the Hayflick limit of its predecessor minus 1. If a cell shows Hayflick limit of 1 or less during testing, it is marked with light blue and will die at a Hayflick limit of 0 or less.

Immune Attack Another state a cell can show is the state of being recognized and attacked by the immune system. Proposing that immune system eliminates cells detected as targets, a cell in this constant state is coded with the color dark gray with a red contour and will be dead after 40 ticks. The probability of immune attack depends on two "hallmarks" (Hanahan and Weinberg, 2000, 2011). First, the intensity of "avoiding immune destruction" determines interval and probability of testing. The test interval ranges between 90 and 10 ticks. Second, the intensity of "genome instability and mutation" further alters the likelihood of testing with a range of 1%–10%, presuming that probability of immune system detection rises with the number of mutations.

THE SIMULATION PROCESS Every simulation round has a defined starting point, both temporal and local. The time unit of the simulation is the "tick," which defines one update loop. Fifty ticks in the simulation equal one day in real time. The duration of one "tick" thereby depends on the power of the processor, which should usually result in a number of about 50 ticks per second. One update loop consists of chronologically determined assessment and subsequent consequences. These conditional links underlie distinct test mechanisms, algorithms, mainly if-then instructions, and probability ranges.

So, at the beginning of the simulation, when the program starts, the first cell evolves containing the characteristics mentioned above, calculated by the initial settings.

In each following tick, the below-mentioned assessment process is performed.

(1) Pathway buttons
The current size of pathway buttons is measured and evaluated against their proportion of entirety of pathway buttons, and the core color of evolving cell or cells is calculated by the proportional summation of distinct color values (Weber et al., 2011).

(2) Mutation and entropy
If the "Mutation" button is set "on," a countdown, starting at "50," is decremented at each tick. If it becomes "0," a random pathway is set at the highest dosage or probability, symbolizing randomly acquired mutations of one of the cellular pathways.
The user cannot adjust pathway buttons manually for as long as they are set to maximum. Automatically set adjustment is removed post 50 ticks, and another random pathway is fixed at the highest dosage or probability.
When "Entropy" button is switched on, the pathways altered once during the current round of the simulation, which have been saved in a background list, will stay permanently altered. Maximally, three pathways can be mutated at once. If a fourth pathway is mutated, the first is ignored.
In the simulation's latest version, this option has been temporarily abandoned due to the insufficient validation.

(3) Cell division
Each cell is tested individually concerning divisibility, and only cells not evolved in this tick can go through cell division.
The event of a division depends on the interval determined by the current state of the pathway "stimulating proliferative signaling." Only when the predetermined interval expires, the value is "0." The probability division occurring is further determined by the dosage of the cell's ability to evade "growth suppressors," which is presumed to impact growth control. It ranges between 0.5 and 1, where the higher the probability, the more that a cell division is probable. A fourth parameter influencing the probability of cell division is the "tumor-promoting inflammation." Here too, the probability of inflammation and cell division are directly proportional. The underlying hypothesis postulates that this parameter supports tumor growth by increasing cell division rate (Hanahan and Weinberg, 2011).
If new cells evolve by the division of mother cells, they are equipped with the characteristics described above. Thereby, qualities are given to all new cells at the time of testing according to the currently set parameters, except for the Hayflick limit and position, which are predefined by the mother cell.

(4) Spatial structure
In the new *2D* simulation, an improvement technique is applied to avoid cell stacking and only permits cell placement where free space is available. This technique consists of a collision process, in which it is tested at each tick if the distance between the centers of two cells has a minimal value of $1.2 \times r$ with r = radius of a cell. This way, every cell is

compared with every other cell existing in the current process (listed in the table of all cells). If two cells do not fulfill this criterion, the compared cell dies so that the new cell survives.

So, through selective mechanism, more cells can grow in areas with more available space such as at the edge of the tumor.

(5) Recognition and destruction of cancer cells by immune system cells
The probability of the cell being attacked by immune system is calculated. It depends on the interval determined by the ability to avoid immune destruction and probability alteration according to the likelihood of "genome instability and mutation." For exact probability calculation, see paragraph "immune attack."

(6) Replication potential
The Hayflick limit of the cell is tested.

(7) Cell death
The probability of cell death is calculated. A cell is eliminated if placed outside the petri dish/primary tumor, if Hayflick limit counts <1, or if ticks determined via "immune attack" reach 0. In reality, this artificial loss of cells would lead to metastasis, an important phenomenon not adequately visualized in our model.

In the simulation's background, a list of all living cells is created simultaneously where each cell is considered an object with a given position determined by its consecutive appearance and is saved with its specific characteristics. As a novel feature to improve transparency, a "lab report bar" is added to the simulation's surface showing the current numbers of total cells, current cells, and dead cells and the growth rate.

THE VALIDATION PROCESS To implement the suggested simulation in oncology, a validation of the depicted processes is inevitable. The validation process was approached in two phases:

(1) Balancing
The first step proceeded during the development of the simulation is via the "balancing" tool (Schell, 2015). In our case, the literal primary basis was the review above by Hanahan and Weinberg (2011), which was complemented by the present literature review on carcinogenesis. The iterative "balancing" was performed by a team of experts in the fields of game design, information theory, cognitive science, and medicine/oncology.

(2) Regression analysis
The second phase contained a statistical analysis of individual correlations by computational regression analyses. More precisely, the function for ridge regression was applied (Pedregosa et al., 2011; Suthaharan, 2016).

Evolutionary sections of the simulation were eliminated here as they were insufficient for this type of analysis.

For the analysis, pathways and cell characteristics were parameterized so that their correlations could be investigated.

All pathways were set to an average level; the pathways to be altered were set to the maximum level. During the simulation, the parameterized cell characteristics were measured at predefined points in time. Using this approach, 175 possibilities plus one standard constellation were tested, six times each so that 1050 experiments were performed. All tests taken together correspond to 365 years in real time.

The results were analyzed utilizing the equation of ridge regression (Pedregosa et al., 2011). Out of the immense amount of generated data, we focused on the results concerning the following aspects: maximum number of current cells, maximum number of all cells over time, and maximum number of cells killed by the immune system. All of these showed both valid results and unexpected correlations and outcomes. The term "unexpected" here means not directly determined by the code.

The two pathways associated with a high number of cells are "signaling," standing for "sustaining proliferative signaling," and "growth," representing "evading growth suppressors."

The qualitative outcome appears to be in concordance with expected results based on current clinical and wet-lab literature.

According to the correlation coefficients shown in Fig. 4, "sustaining proliferative signaling" and "evading growth suppressors" show by far the most significant positive correlation. This finding corresponds to the assertion of Hanahan and Weinberg, who consider "sustaining proliferative signaling" to be "arguably the most fundamental trait of cancer cells" (Hanahan and Weinberg, 2011). The capacity of "evading growth suppressors" is closely linked to mutations in tumor suppressor genes (Vogelstein et al., 2013; Vogelstein and Kinzler, 2015).

A surprising and unexpected result is the large negative coefficient associated with the characteristic of "metastasis" standing for "invasion and metastasis." At first sight, it appears counterintuitive that the ability to invade and disseminate, which is notoriously affiliated with high-grade malignancy and aggressive growth, has an adverse effect on the maximum number of current cancer cells and related tumor size. However, since the number of cells

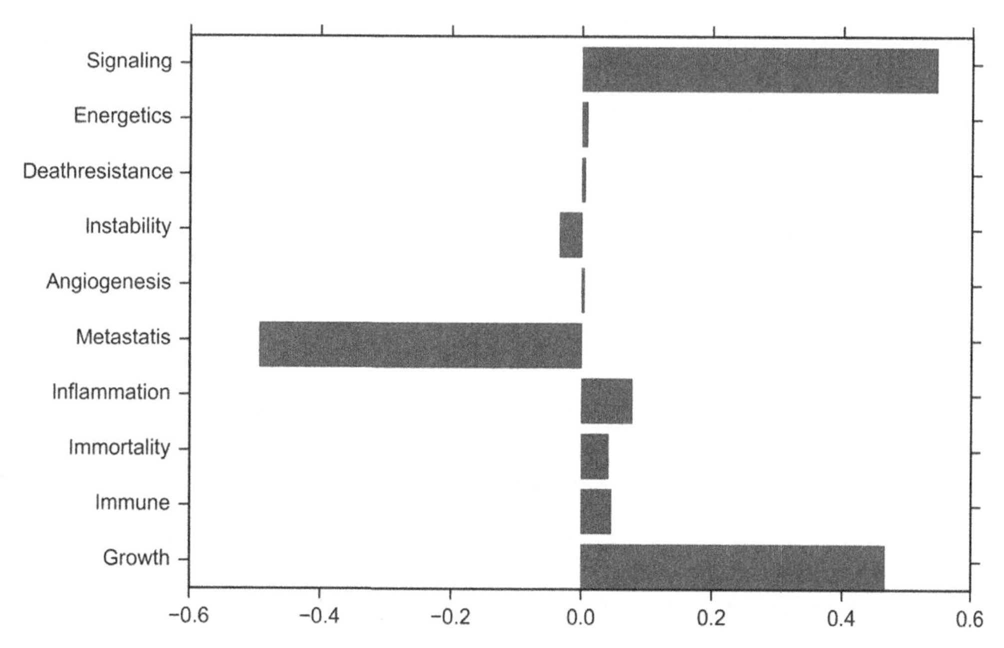

FIG. 4 Coefficients in linear regression of currentCellsMax (Georg and Lau, 2016).

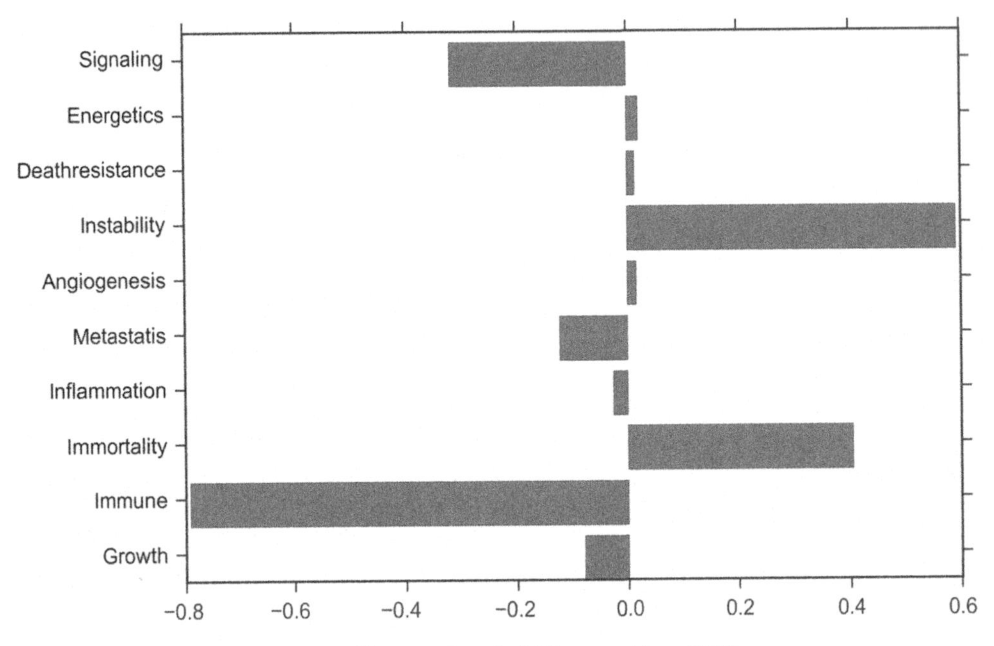

FIG. 5 Coefficients in linear regression of immuneAttacked (Georg and Lau, 2016).

in our simulation corresponds to the number of cells in the primary lesion, cells leaving the primary tumor are not counted in this model.

The rest of the regression results do not show significantly positive or negative coefficients, indicating their less-pronounced impact on cell population size.

One possible way for a cell to die is to be attacked by the immune system.

The number of cells, which are attacked by the immune system in each experiment, is represented by the value "immuneAttacked."

Fig. 5 shows all experimental runs sorted by the number of cells killed by the immune system. The x-axis is nonlinear here because one unit equals one experiment. Nevertheless, the x-axis shows the number of cells attacked by the immune system ranging from 0 to 8979 cells.

The linear regression of the results concerning the number of cells attacked by the immune system indicates that both "instability," standing for genome instability and mutation, and "immortality," standing for enabling replicative immortality, have a positive impact on a cell's probability to be attacked by the immune system, while "signaling," "metastasis," and "immune," meaning avoiding immune destruction, are negatively correlated. In light of the fact that both "genome instability and mutation" and "enabling replicative immortality," despite short telomeres, cause chromosomes with high amounts of genetic failure and defects, it seems plausible that these features increase the probability of the altered cell to be detected as foreign and subsequently eliminated.

3.1.2 *Simulating Hematopoiesis and Acute Myeloid Leukemia*

The details of this model have been described previously (Groten et al., 2016a).

NORMAL HEMATOPOIESIS

The unexpected events occurring in the first two simulations led us to proceed to another independently developed model and simulation depicting hematopoiesis. In hematopoiesis, normal tissue homeostasis could be modeled more easily with the cells production site, the bone marrow, and the offspring getting into circulation leaving the primary site. These results could be validated against known clinical parameters: peripheral blood counts, hematopoietic growth factor effects, and bone marrow cellularity under normal conditions and after leukemic transformation.

Fig. 6 shows the simulation's visible surface. There are nine different figures. The figures in the first row represent the state and differentiation process of erythrocytes, while the second row shows the behavior of granulocytes, and the third shows the development of thrombocytes. The first column depicts the number of fully differentiated cells in the peripheral blood over time, while the second shows the corresponding number of total (dark gray) and free (light gray) transmitters in the bone marrow over time. These graphs are built with the precision of one dot per tick. A bar chart in the last column demonstrates the number of progenitor cells at distinct differentiation levels at the current state, with higher ciphers standing for higher differentiation. The fourth row shows three additional charts. On the left side, there is a description of the number of needed time in milliseconds for each step. The middle represents (dark gray) the total number of cells in the bone marrow compared with the number of cells able to divide (light gray). On the right sight, a pie chart gives an overview of the relative distribution of different progenitor cells in the bone marrow.

SIMULATING LEUKEMOGENESIS: ACUTE MYELOID LEUKEMIA

To simulate the development of acute myeloid leukemia (*AML*), we programmed a new cancer cell type for each differentiation level. They contain an arbitrary combination of up to three different alterations, independence from the Hayflick limit (telomere shortening = 0), independence from transmitters (transmitter dependence = false), and a block of differentiation (as the next differentiation step, the same cell type is determined). Also, for statistical reasons, the cancer state is turned *ON*.

Similar to the validation process of the simulation of carcinogenesis, exemplary experiments have been simulated in this hematopoietic model.

Aiming to investigate the impact of all three proposed pathways of *AML*, both as single parameters and in combination of two or three, leukemia has been simulated in several experimental runs for 10,000 ticks each, which equals 100 days in real time. While the activation of one or two of the three pathways yielded unexpected results (Groten et al., 2016b), the activation of all three pathways led to the development of "leukemia" in this model with all characteristics observed in patients.

4 THE HALLMARK CONCEPT REVISITED

The simulation model depicted in Fig. 6 and performed experiments confirm the hallmark concept extracted from the literature review to a large extent. Furthermore, our findings as shown in Table 1 indicate that a reduction similar to the classification presented by Vogelstein et al. can be reasonable, supplemented by an additional category for differentiation block and acquisition of stem cell features.

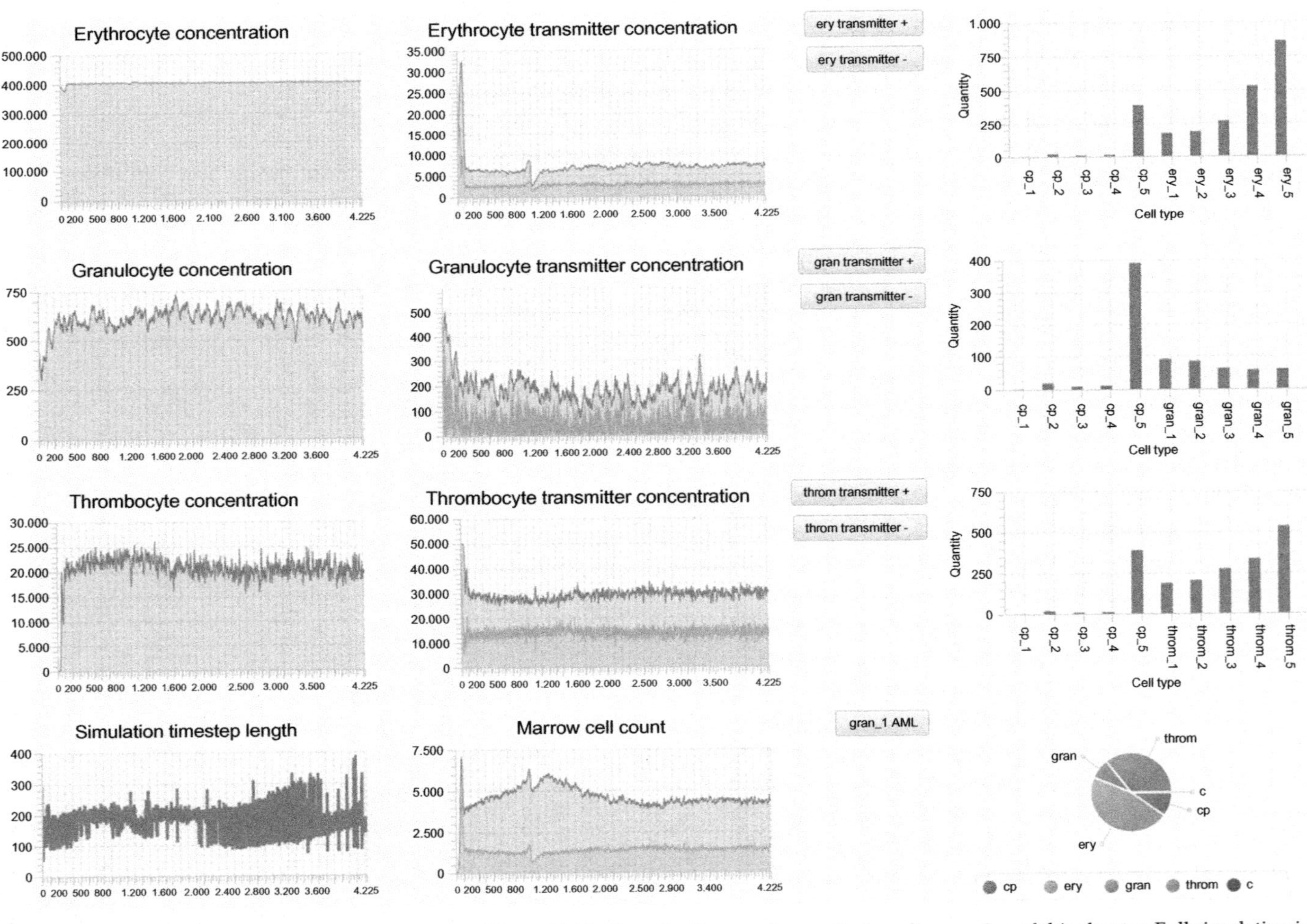

FIG. 6 Surface of the simulation of hematopoiesis (Worm, 2016). For color figure, please refer to online version of this chapter. Full simulation is provided via http://hem-model.psiori.com/hema_simulation.

TABLE 1 The Hallmark Concept Revisited, 1-4b: Relevance Range After Simulating Carcinogenesis, I-III: Final Relevance Range after Simulating Hematopoiesis (Synthesis) (Hanahan and Weinberg, 2011; Vogelstein et al., 2013; Groten et al., 2016b; Torrente et al., 2016)

Hanahan and Weinberg (2011)	Vogelstein et al. (2013)	Groten et al. (2016b) Exp 1	Exp 2	Exp 3	Exp 4	Torrente et al. (2016)
Hallmarks of Cancer	**Core Cellular Processes**	**CurrentCellsMax (Primary Tumor)**	**AllCells**	**ImmuneAttacked**	**Leukemia**	**Genes Overexpressed in Cancer**
Evading growth suppressors	Cell survival	1	1			EMR2
Sustaining proliferative signaling		2	2		I	PTP4A3, RGS1
						RPS6KB1
Tumor-promoting inflammation		3	4	4		TREM2, MAP3K12
Evading immune destruction		4a				TDO2, ANXA11
Enabling replicative immortality	Genome maintenance	4b	3	2	II	?
Genome instability and mutation				1		DDX11, BLM, NUDT1
Activating invasion and metastasis	Cell fate					PTP4A3
New: stem cell properties, block of differentiation, MET					III	LEF1
Resisting cell death		Low effect		None		
Inducing angiogenesis						
Reprogramming energy metabolism						

These are the alterations in the "cell survival" and "genome maintenance," which dominate cell growth in carcinogenesis simulation. These were transferred to the simulation of hematopoiesis. A hallmark presumably missing in the first simulation and the common concepts of cancer traits is the lack of differentiation and maintenance of stem cell features of a cancer cell. Applying the lack of differentiation to the simulation as a third cellular alteration, one can simulate leukemogenesis with largely realistic and plausible results. This result corresponds to the suggestions of Vogelstein et al. (2013) and Vogelstein and Kinzler (2015) and the distribution of genes found overexpressed in different cancer types (Torrente et al., 2016).

5 DISCUSSION

To address the rising need for innovative research approaches (Lowy and Collins, 2016), the overall aim of this project was to investigate the essential phenotypic "hallmarks" of a cancer cell, oriented to the "hallmarks of cancer" suggested by Hanahan and Weinberg (2000, 2011) and extended by findings of pertinent literature about cancer history, cancer hallmarks, genetic hallmarks, cancer therapy, biological and *somatic evolution*, entropy, and chance and recent research objectives. Based on this literature research, we developed an in silico simulation model through a hallmark synthesis and vice versa revisited the identified characteristics via the modeled simulation. We thereby focused on visualization and interactivity of the simulation, aiming for a novel tool for teaching and basic, clinical, and therapeutic research. Implementing this feature utilizing a simulation model, we provide a model, which extends the investigative breadth of previous analytic models. Certainly, other simulation models have been developed in cancer research, for example, to simulate targeted therapies or to investigate the correlation between spatial cancer-cell expansion and tumor morphology. In comparison with our model, the simulation models of targeted therapy provide limited variable and independent parameters and need extensive mathematical descriptions and rules. Despite their complexity, they do not allow the investigation of a high number of parameters nor end points. Especially, the possibility to investigate interactions of parameters is restricted. In contrast, in our model, the true values of parameters result from the simulated dynamic processes and interactions, made possible by simulation of single pathways by iteratively adapted algorithms.

Qualitative validity could be shown by extensive balancing and regression analysis along with reviewing pertinent literature. This way, both simulations have shown considerable coherence and plausibility. Certainly, one could argue that the simulation results could mainly be explained as a self-fulfilling prophecy. However, first, this effect is a sign for inner coherence of a system, in which only single parameters were defined at the beginning and autonomous interactions are still possible, and second, we could still observe undetermined and unexpected outcomes, like the negative impact of "metastasis" on the maximum number of current cells in the primary tumor in the simulation of carcinogenesis.

Even though obvious results provide a certain qualitative validity, an additional validation process called system identification is still inevitable to declare its validity.

Although the full validity of the simulation can be doubted for the reasons mentioned above, we would like to take these results seriously, since the majority can be explained even though they seem counterintuitive at first sight. Our findings indicate that the 10 "hallmarks" proposed by Hanahan and Weinberg can be clustered in two different groups, "growth/apoptosis balance" and "genetic fidelity/immortality," and that carcinogenesis requires just one alteration in each pathway group. Modeling hematopoiesis finally revealed one missing hallmark capability, "block of differentiation," which we have not specifically addressed in the carcinogenesis model, which starts at time 0 already with a cancer-cell population. After having reviewed the literature on cancer evolution, we propose to assign this feature to the broader term "stem cell features" as they largely correlate with earlier suggestions by Vogelstein et al. (2013) and Vogelstein and Kinzler (2015). In their earlier work, they assumed cancer cells to contain alterations in the three core cellular processes: "cell survival," "cell fate," and "genomic maintenance." Out of these three processes, two directly correlate to our suggestions, and their third proposal, "cell fate," primarily describes capability of metastasis, which we would like to extend to "stem cell features," including the "block of differentiation." For this, the assertion results from our simulations and is, according to recent findings, a fundamental trait of cancer cells, which is a prerequisite for several other pathways (Jordan et al., 2006; Gupta et al., 2009). The fact that our simulation of hematopoiesis depends on three different pathway alterations to result in an overt *AML* corresponds to recent suggestions of Vogelstein et al. that three driver mutations are sufficient to initiate the majority of malignancies (Vogelstein and Kinzler, 2015).

In combination with machine-learning tools, autonomous self-learning systems, our simulations promise to contribute to a novel type of evidence and hypothesis generation in cancer research with full exploitation of computational power. The possibly enormous impact of such an approach on the current oncologic research landscape clearly merits an intensive evaluation of tools of artificial intelligence to better understand the process of carcinogenesis. Given such a powerful tool to investigate multiparametric processes in time-lapse experiments, one might no longer be able to justify a reductionist approach, which might not be sufficient when it comes to individual human beings. The same applies to the widely spread assertion that *EBM* is the best way to address huge amounts of data. In contrast, *PM* might be the better way here, enabled by encompassing simulation models to clinical challenges (Sugarman, 2012). Furthermore, in light of our final synthesis of hallmark capabilities and the ability to investigate processes and alterations at any arbitrary point in time and therefore observe a sequence of events via a simulation, one can also doubt the common end points in current cancer research. These end points usually focus on remission induction and treatment-free survival. However, in light of recent findings like the "evolutionary double-bind" (Willyard, 2016), it might be overdue to define new end points with respect to possibilities like living with cancer, that is, overall survival irrespective of remission rates.

Acknowledgments

I would like to thank my thesis advisor Prof. Dr. Christoph Borner. I would further like to acknowledge the essential programming and design work by Maximilian Georg and Oliver Worm. Furthermore, I deeply appreciate the professional support, the challenging discussions, and the expertise in transforming medical and biological concepts into algorithms and highly instructive computer-assisted visualizations of Dr. Boris Lau and Dr. Sascha Lange. Last but not least, I would like to appreciate the generous scholarship for this project provided by the BioThera Foundation (05/2017), Freiburg, Germany.

References

Almendro, V., Marusyk, A., Polyak, K., 2013. Cellular heterogeneity and molecular evolution in cancer. Annu. Rev. Pathol. Mech. Dis. 8, 277–302. https://doi.org/10.1146/annurev-pathol-020712-163923.

Alshalalfa, M., et al., 2013. Coordinate MicroRNA-mediated regulation of protein complexes in prostate cancer. PLoS One 8 (12), e84261. https://doi.org/10.1371/journal.pone.0084261.

Batut, B., et al., 2013. In silico experimental evolution: a tool to test evolutionary scenarios. BMC Bioinform. 14 (Suppl. 15), S11. https://doi.org/10.1186/1471-2105-14-S15-S11.

Beerenwinkel, N., et al., 2015. Cancer evolution: mathematical models and computational inference. Syst. Biol. 64 (1), e1–25. https://doi.org/10.1093/sysbio/syu081.

Blokh, D., Stambler, I., 2017. The application of information theory for the research of aging and aging-related diseases. Prog. Neurobiol 157, 158–173. https://doi.org/10.1016/j.pneurobio.2016.03.005.

Blokh, D., et al., 2007. The information-theory analysis of Michaelis-Menten constants for detection of breast cancer. Cancer Detect. Prev. 31 (6), 489–498. https://doi.org/10.1016/j.cdp.2007.10.010.

Cairns, J., 1975. Mutation selection and the natural history of cancer. Nature 255, cp1. https://doi.org/10.1038/255197a0.

De Carvalho, D.D., et al., 2012. DNA methylation screening identifies driver epigenetic events of cancer cell survival. Cancer Cell 21, 655–667. https://doi.org/10.1016/j.ccr.2012.03.045.

Fan, R., et al., 2011. Entropy-based information gain approaches to detect and to characterize gene-gene and gene-environment interactions/correlations of complex diseases. Genet. Epidemiol. 35 (7), 706–721. https://doi.org/10.1002/gepi.20621.

Geoerg, M., Lau, B., 2016. Personal communication.

Georg, M., Groten, J., Mertelsmann, R., 2016. Personal communication.

Greaves, M., Maley, C.C., 2012. Clonal evolution in cancer. Nature 481 (7381), 306–313. https://doi.org/10.1038/nature10762.

Groten, J., Borner, C., Mertelsmann, R., 2016a. Understanding and controlling cancer: the hallmark concept revisited – chance, evolution and entropy. JOSHA 3 (7), https://doi.org/10.17160/josha.3.7.252.

Groten, J., et al., 2016b. Towards simulating carcinogenesis. JOSHA 3 (7). https://doi.org/10.17160/josha.3.7.253.

Gupta, P.B., Chaffer, C.L., Weinberg, R.A., 2009. Cancer stem cells: mirage or reality? Nat. Med 15 (9), 1010–1012. https://doi.org/10.1038/nm.2032.

Hanahan, D., Weinberg, R.A., 2000. The hallmarks of cancer. Cell 100 (1), 57–70. https://doi.org/10.1007/s00262-010-0968-0.

Hanahan, D., Weinberg, R.A., 2011. Hallmarks of cancer: the next generation. Cell 144, 646–674. https://doi.org/10.1016/j.cell.2011.02.013.

Hayes, D.N., Kim, W.Y., 2015. The next steps in next-gen sequencing of cancer genomes. J. Clin. Investig. 125 (2), 462–468. https://doi.org/10.1172/JCI68339.

Jordan, C.T., Guzman, M.L., Noble, M., 2006. Cancer stem cells. N. Engl. J. Med. 355, 1253–1261. https://doi.org/10.1056/NEJMra061808.

Klein, C.A., 2013. Selection and adaptation during metastatic cancer progression. Nature 501 (7467), 365–372. https://doi.org/10.1038/nature12628.

Komarova, N.L., Wodarz, D., 2014. Targeted Cancer Treatment in Silico Small Molecule Inhibitors and Oncolytic Viruses. Springer Science+Business, New York.

Lowy, D.R., Collins, F.S., 2016. Aiming high—changing the trajectory for cancer. N. Engl. J. Med. 374 (20), 1901–1904. https://doi.org/10.1056/NEJMp1600894.

Luzzatto, L., Pandolfi, P.P., 2015. Causality and chance in the development of cancer. N. Engl. J. Med. 373 (1), 84–88. https://doi.org/10.1056/NEJMsb1502456.

Mackey, M.C., et al., 2015. The utility of simple mathematical models in understanding gene regulatory dynamics. In Silico Biol. 12 (1 & 2), 23–53. https://doi.org/10.3233/ISB-140463.

Merlo, L.M.F., et al., 2006. Cancer as an evolutionary and ecological process. Nature 6, 924–935. https://doi.org/10.1038/nrc2013.

Mertelsmann, R., Georg, M., 2016. Cancer: modeling evolution and natural selection, the "Mitosis Game". JOSHA 3 (1), https://doi.org/10.17160/josha.3.1.100.

Nowell, P.C., 1976. The clonal evolution of tumor cell populations. Science 194 (4260), 23–28. https://doi.org/10.1126/science.959840.

Pedregosa, et al., 2011. Scikit-learn: machine learning in python. J. Mach. Learn. Res. 12, 2825–2830. Available at http://scikit-learn.org/stable/modules/linear_model.html#ridge-regression.

Pelossof, R., et al., 2015. Affinity regression predicts the recognition code of nucleic acid–binding proteins. Nat. Biotechnol. 33 (12), 1242–1249. https://doi.org/10.1038/nbt.3343.

Riedmiller, M., et al., 2009. Reinforcement learning for robot soccer. Auton. Robot. 27 (1), 55–73. https://doi.org/10.1007/s10514-009-9120-4.

Schell, J., 2015. The Art of Game Design: A Book of Lenses, second ed. Available at https://books.google.de/books?hl=de&lr=&id=kRMeBQAAQBAJ&oi=fnd&pg=PP1&dq=balancing+game+design&ots=ACcAYa-i64&sig=XEERO2kFzHEixpJE6ZYdEfnfnks#v=onepage&q=balancing&f=false.

Shay, J.W., Wright, W.E., 2000. Hayflick, his limit, and cellular ageing. Nat. Rev. Mol. Cell Biol. 1, 72–76. https://doi.org/10.1038/35036093.

Sugarman, J., 2012. Questions concerning the clinical translation of cell-based interventions under an innovation pathway. J. Law, Med. Ethics 40 (4), 945–950. https://doi.org/10.1111/j.1748-720X.2012.00723.x.

Suthaharan, S., 2016. Machine learning models and algorithms for big data classification. Integr. Ser. Inf. Syst 36, 1–12. https://doi.org/10.1007/978-1-4899-7641-3.

Tomasetti, C., Vogelstein, B., 2015. Variation in cancer risk among tissues can be explained by the number of stem cell divisions. Science 347 (6217), 78–81.

Torrente, A., et al., 2016. Identification of cancer related genes using a comprehensive map of human gene expression. PLoS One 11 (6), e0157484. https://doi.org/10.1371/journal.pone.0157484.

Vogelstein, B., Kinzler, K.W., 2015. The path to cancer—three strikes and you're out. N. Engl. J. Med. 373 (20), 1895–1898.

Vogelstein, B., et al., 2013. Cancer genome landscapes. Science 339 (6127), 1546–1558. https://doi.org/10.1126/science.1235122.

Waclaw, B., Bozic, I., Pittman, M.E., et al., 2015. A spatial model predicts that dispersal and cell turnover limit intratumour heterogeneity. Nature 525, 261–264. https://doi.org/10.1038/nature14971.

Walker, D.C., Southgate, J., 2009. The virtual cell—a candidate co-ordinator for middle-out modelling of biological systems. Brief. Bioinform. 10 (4), 450–461. https://doi.org/10.1093/bib/bbp010.

Weber, K., et al., 2011. RGB marking facilitates multicolor clonal cell tracking. Nat. Med. 17 (4), 504–509. https://doi.org/10.1038/nm.2338.

Wikipedia, 2016. Blind Man and an Elephant. https://en.wikipedia.org/wiki/Blind_men_and_an_elephant.

Willyard, C., 2016. Cancer: an evolving threat. Nature 532, 166–168.

Worm, O., 2016. Personal communication.

CHAPTER

13

Sepsis: A Challenging Disease With New Promises for Personalized Medicine and Biomarker Discovery

Didem Dayangac-Erden, Mine Durusu-Tanriover
Hacettepe University, Ankara, Turkey

1 INTRODUCTION

Sepsis is defined as a life-threatening organ dysfunction as a result of a dysregulated host response to infection (Singer, 2016). As our understanding in the complex pathophysiology of sepsis has evolved over the years, we now recognize that it is not simply a sequential disequilibrium of inflammation and immunosuppression, but an intricate interplay of several biochemical, immunologic, and pathophysiological abnormalities. Moreover, a substantial variation exists among the septic population, rendering the standardization of the diagnostic and therapeutic approaches nearly impossible in a majority of circumstances. Although the recent years have witnessed a great progress in the immediate management and intensive care treatment of sepsis (Kaukonen et al., 2014), sepsis-associated mortality in the intensive care units on a global basis is still as high as 35.3% in patients with sepsis, with the organ dysfunction being the key factor predicting the outcome (Vincent et al., 2014). Moreover, our understanding of the complex pathophysiology of sepsis could not be translated into targeted therapeutic approaches.

Yet, no effective specific treatment modality exists for sepsis; it is evident that much of the progress in the recent years has been linked to early recognition leading to early antibiotic treatment and appropriate fluid resuscitation. Early recognition of sepsis and differential diagnosis of infection from other causes leading to organ dysfunction and hemodynamic instability may be quite hard in complex and complicated patients with multiple morbidities. Hence, biomarkers and clinical scoring systems may aim to determine an infection, to differentiate between systemic inflammatory response syndrome and sepsis, and to differentiate between sepsis that led organ dysfunction and other etiologies.

Precision Medicine
https://doi.org/10.1016/B978-0-12-805364-5.00013-5

2 EPIDEMIOLOGY AND BURDEN OF DISEASE

Sepsis is a disease with a high clinical and financial burden. It has been reported that 5.2% of the total hospital costs in the United States were attributed to sepsis in 2011 resulting in more than $20 billion expenditure (Torio and Andrews, 2006). The costs of sepsis each year in the United Kingdom were estimated as £7.76 billion (Hex et al., 2017). Epidemiological data regarding the incidence and mortality of sepsis differ depending on the methodology and on the population characteristics; however, studies with a similar methodology showed that incidence of severe sepsis is on the rise. Sepsis is now the leading cause of death in US hospitals and is estimated to account for more than 5 million deaths globally each year (Liu et al., 2014; Fleischmann et al., 2016). Fleischmann and colleagues calculated that the global yearly incidence in adults may be as high as 30.7 million cases of sepsis and 23.8 million cases of severe sepsis, with about 6 million fatalities (Fleischmann et al., 2015). The mortality of severe sepsis seems to be decreasing in the United States as in Australia and New Zealand (Kaukonen et al., 2014; Stoller et al., 2016; Kumar et al., 2011) while still having a mortality rate of approximately 20% in these countries. It should be noted that while the case fatality rate of severe sepsis is decreasing over the years, the total number of cases is increasing leading to a relatively steady state of total mortality in the United States (Kempker and Martin, 2016).

The global epidemiology of sepsis outside the United States is quite variable with a case fatality rate of severe sepsis changing between 22% and 55% (Kempker and Martin, 2016). For instance, the intensive care mortality overall in Europe was 27% (Vincent et al., 2006). The epidemiology of sepsis is much different in resource-limited settings, mainly affecting neonates and their mothers, and unfortunately, there is no significant decrease in these countries. An estimated 1.7 million neonates were affected by sepsis in 2010 in South Asia, sub-Saharan Africa, and Latin America and overall 1.2 million neonate deaths due to severe bacterial infections (Seale et al., 2013).

When evaluating the decreasing mortality of sepsis in developed countries, one should keep in mind that this might well be due to the increased denominator of cases, rather than drastically improved outcomes. So, this should not hinder the enthusiasm in the way we further search for new diagnostic tools and treatment modalities. The clinical outcomes of patients with sepsis are unfavorable even if the patients survive. Long-term disabilities pursue such as weakness and motor loss, long-term cognitive dysfunction, and mental health problems together named as the post-ICU syndrome (Oeyen et al., 2010; Myhren et al., 2010).

3 DEFINITIONS AND DIAGNOSIS OF SEPSIS

The very initial definition of sepsis was made in 1991, and this was the first time when sepsis was defined as the host's systemic inflammatory response syndrome (SIRS) (Table 1) to infection (Bone et al., 1992). The terms "severe sepsis" and "septic shock" were devised to define "sepsis complicated by organ dysfunction" and "sepsis-induced hypotension persisting despite adequate fluid resuscitation," respectively. Afterward, in 2001, new diagnostic criteria were added keeping the same frame of the initial definitions (Levy et al., 2003). However, these definitions had important caveats. First of all, SIRS criteria represented extremely nonspecific parameters overdiagnosing sepsis, while it could skip the diagnosis even though organ

TABLE 1 Systemic Inflammatory Response Syndrome (SIRS) Criteria

Temperature	>38°C or <36°C
Heart rate	>90/min
Respiratory rate	>20/min or $PaCO_2$ <32 mmHg (4.3 kPa)
White blood cell count	>12 000/mm^3 or <4000/mm^3 or >10% immature bands

At Least Two of the Parameters Should Be Present to Diagnose SIRS.
Adapted from Bone, R.C., Balk, R.A., Cerra, F.B., Dellinger, R.P., Fein, A.M., Knaus, W.A., Schein, R.M., Sibbald, W.J., 1992. Definitions for sepsis and organ failure and guidelines for the use of innovative therapies in sepsis. Chest 101, 1644–1655.

dysfunction already developed. Moreover, these definitions were far from explaining the very complex pathobiology of sepsis and describing the clinical context that is affecting the outcome.

As our understanding on the complex pathobiology of sepsis has increased, it became evident that sepsis is not a simple disease that demonstrates a smooth continuum to severe sepsis and septic shock. The degree of organ dysfunction is correlated with mortality, and in this regard, it was obvious that new definitions were needed, particularly for promptly recognizing cases of organ dysfunction. Hence, recently released definitions (The Third International Consensus Definitions for Sepsis and Septic Shock) describe sepsis "as life-threatening organ dysfunction caused by a dysregulated host response to infection" and leave the SIRS criteria simply to aid in the diagnosis of an infectious state (Singer, 2016). Organ dysfunction diagnosis is based on the sequential organ failure assessment (SOFA) score, and it has been accepted that an increase in the SOFA score of two points or more is associated with increased in-hospital mortality. Septic shock is defined as "a subset of sepsis in which particularly profound circulatory, cellular, and metabolic abnormalities are associated with a greater risk of mortality than with sepsis alone." Basically, the terminology "severe sepsis" is left to focus on the organ dysfunction state, and "septic shock" is now accepted not as a continuum but as a subset of sepsis (Table 2). A rapid screening tool to be used in out-of-hospital, emergency department, or general hospital ward settings is a new bedside clinical score termed quickSOFA (qSOFA). Adult patients with (suspected) infection and at least two of the clinical criteria are considered at high risk and require further assessment for evidence of organ dysfunction (Table 2).

As evident, sepsis is a syndrome with no gold standard or validated diagnostic test (Singer, 2016), but we rely on clinical criteria to sort out the diagnosis, yet clinical judgment might not be inferior to the tools we have in hand currently (Quinten et al., 2016). It's clear that a deeper understanding of the pathophysiology of sepsis will necessitate the need to further revise

TABLE 2 Clinical Criteria to Diagnose Sepsis and Septic Shock

Sepsis	Suspected or documented infection AND an acute increase of ≥2 SOFA points (a proxy for organ dysfunction)
Septic shock	Sepsis AND vasopressor therapy needed to elevate MAP ≥65 mmHg AND lactate >2 mmol/L (18 mg/dL) despite adequate fluid resuscitation
Sepsis screening-qSOFA	Respiratory rate ≥22/min, altered mentation, systolic blood pressure ≤100 mmHg

Adapted from Singer, M., Deutschman, C.S., Seymour, C.W., Shankar-Hari, M., Annane, D., Bauer, M., Bellomo, R., Bernard, G.R., Chiche, J.-D., Coopersmith, C.M., 2016. The third international consensus definitions for sepsis and septic shock (sepsis-3). Jama 315, 801–810.

these definitions. Early recognition and proper management of the patients might be better achieved with the integration of certain biomarkers to clinical criteria to generate new sets of diagnostic criteria in the future.

4 MANAGEMENT

The Surviving Sepsis Campaign has shifted the way we understand and manage sepsis with its 2002 guideline prioritizing early recognition, source control, appropriate and timely antibiotic administration, and resuscitation with intravenous fluids and vasoactive drugs in an algorithm, known as the "early goal-directed therapy." Actually, this sepsis resuscitation guideline mainly stemmed from Rivers' study in 2001, which showed a survival benefit in those patients who received a protocolized treatment in the first 6h of early septic shock upon their presentation to the emergency department (Rivers et al., 2001). Central venous pressure, mean arterial pressure, and central venous oxygen saturation were proposed as physiological targets to guide fluid and vasopressor therapy in these "golden hours." However, Rivers' study was criticized because of its small sample size, single center nature, and complexity of the protocol. The early goal-directed therapy was then tested in multicenter studies in different countries such as Protocolized Care for Early Septic Shock (ProCESS), Australasian resuscitation in sepsis evaluation (ARISE), and Protocolized Management in Sepsis (ProMISe) (Yealy et al., 2014; Peake et al., 2014). The recent one of these trials was the ProMISE trial in 56 centers in England with a hypothesis that the 6h early goal-directed therapy resuscitation protocol is superior, in terms of clinical effectiveness and cost-effectiveness, to usual resuscitation in adult patients presenting with early septic shock (Mouncey et al., 2015). As a result, no significant difference was found in terms of mortality at 90 days between those who received 6h of protocolized treatment and those who received usual care. On the other hand, protocolized treatment increased the costs. The findings of ProMISe study were actually the reflection of the proceedings in usual care over the 15 years.

Identification of septic shock in a timely manner and early intravenous antibiotics and adequate fluid resuscitation is the mainstay of this success, rendering the addition of continuous central venous oxygen saturation monitoring and strict protocolization of care useless and costly. The hospital mortality was shown to decrease from 46.5% to 30.5% in Rivers' study with associated care bundles; however, the 90-day mortality of the ProMISe study was already 29.2% in the usual care group (Rivers et al., 2001; Mouncey et al., 2015). ProCESS and ARISE trials also demonstrated a mortality in the usual care group that was lower than the anticipated with no survival benefit of the early goal-directed therapy protocol (The ProCESS Investigators, 2014; ARISE Investigators, ANZICS Clinical Trials Group, 2014). Systemic steroids also did not have a significant effect on mortality or serious adverse events of sepsis syndromes (Volbeda et al., 2015).

Cytokines as possible targets for sepsis therapy have been studied in the 1990s; however, lipopolysaccharide and tumor necrosis factor blockage turn out to be useless (Seeley and Bernard, 2016). Yet, no specific therapeutic strategy is to come, even those who were in practice guidelines in 2008 as activated protein C waned. Specific therapeutic agents are under test such as granulocyte-macrophage colony-3 stimulating factor (NCT00252915), interleukin 7 (NCT02640807), antiprogrammed cell death ligand-1 (NCT02576457), and thymosin

(NCT00711620). It seems that rather than targeting the mediators in the early stages of sepsis, targeting the mediators that take part in the organ injury process is worth studying. This is also important since at the time the patient admits to the emergency room or the intensive care unit, a considerable amount of time has elapsed that hinders the success of therapy against early mediators of sepsis. This is also in parallel to the recent developments in our appraisal of the sepsis pathogenesis and the role of organ failure in the clinical scene.

5 PROGNOSIS

The mortality of sepsis depends on certain factors such as the age, the preexisting illnesses, the inciting agent, the time of recognition, and the initiation of appropriate treatment. For instance, in a cohort of adult patients with invasive pneumococcal disease, the mortality in those with septic shock was 56.9% (Hanada et al., 2016). Sepsis is still the major cause of mortality in patients admitted to EDs or ICUs, with mortality ranging from 35% to 50%. Early recognition, appropriate classification, and proper and specific treatment in the initial phases of sepsis have resulted in improved outcomes in emergency departments and intensive care units (Strehlow et al., 2006).

Prognostic tools help allocate the limited resources more efficiently and guide the patients and families to decide about the intensity of the treatment (Cook et al., 2003). The prognosis is not all about the mortality. It is clear that critical illness is also an important means of morbidity posing a burden of functional loss and brain dysfunction as a result of postintensive care syndrome (PICS) (Pandharipande et al., 2013).

6 PATHOBIOLOGY DRAWS THE ROADMAP FOR BIOMARKER DISCOVERY

Sepsis is not a single disease entity but rather a syndrome provoked and maintained by complex pathobiological changes most of which still remain unelucidated. As once thought, sepsis is not simply a continuum of disease from sepsis to septic shock with a pathophysiology solely based on inflammation. Pathobiology of sepsis is so complicated that inflammatory and antiinflammatory pathways, innate and adaptive immune system, apoptosis, mitochondrial function, translational and transcriptional regulation, and oxidative biology, as well as additional intracellular and extracellular events, are involved at different stages with different kinetics. The view that the complex pathophysiology of immune dysfunction in sepsis may lead to an immune paralysis at some point leading to secondary infections as the main cause of mortality was partially denied by the recent results of the Molecular Diagnosis and Risk Stratification of Sepsis (MARS) program (Hotchkiss et al., 2013; Van Vught et al., 2016). The genomic response of patients with sepsis was consistent with immune suppression at the onset of secondary infection. However, those who had secondary infections and higher mortality rates indeed had higher disease severity scores on admission.

An excellent review by van der Poll and colleagues discussed the key mechanisms of the immune derangements that take place in sepsis patients (Van der Poll et al., 2017). The encounter of the infectious pathogen with the host innate immune system through pathogen-associated molecular patterns (PAMPs) and pattern recognition receptors (PRRs) might result in the clearance of the pathogen or an uncontrolled infection. The homeostasis of the immune system is severely disrupted in patients with sepsis in two extreme directions, excessive inflammation and immune suppression in different levels varying among individuals. Even though the infection is treated timely and properly, the aberrant immune response might persist. The release of damage-associated molecular patterns (DAMPs) can activate many of the PRRs that also recognize PAMPs, and this leads into a self-perpetuating vicious cycle.

An emerging concept in sepsis is the microvascular dysfunction. The endothelial surface layer-glycocalyx integrity is disrupted in sepsis leading to organ injury, edema, and inflammation (Colbert and Schmidt, 2016). Molecules such as circulating glycocalyx elements and angiopoietins have arisen as potential new biomarkers. Additionally, the organ injury and dysfunction are mainly thought to be related with altered cellular metabolism leading to impaired mitochondrial respiratory chain enzymes and oxidative phosphorylation.

7 CLASSICAL AND NOVEL BIOMARKERS

A biomarker can be defined as a "characteristic that can be objectively measured and evaluated as an indicator of normal biological processes, pathological processes, or pharmacological responses to a therapeutic intervention" (Colburn et al., 2001). Sankar and Webster reported in 2013 that 178 biomarkers have been described in the literature (Sankar and Webster, 2013). These ranged from specimen cultures and fractions of leukocytes to more specified biomarkers including cytokines, cell-surface markers, prohormones, enzymes, acute-phase proteins, coagulation factors, apoptosis mediators, receptors, and mediators. However, no single, widely accepted biomarker yet exists to diagnose sepsis with a clear-cut threshold. Moreover, the ultimate goal of developing targeted, personalized therapies to modulate the dysfunctional host immunologic response still remains to be approached. The ideal sepsis biomarker should have fast kinetics, high sensitivity and specificity, fully automated technology, short turnaround time, and low cost and should be available as a point-of-care test as Cohen and his colleagues specified (Cohen et al., 2015). As no single biomarker seems adequate for the diagnosis, monitorization, and prognostication of sepsis in all patients and circumstances, it is logical that an array of several biomarkers or clinical scoring systems that integrate certain biomarkers as point-of-care tests may serve in decision-making processes.

7.1 Microbiological Documentation

Microbiological cultures are the mainstay of diagnosis for an infection and hence required to document the etiology of sepsis appropriately. However, blood cultures are positive only in 30%–40% of severe sepsis and septic shock episodes (Bochud et al., 2004). This may be in part due to the intermittent presence of the microorganism in blood, the suboptimal performance of blood cultures for fastidious bacteria, and the initiation of antimicrobial therapy before the blood samples are obtained. Moreover, the true positivity of blood cultures can be as low as 5%–10% further increasing the uncertainty and cost of diagnosis sepsis (Alahmadi et al., 2011).

Recently, several molecular assays have been developed that allow rapid detection of pathogens directly from whole blood or positive blood cultures of sepsis patients with different performances (Cohen et al., 2015). These assays identify not only the pathogen but also the resistance genes. Nonamplified growth-dependent assays use fluorescence-based hybridization with PNA probes and gold nanoparticle technology, while amplified methods mainly utilize pathogen-specific real-time PCR- or broad-based technologies such as multiplex PCR (Riedel and Carroll, 2016). However, it is claimed that improvements are needed to improve DNA extraction procedures, to reduce contamination rates, and to reduce the detection threshold for difficult-to-grow, slow growing, or nongrowing microorganisms. Moreover, none of the methods was proved to have sufficient diagnostic accuracy, specificity, or sensitivity to replace conventional culture techniques. Assays that will allow point-of-care testing might be clinically relevant if fluid-based, automated platforms are developed to detect pathogens.

7.2 C-Reactive Protein

C-reactive protein is an acute-phase protein released from the liver as a response to IL-6 release in inflammation (Gabay and Kushner, 1999). Though it is not specific for infection, the levels of CRP correlate with the severity of infection (Povoa et al., 2005), and CRP is utilized to monitor the success of antimicrobial therapy in sepsis (Schmit and Vincent, 2008). However, CRP has some caveats such as a peak reached at 24–48h and the absence of specificity rendering it an unsuitable biomarker for sepsis. Several studies failed to show an association between the serum levels of CRP and the severity of sepsis (Tziolos et al., 2015).

7.3 Procalcitonin

Procalcitonin is the prohormone of calcitonin and is an inflammatory biomarker that may guide to determine the presence of a bacterial infection. Assicot and colleagues, in 1993, described a substance immunologically identical to procalcitonin that is raised during septic conditions and showed that the serum concentrations were correlated with the severity of microbial invasion (Assicot et al., 1993). Afterward, studies showed that procalcitonin can be integrated into models of diagnosis to improve the predictive power of standard indicators and hence was thought to be a promising biomarker (Harbarth et al., 2001). The serum levels of PCT start to rise as early as 4h after the onset of infection rendering it a useful biomarker for early diagnosis and early prognostication (Giamarellos-Bourboulis et al., 2002). Not only diagnosis of sepsis but also treatment of sepsis needs guidance. In the era of increasing antibiotic resistance, antibiotic stewardship programs aim to help the physicians to deliver the right antibiotic, at the right time and in the right duration to the right patient. Hence, biomarkers might help to combat with antibiotic resistance while guiding in the proper management of the infection that triggers sepsis. It should be kept in mind that PCT should not be used to start initial empiric antibiotic therapy, and antibiotics should be encouraged in all patients with suspicion of sepsis. Rather, PCT can be used to predict the severity of illness and to discontinue antibiotic. The cutoff to start and stop antibiotics is considered as 0.25ng/mL in patients with respiratory tract infections in the emergency room, while in critically ill patients in the ICU, cutoffs to stop antibiotics are higher as 0.5ng/mL. It has been suggested that procalcitonin led algorithms that can result in reduced antibiotic duration and hospital stay without comprising patient safety (Westwood et al., 2015). Moreover, utilization of procalcitonin to guide discontinuation of antibiotics in the follow-up of adult sepsis patients in the intensive care unit was shown to be cost-effective. A recent meta-analysis by Liu and colleagues showed that single PCT concentrations and PCT nonclearance were strongly associated with all-cause mortality in septic patients with a mention on studies that are needed to define the optimal cutoff point and the optimal definition (Liu et al., 2015). However, PCT did not show a significant area under the curve value for predicting in-hospital mortality as Hong and colleagues reported (Hong et al., 2016). A recent Cochrane review found no clear benefit of procalcitonin-guided antimicrobial therapy to minimize mortality, mechanical ventilation, clinical severity, and reinfection or duration of antimicrobial therapy of patients with septic conditions (Andriolo et al., 2017). However, PCT has been shown in various studies to reduce initiation and shortens the length of antibiotic treatment in different disease conditions (Sager et al., 2017). The beneficial role of PCT to de-escalate antibiotics particularly in critically ill patients has been shown in a Dutch study (De Jong et al., 2016). It is of note that this recent study was not included in the Cochrane review. An important consideration in septic patients is that renal impairment and reduced glomerular filtration rate may lower PCT clearance, and levels thus may be higher than expected (Heredia-Rodríguez et al., 2016). It's argued that in the case of acute kidney injury when creatinine ≥2 mg/dL, procalcitonin levels shall not be used as a marker of infection (Heredia-Rodríguez et al., 2016). Moreover, procalcitonin and C-reactive protein are eliminated during continuous renal replacement therapy, rendering them imperfect biomarkers in patients with acute kidney injury (Honore et al., 2015).

7.4 Lactate

Mitochondrial dysfunction leads to an anaerobic mechanism in the context of increased energy requirements. Lactate, in turn, has become the most commonly used biomarker to guide sepsis diagnosis and therapy. The primary sources of increased lactate production in sepsis are considered as tissue hypoperfusion and anaerobic metabolism. Yet, serum lactate elevation might also be linked to other several reasons such as hepatic and renal failure, drugs and toxins, and inborn errors of metabolism (Andersen et al., 2013). Levels ≥4.0 should be alarming for sepsis, even in patients with normal vital signs (Jones, 2013). Thus, lactate should be considered as a very early biomarker of sepsis and has been shown to be associated with mortality (Mikkelsen et al., 2009). Moreover, lactate clearance of 10% and above is deemed as a positive marker of a successful resuscitation in sepsis treatment in association with other clinical markers (Jones et al., 2010).

7.5 Circulating Immunoglobulins

As innate and adaptive immune responses might demonstrate dynamic alterations in the course of sepsis, measurement of circulating levels of immunoglobulins in the blood of patients may be a biomarker to distinguish patients with severe sepsis who might benefit from intravenous immunoglobulins G, A, and M treatment (Bermejo-Martín et al., 2014). It's also claimed that the presence of specific memory B cells and IgG against the infecting microorganism and previous immunosuppression might have consequences in the presepsis period (Bermejo-Martin and Giamarellos-Bourboulis, 2015). Likewise, consumption or production deficit of immunoglobulin isotypes might have synergistic effects on mortality (Shankar-Hari et al., 2015).

7.6 Cytokines

One of the best studied cytokine in sepsis is IL-6, which is a proinflammatory cytokine. It increases rapidly following acute stimulation of the immune cells and has a half-life of less than 6h. Persistently, high levels of IL-6 were demonstrated in patients with sepsis and were found to be associated with a worse prognosis (Jekarl et al., 2013). Peak IL-6 levels were found to be independently associated with increased mortality (Mat-Nor et al., 2016).

7.7 Dendritic Cells

Dendritic cells are thought to have a pivotal role in immune activation. Both myeloid and plasmacytoid cells were found to be lower in pediatric sepsis patients and even lower in complicated sepsis patients (Elsayh et al., 2013). CD86 and CD83 expression on the entire dendritic cells was found to be low accordingly. Survivors had significantly higher dendritic cell ratios and CD86 and CD83 expression. Hence, it was concluded that sepsis is associated with a reduced level of DCs and decreases their maturation. The estimation of dendritic cell number and maturation state may be used as prognostic markers of sepsis.

7.8 Endothelial Biomarkers

Vascular endothelial injury is a proposed mechanism for organ dysfunction (Coletta et al., 2014; Boisrame-Helms et al., 2013), and profound endothelial damage is claimed to be associated with death in sepsis (Johansen et al., 2015). The glycocalyx works to maintain endothelial permeability, to regulate leukocyte migration and to inhibit intravascular coagulation. Protection of the integrity of the glycocalyx requires hyaluronan and syndecan. The injury to the vascular endothelium may contribute to organ dysfunction via increased leakiness, and the severity of sepsis may be correlated with the circulating levels of hyaluronan and syndecan as few studies have shown (Köhler et al., 2011). Dimple and colleagues recently suggested that serial measurements of hyaluronan and syndecan could help in predicting the prognosis and the outcome of patients with sepsis (Anand et al., 2016).

Among the potential endothelial biomarkers are vascular cell adhesion molecule, soluble intercellular adhesion molecule, sE-selectin, plasminogen activator inhibitor, angiopoietins, and soluble fms-like tyrosine kinase and endocan (Palud et al., 2015). Angiopoietin-2 was shown to be higher in patients presenting with organ dysfunction and was associated with a higher mortality. Endocan was suggested as a biomarker for lung injury (Palud et al., 2015) and a valuable diagnostic and prognostic biomarker even at the very initial presentation of the severe sepsis and septic shock (Pauly et al., 2016).

7.9 Endothelins

Endothelin-1 among the endothelin family is a strong vasoconstrictor that is released in response to bacterial endotoxin and certain cytokines. The precursor peptide C-terminal proendothelin-1 (CT-proET-1) is a more stable molecule than endothelin-1 that it allows a stoichiometric measurement of ET-1. It has been shown that CT-proET-1 serum concentrations were increased in critically ill patients, more significantly in those with sepsis

(Buendgens et al., 2017). Moreover, CT-proET-1 serum levels were associated with clinical disease severity scores and organ dysfunction. Being an independent predictor of ICU mortality, CT-proET-1 might be a promising sepsis biomarker.

7.10 Presepsin

Presepsin, a soluble fragment of CD14 involved in pathogen recognition by innate immunity, has recently been shown to be an early biomarker of sepsis (Shozushima et al., 2011) that might be a more specific marker of bacterial infections (Endo et al., 2012; Zhang et al., 2015). Presepsin can be used to predict bacteremia and bacterial DNAemia (Leli et al., 2016). Per with the results of the ALBIOS study, presepsin concentration increased with the severity of organ dysfunction and with inappropriate antibiotic therapy, whereas it decreased in sepsis patients over 7 days who had negative blood cultures or positive blood cultures but treated with appropriate antibiotic (Masson et al., 2015). It seems that presepsin might be a strong candidate to predict mortality and to guide the appropriateness of antibiotic therapy.

7.11 Neutrophil Gelatinase Associated Lipocalin

Neutrophil gelatinase-associated lipocalin (NGAL) is a 25 kDa protein in the lipocalin family that regulates the small-molecule traffic. NGAL has emerged as a new, most extensively studied biomarker in the diagnosis of acute kidney injury in different settings, and it is found to be predictive of acute kidney injury and its severity (Haase-Fielitz et al., 2014). Sepsis-associated acute kidney injury is common and needs to be diagnosed timely in order to improve the prognosis. The pooled sensitivity and specificity of NGAL were 0.83 (95% confidence interval (CI), 0.77–0.88) and 0.57 (95% CI, 0.54–0.61), respectively (Hong et al., 2016). The AUC was 0.86 (standard error (SE) = 0.04) indicating good, but not excellent, diagnostic accuracy. On the other hand, urine NGAL pooled sensitivity and specificity were 0.80 (95% CI, 0.77–0.83) and 0.80 (95% CI, 0.77–0.83), respectively, with an AUC of 0.90 (SE = 0.02). Hence, NGAL seems as a good predictive factor for AKI in sepsis and associated mortality (Hong et al., 2016). The use of NGAL as a biomarker or AKI has some limitations though. Perhaps the most important of all is the level of NGAL that can increase sepsis in patients who have not developed AKI (Lentini et al., 2012). Other biomarkers such as alpha-1-microglobulin and soluble triggering receptor expressed on myeloid cells-1 have also been studied in a small number of studies (Umbro et al., 2016).

7.12 Cholesterol

Cholesterol has an important role in maintaining immunity and protecting from various kinds of inflammation in conditions of critical illness. As the adrenal glands do not store cortisol, cholesterol is the principal precursor for steroid biosynthesis in steroidogenic tissue in times of increased stress. Systemic inflammation leads to a catabolic state, and malnutrition is an important marker for poor prognosis in sepsis patients. Yamano and colleagues demonstrated that the minimum value of serum total cholesterol level was a significant prognostic indicator for those sepsis patients who stayed in the ICU for longer than 2 weeks (Yamano et al., 2016). Metabolomic profiling in sepsis is one of the promising tools for determining metabolic alterations. The data obtained from metabolomics analysis in mice infected with strains of *Streptococcus pneumoniae* of different virulence leading to organ failure showed that the regulation of cholesterol biosynthesis in the liver as a most relevant pathogenetic factor with respect to outcome and clinical course (Weber et al., 2012).

7.13 Genetic Variants

Susceptibility to early onset and late onset of sepsis in adults and newborns can be identified with genome-wide association studies (GWAS). Several studies examined the association of sepsis with single-nucleotide polymorphisms (SNPs) in genes responsible from pathogen recognition and pro- and antiinflammatory responses (tumor necrosis factor, migration inhibitory factor, plasminogen activator-1, and toll-like receptor genes) and showed conflicting results (Chauhan and McGuire, 2008; Arcaroli et al., 2005; Abu-Maziad et al., 2010; Sutherland et al., 2005; Ahrens et al., 2004; Srinivasan et al., 2017). In a recent study conducted in premature infants with sepsis (Srinivasan et al., 2017), Srinivasan and colleagues have shown that SNPs in FOXC2, FOXL1, ELMO1, IRAK2, RALA, IMMP2L, and PIEZO2 genes that are related to innate immunity, alteration of gut, and respiratory epithelial integrity reached significant levels compared with infants without sepsis. Findings from pathway analyses also confirmed that these pathways may be important mechanisms in genetic susceptibility to infection.

7.14 Cell Free DNA

Cell-free DNA is released from necrotic or apoptotic cells and can be considered as a direct marker of apoptosis in case of sepsis (Zeerleder et al., 2003). Clementi and colleagues have shown that cell-free DNA levels are increased in severe sepsis accompanied by acute kidney injury and correlate with caspase-3, IL-6, and IL-18 (Clementi et al., 2016).

7.15 MicroRNA

In recent years, attention has focused on the use of microRNAs (miRNAs) that are short noncoding endogenous RNA molecules as diagnostic, prognostic, and predictive biomarkers of sepsis. Up to now, more than 1000 miRNAs specific to the human genome have been identified, and it is thought that these miRNAs regulate expression of more than 60% of human genes (Friedman et al., 2009; Krol et al., 2010). To date, with the development of high-throughput sequencing technologies and computational approaches, thousands of miRNAs have been screened and identified as potential transcriptional biomarkers. They harbor several advantages compared with conventional protein-based markers due to their small, less complex chemical structure. In recent years, there are a growing number of studies about how miRNA expression regulates disease process and pathogenesis, especially in infectious diseases (Correia et al., 2017). It has been discovered that they play a regulatory role in various biological processes such as host-pathogen interaction, immunobiology, and inflammation. Identification of a prognostic miRNA signature will offer early detection of disease progression, and also new mechanisms and pathways associated with sepsis will be elucidated in the field of sepsis research. The major advantage of miRNA biomarker analyzed from blood is that it can be used as a point-of-care-based test.

In the literature, miR-25, miR-143, miR-146a, miR-15a, miR-16, miR-126, miR-150, and miR-223 were found to be differentially expressed in sepsis and were regarded as diagnostic and/or prognostic biomarkers (Wang et al., 2010, 2012a,b; Ho et al., 2016; Vasilescu et al., 2009; Roderburg et al., 2013). Some other miRNAs have been found to be related to complications associated with sepsis. miR-122 was associated with liver injury in septic patients (Leelahavanichkul et al., 2015; Roderburg et al., 2015). miR-574-5p and miR-155 were associated with the development of septic shock and respiratory failure (Wang et al., 2012; Liu et al., 2016).

8 COMPOSITE SYSTEMS AND POINT-OF-CARE TESTS

Several composite systems have been studied in sepsis. These systems integrate clinical and laboratory data to yield scoring systems. Acute physiology and chronic health evaluation (APACHE) and SOFA are the most widely used systems that delineate the severity of the patient and the severity of organ dysfunction, respectively. Certain systems evaluated the addition of other parameters to these scores. For instance, infection probability score that consists of heart rate, respiratory rate body temperature, white blood cell count, C-reactive protein, and SOFA score (Bota et al., 2003) has been used to exclude infection in SIRS patients. As a result, it was found that the infection probability score among patients with or without infection was no different (Ratzinger et al., 2013). Some other studies found improvement in the diagnostic or prediction accuracy of parameters when taken together such as addition of peak IL-6 concentration to SOFA score improved risk assessment for the prediction of mortality (Mat-Nor et al., 2016) or combination of C-reactive protein, procalcitonin, and neutrophil CD64 improved the diagnostic accuracy of sepsis (Bauer et al., 2016). However, many studies are disappointing about generating promising data. Certain molecules that increase in the inflammatory process were studied as candidate biomarkers still ending at PCT and CRP as the combination of biomarkers with the highest discriminative ability for bacterial sepsis (Han et al., 2015). A multimarker approach utilizing PCT, presepsin, galectin-3, and soluble suppression of tumorigenicity 2 was shown to predict 30-day mortality of patients with sepsis (Kim et al., 2017).

9 DATA MANAGEMENT

Sepsis is a leading health problem that requires individualized diagnosis and treatment, and the possibility of incorporating novel biomarkers in the sepsis management algorithm is now more exciting than ever. Such a proceeding will change the lives of millions. Clinical decision support systems are computerized tools that aim to help clinicians in their clinical decision-making and patient management processes. These systems may derive the data from the electronic health-care records of the patients. Recently, clinical decision support systems that utilize vital sign data to screen for sepsis and septic shock are developed that rely on machine learning flow (Calvert et al., 2017).

Another aspect of data management in the area of sepsis research is the management of the "big data." The big data obtained from each individual's genetic profile, environment, and nutrition combined with omics technologies (genomics, transcriptomics, proteomics, and metabolomics) need to be linked and interpreted by bioinformatics analysis. This high-level computational analysis will combine all of the system biology approaches and help to identify the key molecules.

10 PERSPECTIVE

Sepsis is one of the oldest diseases described in history, and it seems that it will have an impact in the future as it is one of the major reasons for organ dysfunction, death, and functional disability. Evolving from antimicrobials and supportive care, the science is now trying to enlighten the very complicated pathobiological pathways aiming for specific therapeutic agents. Even though decreased mortality rates in the last 10–15 years are promising, it seems that there is still a long way to go.

The complex interactions of the host-microorganism-environment render sepsis a strong candidate for personalized medicine approaches. Heterogeneous pathobiological pathways either predispose to or protect individuals from sepsis while also determining the course of the disease. There is enormous interindividual variability in sepsis cases alongside with a very rapidly fluctuating disease course within the history of the individual case.

Biomarker research should be based upon the need to stratify patients according to the dominant immunopathophysiological mechanism, so that personalized and targeted therapies can reach the patient. At the moment, biomarkers that can be integrated into diagnostic, prognostic algorithms and useful stratification systems and that can be used as drug targets are lacking (Pierrakos and Vincent, 2010). Moreover, biomarkers are needed to classify and enroll patients in clinical trials that aim to develop or study targeted therapies for sepsis. It's the era for precision medicine and hence the time to chase "ideal" biomarkers.

Host and pathogen factors like host DNA, host gene expression, secreted/circulating proteins, and type of microorganism contribute to sepsis outcome. Depending on the genotype of the patient, "companion diagnostic tests" should be developed, and each patient's genotype should match with related treatment. This makes sepsis a prototype disease for personalized medicine. Different combinations of biomarkers are currently being studied to provide molecular signatures; also integrating different omics technologies will help to develop a personalized treatment approach for sepsis. As the pathobiological proximal pathways are enlightened, "omics" and cellular assays together with bioinformatics to produce the "big data" shall guide the ever-changing management strategies of sepsis.

References

Abu-Maziad, A., Schaa, K., Bell, E.F., Dagle, J.M., Cooper, M., Marazita, M.L., Murray, J.C., 2010. Role of polymorphic variants as genetic modulators of infection in neonatal sepsis. Pediatr. Res. 68, 323–329.

Ahrens, P., Kattner, E., Kohler, B., Hartel, C., Seidenberg, J., Segerer, H., Moller, J., Gopel, W., The Genetic Factors in Neonatology Study Group, 2004. Mutations of genes involved in the innate immune system as predictors of sepsis in very low birth weight infants. Pediatr. Res. 55, 652–656.

Alahmadi, Y., Aldeyab, M., McElnay, J., Scott, M., Elhajji, F.D., Magee, F., Dowds, M., Edwards, C., Fullerton, L., Tate, A., 2011. Clinical and economic impact of contaminated blood cultures within the hospital setting. J. Hosp. Infect. 77, 233–236.

Anand, D., Ray, S., Srivastava, L.M., Bhargava, S., 2016. Evolution of serum hyaluronan and syndecan levels in prognosis of sepsis patients. Clin. Biochem. 49, 768–776.

Andersen, L.W., Mackenhauer, J., Roberts, J.C., Berg, K.M., Cocchi, M.N., Donnino, M.W., 2013. In: Etiology and therapeutic approach to elevated lactate levels. Mayo Clinic Proceedings. Elsevier, pp. 1127–1140.

Andriolo, B.N., Andriolo, R.B., Salomão, R., Atallah, Á.N., 2017. Effectiveness and safety of procalcitonin evaluation for reducing mortality in adults with sepsis, severe sepsis or septic shock. The Cochrane Library.

Arcaroli, J., Fessler, M.B., Abraham, E., 2005. Genetic polymorphisms and sepsis. Shock 24, 300–312.

ARISE Investigators, ANZICS Clinical Trials Group, 2014. Goal-directed resuscitation for patients with early septic shock. N. Engl. J. Med. 2014, 1496–1506.

Assicot, M., Bohuon, C., Gendrel, D., Raymond, J., Carsin, H., Guilbaud, J., 1993. High serum procalcitonin concentrations in patients with sepsis and infection. Lancet 341, 515–518.

Bauer, P.R., Kashyap, R., League, S.C., Park, J.G., Block, D.R., Baumann, N.A., Algeciras-Schimnich, A., Jenkins, S.M., Smith, C.Y., Gajic, O., 2016. Diagnostic accuracy and clinical relevance of an inflammatory biomarker panel for sepsis in adult critically ill patients. Diagn. Microbiol. Infect. Dis. 84, 175–180.

Bermejo-Martin, J.F., Giamarellos-Bourboulis, E.J., 2015. Endogenous immunoglobulins and sepsis: new perspectives for guiding replacement therapies. Int. J. Antimicrob. Agents 46, S25–S28.

Bermejo-Martín, J., Rodriguez-Fernandez, A., Herrán-Monge, R., Andaluz-Ojeda, D., Muriel-Bombín, A., Merino, P., García-García, M., Citores, R., Gandia, F., Almansa, R., 2014. Immunoglobulins IgG1, IgM and IgA: a synergistic team influencing survival in sepsis. J. Intern. Med. 276, 404–412.

Bochud, P.-Y., Bonten, M., Marchetti, O., Calandra, T., 2004. Antimicrobial therapy for patients with severe sepsis and septic shock: an evidence-based review. Crit. Care Med. 32, S495–S512.

Boisrame-Helms, J., Kremer, H., Schini-Kerth, V., Meziani, F., 2013. Endothelial dysfunction in sepsis. Curr. Vasc. Pharmacol. 11, 150–160.

Bone, R.C., Balk, R.A., Cerra, F.B., Dellinger, R.P., Fein, A.M., Knaus, W.A., Schein, R.M., Sibbald, W.J., 1992. Definitions for sepsis and organ failure and guidelines for the use of innovative therapies in sepsis. Chest 101, 1644–1655.

Bota, D.P., Mélot, C., Ferreira, F.L., Vincent, J.-L., 2003. Infection Probability Score (IPS): a method to help assess the probability of infection in critically ill patients. Crit. Care Med. 31, 2579–2584.

Buendgens, L., Yagmur, E., Bruensing, J., Herbers, U., Baeck, C., Trautwein, C., Koch, A., Tacke, F., 2017. C-terminal proendothelin-1 (CT-proET-1) is associated with organ failure and predicts mortality in critically ill patients. J. Intensive Care 5, 25.

Calvert, J., Hoffman, J., Barton, C., Shimabukuro, D., Ries, M., Chettipally, U., Kerem, Y., Jay, M., Mataraso, S., Das, R., 2017. Cost and mortality impact of an algorithm-driven sepsis prediction system. J. Med. Econ. 20, 646–651.

Chauhan, M., McGuire, W., 2008. Interleukin-6 (-174C) polymorphism and the risk of sepsis in very low birth weight infants: meta-analysis. Arch. Dis. Child. Fetal Neonatal Ed. 93, F427–9.

Clementi, A., Virzì, G.M., Brocca, A., Pastori, S., De Cal, M., Marcante, S., Granata, A., Ronco, C., 2016. The role of cell-free plasma DNA in critically Ill patients with sepsis. Blood Purif. 41, 34–40.

Cohen, J., Vincent, J.-L., Adhikari, N.K., Machado, F.R., Angus, D.C., Calandra, T., Jaton, K., Giulieri, S., Delaloye, J., Opal, S., 2015. Sepsis: a roadmap for future research. Lancet Infect. Dis. 15, 581–614.

Colbert, J.F., Schmidt, E.P., 2016. Endothelial and microcirculatory function and dysfunction in sepsis. Clin. Chest Med. 37, 263–275.

Colburn, W., Degruttola, V.G., Demets, D.L., Downing, G.J., Hoth, D.F., Oates, J.A., Peck, C.C., Schooley, R.T., Spilker, B.A., Woodcock, J., 2001. Biomarkers and surrogate endpoints: preferred definitions and conceptual framework. Biomarkers Definitions Working Group. Clin. Pharmacol. Ther. 69, 89–95.

Coletta, C., Módis, K., Oláh, G., Brunyánszki, A., Herzig, D.S., Sherwood, E.R., Ungvári, Z., Szabo, C., 2014. Endothelial dysfunction is a potential contributor to multiple organ failure and mortality in aged mice subjected to septic shock: preclinical studies in a murine model of cecal ligation and puncture. Crit. Care 18, 511.

Cook, D., Rocker, G., Marshall, J., Sjokvist, P., Dodek, P., Griffith, L., Freitag, A., Varon, J., Bradley, C., Levy, M., 2003. Withdrawal of mechanical ventilation in anticipation of death in the intensive care unit. N. Engl. J. Med. 349, 1123–1132.

Correia, C.N., Nalpas, N.C., Mcloughlin, K.E., Browne, J.A., Gordon, S.V., Machugh, D.E., Shaughnessy, R.G., 2017. Circulating microRNAs as potential biomarkers of infectious disease. Front. Immunol. 8, 118.

De Jong, E., Van Oers, J.A., Beishuizen, A., Vos, P., Vermeijden, W.J., Haas, L.E., Loef, B.G., Dormans, T., Van Melsen, G.C., Kluiters, Y.C., 2016. Efficacy and safety of procalcitonin guidance in reducing the duration of antibiotic treatment in critically ill patients: a randomised, controlled, open-label trial. Lancet Infect. Dis. 16, 819–827.

Elsayh, K.I., Zahran, A.M., Mohamad, I.L., Aly, S.S., 2013. Dendritic cells in childhood sepsis. J. Crit. Care 28, 881.e7–881.e13.

Endo, S., Suzuki, Y., Takahashi, G., Shozushima, T., Ishikura, H., Murai, A., Nishida, T., Irie, Y., Miura, M., Iguchi, H., 2012. Usefulness of presepsin in the diagnosis of sepsis in a multicenter prospective study. J. Infect. Chemother. 18, 891–897.

Fleischmann, C., Scherag, A., Adhikari, N., Hartog, C., Tsaganos, T., Schlattmann, P., Angus, D., Reinhart, K., 2015. Global burden of sepsis: a systematic review. Crit. Care 19, P21.

Fleischmann, C., Scherag, A., Adhikari, N.K., Hartog, C.S., Tsaganos, T., Schlattmann, P., Angus, D.C., Reinhart, K., International Forum of Acute Care Trialists, 2016. Assessment of global incidence and mortality of hospital-treated sepsis. Current estimates and limitations. Am. J. Respir. Crit. Care Med. 193, 259–272.

Friedman, R.C., Farh, K.K.H., Burge, C.B., Bartel, D.P., 2009. Most mammalian mRNAs are conserved targets of microRNAs. Genome Res. 19, 92–105.

Gabay, C., Kushner, I., 1999. Acute-phase proteins and other systemic responses to inflammation. N. Engl. J. Med. 340, 448–454.

Giamarellos-Bourboulis, E.J., Mega, A., Grecka, P., Scarpa, N., Koratzanis, G., Thomopoulos, G., Giamarellou, H., 2002. Procalcitonin: a marker to clearly differentiate systemic inflammatory response syndrome and sepsis in the critically ill patient? Intensive Care Med. 28, 1351–1356.

Haase-Fielitz, A., Haase, M., Devarajan, P., 2014. Neutrophil gelatinase-associated lipocalin as a biomarker of acute kidney injury: a critical evaluation of current status. Ann. Clin. Biochem. 51, 335–351.

Han, J.H., Nachamkin, I., Coffin, S.E., Gerber, J.S., Fuchs, B., Garrigan, C., Han, X., Bilker, W.B., Wise, J., Tolomeo, P., 2015. Use of a combination biomarker algorithm to identify medical intensive care unit patients with suspected sepsis at very low likelihood of bacterial infection. Antimicrob. Agents Chemother. 59, 6494–6500.

Hanada, S., Iwata, S., Kishi, K., Morozumi, M., Chiba, N., Wajima, T., Takata, M., Ubukata, K., Invasive Pneumococcal Diseases Surveillance Study Group, 2016. Host factors and biomarkers associated with poor outcomes in adults with invasive pneumococcal disease. PLoS One 11, e0147877.

Harbarth, S., Holeckova, K., Froidevaux, C., Pittet, D., Ricou, B., Grau, G.E., Vadas, L., Pugin, J., 2001. Diagnostic value of procalcitonin, interleukin-6, and interleukin-8 in critically ill patients admitted with suspected sepsis. Am. J. Respir. Crit. Care Med. 164, 396–402.

Heredia-Rodríguez, M., Bustamante-Munguira, J., Fierro, I., Lorenzo, M., Jorge-Monjas, P., Gómez-Sánchez, E., Álvarez, F.J., Bergese, S.D., Eiros, J.M., Bermejo-Martin, J.F., 2016. Procalcitonin cannot be used as a biomarker of infection in heart surgery patients with acute kidney injury. J. Crit. Care 33, 233–239.

Hex, N., Retzler, J., Bartlett, C., Arber, M., 2017. The cost of sepsis care in the UK, YHEC Sepsis Report. Whitewater Charitable Trust.

Ho, J., Chan, H., Wong, S.H., Wang, M.H., Yu, J., Xiao, Z., Liu, X., Choi, G., Leung, C.C., Wong, W.T., Li, Z., Gin, T., Chan, M.T., Wu, W.K., 2016. The involvement of regulatory non-coding RNAs in sepsis: a systematic review. Crit. Care 20, 383.

Hong, D.Y., Kim, J.W., Paik, J.H., Jung, H.M., Baek, K.J., Park, S.O., Lee, K.R., 2016. Value of plasma neutrophil gelatinase-associated lipocalin in predicting the mortality of patients with sepsis at the emergency department. Clin. Chim. Acta 452, 177–181.

Honore, P.M., Jacobs, R., Hendrickx, I., De Waele, E., Van Gorp, V., Spapen, H.D., 2015. "Biomarking" infection during continuous renal replacement therapy: still relevant? Crit. Care 19, 232.

Hotchkiss, R.S., Monneret, G., Payen, D., 2013. Sepsis-induced immunosuppression: from cellular dysfunctions to immunotherapy. Nat. Rev. Immunol. 13, 862–874.

Jekarl, D.W., Lee, S.-Y., Lee, J., Park, Y.-J., Kim, Y., Park, J.H., Wee, J.H., Choi, S.P., 2013. Procalcitonin as a diagnostic marker and IL-6 as a prognostic marker for sepsis. Diagn. Microbiol. Infect. Dis. 75, 342–347.

Johansen, M.E., Johansson, P.I., Ostrowski, S.R., Bestle, M.H., Hein, L., Jensen, A.L., Søe-Jensen, P., Andersen, M.H., Steensen, M., Mohr, T., 2015. In: Profound endothelial damage predicts impending organ failure and death in sepsis. Seminars in Thrombosis and Hemostasis. Thieme Medical Publishers, pp. 016–025.

Jones, A.E., 2013. Lactate clearance for assessing response to resuscitation in severe sepsis. Acad. Emerg. Med. 20, 844–847.

Jones, A.E., Shapiro, N.I., Trzeciak, S., Arnold, R.C., Claremont, H.A., Kline, J.A., Emergency Medicine Shock Research Network (EMShockNet) Investigators, 2010. Lactate clearance vs central venous oxygen saturation as goals of early sepsis therapy: a randomized clinical trial. JAMA 303, 739–746.

Kaukonen, K., Bailey, M., Suzuki, S., Pilcher, D., Bellomo, R., 2014. Mortality related to severe sepsis and septic shock among critically ill patients in Australia and New Zealand, 2000–2012. JAMA 311, 1308–1316.

Kempker, J.A., Martin, G.S., 2016. The changing epidemiology and definitions of sepsis. Clin. Chest Med. 37, 165.

Kim, H., Hur, M., Moon, H.-W., Yun, Y.-M., Di Somma, S., 2017. Multi-marker approach using procalcitonin, presepsin, galectin-3, and soluble suppression of tumorigenicity 2 for the prediction of mortality in sepsis. Ann. Intensive Care 7, 27.

Köhler, M., Kaufmann, I., Briegel, J., Jacob, M., Goeschl, J., Rachinger, W., Thiel, M., Rehm, M., 2011. The endothelial glycocalyx degenerates with increasing sepsis severity. Crit. Care 15, P22.

Krol, J., Loedige, I., Filipowicz, W., 2010. The widespread regulation of microrna biogenesis, function and decay. Nat. Rev. Genet. 11, 597–610.

Kumar, G., Kumar, N., Taneja, A., Kaleekal, T., Tarima, S., McGinley, E., Jimenez, E., Mohan, A., Khan, R.A., Whittle, J., 2011. Nationwide trends of severe sepsis in the 21st century (2000–2007). Chest J. 140, 1223–1231.

Leelahavanichkul, A., Somparn, P., Panich, T., Chancharoenthana, W., Wongphom, J., Pisitkun, T., Hirankarn, N., Eiam-Ong, S., 2015. Serum miRNA-122 in acute liver injury induced by kidney injury and sepsis in CD-1 mouse models. Hepatol. Res. 45, 1341–1352.

Leli, C., Ferranti, M., Marrano, U., Al Dhahab, Z.S., Bozza, S., Cenci, E., Mencacci, A., 2016. Diagnostic accuracy of presepsin (sCD14-ST) and procalcitonin for prediction of bacteraemia and bacterial DNAaemia in patients with suspected sepsis. J. Med. Microbiol. 65, 713–719.

Lentini, P., De Cal, M., Clementi, A., D'angelo, A., Ronco, C., 2012. Sepsis and Aki in ICU patients: the role of plasma biomarkers. Crit. Care Res. Practice 2012, 856401.

Levy, M.M., Fink, M.P., Marshall, J.C., Abraham, E., Angus, D., Cook, D., Cohen, J., Opal, S.M., Vincent, J.L., Ramsay, G., Conf, I.S.D., 2003. 2001 SCCM/ESICM/ACCP/ATS/SIS international sepsis definitions conference. Crit. Care Med. 31, 1250–1256.

Liu, V., Escobar, G.J., Greene, J.D., Soule, J., Whippy, A., Angus, D.C., Iwashyna, T.J., 2014. Hospital deaths in patients with sepsis from 2 independent cohorts. J. Am. Med. Assoc. 312, 90–92.

Liu, S., Liu, C., Wang, Z., Huang, J., Zeng, Q., 2016. MicroRNA-23a-5p acts as a potential biomarker for sepsis-induced acute respiratory distress syndrome in early stage. Cell. Mol. Biol. 62, 31–37.

Liu, D., Su, L., Han, G., Yan, P., Xie, L., 2015. Prognostic value of procalcitonin in adult patients with sepsis: a systematic review and meta-analysis. PLoS One 10, e0129450.

Masson, S., Caironi, P., Fanizza, C., Thomae, R., Bernasconi, R., Noto, A., Oggioni, R., Pasetti, G.S., Romero, M., Tognoni, G., 2015. Circulating presepsin (soluble CD14 subtype) as a marker of host response in patients with severe sepsis or septic shock: data from the multicenter, randomized ALBIOS trial. Intensive Care Med. 41, 12–20.

Mat-Nor, M.B., Ralib, A.M., Abdulah, N.Z., Pickering, J.W., 2016. The diagnostic ability of procalcitonin and interleukin-6 to differentiate infectious from noninfectious systemic inflammatory response syndrome and to predict mortality. J. Crit. Care 33, 245–251.

Mikkelsen, M.E., Miltiades, A.N., Gaieski, D.F., Goyal, M., Fuchs, B.D., Shah, C.V., Bellamy, S.L., Christie, J.D., 2009. Serum lactate is associated with mortality in severe sepsis independent of organ failure and shock. Crit. Care Med. 37, 1670–1677.

Mouncey, P.R., Osborn, T.M., Power, G.S., Harrison, D.A., Sadique, M.Z., Grieve, R.D., Jahan, R., Tan, J.C., Harvey, S.E., Bell, D., 2015. Protocolised Management In Sepsis (ProMISe): a multicentre randomised controlled trial of the clinical effectiveness and cost-effectiveness of early, goal-directed, protocolised resuscitation for emerging septic shock. Health Technol. Assess. 19, 1–150.

Myhren, H., Ekeberg, O., Stokland, O., 2010. Health-related quality of life and return to work after critical illness in general intensive care unit patients: a 1-year follow-up study. Crit. Care Med. 38, 1554–1561.

Oeyen, S.G., Vandijck, D.M., Benoit, D.D., Annemans, L., Decruyenaere, J.M., 2010. Quality of life after intensive care: a systematic review of the literature. Crit. Care Med. 38, 2386–2400.

Palud, A., Parmentier-Decrucq, E., Pastre, J., Caires, N.D.F., Lassalle, P., Mathieu, D., 2015. Evaluation of endothelial biomarkers as predictors of organ failures in septic shock patients. Cytokine 73, 213–218.

Pandharipande, P.P., Girard, T.D., Jackson, J.C., Morandi, A., Thompson, J.L., Pun, B.T., Brummel, N.E., Hughes, C.G., Vasilevskis, E.E., Shintani, A.K., 2013. Long-term cognitive impairment after critical illness. N. Engl. J. Med. 369, 1306–1316.

Pauly, D., Hamed, S., Behnes, M., Lepiorz, D., Lang, S., Akin, I., Borggrefe, M., Bertsch, T., Hoffmann, U., 2016. Endothelial cell-specific molecule-1/endocan: diagnostic and prognostic value in patients suffering from severe sepsis and septic shock. J. Crit. Care 31, 68–75.

Peake, S.L., Delaney, A., Bailey, M., Bellomo, R., Cameron, P.A., Cooper, D.J., Higgins, A.M., Holdgate, A., Howe, B.D., Webb, S.A.R., Williams, P., ARISE Investigators, ANZICS Clinical Trials Group, 2014. Goal-directed resuscitation for patients with early septic shock. N. Engl. J. Med. 371, 1496–1506.

Pierrakos, C., Vincent, J.-L., 2010. Sepsis biomarkers: a review. Crit. Care 14, R15.
Povoa, P., Coelho, L., Almeida, E., Fernandes, A., Mealha, R., Moreira, P., Sabino, H., 2005. C-reactive protein as a marker of infection in critically ill patients. Clin. Microbiol. Infect. 11, 101–108.
Quinten, V.M., Van Meurs, M., Ter Maaten, J.C., Ligtenberg, J.J.M., 2016. Trends in vital signs and routine biomarkers in patients with sepsis during resuscitation in the emergency department: a prospective observational pilot study. BMJ Open 6, 1–9.
Ratzinger, F., Schuardt, M., Eichbichler, K., Tsirkinidou, I., Bauer, M., Haslacher, H., Mitteregger, D., Binder, M., Burgmann, H., 2013. Utility of sepsis biomarkers and the infection probability score to discriminate sepsis and systemic inflammatory response syndrome in standard care patients. PLoS One 8, e82946.
Riedel, S., Carroll, K.C., 2016. Early identification and treatment of pathogens in sepsis: molecular diagnostics and antibiotic choice. Clin. Chest Med. 37, 191–207.
Rivers, E., Nguyen, B., Havstad, S., Ressler, J., Muzzin, A., Knoblich, B., Peterson, E., Tomlanovich, M., 2001. Early goal-directed therapy in the treatment of severe sepsis and septic shock. N. Engl. J. Med. 345, 1368–1377.
Roderburg, C., Benz, F., Vargas Cardenas, D., Koch, A., Janssen, J., Vucur, M., Gautheron, J., Schneider, A.T., Koppe, C., Kreggenwinkel, K., Zimmermann, H.W., Luedde, M., Trautwein, C., Tacke, F., Luedde, T., 2015. Elevated miR-122 serum levels are an independent marker of liver injury in inflammatory diseases. Liver Int. 35, 1172–1184.
Roderburg, C., Luedde, M., Cardenas, D.V., Vucur, M., Scholten, D., Frey, N., Koch, A., Trautwein, C., Tacke, F., Luedde, T., 2013. Circulating microRNA-150 serum levels predict survival in patients with critical illness and sepsis. PLoS One 8, e54612.
Sager, R., Kutz, A., Mueller, B., Schuetz, P., 2017. Procalcitonin-guided diagnosis and antibiotic stewardship revisited. BMC Med. 15, 15.
Sankar, V., Webster, N.R., 2013. Clinical application of sepsis biomarkers. J. Anesth. 27, 269–283.
Schmit, X., Vincent, J.L., 2008. The time course of blood C-reactive protein concentrations in relation to the response to initial antimicrobial therapy in patients with sepsis. Infection 36, 213–219.
Seale, A.C., Blencowe, H., Zaidi, A., Ganatra, H., Syed, S., Engnnann, C., Newton, C.R., Vergnano, S., Stoll, B.J., Cousens, S.N., Lawn, J.E., Tea, N.I.E., 2013. Neonatal severe bacterial infection impairment estimates in South Asia, sub-Saharan Africa, and Latin America for 2010. Pediatr. Res. 74, 73–85.
Seeley, E.J., Bernard, G.R., 2016. Therapeutic targets in sepsis: past, present, and future. Clin. Chest Med. 37, 181–189.
Shankar-Hari, M., Culshaw, N., Post, B., Tamayo, E., Andaluz-Ojeda, D., Bermejo-Martín, J.F., Dietz, S., Werdan, K., Beale, R., Spencer, J., 2015. Endogenous IgG hypogammaglobulinaemia in critically ill adults with sepsis: systematic review and meta-analysis. Intensive Care Med. 41, 1393–1401.
Shozushima, T., Takahashi, G., Matsumoto, N., Kojika, M., Endo, S., Okamura, Y., 2011. Usefulness of presepsin (sCD14-ST) measurements as a marker for the diagnosis and severity of sepsis that satisfied diagnostic criteria of systemic inflammatory response syndrome. J. Infect. Chemother. 17, 764–769.
Singer, M., 2016. The new sepsis consensus definitions (sepsis-3): the good, the not-so-bad, and the actually-quite-pretty. Intensive Care Med. 42, 2027–2029.
Srinivasan, L., Page, G., Kirpalani, H., Murray, J.C., Das, A., Higgins, R.D., Carlo, W.A., Bell, E.F., Goldberg, R.N., Schibler, K., Sood, B.G., Stevenson, D.K., Stoll, B.J., Van Meurs, K.P., Johnson, K.J., Levy, J., McDonald, S.A., Zaterka-Baxter, K.M., Kennedy, K.A., Sanchez, P.J., Duara, S., Walsh, M.C., Shankaran, S., Wynn, J.L., Cotten, C.M., Eunice Kennedy Shriver National Institute of Child Health and Human Development Neonatal Research Network, 2017. Genome-wide association study of sepsis in extremely premature infants. Arch. Dis. Child. Fetal Neonatal Ed. 102, F439–F445.
Srinivasan, L., Swarr, D.T., Sharma, M., Cotten, C.M., Kirpalani, H., 2017. systematic review and meta-analysis: gene association studies in neonatal sepsis. Am. J. Perinatol. 34, 684–692.
Stoller, J., Halpin, L., Weis, M., Aplin, B., Qu, W., Georgescu, C., Nazzal, M., 2016. Epidemiology of severe sepsis: 2008–2012. J. Crit. Care 31, 58–62.
Strehlow, M.C., Emond, S.D., Shapiro, N.I., Pelletier, A.J., Camargo, C.A., 2006. National study of emergency department visits for sepsis, 1992 to 2001. Ann. Emerg. Med. 48, 326–331.
Sutherland, A.M., Walley, K.R., Russell, J.A., 2005. Polymorphisms in CD14, mannose-binding lectin, and Toll-like receptor-2 are associated with increased prevalence of infection in critically ill adults. Crit. Care Med. 33, 638–644.
The ProCESS Investigators, 2014. A randomized trial of protocol-based care for early septic shock. N. Engl. J. Med. 2014, 1683–1693.
Torio, C.M., Andrews, R.M., 2006. National Inpatient Hospital Costs: The Most Expensive Conditions by Payer, 2011: Statistical Brief #160. Agency for Health Care Policy and Research, Rockville, MD.

Tziolos, N., Kotanidou, A., Orfanos, S.E., 2015. Biomarkers in infection and sepsis: Can they really indicate final outcome? Int. J. Antimicrob. Agents 46, S29–S32.

Umbro, I., Gentile, G., Tinti, F., Muiesan, P., Mitterhofer, A.P., 2016. Recent advances in pathophysiology and biomarkers of sepsis-induced acute kidney injury. J. Infect. 72, 131–142.

Van der Poll, T., Van De Veerdonk, F.L., Scicluna, B.P., Netea, M.G., 2017. The immunopathology of sepsis and potential therapeutic targets. Nat. Rev. Immunol. 17, 407–420.

Van Vught, L.A., Klouwenberg, P.M.K., Spitoni, C., Scicluna, B.P., Wiewel, M.A., Horn, J., Schultz, M.J., NÜrnberg, P., Bonten, M.J., Cremer, O.L., 2016. Incidence, risk factors, and attributable mortality of secondary infections in the intensive care unit after admission for sepsis. JAMA 315, 1469–1479.

Vasilescu, C., Rossi, S., Shimizu, M., Tudor, S., Veronese, A., Ferracin, M., Nicoloso, M.S., Barbarotto, E., Popa, M., Stanciulea, O., Fernandez, M.H., Tulbure, D., Bueso-Ramos, C.E., Negrini, M., Calin, G.A., 2009. MicroRNA fingerprints identify miR-150 as a plasma prognostic marker in patients with sepsis. PLoS One 4, e7405.

Vincent, J.-L., Marshall, J.C., Ñamendys-Silva, S.A., FranÇois, B., Martin-Loeches, I., Lipman, J., Reinhart, K., Antonelli, M., Pickkers, P., Njimi, H., 2014. Assessment of the worldwide burden of critical illness: the Intensive Care Over Nations (ICON) audit. Lancet Respir. Med. 2, 380–386.

Vincent, J.L., Sakr, Y., Sprung, C.L., Ranieri, V.M., Reinhart, K., Gerlach, H., Moreno, R., Carlet, J., Le Gall, J.R., Payen, D., Pati, S.O.A.I., 2006. Sepsis in European intensive care units: results of the SOAP study. Crit. Care Med. 34, 344–353.

Volbeda, M., Wetterslev, J., Gluud, C., Zijlstra, J., Van Der Horst, I., Keus, F., 2015. Glucocorticosteroids for sepsis: systematic review with meta-analysis and trial sequential analysis. Intensive Care Med. 41, 1220–1234.

Wang, H., Meng, K., Chen, W., Feng, D., Jia, Y., Xie, L., 2012. Serum miR-574-5p: a prognostic predictor of sepsis patients. Shock 37, 263–267.

Wang, J.F., Yu, M.L., Yu, G., Bian, J.J., Deng, X.M., Wan, X.J., Zhu, K.M., 2010. Serum miR-146a and miR-223 as potential new biomarkers for sepsis. Biochem. Biophys. Res. Commun. 394, 184–188.

Wang, H.J., Zhang, P.J., Chen, W.J., Feng, D., Jia, Y.H., Xie, L.X., 2012a. Evidence for serum miR-15a and miR-16 levels as biomarkers that distinguish sepsis from systemic inflammatory response syndrome in human subjects. Clin. Chem. Lab. Med. 50, 1423–1428.

Wang, H.J., Zhang, P.J., Chen, W.J., Feng, D., Jia, Y.H., Xie, L.X., 2012b. Four serum microRNAs identified as diagnostic biomarkers of sepsis. J. Trauma Acute Care Surg. 73, 850–854.

Weber, M., Lambeck, S., Ding, N., Henken, S., Kohl, M., Deigner, H.P., Enot, D.P., Igwe, E.I., Frappart, L., Kiehntopf, M., Claus, R.A., Kamradt, T., Weih, D., Vodovotz, Y., Briles, D.E., Ogunniyi, A.D., Paton, J.C., Maus, U.A., Bauer, M., 2012. Hepatic induction of cholesterol biosynthesis reflects a remote adaptive response to pneumococcal pneumonia. FASEB J. 26, 2424–2436.

Westwood, M., Ramaekers, B., Whiting, P., Tomini, F., Joore, M., Armstrong, N., Ryder, S., Stirk, L., Severens, J., Kleijnen, J., 2015. Procalcitonin testing to guide antibiotic therapy for the treatment of sepsis in intensive care settings and for suspected bacterial infection in emergency department settings: a systematic review and cost-effectiveness analysis. Health Technol. Assess. 19, 1–236.

Yamano, S., Shimizu, K., Ogura, H., Hirose, T., Hamasaki, T., Shimazu, T., Tasaki, O., 2016. Low total cholesterol and high total bilirubin are associated with prognosis in patients with prolonged sepsis. J. Crit. Care 31, 36–40.

Yealy, D.M., Kellum, J.A., Huang, D.T., Barnato, A.E., Weissfeld, L.A., Pike, F., Terndrup, T., Wang, H.E., Hou, P.C., Lovecchio, F., Filbin, M.R., Shapiro, N.I., Angus, D.C., Investigators, P., 2014. A randomized trial of protocol-based care for early septic shock. N. Engl. J. Med. 370, 1683–1693.

Zeerleder, S., Zwart, B., Wuillemin, W.A., Aarden, L.A., Groeneveld, A.J., Caliezi, C., Van Nieuwenhuijze, A.E., Van Mierlo, G.J., Eerenberg, A.J., LÄmmle, B., 2003. Elevated nucleosome levels in systemic inflammation and sepsis. Crit. Care Med. 31, 1947–1951.

Zhang, X., Liu, D., Liu, Y.-N., Wang, R., Xie, L.-X., 2015. The accuracy of presepsin (scd14-ST) for the diagnosis of sepsis in adults: a meta-analysis. Crit. Care 19, 323.

Further Reading

Singer, M., Deutschman, C.S., Seymour, C.W., Shankar-Hari, M., Annane, D., Bauer, M., Bellomo, R., Bernard, G.R., Chiche, J.-D., Coopersmith, C.M., 2016. The third international consensus definitions for sepsis and septic shock (sepsis-3). JAMA 315, 801–810.

CHAPTER

14

Asphyxia Diagnosis: An Example of Translational Precision Medicine

Jan Lüddecke
InfanDx AG, Cologne, Germany

1 INTRODUCTION

The objective of precision medicine (PM) is to use the wealth of information provided by the emerging "omics" technologies (genomics, transcriptomics, metabolomics, etc.) to generate more complex disease profiles, precise diagnostic tools, and targeted treatments. A field of high need for precise and fast point-of-care (POC) diagnosis is asphyxia-induced neonatal encephalopathy (NE). This article is written from the perspective of the start-up company InfanDx AG (Cologne, Germany), which focuses on the identification and utilization of small endogenous metabolites as asphyxia markers. InfanDx currently develops assays for the quantification of such markers. The article gives an introduction into the recent achievements of PM in diagnostics and therapy, provides an overview over asphyxia research and recent metabolomics studies, and discusses the possible routes for the development of a POC diagnostic device for timely asphyxia diagnosis.

2 THE PM APPROACH

It is well known that drugs affect different patients in different ways, and in many cases, therapies do not have the desired effect on the patients' health. A study from 2001 revealed a lack of drug efficacy in more than 40% of the patients suffering from conditions including diabetes, migraines, cancer, asthma, arthritis, and Alzheimer's disease (Spear et al., 2001).

The emergence and progression of the "omics" technologies (genomics, transcriptomics, proteomics, metabolomics, etc.) over the past 20 years enabled the generation of more complex disease profiles. The objective of PM is to use the wealth of information generated by these novel technologies to identify the right treatments for the individual needs of the patients. The NIH defines PM as "an emerging approach for disease treatment and prevention

Precision Medicine
https://doi.org/10.1016/B978-0-12-805364-5.00014-7

that takes into account individual variability in environment, lifestyle and genes for each person" (National Research Council (US) Committee on A Framework for Developing a New Taxonomy of Disease, 2011).

PM does not focus on generation of patient-specific drugs or treatments, as the older term personalized medicine might suggest. PM focuses on the precise diagnosis and classification of the patients' diseases and needs in order to apply targeted treatment for the right patient at the right time. With PM, treatment can be concentrated on those patients who will benefit from it while preventing unnecessary treatments, side effects, and costs for patients, which are not susceptible.

In January 2015, former president Obama announced the launch of a new Precision Medicine Initiative, a broad research program making use of the advances in "omics" and computational technologies to build the knowledge basis for PM approaches to cancer and other diseases (Collins and Varmus, 2015). This initiative underlines that the time is right, and we have all the necessary tools at hand to implement this promising concept of disease treatment.

3 EXAMPLES FOR SUCCESSFUL PM APPLICATION

The deeper understanding of molecular mechanisms, signaling pathways, and gene expression patterns revealed that various common cancer types have to be further classified into subtypes, each susceptible to different therapeutic approaches. This knowledge also enabled new advances in cancer prevention, diagnosis, prognosis, and therapeutics. At present, PM primarily makes use of genome and gene expression analysis. The steep decline in sequencing costs has fueled this progress (Choi et al., 2009; Ng et al., 2009). Accordingly, treatment is more and more based on precise genetic analysis and less on tissue type and anatomical origin (Jackson and Chester, 2015). As a result, several gene expression signature assays have been developed to guide treatment and predict outcomes. One example is the neoadjuvant therapy in early-stage breast cancer. Classical clinical-pathological diagnosis resulted in an overtreatment of 85% of the patients (Early Breast Cancer Trialists' Collaborative Group, 2012). Gene expression analyses lead to a deeper understanding of breast cancer and enabled new phenotypic classifications, revealing different cancer subtypes, which respond differently, or not at all, to neoadjuvant therapy. Several gene expression signature assays have been marketed (PAM50, MammaTyper, MammaPrint, Oncotype DX, Endopredict, Genomic Grade Index, and Prosigna); these are tests that identify these cancer types and provide treatment guidance in addition to clinical-pathological parameters (Sinn et al., 2013).

Another early success story of PM at the diagnostic and therapeutic front is the monoclonal humanized antibody trastuzumab (Herceptin and its biosimilars) for breast cancer treatment in patients whose tumors overexpress the oncogene HER2. By binding this growth factor on the cell surface, trastuzumab inhibits tumor cell proliferation. The treatment of HER2 overexpressing breast cancer patients with trastuzumab reduces mortality by one-third and cancer reoccurrence by 40% (Goldenberg, 1999; Moja et al., 2012). The efficacy of Herceptin medication is ensured by screening patients with an in vitro diagnostic (IVD) companion diagnostic test (HercepTest), which identifies the approximately one-third of the breast cancer patients with HER2 overexpression. Since the approval of this first pair of drug and companion IVD in 1998, the interest in discovering biomarkers for the combined prognosis and treatment of diseases has grown strongly. This process has been supported by the FDA, which deems companion diagnostics essential for (1) identifying patients that most likely will benefit from the therapy, (2) identifying patients with increased risk of adverse reaction to the treatment, and (3) monitor the response to the treatment and guide treatment progress (US Food and Drug Administration, 2014). Accordingly, in the past years, a great number of therapeutic products with an accompanying IVD have been approved by FDA (current list of IVD companion diagnostics, www.fda.gov/companiondiagnostics).

One of the most rapidly growing families of cancer drugs are protein kinase inhibitors (PKIs), as many types of cancer are the result of an overactivity of kinases, especially tyrosine kinases. Different compounds have been developed, which selectively inhibit these kinases and by that suppress tumor growth. One success story is the PKI imatinib (Gleevec with companion IVD) in the treatment of chronic-phase chronic myeloid leukemia. It inhibits the BCR-ABL fusion oncoprotein, which shows constitutive tyrosine kinase activity (Druker et al., 2001). In most patients, imatinib treatment results in a complete hematologic response and consequently drastically increased survival rates (Smith, 2011; Goldman and Marin, 2012).

In summary, the progress of PM in the recent years has started the process of shifting cancer treatment away from cytotoxic, nonspecific chemotherapies toward precise patient screening and administration of targeted drugs, resulting in increased efficacy and less side effects. Apart from cancer, also other diseases like obstructive coronary artery disease (CAD) are targeted by PM approaches (Clarke et al., 2015; Vargas et al., 2013).

4 METABOLITES IN PM DIAGNOSTICS

As mentioned above, so far, PM mainly focuses on the utilization of genomics and transcriptomics data for diagnosis and therapeutic target selection. However, a pure genomics approach does not take into account environmental influences on the body and thus only covers a part of the reasons for a disease phenotype. To complete the picture, genomics could be complemented by metabolomics on a number of levels. While genetic risk scores indicate a possible outcome, metabolic phenotyping is the most precise way to assess the current status of the patient and could generate unique insights into the fundamental causes of a disease. By combining the information of the genetic background with the effects of gene expression, environmental influences, diet, and the gut microbiome, a holistic picture of a patient's condition could be generated. The promises of such metabolomics evaluations are the discovery of predictive, prognostic, diagnostic, and surrogate biomarkers for disease conditions; the elucidation of underlying disease mechanisms; the subclassification of disease types for targeted treatment; and the identification of drug response phenotypes (Beger et al., 2016; Everett, 2015).

Recent advances in the development of metabolomic analytic platforms and informatics tools make it possible to quantitate thousands of metabolites from small amounts of different body fluids. These analyses are already being used to identify inborn errors in metabolism (IEMs) in newborns, which can be lethal or cause severe organ damage if not diagnosed and treated promptly (Beger et al., 2016; Lindon and Nicholson, 2014). Another example for predictive metabolomics is the discovery of three prediabetes-specific markers (glycine, lysophosphatidylcholine (LPC), and acetylcarnitine), which foretell impaired glucose tolerance (IGT) and/or type 2 diabetes. Glycine and LPC were shown to be predictors of IGT as much as 7 years prior to the disease onset (Wang-Sattler et al., 2012).

A good example for a POC PM application based on metabolite quantification is the personal glucose meter for the self-treatment of diabetic patients with insulin on an hour-to-hour basis (Martin et al., 2006). Advances in the last years in blood glucose measuring devices enable precise real-time continuous monitoring of blood glucose for up to 14 days to enable precise administration of insulin (Rodbard, 2016). In combination with automated, decision-making insulin pumps, this resembles an artificial pancreas device setup, which has the potential to dramatically improve glycemic regulation, reduce acute and chronic complications of diabetes, and relieve patients of diabetes self-management (Russell, 2015).

5 THE GLOBAL BURDEN OF PERINATAL ASPHYXIA

An incidence with urgent need for precise and fast POC diagnostics is neonatal encephalopathy (NE) as a result of asphyxia. Perinatal asphyxia is defined as the reduction or cutoff of respiratory gas exchange to the fetus during the perinatal period. It can result in stillbirth, neonatal mortality, or severe disabilities. Possible causes for asphyxia are perinatal complications like a traumatic birth, umbilical cord compression, shoulder dystocia, maternal or fetal hemorrhage, or uterine rupture (Fattuoni et al., 2015). Perinatal asphyxia affects about 4 million neonates worldwide every year. About 1 million of these do not survive, which attributes 23% of neonatal deaths worldwide to asphyxia (Lawn et al., 2005). Most of the survivors recover; however, some develop hypoxic-ischemic encephalopathy (HIE), possibly causing severe livelong disabilities (Lawn et al., 2011; Ahearne et al., 2016). Although there are other causes for NE like sepsis, meningitis, or metabolic disorder, 50%–80% can be linked to hypoxic ischemia (Lee et al., 2013). In 2010, about 1.6 per 1000 term live-birth neonates in high-income countries developed NE (about 19,000 cases). In developing countries with high neonatal mortality rates, the number is drastically higher, exceeding 12 per 1000 live births (Lee et al., 2013; Blencowe et al., 2013). Possible outcomes of HIE range from persistent motor, sensory, and cognitive impairment to seizure disorders and cerebral palsy (Bhatti and Kumar, 2014; Maneru et al., 2001). The disability-adjusted life year (DALY) estimates the years of life lost due to illness, disabilities, or early death. One of the highest-ranking single conditions in DALY estimations is birth asphyxia, with 50.2 million DALYs worldwide calculated for 2010 (Lee et al., 2013).

HIE in neonates is a highly complex, multifactor-dependent, excitotoxic process, fundamentally different from stroke injury in adults (Baburamani et al., 2012; Millar et al., 2017). The brain-damaging processes do not only occur at the time of the insult but also continue to evolve during the following hours and days. These processes can roughly be separated in two major phases. The initial cerebral hypoxic-ischemic insult, also termed primary energy failure, leads to a switch to anaerobic metabolism and buildup of lactic acid, accompanied by energy depletion and depolarization of the neurons. This ultimately initiates necrosis and/or apoptosis of the cells. The necrosis-induced disruption of the cells in turn causes additional inflammatory reactions. After a latent period of 6–48h, a secondary energy failure and neuroinflammation are induced through inflammatory gene expression and the production of pro-inflammatory cytokines, proteases, and reactive oxygen species. These processes are the major causes of the clinical manifestations of NE, as they trigger even more severe necrosis and apoptosis reactions, resulting in extensive neuronal injury (Millar et al., 2017; Lorek et al., 1994; Fatemi et al., 2009; Allen and Brandon, 2011). Consequently, the aim of therapeutic approaches is to inhibit the inflammatory processes during the latent phase after the primary insult and to prevent the secondary energy failure; the 6h latent phase thus represents the therapeutic window.

The only approved treatment for HIE is therapeutic hypothermia (TH), the cooling of either the head or the whole body of the infant to 33–34°C for 72h (Shankaran, 2012). Multiple studies have demonstrated that TH is both safe and effective in treating HIE (Tagin et al., 2012; Edwards et al., 2010). A number of mechanisms are thought to be involved in the beneficial effects of TH (Schmitt et al., 2014):

- Cooling reduces the cerebral metabolism.
- TH prevents or interrupts apoptotic pathways by reducing mitochondria dysfunction and caspase activation.
- TH attenuates pro-inflammatory immune response and might induce neuroprotection.
- Disruption of the blood-brain barrier is reduced by mild to moderate TH.
- Generation of free oxygen radicals is reduced, relieving endogenous antioxidants.
- Cold-shock proteins are expressed, believed to exhibit antiapoptotic and neuroprotective function.

The initiation of TH during the first 6h after birth is critical; the sooner the therapy is started, the higher the protective effect (Thoresen, 2000). As an alternative to TH, neuroprotective drugs and drug-induced hypothermia are being evaluated. Advantages could be reduced side effects and lower technical requirements, beneficial especially in less developed countries. Apart from that, a combined therapy might yield overadditive protective effects. So far, drug candidates have shown promising results in animal studies, they are however not ready yet for clinical application (Millar et al., 2017; Schmitt et al., 2014; Dixon et al., 2015).

6 PERINATAL ASPHYXIA—THE DIAGNOSTIC GAP

Despite the availability of an effective therapy for HIE, many neonates do not benefit from it due to a lack of reliable tools to diagnose perinatal asphyxia (Ahearne et al., 2016). One of the standard methods for evaluating the health status of a neonate is the Apgar score, an assessment of five simple criteria including skin color and pulse rate, which are determined 5min after birth. However, the score is of subjective nature and suffers from high variability and poor sensitivity and specificity. The false-positive rate of the Apgar scoring for asphyxia can be as high as 50%–80% (Natarajan et al., 2013; Ruth and Raivio, 1988). Similarly, clinical-neurological examinations, like the Amiel-Tison Neurological Assessment (ATNA) or the Sarnat classification for HIE, are not suitable for outcome prediction at the first day after birth. Clinical examinations of critical infants are also hindered by the need for sedative medication (Murray et al., 2010). Other indicators are the base deficit, blood pH, and lactate levels. Again, the predictive values for NE are low, especially in cases of mild to moderate asphyxia (Ruth and Raivio, 1988; Rorbye et al., 2016). Electroencephalography (EEG) and amplitude-integrated EEG (aEEG) are the gold standard in HIE prediction, even within the 6h time window. Studies have shown that the severity of abnormal EEG and aEEG recordings correlates very well with the encephalopathy severity (Toet et al., 1999). Normal EEG readings have a high predictive value for a normal outcome after 2 years, but EEG measurements require the equipment and especially the high clinical expertise to interpret the results correctly, prerequisites not available in many centers (Murray et al., 2009). The same is true for magnetic resonance imaging (MRI), which has been shown to predict injury severity 2h after the insult but requires expensive equipment and clinical experts for interpretation. Especially, since scans show only very subtle and hard to detect changes within the first few hours after the insult. In addition, the transfer of the neonate to the MRI machine is required (Cady et al., 2008; Takeoka et al., 2002).

Consequently, with the currently available diagnostic options, the number needed to treat (NNT) to prevent one case of death or significant neurodevelopmental disability via TH ranges between 6 and 9 (Wachtel and Hendricks-Munoz, 2011). Additionally, 15%–20% of the neonates are diagnosed with only mild or no encephalopathy by Sarnat classification in the first 6h after birth, and therefore do not undergo TH treatment, developing abnormal short-term outcomes, ranging from abnormal neurological examination at discharge to seizures or death from the progressing asphyxia insult (DuPont et al., 2013).

These studies underline the high medical need for specific diagnostic predictors to identify babies, within 6 h after birth, who are at risk of developing HIE and would benefit from TH.

7 ASPHYXIA MARKERS

Multiple studies have been conducted to identify biomarkers suitable to fill this diagnostic gap. Some focused on neuronal and inflammatory proteins to predict the severity of HIE (Lv et al., 2015). In a study from 2013, cord blood samples from 130 neonates were analyzed. Interleukin-16 (IL-16) in combination with the 10min Apgar score was found to differentiate between neonates with normal to mild abnormal EEG and moderate to severely abnormal EEG results better than current markers (Walsh et al., 2013). In a similar study from 2014, cord blood and umbilical artery blood proteins were compared in 17 neonates with moderate HIE and 3 neonates with severe HIE. It was found that at birth, neuronal glial fibrillary acidic protein (GFAP) and ubiquitin carboxyl-terminal hydrolase L1 (UCH-L1) concentrations increased with the severity of HIE (Chalak et al., 2014). Other recent studies focused on the use of circulating microRNAs (miRNAs) as HIE markers (Looney et al., 2015). However, as described above, perinatal asphyxia is a complex disease with a variety of different injury types involved, dependent on the duration of the insult. HIE develops over time and different markers show different kinetics. Hence, single markers will never suffice to diagnose all possible cases correctly. The strength of metabolome analyses is the ability to screen a wide variety of possible markers from different pathways at once. Consequently, the majority of the studies focused on the evaluation of plasma and urine metabolites as possible asphyxia and HIE markers (Fattuoni et al., 2015; Denihan et al., 2015). Table 1 summarizes studies focused on neonatal asphyxia and HIE metabolites of the recent years.

A problem in studying HIE in neonates is the low prevalence of the disease. In 2006, a study by Chu et al. recruited 256 neonates in three hospitals (Chu et al., 2006). Only 11 of these

TABLE 1 Recent Studies Aiming to Discover Metabolite Markers for Neonatal Asphyxia or HIE in Human or Animal Urine and Blood Samples

Author (Year)	Sample Type	*n*	Method	Study Aim	Marker Findings
Kuligowski et al. (2017)	Plasma (piglets)	32	LC-TOFMS	Hypoxia severity assessment	Predictive score, based on choline, 6.8-dihydroxypurine, hypoxanthine
Sanchez-Illana et al. (2017)	Plasma, urine (piglets)	32	LC-MS/MS	Hypoxia severity assessment	Improved prediction by choline and related biomarkers in combination with lactate
Solberg et al. (2016)	Plasma (piglets)	32	LC-TOFMS	Hypoxia severity assessment	Choline, purine catabolism intermediates
Sachse et al. (2016)	Plasma, urine (piglets)	125	^{1}H NMR	Asphyxia severity assessment, resuscitation methods	Lactate, pyruvate Plasma samples superior to urine
Ahearne et al. (2016)	Cord blood (human)	31	^{1}H NMR	HIE severity prediction	Succinate, glycerol, 3-hydroxybutyrate, O-phosphocholine
Longini et al. (2015)	Urine (human)	14	^{1}H NMR	Hypoxia severity assessment	Lactate, glucose, TMAO, threonine, 3-hydroxyisovalerate
Reinke et al. (2013)	Cord blood (human)	59	^{1}H NMR	HIE severity prediction	Acetone, 3-hydroxybutyrate, glycerol, succinate
Skappak et al. (2013)	Urine (piglets)	32	^{1}H NMR	Hypoxia assessment	Lactate, hippurate, betaine, valine, asparagine (13 out of 50 measured metabolites specific for asphyxia)
Walsh et al. (2012)	Cord blood (human)	142	LC-MS/MS	HIE severity prediction	148 Measured metabolites analyzed, 9 specific for HIE (amino acids, acylcarnitines, and glycerophospholipids)
Beckstrom et al. (2011)	Plasma (macaques)	24	2D GC-TOFMS	Asphyxia assessment	10 out of 50 metabolites significantly increased in asphyxia. Lactate, creatine, succinic acid, malate
Solberg et al. (2010)	Plasma (piglets)	33	FIA-MS/MS LC-MS/MS	Hypoxia severity assessment	Alanine/BCAA, glycine/BCAA, Krebs cycle intermediates
Chu et al. (2006)	Urine (human)	256	GC-MS	Asphyxia assessment	Glutarate, methylmalonate, 3-hydroxy-butyrate, and orotate

LC-TOFMS, liquid chromatography time-of-flight mass spectrometry; *LC-MS/MS*, liquid chromatography-tandem mass spectrometry; *^{1}H NMR*, proton nuclear magnetic resonance spectroscopy; *2D GC-TOFMS*, two-dimensional gas chromatography time-of-flight mass spectrometry; *FIA-MS/MS*, flow injection analysis-tandem mass spectrometry; *TMAO*, trimethylamine N-oxide; *BCAA*, branched chain amino acids.

were diagnosed with severe asphyxia. In urine samples, eight organic acids were identified, which could act as candidates for the prognosis of HIE outcomes. Urine samples have the advantages to be easily obtainable by noninvasive means, in contrast to plasma samples or lactate measurements. On the other hand, a later study comparing urine and plasma samples of 125 piglets only found poor correlation. The relatively big variations of concentrations along with difficult normalization are well-known limitations when analyzing this matrix. Thus, the authors caution about using urine for real-time monitoring of acute conditions such as hypoxia, as urine samples are delayed and are dependent on renal function and individual clearance patterns of the metabolites (Sachse et al., 2016).

In 2010, Solberg et al. screened a broad range of endogenous intermediates in plasma from 33 asphyxiated piglets (Solberg et al., 2010). They found that lactate as well as pH and base deficit did not correlate well with hypoxia duration. In contrast, ratios of alanine or glycine to branched chain amino acids (BCAA) showed a good correlation with hypoxia duration. The finding that resuscitation with 100% oxygen delays cellular recovery was not confirmed in a study from 2016, examining 125 piglets in terms of different resuscitation and chest compression methods (Sachse et al., 2016).

A first human study with plasma from umbilical cord blood, which can also easily be obtained, was conducted by Walsh et al. (2012). Out of five different classes (acylcarnitines, glycerophospholipids, sphingolipids, amino acids, and biogenic amines), 148 metabolites were quantified from samples of 142 infants. Of these, 31 infants were confirmed HIE cases, 40 were asphyxiated without HIE, and 71 were matched controls. It was found that three distinct metabolite classes were disrupted by asphyxia: amino acids, acylcarnitines, and phosphatidylcholines. Fourteen metabolites were found to differentiate between infants with confirmed HIE and those that were asphyxiated without developing HIE.

A similar study was conducted by Reinke et al. in 2013 with a focus on organic acids and carbohydrates of the central energy metabolism. From 100 recruited neonates, 41 had to be excluded due to insufficient sample quantity, missing EEG or clinical data or alternate diagnosis. The remaining 59 candidates consisted of 13 mild, 6 moderate, and 6 severe HIE cases and 34 asphyxiated infants without clinical-neurological signs. Eighteen metabolites were found to discriminate between asphyxia candidates versus matched controls, and 13 were found to discriminate between HIE versus controls.

Sánchez-Illana et al. reevaluated a group of previously discovered markers (choline, betaine, cytidine, and uridine) in 2017 in plasma and urine samples of 32 piglets (Sanchez-Illana et al., 2017). By combining them with lactate measurements, they could increase the predictive performance and precision for hypoxia in contrast to lactate measurements alone. However, this was only true for samples collected up to 2h after reoxygenation; 9h samples only showed poor predictive capacity. Similar results were obtained by Kuligowski et al., who evaluated the predictive capacity of choline, 6.8-dihydroxypurine, and hypoxanthine for hypoxia duration in the same study (Kuligowski et al., 2017). Again, an enhanced prediction could be achieved 2h after resuscitation compared with lactate alone.

In summary, only a limited number of studies focusing on the search for perinatal asphyxia and HIE biomarkers from blood and urine have been conducted until now, the majority of which are animal studies. Even though the noninvasive sampling of umbilical cord blood is an established method, the number of human studies is small. This can be explained by the low prevalence of the disease and the impossibility to perform risky novel treatments on infants. As HIE evolves over time, animal studies also have the advantage of a known time point of the hypoxic-ischemic insult. On the other hand, the majority of the animal studies induce neonatal, but not perinatal systemic hypoxia by lowering the fraction of inspired oxygen, which does not include an ischemic component. In the complex process of asphyxia-induced HIE development, these differences, together with general neurodevelopmental differences, could result in marker sets that might not be translatable to humans. In any case, a gold-standard animal model is yet to be found (Yager and Ashwal, 2009).

A challenge in human neonatal studies remains the standardization of recruitment and classification of patients and HIE severity, especially in multicenter trails. HIE classification is often impossible during the first hours after birth, which makes it necessary define broad study inclusion criteria and to collect samples from all infants with a certain HIE risk (Denihan et al., 2015). Nevertheless, promising first results have been generated. Marker candidates have been found in the classes of Krebs cycle intermediates, amino acids, and cell membrane components in both animal and human studies. Comprehensive cord blood studies yielded marker combinations that could successfully discriminate pathological (asphyxia or HIE) from healthy neonates (Reinke et al., 2013; Walsh et al., 2012).

Importantly, no study could find single markers with significant predictive capacities. This underlines the necessity of comprehensive multivariate analysis, a strength of metabolomics.

8 STRATEGIES FOR THE POC DIAGNOSIS OF PERINATAL ASPHYXIA

The goal of the InfanDx AG is to confirm and refine the knowledge gained about the metabolite markers and kinetics involved in asphyxia and to develop a sample-to-answer POC diagnostic device with the capabilities to distinguish healthy neonates from those who would benefit from hypothermia therapy. Standard metabolomics methods like mass spectrometry (MS) are not suited for this task, as they are highly complex and costly. To keep the price low, such a system could consist of disposable cartridges, including a microfluidic system or a lateral-flow stripe, and a reader for the automated evaluation. Several biological capture molecules and countless ligand-binding assay formats come into consideration to detect small molecules, all with specific advantages and drawbacks. Not only the most prominent candidates for detecting target molecules are of course antibodies, but also enzymes have a track record in biosensors. DNA and RNA aptamers are also promising capture molecule candidates, although they have not fully reached the market yet. Possible capture components and POC device setups are described in the following.

9 ANTIBODIES

Today, antibodies are the undisputed gold standard for use in biosensors. The majority of immunoassays make use of immunoglobulin G (IgG) antibodies with a molecular weight of about 150 kDa. A wide set of assay formats are used, ranging from classical colorimetric, fluorescent, or chemiluminescent enzyme-linked immunosorbent assays (ELISAs) to Western blot and lateral-flow assays. Many of these assays can be fully automated to enable high-throughput screening in clinical laboratories (Dinis-Oliveira, 2014). Target molecules can be detected in a wide range of matrices including whole blood, serum, plasma, urine, or swabs from mucosa. Immunoassays are available for the determination of cardiac and cancer biomarkers, hormones, toxins, pathogenic bacteria, viruses, etc. (Justino et al., 2016). A well-known and widely used POC immunoassay is the lateral-flow-based pregnancy test. It detects the peptide hormone human chorionic gonadotropin (hCG), a reliable indicator of pregnancy in urine (Gnoth and Johnson, 2014).

Although antibodies show exceptional affinity and specificity when binding big molecules like proteins, detection of low-molecular-weight targets is more difficult. Many small molecules, like metabolites, do not elicit an immune response, which is important for antibody generation. To generate antibodies against these molecules, also termed haptens, carrier proteins like BSA are used. If the haptens are fused to those carriers, the immune system recognizes the complex as foreign and an immune response and antibody generation is induced. With this strategy, specific antibodies can be generated against almost any target containing a sufficiently unique chemical structure (Shreder, 2000). To date, over 450 antibodies are commercially available against a wide variety of haptens like pesticides, herbicides, insecticides, drugs, and natural compounds (Singh et al., 2006; Gunther et al., 2007).

One aspect reducing the suitability of antibodies for application in POC devices is their limited shelf life and temperature stability. To minimize degradation, antibodies ideally have to be lyophilized and stored in a moisture-free environment (Johnson, 2012). Classical ELISAs also have the disadvantage of being very time-consuming. Different steps of antibody binding and washing can sum up to several hours of overall assay duration, rendering such assays unsuitable for prompt diagnosis. Apart from practical issues of antibody handling, general concerns have been raised about the reliability of immunoassays. Endo- and exogenous influences can interfere with assay reactions, which may lead to misinterpretation of a patient's results (Tate and Ward, 2004). An article from 2015 outlines massive issues in reproducibility of data generated with antibody affinity tools due to high batch-to-batch variability and cross-reactivity. A careful characterization of antibodies and careful quality control of each batch is mandatory to overcome these issues (Baker, 2015).

10 APTAMERS

In contrast to antibodies, aptamers are selected in vitro by a process called systematic evolution of ligands by exponential enrichment (SELEX) (Tuerk and Gold, 1990), which not only is a more ethical approach than immunization of animals but also enables the selection of aptamers against nonimmunogenic or toxic targets (Bruno et al., 2012; Lauridsen et al., 2012). The chemical synthesis of aptamers is also significantly more cost-efficient and reproducible than hybridoma antibody production (Smith et al., 2007). The in vitro synthesis of aptamers also enables straightforward modification with fluorescent or affinity tags. Another advantage of DNA aptamers is a much longer shelf life and thermal stability compared with proteins. In comparison with antibodies, typical aptamers are 10–100 times smaller in size (around 10 kDa), which can increase the sensitivity toward small molecules by enabling construction of high density sensors (Chang et al., 2014).

Aptamer-based assays can be divided into three different categories. Target binding can (1) induce a switching of the aptamer structure, (2) lead to a dissociation or strand displacement reaction, or (3) be used in sandwich approaches, although the latter is not well suited for small-molecule detection (Han et al., 2010). Each of these aptamer categories can be used in wide variety of detection formats, including colorimetry, fluorescence, luminescence and electrochemical detection, mass sensitivity, and nucleic acid amplification (Pfeiffer and Mayer, 2016). During the last 15 years, the number of articles describing novel aptamer assays has increased almost exponentially.

However, aptamers also have disadvantages. In liquid, aptamers are highly susceptible to degradation by DNases/RNases, if they are not modified by artificial bases (Rohloff et al., 2014). Another drawback is that natural DNA/RNA aptamers are limited to four bases, resulting in a much lower potential variability in secondary and tertiary structures in comparison with proteins, which consist of 22 amino acids (Ruscito and DeRosa, 2016). Aptamer selection for detection of small molecules is not trivial due to limitations of the SELEX process. As a result, aptamers selected for low-molecular-weight targets show reduced average affinity than aptamers selected for proteins or cells (McKeague et al., 2015). Nonetheless, several aptamers have been generated for a variety of small targets ranging from hormones and toxins over antibiotics to heavy metals. The Kds published for these aptamers are in the low nM–μM range (Pfeiffer and Mayer, 2016; Ruscito and DeRosa, 2016). A direct comparison of aptamers and antibodies with small molecules (antibiotics, bisphenol A, cocaine, ochratoxin A, and estradiol) revealed that antibodies generally show a higher affinity. This does lead not only to two to three orders of magnitude lower limits of detection (LoDs) but also to a higher specificity in complex matrices (Piro et al., 2016).

Although in theory, aptamers have many advantages over antibodies, and articles have been published demonstrating the general suitability of aptamers for diagnostic and therapeutic purposes; on the clinical and diagnostic market, aptamers are of no relevance yet. The only aptamer that has been approved for clinical treatment (Pegaptanib, Macugen, and FDA approved for neovascular wet age-related macular degeneration in 2004) was outcompeted in 2011 because of the marketing of the more effective treatment with a monoclonal antibody (Colquitt et al., 2008; Gower et al., 2010; Chapman and Beckey, 2006).

11 ENZYMES

Regardless of their high specificity, aptamers and antibodies that recognize specific targets can also bind to molecules with a similar structure. This is especially true for low-complexity small-molecule targets. If the target is a metabolite, an enzymatic assay might be a better solution. Enzymes are known for their high selectivity and sensitive detection kits are available for a wide range of metabolic targets, for example, sugars, alcohols, or amino acids. The most widely used methods for quantifying enzymatic reactions are colorimetric assays, as the change in absorbance can be easily measured. Fluorimetry or luminometry measurements not only are more sensitive but also require instrumentation that is more sophisticated. Finally, electrochemical detection of enzymatic reactions is easily miniaturized and thus ideally suited for use in POC devices (Bisswanger, 2014).

Although plenty of kits for metabolite detection based on enzymatic reactions are available for research use, in the clinical diagnostic sector, the number of enzymatic assays is much lower than the number of immunoassays. Most assays are fully automatable, enabling high-throughput screening. One example is the glycated hemoglobin (HbA1c) quantification in diabetes diagnosis. The enzymatic reaction cascade involves proteinases and a fructosyl peptide oxidase. The final detection step of the assay involves the production of hydrogen peroxide, which reacts with a substrate molecule resulting in a change of the absorption at 660 nm wavelength (Hirokawa et al., 2005; Jaisson et al., 2014).

In kidney function analysis, the glomerular filtration rate marker creatinine is measured via an enzymatic assay, which involves a series of coupled enzymatic reactions including the conversion of creatinine to creatine by a creatininase. Again, the final reaction results in an absorption change (Hoste et al., 2015; Kume et al., 2017).

Other examples are the enzymatic diagnosis of phenylketonuria, an inborn error of metabolism that results in decreased metabolism of phenylalanine (Hassan et al., 2012) or the diagnosis of magnesium deficiency (Bailey et al., 2014). The latter is a very straightforward assay, making use of the dependency of the isocitrate dehydrogenase on magnesium as a cofactor. The more magnesium is available, the higher is the enzymatic activity. Of course, also the detection of blood glucose in personal glucose meters (PGMs) is accomplished by enzymes. Most of today's PGMs make use of glucose oxidase (GOx) and glucose dehydrogenase (GDH) to create an electrochemical signal (Lan et al., 2016).

The degree of complexity of enzymatic assays is highly variable. Relatively, straightforward setups make use of simple reactions, catalyzed by just one enzyme. At the other end of the spectrum complex coupled enzyme reactions, which can consist of several enzymes and their specific cofactors and substrates, might be required to detect a target. In these cases, also the storage of the components, together or separated, becomes more complex. Similar to antibodies, most enzymes are susceptible to thermal degradation if not specifically engineered or from thermophilic organisms. This requires careful selection of storage form and conditions (Linares-Pasten et al., 2014). The activity of enzymes is strongly dependent on factors like the pH, temperature, cofactors, and inhibitors (Bisswanger, 2014). This makes reliable measurements of targets in biological samples like plasma difficult and requires a very careful validation of the assay.

12 POC DEVICES

In many clinical scenarios and especially in the case of neonatal asphyxia, the rapid, reliable, and low-cost confirmation of clinical findings is needed to make timely decisions of patient management, without the involvement the core clinical laboratory. This role is fulfilled by POC in vitro diagnostic devices. As described above, monitoring of single biomarkers is not sufficient. To improve diagnostic efficiency, enhance precision, and reduce costs, multiplexing is necessary. Hence, multiplex POC testing (xPOCT) of several markers from a single sample is inevitable.

The demands on such xPOCT devices are high: They have to (i) generate highly reliable results, in accordance with central laboratory findings (ii) in short turnaround times; (iii) work with low sample volumes, especially in the case of newborn plasma; (iv) be easy to operate by nonexpert staff in stressful situations like in the maternity unit after a difficult birth; (v) have a long shelf life; and (vi) be cost-efficient (Dincer et al., 2017; Spindel and Sapsford, 2014). Examples of methods, which could fulfill these demands, are described in the following sections.

12.1 Paper Based Analytical Devices

Paper Based Analytical Devices (PADs) consist of membranes or cellulose paper with different reaction zones. Once the sample is applied, these devices are self-operating, making use of capillary forces to transport the sample to the different zones, where it reacts with embedded labeling reagents or is caught by capturing molecules. The labeled and captured analytes are then used for quantification by optical systems.

An example for a widespread PAD is the pregnancy test lateral-flow assays (LFA), which produces a simple optical readout, enabling the interpretation of the results also by nonexperts without the need for instrumentation (Li and Macdonald, 2016). To allow quantitative evaluation of several analytes, a readout instrument is required. An example for such a device is the Alere Triage platform, a portable battery-powered fluorimeter that can analyze different LFA cartridges and detect up to 10 analytes at once. One application is the diagnosis of myocardial infarction via an LF-based fluorescence sandwich ELISA (Clark et al., 2002). Similar platforms could also be established by using aptamers as capture molecules.

LFAs have the advantages to be simple to operate and low cost and have a short sample-to-answer time of 15–30 min. LFAs are also very well suited for qualitative naked-eye readouts, even by patients. On the other hand, demands for quantitative and reproducible results are still challenging to be realized with LFAs (Dincer et al., 2017; Sajid et al., 2015).

An enhancement of classical LFAs are microfluidic PADs (μPADs). By designing PADs with a 3D structure, these μPADs enable operations like mixing, splitting, separation, and filtration. Although not on the market yet, μPADs might enable semiquantitative and quantitative evaluations even by patients, if making use of smartphone cameras and appropriate software. Several articles regarding this topic have been published in the recent years (Yang et al., 2017). As an alternative to fluorimetric readouts, the use of enzymes can enable electrochemical signal generation, as shown in a μPAD for glucose, lactate and uric acid detection (Dungchai et al., 2009).

12.2 Microfluidic Systems

Another approach for the realization of xPOCTs is microfluidic systems. The promise of these devices is to miniaturize and combine complex laboratory procedures on small microchips. These lab-on-a-chip (LOC) devices enable much more complex operations than PADs. The flow is usually generated by micropumps, limiting user action to the addition of the sample. Sample volumes can be as small as 1 μL of blood. These points together make microfluidic devices predestined for complex POC applications (Chin et al., 2012).

One example of a successful LOC device is the 1992 approved iSTAT, one of the first microfluidic device marketed, a system consisting of disposable chips and a battery-powered handheld reader. It requires 65–100 μL of whole blood to test for a set of blood parameters such as sodium, potassium, chloride, glucose, hematocrit, gases, coagulation, and cardiac markers.

The signal is generated via thin-film electrodes, which have been coated with ionophores or enzymes. The cartridges also contain a calibration solution enabling the automatic calibration of the device prior to the addition of the sample (Erickson and Wilding, 1993; Peled, 1996). Newer devices like the Alere epoc also allow Wi-Fi communication of the handheld reader with the laboratory information system to exchange patient data (Chin et al., 2012).

An ELISA-based LOC system (Claros 1) was developed by Opko Health (former Claros Diagnostics) for the diagnosis of prostate cancer and is currently being validated in clinical trials for FDA approval (started in January 2017). The protocol realized on these chips uses only 12 μL of finger prick blood and resembles a complete ELISA, including multiple incubation and washing steps. However, in contrast to classical benchtop ELISAs, results are generated in just 10 min. The used plastic cartridges are produced by injection molding, a cheap and easily scalable process (Chin et al., 2012).

Nevertheless, even though many research articles about LOC devices are being published and improvements of individual LOC components are achieved, the main difficulty on the way to a usable product lies in the integration of all those components into a functional and reliable device. The complex requirements of different components like sampling, sample pretreatment, analyte-specific reaction, and signal generation, amplification, and measurement have to be harmonized in a seamless automated system (Chin et al., 2012).

13 OUTLOOK

As described, the translation of a set of endogenous metabolites, characteristic for a distinct pathobiochemical state, into a diagnostic tool that meets a clinical need, remains a great challenge. Starting from an animal model, the InfanDx AG has successfully identified several plasma-derived metabolite marker candidates for the diagnosis and classification of asphyxia in neonates. These candidates are currently confirmed in two large-scale international clinical studies. In parallel, the company is translating clinically confirmed sets of markers into a sensitive and reliable diagnostic POC assay. In fact, the InfanDx AG asphyxia assay represents a showcase for the PM approach. The de novo combination of recent biochemical insights with emerging affordable technological options, such as aptamer-, antibody-, or enzyme-based metabolite quantification, in an xPOCT, will enable the individualized treatment of neonates based on precise diagnostic differentiation. The company will provide a first diagnostic assay based on objective molecular criteria for the identification of asphyctic neonates requiring and benefitting from hypothermia therapy.

References

Ahearne, C.E., Boylan, G.B., Murray, D.M., 2016. Short and long term prognosis in perinatal asphyxia: an update. World J. Clin. Pediatr. 5 (1), 67–74.

Ahearne, C.E., et al., 2016. Early cord metabolite index and outcome in perinatal asphyxia and hypoxic-ischaemic encephalopathy. Neonatology 110 (4), 296–302.

Allen, K.A., Brandon, D.H., 2011. Hypoxic ischemic encephalopathy: pathophysiology and experimental treatments. Newborn Infant Nurs. Rev. 11 (3), 125–133.

Baburamani, A.A., et al., 2012. Vulnerability of the developing brain to hypoxic-ischemic damage: contribution of the cerebral vasculature to injury and repair? Front. Physiol. 3, 424.

Bailey, D., et al., 2014. A next generation enzymatic magnesium assay on the Abbott ARCHITECT chemistry system meets performance goals based on biological variation. Clin. Biochem. 47 (1–2), 142–144.

Baker, M., 2015. Reproducibility crisis: blame it on the antibodies. Nature 521 (7552), 274–276.

Beckstrom, A.C., et al., 2011. Application of comprehensive two-dimensional gas chromatography with time-of-flight mass spectrometry method to identify potential biomarkers of perinatal asphyxia in a non-human primate model. J. Chromatogr. A 1218 (14), 1899–1906.

Beger, R.D., et al., 2016. Metabolomics enables precision medicine: "A White Paper, Community Perspective". Metabolomics 12 (10), 149.

Bhatti, A., Kumar, P., 2014. Systemic effects of perinatal asphyxia. Indian J. Pediatr. 81 (3), 231–233.

Bisswanger, H., 2014. Enzyme assays. Perspect. Sci. 1 (1), 41–55.

Blencowe, H., et al., 2013. Estimates of neonatal morbidities and disabilities at regional and global levels for 2010: introduction, methods overview, and relevant findings from the Global Burden of Disease study. Pediatr. Res. 74 (Suppl. 1), 4–16.

Bruno, J.G., et al., 2012. An aptamer beacon responsive to botulinum toxins. Biosens. Bioelectron. 31 (1), 240–243.

Cady, E.B., et al., 2008. Phosphorus magnetic resonance spectroscopy 2 h after perinatal cerebral hypoxia-ischemia prognosticates outcome in the newborn piglet. J. Neurochem. 107 (4), 1027–1035.

Chalak, L.F., et al., 2014. Biomarkers for severity of neonatal hypoxic-ischemic encephalopathy and outcomes in newborns receiving hypothermia therapy. J. Pediatr. 164 (3), 468–474. e1.

Chang, A.L., et al., 2014. Kinetic and equilibrium binding characterization of aptamers to small molecules using a label-free, sensitive, and scalable platform. Anal. Chem. 86 (7), 3273–3278.

Chapman, J.A., Beckey, C., 2006. Pegaptanib: a novel approach to ocular neovascularization. Ann. Pharmacother. 40 (7-8), 1322–1326.

Chin, C.D., Linder, V., Sia, S.K., 2012. Commercialization of microfluidic point-of-care diagnostic devices. Lab Chip 12 (12), 2118–2134.

Choi, M., et al., 2009. Genetic diagnosis by whole exome capture and massively parallel DNA sequencing. Proc. Natl. Acad. Sci. U. S. A. 106 (45), 19096–19101.

Chu, C.Y., et al., 2006. Metabolomic and bioinformatic analyses in asphyxiated neonates. Clin. Biochem. 39 (3), 203–209.

Clark, T.J., McPherson, P.H., Buechler, K.F., 2002. The triage cardiac panel: cardiac markers for the triage system. Point of Care 1 (1), 42–46.

Clarke, J.L., et al., 2015. The diagnosis of CAD in women: addressing the unmet need—a report from the national expert roundtable meeting. Popul. Health Manag. 18 (2), 86–92.

Collins, F.S., Varmus, H., 2015. A new initiative on precision medicine. N. Engl. J. Med. 372 (9), 793–795.

Colquitt, J.L., et al., 2008. Ranibizumab and pegaptanib for the treatment of age-related macular degeneration: a systematic review and economic evaluation. Health Technol. Assess. 12 (16), iii–iv. ix-201.

Denihan, N.M., Boylan, G.B., Murray, D.M., 2015. Metabolomic profiling in perinatal asphyxia: a promising new field. Biomed. Res. Int. 2015, 254076.

Dincer, C., et al., 2017. Multiplexed point-of-care testing—xPOCT. Trends Biotechnol. 35 (8), 728–742.

Dinis-Oliveira, R.J., 2014. Heterogeneous and homogeneous immunoassays for drug analysis. Bioanalysis 6 (21), 2877–2896.

Dixon, B.J., et al., 2015. Neuroprotective strategies after neonatal hypoxic ischemic encephalopathy. Int. J. Mol. Sci. 16 (9), 22368–22401.

Druker, B.J., et al., 2001. Efficacy and safety of a specific inhibitor of the BCR-ABL tyrosine kinase in chronic myeloid leukemia. N. Engl. J. Med. 344 (14), 1031–1037.

Dungchai, W., Chailapakul, O., Henry, C.S., 2009. Electrochemical detection for paper-based microfluidics. Anal. Chem. 81 (14), 5821–5826.

DuPont, T.L., et al., 2013. Short-term outcomes of newborns with perinatal acidemia who are not eligible for systemic hypothermia therapy. J. Pediatr. 162 (1), 35–41.

Early Breast Cancer Trialists' Collaborative Group, 2012. Comparisons between different polychemotherapy regimens for early breast cancer: meta-analyses of long-term outcome among 100000 women in 123 randomised trials. Lancet 379 (9814), 432–444.

Edwards, A.D., et al., 2010. Neurological outcomes at 18 months of age after moderate hypothermia for perinatal hypoxic ischaemic encephalopathy: synthesis and meta-analysis of trial data. BMJ 340, c363.

Erickson, K.A., Wilding, P., 1993. Evaluation of a novel point-of-care system, the i-STAT portable clinical analyzer. Clin. Chem. 39 (2), 283–287.

Everett, J.R., 2015. Pharmacometabonomics in humans: a new tool for personalized medicine. Pharmacogenomics 16 (7), 737–754.

Fatemi, A., Wilson, M.A., Johnston, M.V., 2009. Hypoxic-ischemic encephalopathy in the term infant. Clin. Perinatol. 36 (4), 835–858. vii.

Fattuoni, C., et al., 2015. Perinatal asphyxia: a review from a metabolomics perspective. Molecules 20 (4), 7000–7016.

Gnoth, C., Johnson, S., 2014. Strips of hope: accuracy of home pregnancy tests and new developments. Geburtshilfe Frauenheilkd. 74 (7), 661–669.

Goldenberg, M.M., 1999. Trastuzumab, a recombinant DNA-derived humanized monoclonal antibody, a novel agent for the treatment of metastatic breast cancer. Clin. Ther. 21 (2), 309–318.

Goldman, J.M., Marin, D., 2012. Is imatinib still an acceptable first-line treatment for CML in chronic phase? Oncology (Williston Park) 26 (10), 901–907.

Gower, E.W., et al., 2010. A cost-effectiveness analysis of three treatments for age-related macular degeneration. Retina 30 (2), 212–221.

Gunther, S., et al., 2007. SuperHapten: a comprehensive database for small immunogenic compounds. Nucleic Acids Res. 35 (Database issue), D906–10.

Han, K., Liang, Z., Zhou, N., 2010. Design strategies for aptamer-based biosensors. Sensors (Basel) 10 (5), 4541–4557.

Hassan, F.A., et al., 2012. Evaluation of the diagnostic efficacy of enzyme colorimetric assay compared to tandem mass spectrometer in neonatal screening for phenylketonuria. Comp. Clin. Pathol. 21 (6), 1509–1513.

Hirokawa, K., Shimoji, K., Kajiyama, N., 2005. An enzymatic method for the determination of hemoglobinA(1C). Biotechnol. Lett. 27 (14), 963–968.

Hoste, L., et al., 2015. Routine serum creatinine measurements: how well do we perform? BMC Nephrol. 16, 21.

Jackson, S.E., Chester, J.D., 2015. Personalised cancer medicine. Int. J. Cancer 137 (2), 262–266.

Jaisson, S., et al., 2014. Analytical performances of a new enzymatic assay for hemoglobin A1c. Clin. Chim. Acta 434, 48–52.

Johnson, M., 2012. Antibody shelf life/how to store antibodies. Mater. Methods.

Justino, C.I., Duarte, A.C., Rocha-Santos, T.A., 2016. Immunosensors in clinical laboratory diagnostics. Adv. Clin. Chem. 73, 65–108.

Kuligowski, J., et al., 2017. Plasma metabolite score correlates with Hypoxia time in a newly born piglet model for asphyxia. Redox Biol. 12, 1–7.

Kume, T., et al., 2017. Evaluation and comparison of Abbott Jaffe and enzymatic creatinine methods: could the old method meet the new requirements? J. Clin. Lab. Anal.

Lan, T., Zhang, J., Lu, Y., 2016. Transforming the blood glucose meter into a general healthcare meter for in vitro diagnostics in mobile health. Biotechnol. Adv. 34 (3), 331–341.

Lauridsen, L.H., et al., 2012. Rapid one-step selection method for generating nucleic acid aptamers: development of a DNA aptamer against alpha-bungarotoxin. PLoS One 7 (7), e41702.

Lawn, J.E., et al., 2005. 4 Million neonatal deaths: when? Where? Why? Lancet 365 (9462), 891–900.

Lawn, J.E., et al., 2011. Setting research priorities to reduce almost one million deaths from birth asphyxia by 2015. PLoS Med. 8 (1), e1000389.

Lee, A.C., et al., 2013. Intrapartum-related neonatal encephalopathy incidence and impairment at regional and global levels for 2010 with trends from 1990. Pediatr. Res. 74 (Suppl. 1), 50–72.

Li, J., Macdonald, J., 2016. Multiplexed lateral flow biosensors: technological advances for radically improving point-of-care diagnoses. Biosens. Bioelectron. 83, 177–192.

Linares-Pasten, J.A., Andersson, M., Karlsson, E.N., 2014. Thermostable glycoside hydrolases in biorefinery technologies. Curr. Biotechnol. 3 (1), 26–44.

Lindon, J.C., Nicholson, J.K., 2014. The emergent role of metabolic phenotyping in dynamic patient stratification. Expert Opin. Drug Metab. Toxicol. 10 (7), 915–919.

Longini, M., et al., 2015. Proton nuclear magnetic resonance spectroscopy of urine samples in preterm asphyctic newborn: a metabolomic approach. Clin. Chim. Acta 444, 250–256.

Looney, A.M., et al., 2015. Downregulation of umbilical cord blood levels of miR-374a in neonatal hypoxic ischemic encephalopathy. J. Pediatr. 167 (2), 269–273. e2.

Lorek, A., et al., 1994. Delayed ("secondary") cerebral energy failure after acute hypoxia-ischemia in the newborn piglet: continuous 48-hour studies by phosphorus magnetic resonance spectroscopy. Pediatr. Res. 36 (6), 699–706.

Lv, H., et al., 2015. Neonatal hypoxic ischemic encephalopathy-related biomarkers in serum and cerebrospinal fluid. Clin. Chim. Acta 450, 282–297.

Maneru, C., et al., 2001. Neuropsychological long-term sequelae of perinatal asphyxia. Brain Inj. 15 (12), 1029–1039.

Martin, S., et al., 2006. Self-monitoring of blood glucose in type 2 diabetes and long-term outcome: an epidemiological cohort study. Diabetologia 49 (2), 271–278.

McKeague, M., et al., 2015. Analysis of in vitro aptamer selection parameters. J. Mol. Evol. 81 (5–6), 150–161.

Millar, L.J., et al., 2017. Neonatal hypoxia ischaemia: mechanisms, models, and therapeutic challenges. Front. Cell. Neurosci. 11, 78.

Moja, L., et al., 2012. Trastuzumab containing regimens for early breast cancer. Cochrane Database Syst. Rev. (4)CD006243.

Murray, D.M., et al., 2009. Early EEG findings in hypoxic-ischemic encephalopathy predict outcomes at 2 years. Pediatrics 124 (3), e459–67.

Murray, D.M., et al., 2010. The predictive value of early neurological examination in neonatal hypoxic-ischaemic encephalopathy and neurodevelopmental outcome at 24 months. Dev. Med. Child Neurol. 52 (2), e55–9.

Natarajan, G., et al., 2013. Apgar scores at 10 min and outcomes at 6–7 years following hypoxic-ischaemic encephalopathy. Arch. Dis. Child. Fetal Neonatal Ed. 98 (6), F473–9.

National Research Council (US) Committee on A Framework for Developing a New Taxonomy of Disease, 2011. Toward Precision Medicine: Building a Knowledge Network for Biomedical Research and a New Taxonomy of Disease. National Academies Press, Washington, DC.

Ng, S.B., et al., 2009. Targeted capture and massively parallel sequencing of 12 human exomes. Nature 461 (7261), 272–276.

Peled, N., 1996. Design and implementation of a microchemistry analyzer. Pure Appl. Chem. 68 (10), 1837–1841.

Pfeiffer, F., Mayer, G., 2016. Selection and biosensor application of aptamers for small molecules. Front. Chem. 4, 25.

Piro, B., et al., 2016. Comparison of electrochemical immunosensors and aptasensors for detection of small organic molecules in environment, food safety, clinical and public security. Biosensors (Basel) 6 (1).

Reinke, S.N., et al., 2013. ^{1}H NMR derived metabolomic profile of neonatal asphyxia in umbilical cord serum: implications for hypoxic ischemic encephalopathy. J. Proteome Res. 12 (9), 4230–4239.

Rodbard, D., 2016. Continuous glucose monitoring: a review of successes, challenges, and opportunities. Diabetes Technol. Ther. 18 (Suppl. 2), S3–S13.

Rohloff, J.C., et al., 2014. Nucleic acid ligands with protein-like side chains: modified aptamers and their use as diagnostic and therapeutic agents. Mol. Ther. Nucleic Acids 3, e201.

Rorbye, C., Perslev, A., Nickelsen, C., 2016. Lactate versus pH levels in fetal scalp blood during labor—using the Lactate Scout System. J. Matern. Fetal Neonatal Med. 29 (8), 1200–1204.

Ruscito, A., DeRosa, M.C., 2016. Small-molecule binding aptamers: selection strategies, characterization, and applications. Front. Chem. 4, 14.

Russell, S.J., 2015. Progress of artificial pancreas devices towards clinical use: the first outpatient studies. Curr. Opin. Endocrinol. Diabetes Obes. 22 (2), 106–111.

Ruth, V.J., Raivio, K.O., 1988. Perinatal brain damage: predictive value of metabolic acidosis and the Apgar score. BMJ 297 (6640), 24–27.

Sachse, D., et al., 2016. The role of plasma and urine metabolomics in identifying new biomarkers in severe newborn asphyxia: a study of asphyxiated newborn pigs following cardiopulmonary resuscitation. PLoS One 11 (8), e0161123.

Sajid, M., Kawde, A.-N., Daud, M., 2015. Designs, formats and applications of lateral flow assay: a literature review. J. Saudi Chem. Soc. 19 (6), 689–705.

Sanchez-Illana, A., et al., 2017. Assessment of phospholipid synthesis related biomarkers for perinatal asphyxia: a piglet study. Sci. Rep. 7, 40315.

Schmitt, K.R., Tong, G., Berger, F., 2014. Mechanisms of hypothermia-induced cell protection in the brain. Mol. Cell. Pediatr. 1 (1), 7.

Shankaran, S., 2012. Therapeutic hypothermia for neonatal encephalopathy. Curr. Treat. Options Neurol. 14 (6), 608–619.

Shreder, K., 2000. Synthetic haptens as probes of antibody response and immunorecognition. Methods 20 (3), 372–379.

Singh, M.K., et al., 2006. HaptenDB: a comprehensive database of haptens, carrier proteins and anti-hapten antibodies. Bioinformatics 22 (2), 253–255.

Sinn, P., et al., 2013. Multigene assays for classification, prognosis, and prediction in breast cancer: a critical review on the background and clinical utility. Geburtshilfe Frauenheilkd. 73 (9), 932–940.

Skappak, C., et al., 2013. Identifying hypoxia in a newborn piglet model using urinary NMR metabolomic profiling. PLoS One 8 (5), e65035.

Smith, B.D., 2011. Imatinib for chronic myeloid leukemia: the impact of its effectiveness and long-term side effects. J. Natl. Cancer Inst. 103 (7), 527–529.

Smith, J.E., et al., 2007. Aptamer-conjugated nanoparticles for the collection and detection of multiple cancer cells. Anal. Chem. 79 (8), 3075–3082.

Solberg, R., et al., 2010. Metabolomic analyses of plasma reveals new insights into asphyxia and resuscitation in pigs. PLoS One 5 (3), e9606.

Solberg, R., et al., 2016. Changes of the plasma metabolome of newly born piglets subjected to postnatal hypoxia and resuscitation with air. Pediatr. Res. 80 (2), 284–292.

Spear, B.B., Heath-Chiozzi, M., Huff, J., 2001. Clinical application of pharmacogenetics. Trends Mol. Med. 7 (5), 201–204.

Spindel, S., Sapsford, K.E., 2014. Evaluation of optical detection platforms for multiplexed detection of proteins and the need for point-of-care biosensors for clinical use. Sensors (Basel) 14 (12), 22313–22341.

Tagin, M.A., et al., 2012. Hypothermia for neonatal hypoxic ischemic encephalopathy: an updated systematic review and meta-analysis. Arch. Pediatr. Adolesc. Med. 166 (6), 558–566.

Takeoka, M., et al., 2002. Diffusion-weighted images in neonatal cerebral hypoxic-ischemic injury. Pediatr. Neurol. 26 (4), 274–281.

Tate, J., Ward, G., 2004. Interferences in immunoassay. Clin. Biochem. Rev. 25 (2), 105–120.

Thoresen, M., 2000. Cooling the newborn after asphyxia—physiological and experimental background and its clinical use. Semin. Neonatol. 5 (1), 61–73.

Toet, M.C., et al., 1999. Amplitude integrated EEG 3 and 6 hours after birth in full term neonates with hypoxic-ischaemic encephalopathy. Arch. Dis. Child. Fetal Neonatal Ed. 81 (1), F19–23.

Tuerk, C., Gold, L., 1990. Systematic evolution of ligands by exponential enrichment: RNA ligands to bacteriophage T4 DNA polymerase. Science 249 (4968), 505–510.

US Food and Drug Administration, 2014. In Vitro Companion Diagnostic Devices—Guidance for Industry and Food and Drug Administration Staff.

Vargas, J., et al., 2013. Use of the Corus(R) CAD gene expression test for assessment of obstructive coronary artery disease likelihood in symptomatic non-diabetic patients. PLoS Curr. 5.

Wachtel, E.V., Hendricks-Munoz, K.D., 2011. Current management of the infant who presents with neonatal encephalopathy. Curr. Probl. Pediatr. Adolesc. Health Care 41 (5), 132–153.

Walsh, B.H., et al., 2012. The metabolomic profile of umbilical cord blood in neonatal hypoxic ischaemic encephalopathy. PLoS One 7 (12), e50520.

Walsh, B.H., et al., 2013. Cord blood proteins and multichannel-electroencephalography in hypoxic-ischemic encephalopathy. Pediatr. Crit. Care Med. 14 (6), 621–630.

Wang-Sattler, R., et al., 2012. Novel biomarkers for pre-diabetes identified by metabolomics. Mol. Syst. Biol. 8, 615.

Yager, J.Y., Ashwal, S., 2009. Animal models of perinatal hypoxic-ischemic brain damage. Pediatr. Neurol. 40 (3), 156–167.

Yang, Y., et al., 2017. Paper-based microfluidic devices: emerging themes and applications. Anal. Chem. 89 (1), 71–91.

CHAPTER

15

Biomarker for Alzheimer's Disease

Dorothee Haas

Furtwangen University, Furtwangen, Germany

1 INTRODUCTION

According to the World Alzheimer Report 2015 provided by the Alzheimer's Disease International (ADI), nearly 46.8 million people in the world suffer from dementia (Wortmann, 2015). Dementia is a term used for many diseases, which occur primarily in the elderly and lead to loss of mental function. Alzheimer's disease (AD) is the most common form of dementia (Olsson et al., 2016). In Germany, around 1.5 million people suffer from dementia, two-thirds are by AD (Hebert et al., 2003). AD is a progressive degenerative form of dementia with diffuse atrophy of the cerebral cortex. The disease was first described in 1906 by the German neurologist Alois Alzheimer.

Less than 1% of all AD cases are the heredity-based form of AD caused by mutations in the genetic material (Selkoe, 1997). Particularly, the genes for the amyloid precursor protein, presenilin 1, and apolipoprotein E are affected. This form is known as the "early onset" of AD and occurs in people before they reach the age of 65 (Cai et al., 2012). Most of the AD cases, however, are sporadic forms and affect people beyond the age of 65. Both the heredity-based and the sporadic form lead to the same symptoms and pathophysiology. Women are affected more often than men, which may be the result of the higher life expectancy of women. This, in turn, correlates with the steadily rising number of people of over 60 years of age. The ADI estimates that 131.5 million people in the world will be suffering from dementia by 2050 (Wortmann, 2015). The increasing number of dementia cases bears a challenge for current health and social systems.

A distinction is made between a preliminary stage, called mild cognitive impairment (MCI), and the manifested stage of AD (Blennow and Hampel, 2003). Patients usually die 5–10 years after the diagnosis of the consequences of the increasing immobility (pneumonia, pulmonary, and embolism) (Lange-Asschenfeldt, 2009; Liu et al., 2004). Presently, AD is incurable. Through medication, only the symptoms can be alleviated.

Although some pathophysiological bases have been discovered, the complexity and entirety of the disease are still unclear. Meanwhile, some important elements of the pathogenesis have been explored, which build the basis for many forms of therapy today.

Precision Medicine
https://doi.org/10.1016/B978-0-12-805364-5.00015-9

Worldwide studies have been published, which deal with biomarkers of AD patients to detect the disease in it's the early stage. They are detectable, before clinical signs are noticed.

The aim of this article is to provide an overview of the current state of biomarker research regarding AD.

2 DEFINITION "BIOMARKER"

Biomarkers were first defined in 1998 by the National Institutes of Health Biomarkers Definitions Working Group: "A characteristic that is objectively measured and evaluated as an indicator of normal biological processes, pathogenic processes, or pharmacologic responses to a therapeutic intervention" (Biomarkers Definitions Working Group, 2001). Diseases are often characterized by different clinical symptoms and make it difficult, therefore, to provide a specific diagnosis and prognosis. Biomarker can detect the disease early and unequivocally and track the disease process. Increasing research in pathophysiology and molecular biology helps us to understand the processes of diseases. Increasingly, a link between disease and DNA is prepared. With the biomarker research, there is hope that biomarkers can be detected through next-generation sequencing.

Biomarkers can be produced biologically, physiologically, or from imaging procedures (Schmitz and Anz, 2008).

Biological biomarkers are obtained from patient samples such as blood, urine, tissue, and cerebrospinal fluid (CSF) (Schmitz and Anz, 2008). A common method is the complete blood count (CBC), which became an integral part of the practical medicine. In this case, values from the blood, such as hematocrit, hemoglobin level, cholesterol levels, or leukocyte number, are read out. These serve as biomarkers and may indicate certain diseases. However, cells, hormones, proteins, and DNA can be used as biological biomarkers too.

Physiological biomarkers include values that can be measured directly at the patient. Examples are the ECG and blood pressure (Schmitz and Anz, 2008).

Lastly, biomarkers can also be obtained from *imaging techniques* such as magnetic resonance imaging (MRI), computed tomography (CT), ultrasound, and positron emission tomography (PET) (Schmitz and Anz, 2008). The picture's features can provide evidence of certain diseases.

Biomarker research provides great potential for the treatment of diseases, and almost every pharmaceutical company conducts biomarker research. Drug-related biomarkers help to track the drug effect in the metabolism of the body. With the marker, the therapy can be individually adapted to each patient. Especially in oncology, where every case develops and proceeds differently, thereby requiring various therapies, appropriate medication can be chosen based on the biomarkers. With advancing biomarker research, the medicine not only is more targeted and effective but also moves increasingly in the direction of personalized medicine.

A good biomarker has certain characteristics. At best, it should have a specificity and sensitivity of at least 80% (Polivka et al., 2016). The measurement procedure should be easy, unequivocal, and cheap. Additionally, the sampling of biomarkers from the patient should be as comfortable as possible (Polivka et al., 2016). A blood sample is, for example, less invasive than a spinal puncture and thereby a common routine examination. Therefore, the biomarker research attaches great value on biomarkers, which can be obtained from the blood.

3 CLASSIFICATION OF BIOMARKERS

A distinction is primarily made between biomarkers that give information about a potential risk of illness and biomarkers used for diagnosis and prognosis of diseases (Mayeux, 2004). They are measurable parameters for the presence (trait) and/or severity (state) of a disease.

3.1 Trait Markers

Biomarkers, which are used to identify at-risk family members, are called trait markers (Mandel et al., 2010). Trait markers are invariable characteristics such as mutations in the genome and do not change as the disease progresses. Therefore, they can display the probability of a future disease.

3.2 State Markers

State markers are measurable parameters, which change with the process of the disease. During the illness the markers can be clearly measured, with the decline of the disease, they are no longer measurable (Schmitz and Anz, 2008). Due to this characteristic, state markers can be divided into different groups according to their use. Diagnostic markers can be used to identify the disease clearly and definitely. Prognostic markers, in turn, can give information about the disease process and about an eventual healing process. Predictive markers, in contrast, are used in the treatment of diseases. They give information about the patient's response to a certain therapy and make the treatment more targeted.

The boundaries of the classification overlap and biomarkers can be often assigned to multiple categories.

4 CLINIC AND PATHOGENESIS OF ALZHEIMER'S DISEASE

AD is a disease, which affects the central nervous system (CNS). The main symptoms of AD include primarily the loss of memory and finally a loss of all cognitive abilities, leading to aphasia, agnosia, apraxia, and disturbance of executive functions (Talwalker, 1996). Due to the loss of cholinergic transmission in neurons in the frontal lobe, declining cognitive abilities arise (Mandel et al., 2010).

Before first symptoms are seen and noticeable as MCI, a pathological change in the brain has already taken place. AD patients show characteristic deposits in the brain. Extracellular deposits of beta-amyloid (Aβ) plaques formed by Aβ peptides and intracellular neurofibrillary tangles caused by tau proteins occurs in the brain (Katzman and Saitoh, 1991). These deposits lead to inflammation and oxidative-toxic processes with subsequent neuronal damage. The plaques are deposited not only in the brain tissue, but can also lead to constriction of cerebral vessels and thus a reduced perfusion in advanced amyloid deposition (Lange-Asschenfeldt, 2009). All these pathological processes result in cell death, causing a cholinergic deficit. The cholinergic deficit correlates with the severity of AD. The mechanism of increased toxic presence of Aβ and tau proteins is not yet completely understood, but an explanation could be pathological alterations in the amyloid precursor protein (APP)-cleaving process.

4.1 Amyloid Precursor Protein and the Production of Aβ

The APP is a membrane-bound protein, which can be found in brain cells. By two consecutive enzymatic cleavages, the protein is proteolytically cleaved into three parts (Miyagawa et al., 2016) (see Fig. 1). First, the β-secretase (also known as β-site amyloid precursor protein-cleaving enzyme 1, BACE1) cuts off an extracellular part from APP (Miyagawa et al.,

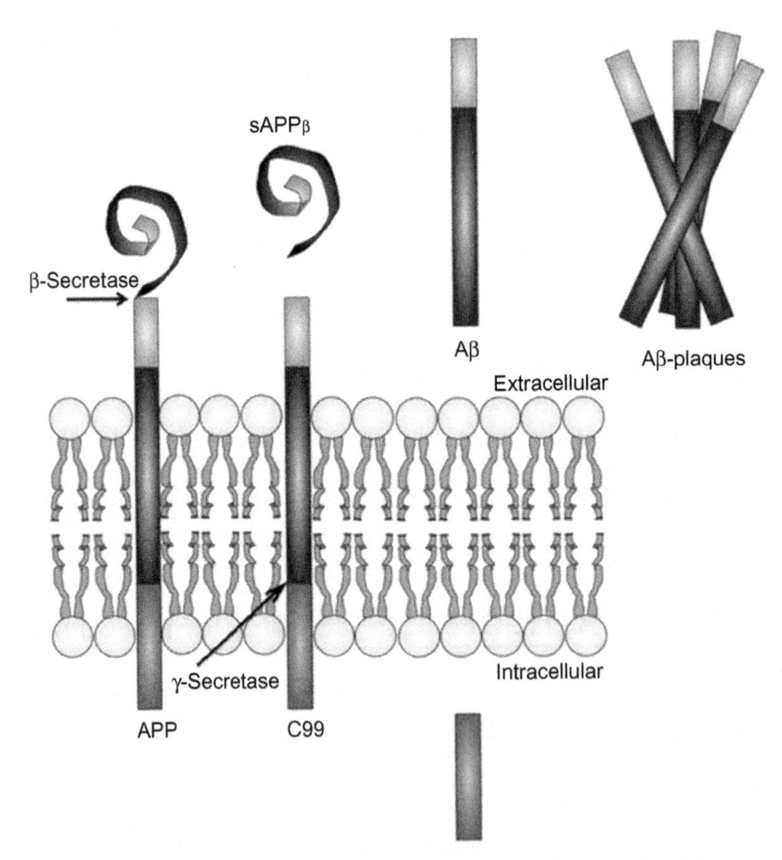

FIG. 1 APP-cleaving process. APP is a membrane-bound protein. β-Secretase separates sAPPβ from APP, which is released extracellularly. γ-Secretase cuts at a specific position, so that Aβ is released in the extracellular space. With an increased presence of Aβ, oligomers are built. These form the amyloid plaques that are deposited in the extracellular space and in the vessels.

2016). The chopped soluble APP fragment, also known as sAPPβ, is released into the extracellular space (Miyagawa et al., 2016). Thereafter, the γ-secretase makes a transmembrane cut through the membrane-bound stub named C99 (Miyagawa et al., 2016). It separates the Aβ from the rest of the APP and releases it extracellularly. Aβ is a 40 (Aβ40) or 42 (Aβ42) amino acid long peptide. While Aβ40 can be dismantled without problems, Aβ42 agglomerates into oligomers, when there is an increased presence.

The excessive production of Aβ42 depends either on increased incidence of APP or on increased activity or concentration of β-secretase or both (Schmitz and Anz, 2008).

4.2 Tau Proteins

Tau proteins or t-tau are parts of the cytoskeleton of neurons, which bind to microtubules and form a stabilizing component. In addition, they control the aggregation of the microtubules. When hyperphosphorylation occurs (then called phospho tau or p-tau proteins), there is an increase in the self-assembly of tau proteins (Gong et al., 2000). The increased aggregation leads to the neurofibrillary tangles, which are deposited in the brain of AD patients.

Several studies have indicated that *oxidative stress* is also a contributing factor of the pathogenesis. Through brain aging and cell death, it results in mitochondrial and peroxisomal dysfunction (Ciavardelli et al., 2016). A result of the dysfunction of cellular organelles and the antioxidant defense system is an increased presence of reactive oxygen species (ROS) and reactive nitrogen species (RNS) (Aksenov et al., 1998). Consequences are lipid peroxidation, proteolysis, and nucleic acid oxidation. These processes result in cell death and promote the course of the disease further (Butterfield et al., 2013).

Besides the two main mechanism of depositing Aβ plaques and neurofibrillary tangles, pathological processes such as imbalances of hormones and neurotransmitters, mitochondrial and synaptic damage, and inflammatory response occur (Reddy et al., 2016). As a result, the focus of the AD biomarker research is not only on deposits production but also on defined metabolic pathways and immunologic processes.

5 CURRENT STATUS OF RESEARCH: POTENTIAL BIOMARKERS FOR ALZHEIMER'S DISEASE

Previously, it was possible to consolidate the suspected AD with the help of memory and behavioral tests such as the mini-mental state evaluation (MMSE), a method to categorize the severity of cognitive impairment of patients with neurodegenerative diseases (Chang et al., 2016). With a score above 25, the patients are classified as normal; below 25 speaks for an impairment of cognitive prowess (Folstein et al., 1975).

Now we know that plaques and neurofibrillary depositions appear in the brain of AD patients. These deposits can be made visible using imaging techniques such as MRI and PET (Schmitz and Anz, 2008). These values belong to the neuroimaging biomarkers of AD. On top of that, the evidence of tau proteins and Aβ42 has been incorporated into current diagnostic research criteria (Schott and Petersen, 2015). These tests, though, are made when the disease is already in its advanced stage. Below genetic, protein, metabolic, immunologic, and transcriptomic biomarkers are listed, which play a main role in the early diagnosis of AD.

5.1 Genetic Biomarkers

A disease disposition is defined as an increased susceptibility to a particular disease due to a genetic trait. These affected genes are called trait biomarkers or trait markers. Some genetic biomarkers have been associated with AD. If there are mutations in these genes, at-risk family members can be identified (Mandel et al., 2010). The mutations can be detected by DNA-sequencing.

The most important trait marker for AD is the gene for *APP*, which is located on chromosome 21. Caused by mutations in the gene for APP or genes encoding the cleaving enzymes, an increased production of Aβ takes place (Schwarz, 2007). Interestingly, humans with trisomy 21 often show the typical Aβ plaques as well (Schwarz, 2007).

The gene for *apolipoprotein E ε4* (APOε4) is located on chromosome 19. APOε4 can be found in the CNS and is responsible for the support and repair of neurons (Mandel et al., 2010). In addition, APOε4 is a lipid-trafficking molecule (Cai et al., 2012). It can bind to both the tau protein and Aβ. Mutations in APOε4 are postulated to be a genetic risk factor for AD (Strittmatter et al., 1993).

The *presenilin 1* (PS1) gene on chromosome 14 and the *presenilin 2* (PS2) gene on chromosome 1 also count to the genetic markers of AD. The main task of presenilins is to regulate γ-secretase. Through changes in this gene, the activity of γ-secretase is modified as well. Through a shift, presenilin 1 effects that the γ-secretase cuts the longer insoluble pathogenic Aβ42 from APP instead of the soluble Aβ40 (Zoltowska et al., 2016).

5.2 Protein Biomarkers

To diagnose AD, nowadays, the CSF levels of t-tau, p-tau, and Aβ42 are already determined. Here, an overall specificity and sensitivity from 85% to 90% are achieved (Blennow and Hampel, 2003). In addition to these hallmarks, more protein biomarkers have been investigated, that reflect the process of the disease.

5.2.1 APP Metabolism

Below are lists of some variants that provide explanations for increased Aβ presence in AD patients.

AB42

CSF levels of Aβ42 are significantly lower in AD patients than controls (Olsson et al., 2016). The aggregation is detectable as a decreased concentration of Aβ42 (Strozyk et al., 2003). Aβ42 is a peptide, which cannot be degraded by proteolysis.

Even if the concentration of Aβ40 in the CSF remains relatively stable, it should be included in the diagnostic (Blennow and Hampel, 2003). The CSF Aβ42/Aβ40 ratio has more meaning than the sole concentration of Aβ42, because the normal Aβ production of individuals is taken into account as well (Lewczuk et al., 2015). The role of Aβ42 in plasma is currently still being debated. It is fact that the plasma concentration of Aβ42 changes during disease process. It is reported that it rises at the beginning of the disease and then slowly decreases as it proliferates (Poljak and Sachdev, 2016). Although changes are measurable, they are not sufficient enough to serve as a biomarker for AD.

Several studies deal with the neurotoxic effect of Aβ, and how it affects neuronal damage. According to Shankar et al. (2007), by binding the receptors of NDMA-type glutamate receptor, the Ca^{2+} flow is changed and this leads to oxidative stress, which ultimately results in synaptic cell death. According to Zhao et al. (2006), oligomeric Aβ has negative effects on p21-activated kinase (PAK), which is involved in cellular signal transduction. Lastly, a few studies discuss the toxic effect of Aβ on cell membranes. It is known that Aβ builds Aβ pores and allows abnormal ion flow in synapses (Kawahara and Kuroda, 2000). All these changes leads to the formation of ROS and cell death.

BACE1

BACE1, also known as β-secretase or β-site APP-cleaving enzyme 1, is mainly expressed in neurons of the brain (Vassar et al., 1999). Under pathological conditions, the β-secretase cuts the APP at a specific point so that the neurotoxic Aβ42 is released extracellularly (Murphy and LeVine, 2010). β-Secretase is encoded by the BACE1 gene. According to Li et al. (2016), either an increased expression of the BACE1 gene or an abnormal function of β-secretase is one of the earliest processes of the pathogenesis of AD. For this reason, BACE1 may serve as a sensitive biomarker to detect AD already in its early stage (Li et al., 2016).

BIN1

BIN1 is an amphiphysin protein that is encoded in humans by the bridging integrator 1 (BIN1) gene (Nicot et al., 2007). BIN1 is mainly found in the CNS, and its tasks include the regulation of endocytose and the endosomal vesicle sorting of membrane proteins (Miyagawa et al., 2016). Additionally, BIN1 regulates the fluidity of proteins that are involved in the pathogenesis of AD (Miyagawa et al., 2016). Several studies discuss the effect of BIN1 on BACE1. Through decreased levels of BIN1, endosomal trafficking is impaired, which results in increased levels of BACE1 (Miyagawa et al., 2016). Through this increased presence of BACE1, the cleaving process occurs more frequently, so that more Aβ is produced (Miyagawa et al., 2016). Sun et al. (2013) reports on significantly increased mRNA and protein levels of BIN1 in the plasma of AD patients compared with healthy control groups, with an achieved sensitivity and specificity of 73% and 75%, respectively.

5.2.2 T-Tau/P-Tau Proteins

In AD patients tau proteins are phosphorylated abnormally (Gong et al., 2000). This hyperphosphorylation produces proteins, which have lost their stabilizing task and assemble into paired helical filaments instead (Gong et al., 2000). This contributes to cell death. While the concentration of t-tau in CSF correlates with the intensity of neuronal degeneration, the concentration of p-tau mirrors the pathogenesis of neurofibrillary tangles (Poljak and Sachdev, 2016). In AD patients the level of t-tau is significantly elevated in CSF and plasma, while the level of p-tau is only elevated in the CSF (Olsson et al., 2016).

A-SYNUCLEIN

α-Synuclein (α-syn) is a protein found in brain cells, which regulates dopamine release (Butler et al., 2016). According to several studies, α-syn induces hyperphosphorylation and aggregation of tau proteins (Duka et al., 2006). It is also known that there is a measurable increased level of α-syn in the CSF of AD patients compared with control groups (Hall et al., 2012). Although the concentration is elevated in other neurodegenerative diseases such as Parkinson's disease (PD), the level is significantly higher in AD patients than in PD patients (Wang et al., 2015). It is possible that there is a positive correlation between the α-syn level in the CSF of AD patients and the reached MMSE score (Korff et al., 2013).

5.2.3 *Neurofilaments*

Neurofilaments are cytoskeletal proteins and components of the axons and neurons (Al-Chalabi and Miller, 2003). They are composed of three subunits, which are named according to their weight: the light (NFL), the medium (NFM), and the heavy subunit (NFH) (Bruno et al., 2012). Each subunit is encoded by a separate gene. The main tasks of the neurofilaments are forging the cell shape and regulate the transport of cellular components within the cytoplasm (Liu et al., 2004). Through imbalance between kinase and phosphatase activities, abnormal hyperphosphorylation of neurofilaments occurs (Gong et al., 2000). It is possible that MAP kinase, GSK-3, and CDK5 pathways are part of this process (Liu et al., 2004). The abnormal hyperphosphorylation of neurofilaments leads to the loss of their stabilizing task and results in axonal damage in the brain and white matter. Subsequently, components of neurofilaments are released and can be detected in the CSF (Bruno et al., 2012). However, the increased concentration of NFL not only is elevated in AD patients but also may be an indicator of other neurological diseases such as dementia or PD.

5.2.4 *sNRG-1*

The endogenous neuregulin 1 (NRG-1), called sNRG-1 in the soluble form, is a protein that is expressed in the human body by the NRG-1 gene (Orr-Urtreger et al., 1993). Although NRG is mainly expressed in cortical neurons, sNRG1 can be found on the surface of white matter astrocytes (Pankonin et al., 2009). Previously, studies showed that the level of sNRG-1 is significantly elevated in the plasma of AD patients, because the corresponding receptors are occupied by presenilin (Chang et al., 2016). Unlike Aβ and α-syn, whose plasma concentrations in AD patients are not significantly different than in the plasma of healthy subjects, sNRG-1 is suitable as a plasma-sensitive biomarker (Chang et al., 2016). On top of that, there is a visible correlation between the sNRG-1 concentration and MMSE scoring. The lower the MMSE score of the AD patients is, the higher the sNRG-1 concentration (Chang et al., 2016).

5.2.5 *YKL-40*

YKL-40 (also known as chitinase-3-like protein-1) is a glycoprotein, which is mainly expressed in astrocytes (Bonneh-Barkay et al., 2010). YKL-40 is a marker for inflammatory processes and activated astrocytes, which can be measured in the CSF (Olsson et al., 2016). AD patients have significantly higher YKL-40 levels in the CSF compared with controls (Rosén et al., 2014). Even if YKL-40 levels correlate with t-tau and p-tau levels and MMSE scores, it is not a specific biomarker for AD, because it only reflects the inflammatory progress. However, YKL-40 is suitable as a marker for clinical drug trials to give information about neurodegeneration and glial activation independently of tau and Aβ (Olsson et al., 2016).

5.3 Metabolic Biomarkers

AD has an influence on the metabolism. Several studies show that there are significant imbalances in the level of some metabolites measurable, contributing to neurological symptoms and deposits. Since the pathological metabolic pathways are involved in the pathophysiology, they may provide targets for therapeutic strategies. Below are listed metabolic biomarkers, which arise during disturbances in AD patients' metabolism.

5.3.1 *Glucose Metabolism and Plasma Acylcarnitines*

Several studies deal with the idea that excess body weight during middle age leads to an increased risk of developing AD (Cai et al., 2012). A healthy lifestyle and exercising, in turn, reduces the risk (Cyna et al., 2017). However, it is known that hyperglycemia has toxic effects on brain cells. According to Cai et al. (2012), deficits in cerebral glucose metabolism occur in AD, which leads to hyperglycemia and promote the pathogenesis further.

Effects of hyperglycemia include the activation of polyol pathway, the formation of advanced glycation end products (AGE), the activation of protein kinase C and an increased incidence of ROS (Biessels et al., 2002). There is also evidence that hyperglycemia results in altered functions of neurotransmission in animal models, such as changed levels of acetylcholine, serotonin, dopamine, and norepinephrine (Biessels et al., 2002; Kamal et al., 1999; Ramakrishnan et al., 2004; Welsh and Wecker, 1991).

Self-oxidation of glucose and increased level of ROS lead to oxidative stress, lipid peroxidation, and dysfunction of cell organelles. Due to the mitochondria dysfunction, it is known that ATP synthesis and oxygen consumption are impaired and consequently contribute to the pathogenesis of AD (Ohta and Ohsawa, 2006). AD-related mitochondria deficits affect the liver as well. Acylcarnitine is a transport form of fatty acids. Plasma acylcarnitines are an indirect indicator of hepatic fatty acid beta-oxidation (Schooneman et al., 2015). A decreased plasma concentration of acylcarnitines can be measured in AD patients compared with controls (Ciavardelli et al., 2016).

5.3.2 *Insulin Metabolism*

A higher rate of insulin receptors can be found in the brain of AD patients (Cai et al., 2012). As there occurs an increased level of insulin as well, conclusions on insulin resistance can be drawn (Cai et al., 2012). The enhanced level of insulin leads to an impaired level of insulin-degrading enzyme (IDE), which is responsible for the reduction of insulin and, intriguingly, for the reduction of Aβ (Cai et al., 2012).

Interestingly, in patients with diabetes mellitus type 1 and 2, significant memory impairment was observed compared with control groups (Cai et al., 2012). According to Luchsinger et al. (2004), patients with hyperinsulinemia have a doubled risk of developing AD. It is known, that there is a correlation between AD and the insulin metabolism, but until now, there is no suitable marker, which reflects the impaired insulin pathway.

5.3.3 *Cholesterol Metabolism and 24-Hydroxycholesterol/27-Hydroxycholesterol*

It is known that the cholesterol metabolism is affected by AD. According to Wang et al. (2016), 24-hydroxycholesterol and 27-hydroxycholesterol levels are elevated in the CSF of AD patients compared with healthy controls. Changes in the APOε4 occur both in the heredity-based and in sporadic forms of AD. APOε4 is known as a regulator of cholesterol uptake (Mahley, 1988). Due to alterations and limited function of this protein, it results in a higher level of 24-hydroxycholesterol and 27-hydroxycholesterol (Wang et al., 2016). According to Polivka et al. (2016), 24-hydroxycholesterol is elevated in plasma as well. Alterations in APOε4 are also a risk factor for diabetes mellitus or hypercholesterolemia (Peila et al., 2002). Besides Aβ42, t-tau, and p-tau, 24-hydroxycholesterol and 27-hydroxycholesterol are suitable as sensitive biomarkers for AD (Wang et al., 2016).

5.4 Immunologic Biomarker

There is increasing evidence that the immune system contributes to the pathogenesis of AD. According to Wu and Li (2016), autoantibodies against some molecules that are involved in AD, have been found. Several studies suggest that autoantibodies have the potential to serve as biomarkers for AD. Autoantibodies are antibodies that are formed against endogenous structures, which have two important tasks: They stimulate cells to apoptosis and prevent inflammatory processes (Wu and Li, 2016). That means that the detected autoantibodies can be both involved in the pathogenesis, but can also assume a protective function.

It can be stated that there is a definite correlation between autoantibodies against Aβ and AD. In future studies value should be put on finding a method, which makes it possible to detect and distinguish the bound and unbound autoantibodies. Then, the exact relationship between the pathogenesis of AD and the formation of the antibodies can be explored. Particularly autoantibodies serve as biomarkers, because they can be obtained from the patient serum and can be detected with relatively simple and cheap molecular biological methods.

5.4.1 Autoantibodies Against Aβ

Many studies report about antibodies against Aβ. Autoantibodies circulate both free and antigen-bound. That makes it difficult to determine the accurate autoantibody concentration, and therefore, the results of the studies are varied (Wu and Li, 2016). Several studies report about significantly lower levels of Aβ autoantibodies in the serum of AD patients than healthy control groups (Du et al., 2001). In contrast, according to Mruthinti et al. (2004), there are significantly higher levels of Aβ autoantibodies than in control groups. In both studies, bound or unbound characteristics were disregarded (Wu and Li, 2016).

5.4.2 Autoantibodies Against Anti-CAPS

Recent studies report about low-molecular-weight oligomeric cross-linked Aβ protein species (CAPS) and anti-CAPS antibodies (Moir et al., 2005). Anti-CAPS antibody levels in the plasma of AD patients are low compared with control groups (Polivka et al., 2016).

5.5 Transcriptomic Biomarkers

The latest research shows that microRNAs (miRNAs) are involved in the pathological processes of AD such as synaptic and mitochondrial damage and the deposits of Aβ and tau proteins (Reddy et al., 2016). RNA occurs in the body as different types, each with different functions and application areas. miRNA regulates gene expression typically by blocking the translation of particular mRNAs (Alberts et al., 2011). By up- or downregulation, miRNA affects the pathological processes and has thereby a toxic or protective effect (Reddy et al., 2016).

One class of miRNA, miR-124, is downregulated, so that the expression of its targeted mRNA is elevated, which is involved in the regulation of the APP splicing process (Smith et al., 2011). miR-124 is known for its function in regulating cleaving enzyme BACE1 (Fang et al., 2012). By downregulation of miR-124, BACE1 is activated, so that more Aβ is produced (Fang et al., 2012). Previously, fluctuations of particular miRNA types in plasma were associated with AD (Kumar and Reddy, 2016). This makes it possible to dissociate healthy humans from AD patients and probably make a statement about whether a person will develop AD (Reddy et al., 2016).

Several studies report about the diagnostic potential of miR-191-5p and miR-342-3p as blood-based biomarkers for AD. According to Tan et al. (2014), miR-342-3p has the best sensitivity of about 81.5% and specificity of about 70.1%, and there is a correlation to the MMSE score visible. In contrast, according to Kumar and Reddy (2016), miR-191-5p shows the largest measurable change compared with other types of miRNA.

However, the knowledge to use miRNA as biomarkers is currently not large enough so that the traditional hallmarks, such as tau proteins and Aβ, have more potential to serve as biomarkers for AD.

5.6 Overview

Table 1 lists all the biomarkers, which were discussed in this review and can be measured in the patient's CSF or blood. Genetic biomarkers, which can be detected by DNA sequencing, are not mentioned.

6 BARRIERS BETWEEN THE BRAIN AND THE BLOOD CIRCULATION

In several studies, proteins and metabolites were found to be sensitive biomarkers for AD both in blood and in the CSF. Nevertheless, a distinction between the biomarkers in the CSF and blood must be made. An important boundary for the molecules represents the blood-brain barrier (BBB).

The BBB builds a border between the blood circulation and the CNS. It is composed of tight capillary endothelium, basal membranes, and astrocytes (Ueno et al., 2016). The capillary endothelium is not fenestrated (Ueno et al., 2016). Ions and polar molecules can only cross the BBB if there is a related transporter. Fat-soluble substances pass the barrier unimpeded. The BBB creates a protective barrier around the brain so that the sensitive structures are spared during temporary electrolyte changes. However, the BBB makes it difficult to treat diseases in the brain with medication, since most drugs consisting of heavy and complex structures cannot cross the BBB.

A further important barrier is the blood-cerebrospinal fluid barrier (BCB). A part of the BCB is the choroid plexus, which is a vascular construct in the brain and responsible for the production of cerebrospinal fluid (Chodobski and Zheng, 2005). The CSF flows through the ventricles and spinal cord. In contrast to the BBB, BCB is fenestrated so that small hydrophilic

TABLE 1 Overview of Protein, Metabolic, Immunologic, and Transcriptomic AD Biomarkers

		Measurable Alteration	
Type	**Biomarker**	**CSF**	**Plasma/Serum**
Protein markers	Aβ42	↓ (Olsson et al., 2016)	– (Olsson et al., 2016)
	BACE1	na	na
	BIN1	↓ (Miyagawa et al., 2016)	↑ (Sun et al., 2013)
	T-tau	↑ (Olsson et al., 2016)	↑ (Olsson et al., 2016)
	P-tau	↑ (Olsson et al., 2016)	– (Olsson et al., 2016)
	α-Synuclein	↑ (Hall et al., 2012)	na
	NFL	↑ (Liu et al., 2004)	na
	sNRG-1	na	↑ (Chang et al., 2016)
	YKL-40	↑ (Rosén et al., 2014)	na
Metabolic markers	Acylcarnitines	na	↓ (Ciavardelli et al., 2016)
	24-Hydroxycholesterol	↑ (Wang et al., 2016)	↑ (Polivka et al. (2016)
	27-Hydroxycholesterol	↑ (Wang et al., 2016)	na
Immunologic markers	Autoantibodies against Aβ42	na	↓ (Du et al., 2001) ↑ (Mruthinti et al., 2004)
	Autoantibodies against CAPS	na	↓ (Polivka et al., 2016)
Transcriptomic markers	miR-124	↓ (Smith et al., 2011)	na
	miR-191-5p	na	↑ (Kumar and Reddy, 2016)
	miR-342-3p	na	↑ (Tan et al., 2014)

The measurable changes for each biomarker are listed: elevated levels (↑), decreased levels (↓), steady level (–), and not assessed (na).

molecules and peptides can pass through the barrier. By passive diffusion, CSF proteins can get into the serum, which happens along the nerve sheaths in the spinal cord (Reiber and Peter, 2001). This allows some molecules to cross the BCB and, therefore, can be found in the blood. In AD, specific molecules and proteins are released into the CSF, when axons are damaged. It is known that choroid plexus function changes with disease and aging, such as in the pathogenesis of AD (Chodobski and Zheng, 2005). This explains the difference in CSF and plasma molecules between AD patients and healthy controls.

The concentration of tau proteins is elevated in both the CSF and the plasma. Triggered by neuronal cell death and high intracellular tau concentration, tau is released extracellularly in the CSF (Tarasoff-Conway et al., 2015). Transporters that transport tau through the BBB have not yet been found (Ueno et al., 2016). It is known that tau proteins are small proteins with a weight around 40 kDa and have a large proportion of basic amino acids, so they are very hydrophilic. It could be possible, due to this hydrophilic property, that they can easily pass the barrier and enter the plasma. Therefore, the levels in both CSF and plasma are altered.

7 OUTLOOK

Due to the fact that the number of AD cases will rapidly increase in the years ahead, it becomes more and more important to find medication for AD. Previously, AD could not be treated successfully, because the mechanisms and processes of the disease were not fully understood.

Through a progression in biomarker research, the mechanisms of pathophysiology are enlightened step by step. This makes it possible to find more targets for drugs. Hitherto, the symptoms of AD are treated with cholinesterase inhibitors with only moderate success. Cholinesterase inhibitors try to compensate the cholinergic deficit in the synapses, which is caused by cell death, by preventing the degradation of the neurotransmitter acetylcholine and keep the concentration high in the synaptic cleft. Nevertheless, only the symptoms are treated, and the progression of the disease is not halted.

Researchers are currently working on new treatment methods. A promising method could be the vaccination against antibodies. Several studies deal with autoantibodies, which are directed against structures in the brain and thus perpetuate the process of the disease. Vaccination against these antibodies could slow down the process so that the disease cannot proliferate.

Another method could be the application of β-secretase inhibitors. By elucidating the cutting mechanisms of the APP, it has been found out that more Aβ is prepared by an increased activity or an increased incidence of cleaving enzyme. By inhibiting the cutting enzyme, the concentration of Aβ is kept low. Therefore, Aβ cannot unfold its neurotoxic feature and less amyloid plaques are formed.

In recent years, it has been found that changes in metabolic pathways in AD occur. It was shown that the insulin and glucose metabolism were adversely affected, for example. In animal models, beneficial effects of peroxisome proliferator-activated receptor (PPAR) agonists have been observed (Cai et al., 2012). Both the glucose and insulin metabolism are stimulated, and the expression of inflammatory substances is prevented.

In recent years, many biomarkers have been found that can be collected from the CSF and plasma of AD patients. Biomarkers make it possible to identify the disease unambiguously, besides the evidence of plaques and neurofibrillary tangle bundles provided by imaging procedures and tests that check the mental abilities. They can also give information about how far the disease has progressed. Nowadays, biomarkers from the CSF are being applied in practice for the diagnosis of AD by determining the concentrations of t-tau, p-tau, and Aβ42. However, the research should focus on biomarkers specifically from the plasma. Blood-based biomarkers are easier and less expensive to analyze and can be extracted by a less invasive procedure.

References

Aksenov, M.Y., Tucker, H.M., Nair, P., Aksenova, M.V., Butterfield, D.A., Estus, S., Markesbery, W.R., 1998. The expression of key oxidative stress-handling genes in different brain regions in Alzheimer's disease. J. Mol. Neurosci. 11 (2), 151–164. https://doi.org/10.1385/JMN:11:2:151.

Alberts, B., Schäfer, U., Häcker, B. (Eds.), 2011. Molekularbiologie der Zelle, fifth ed. Wiley-VCH, Weinheim. 1928 pp.

Al-Chalabi, A., Miller, C.C.J., 2003. Neurofilaments and neurological disease. Bioessays 25 (4), 346–355. https://doi.org/10.1002/bies.10251.

Biessels, G.J., van der Heide, Lars, P., Kamal, A., Bleys, R.L.A.W., Gispen, W.H., 2002. Ageing and diabetes: implications for brain function. Eur. J. Pharmacol. 441 (1–2), 1–14.

Biomarkers Definitions Working Group, 2001. Biomarkers and surrogate endpoints: preferred definitions and conceptual framework. Clin. Pharmacol. Ther. 69 (3), 89–95. https://doi.org/10.1067/mcp.2001.113989.

Blennow, K., Hampel, H., 2003. CSF markers for incipient Alzheimer's disease. Lancet Neurol. 2 (10), 605–613. https://doi.org/10.1016/S1474-4422(03)00530-1.

Bonneh-Barkay, D., Wang, G., Starkey, A., Hamilton, R.L., Wiley, C.A., 2010. In vivo CHI3L1 (YKL-40) expression in astrocytes in acute and chronic neurological diseases. J. Neuroinflammation 7, 34. https://doi.org/10.1186/1742-2094-7-34.

Bruno, D., Pomara, N., Nierenberg, J., Ritchie, J.C., Lutz, M.W., Zetterberg, H., Blennow, K., 2012. Levels of cerebrospinal fluid neurofilament light protein in healthy elderly vary as a function of TOMM40 variants. Exp. Gerontol. 47 (5), 347–352. https://doi.org/10.1016/j.exger.2011.09.008.

Butler, B., Sambo, D., Khoshbouei, H., 2016. Alpha-synuclein modulates dopamine neurotransmission. J. Chem. Neuroanat. https://doi.org/10.1016/j.jchemneu.2016.06.001.

Butterfield, D.A., Swomley, A.M., Sultana, R., 2013. Amyloid beta-peptide (1–42)-induced oxidative stress in Alzheimer disease: importance in disease pathogenesis and progression. Antioxid. Redox Signal. 19 (8), 823–835. https://doi.org/10.1089/ars.2012.5027.

Cai, H., Cong, W.-n., Ji, S., Rothman, S., Maudsley, S., Martin, B., 2012. Metabolic dysfunction in Alzheimer's disease and related neurodegenerative disorders. Curr. Alzheimer Res. 9 (1), 5–17.

Chang, K.-A., Shin, K.Y., Nam, E., Lee, Y.-B., Moon, C., Suh, Y.-H., Lee, S.H., 2016. Plasma soluble neuregulin-1 as a diagnostic biomarker for Alzheimer's disease. Neurochem. Int. 97, 1–7. https://doi.org/10.1016/j.neuint.2016.04.012.

Chodobski, A., Zheng, W. (Eds.), 2005. The Blood-Cerebrospinal Fluid Barrier. Taylor & Francis, Boca Raton. 629 pp.

Ciavardelli, D., Piras, F., Consalvo, A., Rossi, C., Zucchelli, M., Di Ilio, C., Frazzini, V., Caltagirone, C., Spalletta, G., Sensi, S.L., 2016. Medium-chain plasma acylcarnitines, ketone levels, cognition, and gray matter volumes in healthy elderly, mildly cognitively impaired, or Alzheimer's disease subjects. Neurobiol. Aging 43, 1–12. https://doi.org/10.1016/j.neurobiolaging.2016.03.005.

Cyna, M., Lynch, M., Oore, J.J., Nagy, P.M., Aubert, I., 2017. The benefits of exercise and metabolic interventions for the prevention and early treatment of Alzheimer's disease. Curr. Alzheimer Res. 14, 47–60.

Du, Y., Dodel, R., Hampel, H., Buerger, K., Lin, S., Eastwood, B., Bales, K., Gao, F., Moeller, H.J., Oertel, W., Farlow, M., Paul, S., 2001. Reduced levels of amyloid beta-peptide antibody in Alzheimer disease. Neurology 57 (5), 801–805.

Duka, T., Rusnak, M., Drolet, R.E., Duka, V., Wersinger, C., Goudreau, J.L., Sidhu, A., 2006. Alpha-synuclein induces hyperphosphorylation of Tau in the MPTP model of parkinsonism. FASEB J. 20 (13), 2302–2312. https://doi.org/10.1096/fj.06-6092com.

Fang, M., Wang, J., Zhang, X., Geng, Y., Hu, Z., Rudd, J.A., Ling, S., Chen, W., Han, S., 2012. The miR-124 regulates the expression of BACE1/beta-secretase correlated with cell death in Alzheimer's disease. Toxicol. Lett. 209 (1), 94–105. https://doi.org/10.1016/j.toxlet.2011.11.032.

Folstein, M.F., Folstein, S.E., McHugh, P.R., 1975. "Mini-mental state." a practical method for grading the cognitive state of patients for the clinician. J. Psychiatr. Res. 12 (3), 189–198.

Gong, C.X., Lidsky, T., Wegiel, J., Zuck, L., Grundke-Iqbal, I., Iqbal, K., 2000. Phosphorylation of microtubule-associated protein tau is regulated by protein phosphatase 2A in mammalian brain. Implications for neurofibrillary degeneration in Alzheimer's disease. J. Biol. Chem. 275 (8), 5535–5544.

Hall, S., Ohrfelt, A., Constantinescu, R., Andreasson, U., Surova, Y., Bostrom, F., Nilsson, C., Hakan, W., Decraemer, H., Nagga, K., Minthon, L., Londos, E., Vanmechelen, E., Holmberg, B., Zetterberg, H., Blennow, K., Hansson, O., 2012. Accuracy of a panel of 5 cerebrospinal fluid biomarkers in the differential diagnosis of patients with dementia and/or parkinsonian disorders. Arch. Neurol. 69 (11), 1445–1452. https://doi.org/10.1001/archneurol.2012.1654.

Hebert, L.E., Scherr, P.A., Bienias, J.L., Bennett, D.A., Evans, D.A., 2003. Alzheimer disease in the US population: prevalence estimates using the 2000 census. Arch. Neurol. 60 (8), 1119–1122. https://doi.org/10.1001/archneur.60.8.1119.

Kamal, A., Biessels, G.J., Urban, I.J., Gispen, W.H., 1999. Hippocampal synaptic plasticity in streptozotocin-diabetic rats: impairment of long-term potentiation and facilitation of long-term depression. Neuroscience 90 (3), 737–745.

Katzman, R., Saitoh, T., 1991. Advances in Alzheimer's disease. FASEB J. 5 (3), 278–286.

Kawahara, M., Kuroda, Y., 2000. Molecular mechanism of neurodegeneration induced by Alzheimer's beta-amyloid protein: channel formation and disruption of calcium homeostasis. Brain Res. Bull. 53 (4), 389–397.

Korff, A., Liu, C., Ginghina, C., Shi, M., Zhang, J., 2013. Alpha-synuclein in cerebrospinal fluid of Alzheimer's disease and mild cognitive impairment. J. Alzheimers Dis. 36 (4), 679–688. https://doi.org/10.3233/JAD-130458.

Kumar, S., Reddy, P.H., 2016. Are circulating microRNAs peripheral biomarkers for Alzheimer's disease? Biochim. Biophys. Acta 1862 (9), 1617–1627. https://doi.org/10.1016/j.bbadis.2016.06.001.

Lange-Asschenfeldt, C., 2009. Pathophysiologie, Diagnostik, Therapie der Alzheimer-Demenz. Fortbildungstelegramm Pharm. 3, 1–17.

Lewczuk, P., Lelental, N., Spitzer, P., Maler, J.M., Kornhuber, J., 2015. Amyloid-beta 42/40 cerebrospinal fluid concentration ratio in the diagnostics of Alzheimer's disease: validation of two novel assays. J. Alzheimers Dis. 43 (1), 183–191. https://doi.org/10.3233/JAD-140771.

Li, L., Luo, J., Chen, D., Tong, J.-B., Zeng, L.-P., Cao, Y.-Q., Xiang, J., Luo, X.-G., Shi, J.-M., Wang, H., Huang, J.-F., 2016. BACE1 in the retina: a sensitive biomarker for monitoring early pathological changes in Alzheimer's disease. Neural Regen. Res. 11 (3), 447–453. https://doi.org/10.4103/1673-5374.179057.

Liu, Q., Xie, F., Siedlak, S.L., Nunomura, A., Honda, K., Moreira, P.I., Zhua, X., Smith, M.A., Perry, G., 2004. Neurofilament proteins in neurodegenerative diseases. Cell. Mol. Life Sci. 61 (24), 3057–3075. https://doi.org/10.1007/s00018-004-4268-8.

Luchsinger, J.A., Tang, M.-X., Shea, S., Mayeux, R., 2004. Hyperinsulinemia and risk of Alzheimer disease. Neurology 63 (7), 1187–1192.

Mahley, R.W., 1988. Apolipoprotein E: cholesterol transport protein with expanding role in cell biology. Science (New York, N.Y.) 240 (4852), 622–630.

Mandel, S.A., Morelli, M., Halperin, I., Korczyn, A.D., 2010. Biomarkers for prediction and targeted prevention of Alzheimer's and Parkinson's diseases: evaluation of drug clinical efficacy. EPMA J. 1 (2), 273–292. https://doi.org/10.1007/s13167-010-0036-z.

Mayeux, R., 2004. Biomarkers: potential uses and limitations. NeuroRx 1 (2), 182–188.

Miyagawa, T., Ebinuma, I., Morohashi, Y., Hori, Y., Young Chang, M., Hattori, H., Maehara, T., Yokoshima, S., Fukuyama, T., Tsuji, S., Iwatsubo, T., Prendergast, G.C., Tomita, T., 2016. BIN1 regulates BACE1 intracellular trafficking and amyloid-beta production. Hum. Mol. Genet. 25 (14), 2948–2958. https://doi.org/10.1093/hmg/ddw146.

Moir, R.D., Tseitlin, K.A., Soscia, S., Hyman, B.T., Irizarry, M.C., Tanzi, R.E., 2005. Autoantibodies to redox-modified oligomeric Abeta are attenuated in the plasma of Alzheimer's disease patients. J. Biol. Chem. 280 (17), 17458–17463. https://doi.org/10.1074/jbc.M414176200.

Mruthinti, S., Buccafusco, J.J., Hill, W.D., Waller, J.L., Jackson, T.W., Zamrini, E.Y., Schade, R.F., 2004. Autoimmunity in Alzheimer's disease: increased levels of circulating IgGs binding Abeta and RAGE peptides. Neurobiol. Aging 25 (8), 1023–1032. https://doi.org/10.1016/j.neurobiolaging.2003.11.001.

Murphy, M.P., LeVine 3rd, H., 2010. Alzheimer's disease and the amyloid-beta peptide. J. Alzheimers Dis. 19 (1), 311–323. https://doi.org/10.3233/JAD-2010-1221.

Nicot, A.-S., Toussaint, A., Tosch, V., Kretz, C., Wallgren-Pettersson, C., Iwarsson, E., Kingston, H., Garnier, J.-M., Biancalana, V., Oldfors, A., Mandel, J.-L., Laporte, J., 2007. Mutations in amphiphysin 2 (BIN1) disrupt interaction with dynamin 2 and cause autosomal recessive centronuclear myopathy. Nat. Genet. 39 (9), 1134–1139. https://doi.org/10.1038/ng2086.

Ohta, S., Ohsawa, I., 2006. Dysfunction of mitochondria and oxidative stress in the pathogenesis of Alzheimer's disease: on defects in the cytochrome c oxidase complex and aldehyde detoxification. J. Alzheimers Dis. 9 (2), 155–166.

Olsson, B., Lautner, R., Andreasson, U., Ohrfelt, A., Portelius, E., Bjerke, M., Holtta, M., Rosen, C., Olsson, C., Strobel, G., Wu, E., Dakin, K., Petzold, M., Blennow, K., Zetterberg, H., 2016. CSF and blood biomarkers for the diagnosis of Alzheimer's disease: a systematic review and meta-analysis. Lancet Neurol. 15 (7), 673–684. https://doi.org/10.1016/S1474-4422(16)00070-3.

Orr-Urtreger, A., Trakhtenbrot, L., Ben-Levy, R., Wen, D., Rechavi, G., Lonai, P., Yarden, Y., 1993. Neural expression and chromosomal mapping of Neu differentiation factor to 8p12-p21. Proc. Natl. Acad. Sci. U. S. A. 90 (5), 1867–1871.

Pankonin, M.S., Sohi, J., Kamholz, J., Loeb, J.A., 2009. Differential distribution of neuregulin in human brain and spinal fluid. Brain Res. 1258, 1–11. https://doi.org/10.1016/j.brainres.2008.12.047.

Peila, R., Rodriguez, B.L., Launer, L.J., 2002. Type 2 diabetes, APOE gene, and the risk for dementia and related pathologies: the Honolulu-Asia aging study. Diabetes 51 (4), 1256–1262.

Polivka, J., Polivka Jr., J., Krakorova, K., Peterka, M., Topolcan, O., 2016. Current status of biomarker research in neurology. EPMA J. 7, 14. https://doi.org/10.1186/s13167-016-0063-5.

Poljak, A., Sachdev, P.S., 2016. Plasma amyloid beta peptides: an Alzheimer's conundrum or a more accessible Alzheimer's biomarker? Expert. Rev. Neurother. 1–3. https://doi.org/10.1080/14737175.2016.1217156.

Ramakrishnan, R., Sheeladevi, R., Suthanthirarajan, N., 2004. PKC-alpha mediated alterations of indoleamine contents in diabetic rat brain. Brain Res. Bull. 64 (2), 189–194. https://doi.org/10.1016/j.brainresbull.2004.07.002.

Reddy, P.H., Tonk, S., Kumar, S., Vijayan, M., Kandimalla, R., Kuruva, C.S., Reddy, A.P., 2016. A critical evaluation of neuroprotective and neurodegenerative MicroRNAs in Alzheimer's disease. Biochem. Biophys. Res. Commun. https://doi.org/10.1016/j.bbrc.2016.08.067.

Reiber, H., Peter, J.B., 2001. Cerebrospinal fluid analysis: disease-related data patterns and evaluation programs. J. Neurol. Sci. 184 (2), 101–122.

Rosén, C., Andersson, C.H., Andreasson, U., Molinuevo, J.L., Bjerke, M., Rami, L., Lladó, A., Blennow, K., Zetterberg, H., 2014. Increased levels of chitotriosidase and YKL-40 in cerebrospinal fluid from patients with Alzheimer's disease. Dement. Geriatr. Cogn. Disord. 4 (2), 297–304. https://doi.org/10.1159/000362164.

Schmitz, G., Anz, D. (Eds.), 2008. Biomarker: Bedeutung für medizinischen Fortschritt und Nutzenbewertung; mit 14 Tabellen. Schattauer, Stuttgart. 168 pp.

Schooneman, M.G., Ten Have, G.A.M., van Vlies, N., Houten, S.M., Deutz, N.E.P., Soeters, M.R., 2015. Transorgan fluxes in a porcine model reveal a central role for liver in acylcarnitine metabolism. Am. J. Physiol. Endocrinol. Metab. 309 (3), E256–64. https://doi.org/10.1152/ajpendo.00503.2014.

Schott, J.M., Petersen, R.C., 2015. New criteria for Alzheimer's disease: which, when and why? Brain J. Neurol. 138 (Pt 5), 1134–1137. https://doi.org/10.1093/brain/awv055.

Schwarz, S. (Ed.), 2007. Pathophysiologie: Molekulare, zelluläre, systemische Grundlagen von Erkrankungen; 398 Tabellen. Maudrich, Wien.

Selkoe, D.J., 1997. Alzheimer's disease: genotypes, phenotypes, and treatments. Science (New York, N.Y.) 275 (5300), 630–631.

Shankar, G.M., Bloodgood, B.L., Townsend, M., Walsh, D.M., Selkoe, D.J., Sabatini, B.L., 2007. Natural oligomers of the Alzheimer amyloid-beta protein induce reversible synapse loss by modulating an NMDA-type glutamate receptor-dependent signaling pathway. J. Neurosci. Off. J. Soc. Neurosci. 27 (11), 2866–2875. https://doi.org/10.1523/JNEUROSCI.4970-06.2007.

Smith, P., Al Hashimi, A., Girard, J., Delay, C., Hebert, S.S., 2011. In vivo regulation of amyloid precursor protein neuronal splicing by microRNAs. J. Neurochem. 116 (2), 240–247. https://doi.org/10.1111/j.1471-4159.2010.07097.x.

Strittmatter, W.J., Saunders, A.M., Schmechel, D., Pericak-Vance, M., Enghild, J., Salvesen, G.S., Roses, A.D., 1993. Apolipoprotein E: high-avidity binding to beta-amyloid and increased frequency of type 4 allele in late-onset familial Alzheimer disease. Proc. Natl. Acad. Sci. U. S. A. 90 (5), 1977–1981.

Strozyk, D., Blennow, K., White, L.R., Launer, L.J., 2003. CSF Abeta 42 levels correlate with amyloid-neuropathology in a population-based autopsy study. Neurology 60 (4), 652–656.

Sun, L., Tan, M.-S., Hu, N., Yu, J.-T., Tan, L., 2013. Exploring the value of plasma BIN1 as a potential biomarker for alzheimer's disease. J. Alzheimers Dis. 37 (2), 291–295. https://doi.org/10.3233/JAD-130392.

Talwalker, S., 1996. The cardinal features of cognitive and noncognitive dysfunction and the differential efficacy of tacrine in Alzheimer's disease patients. J. Biopharm. Stat. 6 (4), 443–456. https://doi.org/10.1080/10543409608835155.

Tan, L., Yu, J.-T., Tan, M.-S., Liu, Q.-Y., Wang, H.-F., Zhang, W., Jiang, T., Tan, L., 2014. Genome-wide serum microRNA expression profiling identifies serum biomarkers for Alzheimer's disease. J. Alzheimers Dis. 40 (4), 1017–1027. https://doi.org/10.3233/JAD-132144.

Tarasoff-Conway, J.M., Carare, R.O., Osorio, R.S., Glodzik, L., Butler, T., Fieremans, E., Axel, L., Rusinek, H., Nicholson, C., Zlokovic, B.V., Frangione, B., Blennow, K., Menard, J., Zetterberg, H., Wisniewski, T., de Leon, M.J., 2015. Clearance systems in the brain-implications for Alzheimer disease. Nat. Rev. Neurol. 11 (8), 457–470. https://doi.org/10.1038/nrneurol.2015.119.

Ueno, M., Chiba, Y., Murakami, R., Matsumoto, K., Kawauchi, M., Fujihara, R., 2016. Blood-brain barrier and blood-cerebrospinal fluid barrier in normal and pathological conditions. Brain Tumor Pathol. 33 (2), 89–96. https://doi.org/10.1007/s10014-016-0255-7.

Vassar, R., Bennett, B.D., Babu-Khan, S., Kahn, S., Mendiaz, E.A., Denis, P., Teplow, D.B., Ross, S., Amarante, P., Loeloff, R., Luo, Y., Fisher, S., Fuller, J., Edenson, S., Lile, J., Jarosinski, M.A., Biere, A.L., Curran, E., Burgess, T., Louis, J.C., Collins, F., Treanor, J., Rogers, G., Citron, M., 1999. Beta-secretase cleavage of Alzheimer's amyloid precursor protein by the transmembrane aspartic protease BACE. Science (New York, N.Y.) 286 (5440), 735–741.

Wang, H.-L., Wang, Y.-Y., Liu, X.-G., Kuo, S.-H., Liu, N., Song, Q.-Y., Wang, M.-W., 2016. Cholesterol, 24-hydroxycholesterol, and 27-hydroxycholesterol as surrogate biomarkers in cerebrospinal fluid in mild cognitive impairment and Alzheimer's disease: a meta-analysis. J. Alzheimers Dis. 51 (1), 45–55. https://doi.org/10.3233/JAD-150734.

Wang, Z.-Y., Han, Z.-M., Liu, Q.-F., Tang, W., Ye, K., Yao, Y.-Y., 2015. Use of CSF alpha-synuclein in the differential diagnosis between Alzheimer's disease and other neurodegenerative disorders. Int. Psychogeriatr. 27 (9), 1429–1438. https://doi.org/10.1017/S1041610215000447.

Welsh, B., Wecker, L., 1991. Effects of streptozotocin-induced diabetes on acetylcholine metabolism in rat brain. Neurochem. Res. 16 (4), 453–460.

Wortmann, M., 2015. World Alzheimer report 2014: dementia and risk reduction. Alzheimers Dement. 11 (7), P837. https://doi.org/10.1016/j.jalz.2015.06.1858.

Wu, J., Li, L., 2016. Autoantibodies in Alzheimer's disease: potential biomarkers, pathogenic roles, and therapeutic implications. J. Biomed. Res. 30, 361–372. https://doi.org/10.7555/JBR.30.20150131.

Zhao, L., Ma, Q.-L., Calon, F., Harris-White, M.E., Yang, F., Lim, G.P., Morihara, T., Ubeda, O.J., Ambegaokar, S., Hansen, J.E., Weisbart, R.H., Teter, B., Frautschy, S.A., Cole, G.M., 2006. Role of p21-activated kinase pathway defects in the cognitive deficits of Alzheimer disease. Nat. Neurosci. 9 (2), 234–242. https://doi.org/10.1038/nn1630.

Zoltowska, K.M., Maesako, M., Berezovska, O., 2016. Interrelationship between changes in the amyloid beta 42/40 ratio and presenilin 1 conformation. Mol. Med. 22, 329–337. https://doi.org/10.2119/molmed.2016.00127.

Index

Note: Page numbers followed by *f* indicate figures, and *t* indicate tables.

N

R

S

Printed in the United States
By Bookmasters